T0203064

Lecture Notes in Computer Science

Lecture Notes in Artificial Intelligence 13995

Founding Editor

Jörg Siekmann

The series Lecture Notes in Artificial Intelligence (LNAI) was established in 1988 as a topical subseries of LNCS devoted to artificial intelligence.

The series publishes state-of-the-art research results at a high level. As with the LNCS mother series, the mission of the series is to serve the international R & D community by providing an invaluable service, mainly focused on the publication of conference and workshop proceedings and postproceedings.

Ngoc Thanh Nguyen · Siridech Boonsang ·
Hamido Fujita · Bogumiła Hnatkowska ·
Tzung-Pei Hong · Kitsuchart Pasupa ·
Ali Selamat
Editors

Intelligent Information and Database Systems

15th Asian Conference, ACIIDS 2023
Phuket, Thailand, July 24–26, 2023
Proceedings, Part I

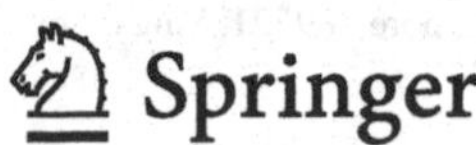

Editors
Ngoc Thanh Nguyen
Wrocław University of Science and Technology
Wrocław, Poland

Hamido Fujita
Iwate Prefectural University Iwate
Iwate, Japan

Tzung-Pei Hong
National University of Kaohsiung
Kaohsiung, Taiwan

Ali Selamat
Malaysia Japan International Institute of Technology
Kuala Lumpur, Malaysia

Siridech Boonsang
King Mongkut's Institute of Technology Ladkrabang
Bangkok, Thailand

Bogumiła Hnatkowska
Wroclaw University of Science and Technology
Wrocław, Poland

Kitsuchart Pasupa
King Mongkut's Institute of Technology Ladkrabang
Bangkok, Thailand

ISSN 0302-9743 ISSN 1611-3349 (electronic)
Lecture Notes in Artificial Intelligence
ISBN 978-981-99-5833-7 ISBN 978-981-99-5834-4 (eBook)
https://doi.org/10.1007/978-981-99-5834-4

LNCS Sublibrary: SL7 – Artificial Intelligence

This Springer imprint is published by the registered company Springer Nature Singapore Pte Ltd.
The registered company address is: 152 Beach Road, #21-01/04 Gateway East, Singapore 189721, Singapore

Paper in this product is recyclable.

Preface

ACIIDS 2023 was the 15th event in a series of international scientific conferences on research and applications in the field of intelligent information and database systems. The aim of ACIIDS 2023 was to provide an international forum for research workers with scientific backgrounds in the technology of intelligent information and database systems and their various applications. The conference was hosted by King Mongkut's Institute of Technology Ladkrabang, Thailand, and jointly organized by Wrocław University of Science and Technology, Poland, in cooperation with IEEE SMC Technical Committee on Computational Collective Intelligence, European Research Center for Information Systems (ERCIS), University of Newcastle (Australia), Yeungnam University (South Korea), International University - Vietnam National University HCMC (Vietnam), Leiden University (The Netherlands), Universiti Teknologi Malaysia (Malaysia), Nguyen Tat Thanh University (Vietnam), BINUS University (Indonesia), and Vietnam National University, Hanoi (Vietnam). ACIIDS 2023 occurred in Phuket, Thailand, on July 24–26, 2023.

The ACIIDS conference series is already well established. The first two events, ACIIDS 2009 and ACIIDS 2010, took place in Dong Hoi City and Hue City in Vietnam, respectively. The third event, ACIIDS 2011, occurred in Daegu (South Korea), followed by the fourth, ACIIDS 2012, in Kaohsiung (Taiwan). The fifth event, ACIIDS 2013, was held in Kuala Lumpur (Malaysia), while the sixth event, ACIIDS 2014, was held in Bangkok (Thailand). The seventh event, ACIIDS 2015, occurred in Bali (Indonesia), followed by the eighth, ACIIDS 2016, in Da Nang (Vietnam). The ninth event, ACIIDS 2017, was organized in Kanazawa (Japan). The 10th jubilee conference, ACIIDS 2018, was held in Dong Hoi City (Vietnam), followed by the 11th event, ACIIDS 2019, in Yogyakarta (Indonesia). The 12th and 13th events were planned to be on-site in Phuket (Thailand). However, the global pandemic relating to COVID-19 resulted in both editions of the conference being held online in virtual space. ACIIDS 2022 was held in Ho Chi Minh City as a hybrid conference, and it restarted in-person meetings at conferences.

These two volumes contain 65 peer-reviewed papers selected for presentation from 224 submissions. Papers included in this volume cover the following topics: data mining and machine learning methods, advanced data mining techniques and applications, intelligent and contextual systems, natural language processing, network systems and applications, computational imaging and vision, decision support, control systems, and data modeling and processing for industry 4.0.

The accepted and presented papers focus on new trends and challenges facing the intelligent information and database systems community. The presenters showed how research work could stimulate novel and innovative applications. We hope you find these results valuable and inspiring for future research work. We would like to express our sincere thanks to the honorary chairs for their support: Arkadiusz Wójs (Rector of Wrocław University of Science and Technology, Poland), Moonis Ali (Texas State

University, President of International Society of Applied Intelligence, USA), Komsan Maleesee (President of King Mongkut's Institute of Technology Ladkrabang, Thailand).

We thank the keynote speakers for their world-class plenary speeches: Saman K. Halgamuge from The University of Melbourne (Australia), Jerzy Stefanowski from Poznań University of Technology (Poland), Siridech Boonsang from King Mongkut's Institute of Technology Ladkrabang (Thailand), and Masaru Kitsuregawa from The University of Tokyo (Japan).

We cordially thank our main sponsors, King Mongkut's Institute of Technology Ladkrabang (Thailand), Wrocław University of Science and Technology (Poland), IEEE SMC Technical Committee on Computational Collective Intelligence, European Research Center for Information Systems (ERCIS), University of Newcastle (Australia), Yeungnam University (South Korea), Leiden University (The Netherlands), Universiti Teknologi Malaysia (Malaysia), BINUS University (Indonesia), Vietnam National University (Vietnam), and Nguyen Tat Thanh University (Vietnam). Our special thanks go to Springer for publishing the proceedings and to all the other sponsors for their kind support.

Our special thanks go to the program chairs, the special session chairs, the organizing chairs, the publicity chairs, the liaison chairs, and the Local Organizing Committee for their work towards the conference. We sincerely thank all the members of the International Program Committee for their valuable efforts in the review process, which helped us to guarantee the highest quality of the selected papers for the conference. We cordially thank all the authors and other conference participants for their valuable contributions. The conference would not have been possible without their support. Thanks are also due to the many experts who contributed to the event being a success.

July 2023

Ngoc Thanh Nguyen
Siridech Boonsang
Hamido Fujita
Bogumiła Hnatkowska
Tzung-Pei Hong
Kitsuchart Pasupa
Ali Selamat

Organization

Honorary Chairs

Arkadiusz Wójs	Rector of Wrocław University of Science and Technology, Poland
Moonis Ali	Texas State University, President of International Society of Applied Intelligence, USA
Komsan Maleesee	President of King Mongkut's Institute of Technology Ladkrabang, Thailand

General Chairs

Ngoc Thanh Nguyen	Wrocław University of Science and Technology, Poland
Suphamit Chittayasothorn	King Mongkut's Institute of Technology Ladkrabang, Thailand

Program Chairs

Hamido Fujita	Iwate Prefectural University, Japan
Tzung-Pei Hong	National University of Kaohsiung, Taiwan
Ali Selamat	Universiti Teknologi Malaysia, Malaysia
Siridech Boonsang	King Mongkut's Institute of Technology Ladkrabang, Thailand
Kitsuchart Pasupa	King Mongkut's Institute of Technology Ladkrabang, Thailand

Steering Committee

Ngoc Thanh Nguyen (Chair)	Wrocław University of Science and Technology, Poland
Longbing Cao	University of Science and Technology Sydney, Australia
Suphamit Chittayasothorn	King Mongkut's Institute of Technology Ladkrabang, Thailand
Ford Lumban Gaol	Bina Nusantara University, Indonesia

Tzung-Pei Hong	National University of Kaohsiung, Taiwan
Dosam Hwang	Yeungnam University, South Korea
Bela Stantic	Griffith University, Australia
Geun-Sik Jo	Inha University, South Korea
Hoai An Le Thi	University of Lorraine, France
Toyoaki Nishida	Kyoto University, Japan
Leszek Rutkowski	Częstochowa University of Technology, Poland
Ali Selamat	Universiti Teknologi Malaysia, Malaysia
Edward Szczerbicki	University of Newcastle, Australia

Special Session Chairs

Bogumiła Hnatkowska	Wrocław University of Science and Technology, Poland
Arit Thammano	King Mongkut's Institute of Technology Ladkrabang, Thailand
Krystian Wojtkiewicz	Wrocław University of Science and Technology, Poland

Doctoral Track Chairs

Marek Krótkiewicz	Wrocław University of Science and Technology, Poland
Nont Kanungsukkasem	King Mongkut's Institute of Technology Ladkrabang, Thailand

Liaison Chairs

Sirasit Lochanachit	King Mongkut's Institute of Technology Ladkrabang, Thailand
Ford Lumban Gaol	Bina Nusantara University, Indonesia
Quang-Thuy Ha	VNU-University of Engineering and Technology, Vietnam
Mong-Fong Horng	National Kaohsiung University of Applied Sciences, Taiwan
Dosam Hwang	Yeungnam University, South Korea
Le Minh Nguyen	Japan Advanced Institute of Science and Technology, Japan
Ali Selamat	Universiti Teknologi Malaysia, Malaysia

Organizing Chairs

Kamol Wasapinyokul	King Mongkut's Institute of Technology Ladkrabang, Thailand
Krystian Wojtkiewicz	Wrocław University of Science and Technology, Poland

Publicity Chairs

Marcin Jodłowiec	Wrocław University of Science and Technology, Poland
Rafał Palak	Wrocław University of Science and Technology, Poland
Nat Dilokthanakul	King Mongkut's Institute of Technology Ladkrabang, Thailand

Finance Chair

Pattanapong Chantamit-O-Pas	King Mongkut's Institute of Technology Ladkrabang, Thailand

Webmaster

Marek Kopel	Wrocław University of Science and Technology, Poland

Local Organizing Committee

Taravichet Titijaroonroj	King Mongkut's Institute of Technology Ladkrabang, Thailand
Praphan Pavarangkoon	King Mongkut's Institute of Technology Ladkrabang, Thailand
Natthapong Jungteerapanich	King Mongkut's Institute of Technology Ladkrabang, Thailand
Putsadee Pornphol	Phuket Rajabhat University, Thailand
Patient Zihisire Muke	Wrocław University of Science and Technology, Poland

Thanh-Ngo Nguyen	Wrocław University of Science and Technology, Poland
Katarzyna Zombroń	Wrocław University of Science and Technology, Poland
Kulwadee Somboonviwat	Kasetsart University Sriracha, Thailand

Keynote Speakers

Saman K. Halgamuge	University of Melbourne, Australia
Jerzy Stefanowski	Poznań University of Technology, Poland
Siridech Boonsang	King Mongkut's Institute of Technology Ladkrabang, Thailand
Masaru Kitsuregawa	University of Tokyo, Japan

Special Sessions Organizers

ADMTA 2023: Special Session on Advanced Data Mining Techniques and Applications

Chun-Hao Chen	National Kaohsiung University of Science and Technology, Taiwan
Bay Vo	Ho Chi Minh City University of Technology, Vietnam
Tzung-Pei Hong	National University of Kaohsiung, Taiwan

AINBC 2023: Special Session on Advanced Data Mining Techniques and Applications

Andrzej W. Przybyszewski	University of Massachusetts Medical School, USA
Jerzy P. Nowacki	Polish-Japanese Academy of Information Technology, Poland

CDF 2023: Special Session on Computational Document Forensics

Jean-Marc Ogier	La Rochelle Université, France
Mickaël Coustaty	La Rochelle Université, France
Surapong Uttama	Mae Fah Luang University, Thailand

CSDT 2023: Special Session on Cyber Science in Digital Transformation

Dariusz Szostek	University of Silesia in Katowice, Poland
Jan Kozak	University of Economics in Katowice, Poland
Paweł Kasprowski	Silesian University of Technology, Poland

CVIS 2023: Special Session on Computer Vision and Intelligent Systems

Van-Dung Hoang	Ho Chi Minh City University of Technology and Education, Vietnam
Dinh-Hien Nguyen	University of Information Technology, VNU-HCM, Vietnam
Chi-Mai Luong	Vietnam Academy of Science and Technology, Vietnam

DMPCPA 2023: Special Session on Data Modelling and Processing in City Pollution Assessment

Hoai Phuong Ha	UiT The Arctic University of Norway, Norway
Manuel Nuñez	Universidad Complutense de Madrid, Spain
Rafał Palak	Wrocław University of Science and Technology, Poland
Krystian Wojtkiewicz	Wrocław University of Science and Technology, Poland

HPC-ComCon 2023: Special Session on HPC and Computing Continuum

Pascal Bouvry	University of Luxembourg, Luxembourg
Johnatan E. Pecero	University of Luxembourg, Luxembourg
Arijit Roy	Indian Institute of Information Technology, Sri City, India

LRLSTP 2023: Special Session on Low Resource Languages Speech and Text Processing

Ualsher Tukeyev	Al-Farabi Kazakh National University, Kazakhstan
Orken Mamyrbayev	Institute of Information and Computational Technologies, Kazakhstan

Senior Program Committee

Ajith Abraham	Machine Intelligence Research Labs, USA
Jesús Alcalá Fernández	University of Granada, Spain
Lionel Amodeo	University of Technology of Troyes, France
Ahmad Taher Azar	Prince Sultan University, Saudi Arabia
Thomas Bäck	Leiden University, The Netherlands
Costin Badica	University of Craiova, Romania
Ramazan Bayindir	Gazi University, Turkey
Abdelhamid Bouchachia	Bournemouth University, UK
David Camacho	Universidad Autónoma de Madrid, Spain
Leopoldo Eduardo Cardenas-Barron	Tecnológico de Monterrey, Mexico
Oscar Castillo	Tijuana Institute of Technology, Mexico
Nitesh Chawla	University of Notre Dame, USA
Rung-Ching Chen	Chaoyang University of Technology, Taiwan
Shyi-Ming Chen	National Taiwan University of Science and Technology, Taiwan
Simon Fong	University of Macau, China
Hamido Fujita	Iwate Prefectural University, Japan
Mohamed Gaber	Birmingham City University, UK
Marina L. Gavrilova	University of Calgary, Canada
Daniela Godoy	ISISTAN Research Institute, Argentina
Fernando Gomide	University of Campinas, Brazil
Manuel Grana	University of the Basque Country, Spain
Claudio Gutierrez	Universidad de Chile, Chile
Francisco Herrera	University of Granada, Spain
Tzung-Pei Hong	National University of Kaohsiung, Taiwan
Dosam Hwang	Yeungnam University, South Korea
Mirjana Ivanovic	University of Novi Sad, Serbia
Janusz Jeżewski	Institute of Medical Technology and Equipment ITAM, Poland
Piotr Jędrzejowicz	Gdynia Maritime University, Poland
Kang-Hyun Jo	University of Ulsan, South Korea
Janusz Kacprzyk	Systems Research Institute, Polish Academy of Sciences, Poland
Nikola Kasabov	Auckland University of Technology, New Zealand
Muhammad Khurram Khan	King Saud University, Saudi Arabia
Frank Klawonn	Ostfalia University of Applied Sciences, Germany
Joanna Kolodziej	Cracow University of Technology, Poland
Józef Korbicz	University of Zielona Gora, Poland
Ryszard Kowalczyk	Swinburne University of Technology, Australia

Bartosz Krawczyk	Virginia Commonwealth University, USA
Ondrej Krejcar	University of Hradec Králové, Czech Republic
Adam Krzyzak	Concordia University, Canada
Mark Last	Ben-Gurion University of the Negev, Israel
Hoai An Le Thi	University of Lorraine, France
Kun Chang Lee	Sungkyunkwan University, South Korea
Edwin Lughofer	Johannes Kepler University Linz, Austria
Nezam Mahdavi-Amiri	Sharif University of Technology, Iran
Yannis Manolopoulos	Open University of Cyprus, Cyprus
Klaus-Robert Müller	Technical University of Berlin, Germany
Saeid Nahavandi	Deakin University, Australia
Grzegorz J Nalepa	AGH University of Science and Technology, Poland
Ngoc-Thanh Nguyen	Wrocław University of Science and Technology, Poland
Dusit Niyato	Nanyang Technological University, Singapore
Manuel Núñez	Universidad Complutense de Madrid, Spain
Jeng-Shyang Pan	Fujian University of Technology, China
Marcin Paprzycki	Systems Research Institute, Polish Academy of Sciences, Poland
Hoang Pham	Rutgers University, USA
Tao Pham Dinh	INSA Rouen, France
Radu-Emil Precup	Politehnica University of Timisoara, Romania
Leszek Rutkowski	Częstochowa University of Technology, Poland
Jürgen Schmidhuber	Swiss AI Lab IDSIA, Switzerland
Björn Schuller	University of Passau, Germany
Ali Selamat	Universiti Teknologi Malaysia, Malaysia
Andrzej Skowron	Warsaw University, Poland
Jerzy Stefanowski	Poznań University of Technology, Poland
Edward Szczerbicki	University of Newcastle, Australia
Ryszard Tadeusiewicz	AGH University of Science and Technology, Poland
Muhammad Atif Tahir	National University of Computing & Emerging Sciences, Pakistan
Bay Vo	Ho Chi Minh City University of Technology, Vietnam
Dinh Duc Anh Vu	Vietnam National University HCMC, Vietnam
Lipo Wang	Nanyang Technological University, Singapore
Junzo Watada	Waseda University, Japan
Michał Woźniak	Wrocław University of Science and Technology, Poland
Farouk Yalaoui	University of Technology of Troyes, France

Sławomir Zadrożny — Systems Research Institute, Polish Academy of Sciences, Poland
Zhi-Hua Zhou — Nanjing University, China

Program Committee

Muhammad Abulaish — South Asian University, India
Bashar Al-Shboul — University of Jordan, Jordan
Toni Anwar — Universiti Teknologi PETRONAS, Malaysia
Taha Arbaoui — University of Technology of Troyes, France
Mehmet Emin Aydin — University of the West of England, UK
Amelia Badica — University of Craiova, Romania
Kambiz Badie — ICT Research Institute, Iran
Hassan Badir — École Nationale des Sciences Appliquées de Tanger, Morocco
Zbigniew Banaszak — Warsaw University of Technology, Poland
Dariusz Barbucha — Gdynia Maritime University, Poland
Maumita Bhattacharya — Charles Sturt University, Australia
Leon Bobrowski — Białystok University of Technology, Poland
Bülent Bolat — Yildiz Technical University, Turkey
Mariusz Boryczka — University of Silesia in Katowice, Poland
Urszula Boryczka — University of Silesia in Katowice, Poland
Zouhaier Brahmia — University of Sfax, Tunisia
Stéphane Bressan — National University of Singapore, Singapore
Peter Brida — University of Žilina, Slovakia
Piotr Bródka — Wrocław University of Science and Technology, Poland
Grażyna Brzykcy — Poznań University of Technology, Poland
Robert Burduk — Wrocław University of Science and Technology, Poland
Aleksander Byrski — AGH University of Science and Technology, Poland
Dariusz Ceglarek — WSB University in Poznań, Poland
Somchai Chatvichienchai — University of Nagasaki, Japan
Chun-Hao Chen — Tamkang University, Taiwan
Leszek J. Chmielewski — Warsaw University of Life Sciences, Poland
Kazimierz Choroś — Wrocław University of Science and Technology, Poland
Kun-Ta Chuang — National Cheng Kung University, Taiwan
Dorian Cojocaru — University of Craiova, Romania
Jose Alfredo Ferreira Costa — Federal University of Rio Grande do Norte (UFRN), Brazil

Ireneusz Czarnowski	Gdynia Maritime University, Poland
Piotr Czekalski	Silesian University of Technology, Poland
Theophile Dagba	University of Abomey-Calavi, Benin
Tien V. Do	Budapest University of Technology and Economics, Hungary
Rafał Doroz	University of Silesia in Katowice, Poland
El-Sayed M. El-Alfy	King Fahd University of Petroleum and Minerals, Saudi Arabia
Keiichi Endo	Ehime University, Japan
Sebastian Ernst	AGH University of Science and Technology, Poland
Nadia Essoussi	University of Carthage, Tunisia
Usef Faghihi	Université du Québec à Trois-Rivières, Canada
Dariusz Frejlichowski	West Pomeranian University of Technology, Szczecin, Poland
Blanka Frydrychova Klimova	University of Hradec Králové, Czech Republic
Janusz Getta	University of Wollongong, Australia
Daniela Gifu	University "Alexandru loan Cuza" of Iași, Romania
Gergo Gombos	Eötvös Loránd University, Hungary
Manuel Grana	University of the Basque Country, Spain
Janis Grundspenkis	Riga Technical University, Latvia
Dawit Haile	Addis Ababa University, Ethiopia
Marcin Hernes	Wrocław University of Business and Economics, Poland
Koichi Hirata	Kyushu Institute of Technology, Japan
Bogumiła Hnatkowska	Wrocław University of Science and Technology, Poland
Bao An Mai Hoang	Vietnam National University HCMC, Vietnam
Huu Hanh Hoang	Posts and Telecommunications Institute of Technology, Vietnam
Van-Dung Hoang	Quang Binh University, Vietnam
Jeongkyu Hong	Yeungnam University, South Korea
Yung-Fa Huang	Chaoyang University of Technology, Taiwan
Maciej Huk	Wrocław University of Science and Technology, Poland
Kha Tu Huynh	Vietnam National University HCMC, Vietnam
Sanjay Jain	National University of Singapore, Singapore
Khalid Jebari	LCS Rabat, Morocco
Joanna Jędrzejowicz	University of Gdańsk, Poland
Przemysław Juszczuk	University of Economics in Katowice, Poland
Krzysztof Juszczyszyn	Wrocław University of Science and Technology, Poland

Mehmet Karaata	Kuwait University, Kuwait
Rafał Kern	Wrocław University of Science and Technology, Poland
Zaheer Khan	University of the West of England, UK
Marek Kisiel-Dorohinicki	AGH University of Science and Technology, Poland
Attila Kiss	Eötvös Loránd University, Hungary
Shinya Kobayashi	Ehime University, Japan
Grzegorz Kołaczek	Wrocław University of Science and Technology, Poland
Marek Kopel	Wrocław University of Science and Technology, Poland
Jan Kozak	University of Economics in Katowice, Poland
Adrianna Kozierkiewicz	Wrocław University of Science and Technology, Poland
Dalia Kriksciuniene	Vilnius University, Lithuania
Dariusz Król	Wrocław University of Science and Technology, Poland
Marek Krótkiewicz	Wrocław University of Science and Technology, Poland
Marzena Kryszkiewicz	Warsaw University of Technology, Poland
Jan Kubicek	VSB -Technical University of Ostrava, Czech Republic
Tetsuji Kuboyama	Gakushuin University, Japan
Elżbieta Kukla	Wrocław University of Science and Technology, Poland
Marek Kulbacki	Polish-Japanese Academy of Information Technology, Poland
Kazuhiro Kuwabara	Ritsumeikan University, Japan
Annabel Latham	Manchester Metropolitan University, UK
Tu Nga Le	Vietnam National University HCMC, Vietnam
Yue-Shi Lee	Ming Chuan University, Taiwan
Florin Leon	Gheorghe Asachi Technical University of Iasi, Romania
Chunshien Li	National Central University, Taiwan
Horst Lichter	RWTH Aachen University, Germany
Igor Litvinchev	Nuevo Leon State University, Mexico
Doina Logofatu	Frankfurt University of Applied Sciences, Germany
Lech Madeyski	Wrocław University of Science and Technology, Poland
Bernadetta Maleszka	Wrocław University of Science and Technology, Poland

Marcin Maleszka	Wrocław University of Science and Technology, Poland
Tamás Matuszka	Eötvös Loránd University, Hungary
Michael Mayo	University of Waikato, New Zealand
Héctor Menéndez	University College London, UK
Jacek Mercik	WSB University in Wrocław, Poland
Radosław Michalski	Wrocław University of Science and Technology, Poland
Peter Mikulecky	University of Hradec Králové, Czech Republic
Miroslava Mikusova	University of Žilina, Slovakia
Marek Milosz	Lublin University of Technology, Poland
Jolanta Mizera-Pietraszko	Opole University, Poland
Dariusz Mrozek	Silesian University of Technology, Poland
Leo Mrsic	IN2data Ltd Data Science Company, Croatia
Agnieszka Mykowiecka	Institute of Computer Science, Polish Academy of Sciences, Poland
Pawel Myszkowski	Wrocław University of Science and Technology, Poland
Huu-Tuan Nguyen	Vietnam Maritime University, Vietnam
Le Minh Nguyen	Japan Advanced Institute of Science and Technology, Japan
Loan T. T. Nguyen	Vietnam National University HCMC, Vietnam
Quang-Vu Nguyen	Korea-Vietnam Friendship Information Technology College, Vietnam
Thai-Nghe Nguyen	Cantho University, Vietnam
Thi Thanh Sang Nguyen	Vietnam National University HCMC, Vietnam
Van Sinh Nguyen	Vietnam National University HCMC, Vietnam
Agnieszka Nowak-Brzezińska	University of Silesia in Katowice, Poland
Alberto Núñez	Universidad Complutense de Madrid, Spain
Mieczysław Owoc	Wrocław University of Business and Economics, Poland
Panos Patros	University of Waikato, New Zealand
Maciej Piasecki	Wrocław University of Science and Technology, Poland
Bartłomiej Pierański	Poznań University of Economics and Business, Poland
Dariusz Pierzchała	Military University of Technology, Poland
Marcin Pietranik	Wrocław University of Science and Technology, Poland
Elias Pimenidis	University of the West of England, UK
Jaroslav Pokorný	Charles University in Prague, Czech Republic
Nikolaos Polatidis	University of Brighton, UK
Elvira Popescu	University of Craiova, Romania

Piotr Porwik	University of Silesia in Katowice, Poland
Petra Poulova	University of Hradec Králové, Czech Republic
Małgorzata Przybyła-Kasperek	University of Silesia in Katowice, Poland
Paulo Quaresma	Universidade de Évora, Portugal
David Ramsey	Wrocław University of Science and Technology, Poland
Mohammad Rashedur Rahman	North South University, Bangladesh
Ewa Ratajczak-Ropel	Gdynia Maritime University, Poland
Sebastian A. Rios	University of Chile, Chile
Keun Ho Ryu	Chungbuk National University, South Korea
Daniel Sanchez	University of Granada, Spain
Rafał Scherer	Częstochowa University of Technology, Poland
Yeong-Seok Seo	Yeungnam University, South Korea
Donghwa Shin	Yeungnam University, South Korea
Andrzej Siemiński	Wrocław University of Science and Technology, Poland
Dragan Simic	University of Novi Sad, Serbia
Bharat Singh	Universiti Teknologi PETRONAS, Malaysia
Paweł Sitek	Kielce University of Technology, Poland
Adam Słowik	Koszalin University of Technology, Poland
Vladimir Sobeslav	University of Hradec Králové, Czech Republic
Kamran Soomro	University of the West of England, UK
Zenon A. Sosnowski	Białystok University of Technology, Poland
Bela Stantic	Griffith University, Australia
Stanimir Stoyanov	University of Plovdiv "Paisii Hilendarski", Bulgaria
Ja-Hwung Su	Cheng Shiu University, Taiwan
Libuse Svobodova	University of Hradec Králové, Czech Republic
Jerzy Swiątek	Wrocław University of Science and Technology, Poland
Andrzej Swierniak	Silesian University of Technology, Poland
Julian Szymański	Gdańsk University of Technology, Poland
Yasufumi Takama	Tokyo Metropolitan University, Japan
Zbigniew Telec	Wrocław University of Science and Technology, Poland
Dilhan Thilakarathne	Vrije Universiteit Amsterdam, The Netherlands
Diana Trandabat	University "Alexandru loan Cuza" of Iași, Romania
Maria Trocan	Institut Superieur d'Electronique de Paris, France
Krzysztof Trojanowski	Cardinal Stefan Wyszyński University in Warsaw, Poland
Ualsher Tukeyev	al-Farabi Kazakh National University, Kazakhstan

Olgierd Unold	Wrocław University of Science and Technology, Poland
Jørgen Villadsen	Technical University of Denmark, Denmark
Thi Luu Phuong Vo	Vietnam National University HCMC, Vietnam
Wahyono Wahyono	Universitas Gadjah Mada, Indonesia
Paweł Weichbroth	Gdańsk University of Technology, Poland
Izabela Wierzbowska	Gdynia Maritime University, Poland
Krystian Wojtkiewicz	Wrocław University of Science and Technology, Poland
Xin-She Yang	Middlesex University London, UK
Tulay Yildirim	Yildiz Technical University, Turkey
Drago Zagar	University of Osijek, Croatia
Danuta Zakrzewska	Łódź University of Technology, Poland
Constantin-Bala Zamfirescu	Lucian Blaga University of Sibiu, Romania
Katerina Zdravkova	Ss. Cyril and Methodius University in Skopje, Macedonia
Vesna Zeljkovic	Lincoln University, USA
Jianlei Zhang	Nankai University, China
Zhongwei Zhang	University of Southern Queensland, Australia
Adam Ziębiński	Silesian University of Technology, Poland

Contents – Part I

Case-Based Reasoning and Machine Comprehension

Computer Vision

Data Mining and Machine Learning

Contents – Part II

Knowledge Integration and Analysis

Case-Based Reasoning and Machine Comprehension

On the Improvement of the Reasoning Cycle in Case-Based Reasoning

Fateh Boulmaiz[1(✉)], Patrick Reignier[1], and Stephane Ploix[2]

[1] University Grenoble Alpes, INP Grenoble, LIG, 38000 Grenoble, France
{fateh.boulmaiz,patrick.reignier}@univ-genoble-alpes.fr

[2] University Grenoble Alpes, INP Grenoble, G-SCOP, 38000 Grenoble, France
stephane.ploix@univ-genoble-alpes.fr

Abstract. Case-based reasoning (CBR) is undoubtedly by far one of the most intuitive artificial intelligence problem-solving approaches. It is inherent to the reuse of existing experience that includes solutions to problems or mechanisms to derive these solutions. Unfortunately, existing CBR systems are lacking in generality as the adaptation process is usually driven by the application domain and heavily based on the domain expert's knowledge. Moreover, existing CBR systems perform poorly as they deal with each step of the CBR methodology separately and independently from the other steps. This work is an effort to lay a primary foundation for a generic case-based reasoning framework that relies on a domain-independent approach. Each step of the reasoning process is conceived to fulfil not only specific requirements of that stage but also to sustain other stages' processes. The experimental results of the application of our approach in a sober consumption energy system in buildings show the effectiveness of our approach.

Keywords: Case based reasoning · Similarity · Compositional adaptation · Local model · Genetic algorithm

1 Introduction

Case-based reasoning is a discipline of artificial intelligence science that is increasingly adopted for extremely diverse applications in domains as varied as engineering [13], medicine [2], finance [14], etc. The CBR approach is akin to human reasoning that uses past problem-solving experiences to address a target (new) problem. Previous experiences are organized in a case base in the form of cases called source cases.

Since its early days, the CBR approach has enjoyed significant enthusiasm within the artificial intelligence community, as it was broadly considered to be a successful approach for solving many of the issues posed, for so long time, by the classical rule-based approach and alternative techniques such as neural networks or regression models. This is motivated by many considerations. Firstly, CBR approach can reduce the knowledge acquisition bottleneck that

N. T. Nguyen et al. (Eds.): ACIIDS 2023, LNAI 13995, pp. 3–16, 2023.
https://doi.org/10.1007/978-981-99-5834-4_1

has plagued machine learning approaches since their inception. Second, CBR may facilitate the maintenance of the system (e.g., deleting data or modifying reasoning rules) by overcoming the knock-on effect challenging some rule-based approaches. Finally, unlike other machine learning approaches, a CBR system can reason by providing reasonable solutions even if it lacks a sufficiently large and correct data set to guarantee optimal learning. Indeed, the focus on situation-specific knowledge as a reference resource for reasoning considerably facilitates the knowledge acquisition process.

Despite the benefits afforded by the CBR approach, the latter has introduced new challenges of its own. It has facilitated the acquisition of knowledge on a certain level since cases are easy to learn, however, this is just one category of knowledge needed for CBR. Additional types of knowledge are required to accomplish the reasoning process including domain description, similarity knowledge, and adaptation knowledge. Each type of knowledge involves a dedicated learning process. Unlike domain knowledge and case knowledge, the similarity and adaptation knowledge available initially are difficult to model, imprecise, or incomplete, moreover, they may evolve over time. Therefore, it is important to set up mechanisms to support the acquisition of this knowledge to refine it and make it evolve as the system is being used. This highlights the challenge of managing these knowledge bases from their design through the acquisition process to their maintenance. Past works have addressed this problem by considering each type of knowledge autonomously. However, these knowledge are closely related and should not be considered independently. For instance, adaptation knowledge is often used in the retrieval stage, to decrease the adaptation effort.

Moreover, several efforts have focused on similarity estimation in the retrieval process by proposing approaches involving in-depth domain knowledge or by specifying static similarity knowledge (a taxonomy is proposed in [5]), such that this knowledge may be applied to all target problems. Unfortunately, static knowledge may be relevant for some target problems but not for others, resulting in fluctuating performance of the retrieval process even in the same domain. Furthermore, despite the vast literature on the topic (see for instance [12]), existing adaptation approaches are heavily dependent on the application domain which makes mass deployment of existing systems difficult to envisage and the maintenance processes challenging. A comprehensive generic adaptation method has yet to be formulated. Currently, adaptation heavily relies on the application domain, requiring domain experts to provide adaptation rules. A domain-independent framework has been proposed, with the variational method [7] being the most well-formulated approach. This method evaluates the influence of source problem descriptors on source solution descriptors and measures variations between the source and target problems to compute a target solution. While the native technique requires adaptation knowledge from domain experts, an extended approach [11] incorporates learning adaptation knowledge from other domain knowledge sources. Additionally, knowledge discovery techniques [4] from data mining have been suggested for generating adaptation rules. However, accurately identifying influence relationships in complex systems, especially with delayed impact phenomena, remains challenging. An alternative perspective [8] argues for fully automated learning of adaptation knowledge, based on the assumption that differences between cases in the case base reflect differences between future problems and existing cases. This involves pairwise comparisons and the formulation of adaptation rules based on differences in problem features and solutions. However, this method shifts the maintenance of adaptation knowledge to a rules base, introducing additional maintenance issues.

To address the previously mentioned issues, we present in this study an efficient framework for the CBR approach based on a holistic methodology. Specifically, the main contributions of this research are summed up as follows:

- propose a more refined formalism than the traditional one (problem, solution) to represent a case.
- develop an approach based on a genetic algorithm for features weighting.
- provide a method to identify similar cases by defining appropriate metrics based on the new formalism and the statistical metric F1-score.
- suggest a novel adaptation strategy aiming at developing an efficient reuse process that is as independent as possible from the application domain.

Our CBR methodology is presented in Sect. 2. Section 3 reports on an evaluation of our approach by considering a case study with real-world data. Section 4 provides conclusion and future improvements.

2 A More Domain Knowledge Independent Approach

This section presents a new methodology to address the findings outlined in the introduction. Particularly, we present a new formalism to represent a case, followed by a description of a novel method to evaluate the similarity and finally we develop a new domain-independent approach for the adaptation process.

2.1 Case Structure

Existing CBR studies most often adopt the case structure originally proposed in [10], where the case knowledge is divided into a specification of the problem and a description of the solution. The problem part describes the objectives to be achieved, whereas the solution part includes the description of the solution provided by the reasoning process. In contrast to earlier systems and for the design requirements of the different phases of our approach, we propose a new structure for a case based on a finer classification of the application domain phenomena, that are assumed to be described in a language $\mathfrak{L}_D$, as:

- *context variables (features)*: Model the phenomena undergone by the environment of the application domain, they are phenomena over which one does not have control. For instance, the weather conditions in an energy management system (EMS).
- *action variables*: Represent the controllable phenomena. For instance, adjusting the set-point of a heating system in an EMS.

- *effect variables*: Model the state of the system after applying the actions. For instance, the temperature inside a room after opening the windows in an EMS.

We suppose that the context, action and effect variables are described in languages $\mathcal{L}_C$, $\mathcal{L}_A$ and $\mathcal{L}_E$ respectively, thus $\mathcal{L}_D = \mathcal{L}_C \cup \mathcal{L}_A \cup \mathcal{L}_E$.

Definition 1. (Case). *A case is a tuple of three components* $(\mathcal{C}, \mathcal{A}, \mathcal{E}) \in \mathcal{L}_C \times \mathcal{L}_A \times \mathcal{L}_E$ *which assumes the existence of a relation* $\mathcal{R} : \mathcal{C} \times \mathcal{A} \to \mathcal{E}$, *meaning that* $\mathcal{E}$ *is the consequence of application of* $\mathcal{A}$ *to* $\mathcal{C}$.

Considering the revised structure of a case introduced in Definition 1, the guiding hypothesis of the CBR approach (Similar problems produce similar solutions), is reformulated by Hypothesis 1.

Hypothesis 1. (Consistency). *Carrying out the same actions in similar contexts generates similar effects.*

In the following, we assume that each variable $\mathcal{Q} \in \{context, action, effect\}$ is expressed by a descriptor $d_{\mathcal{Q}} = (v_{\mathcal{Q}}, R_{\mathcal{Q}})$, where $v_{\mathcal{Q}}$ is a unique attribute associated with a sub-language $R_{\mathcal{Q}} \in \{\mathcal{L}_C, \mathcal{L}_A, \mathcal{L}_E\}$ and $R_{\mathcal{Q}}$ could take several forms. e.g., an atomic value, in the form of constraints, or more generally, in the form of a vector $\{x_1, x_2, ..., x_{n'}\} \subseteq \mathcal{L}_{\mathcal{Q}}$. In our approach, we adopt the last representation (vector representation) which is particularly appropriated for weak theory domains.

2.2 Retrieving Similar Cases

The main purpose of the retrieval stage in CBR is to identify relevant cases for the process of current problem resolution. In the following, we first introduce some definitions and notations for the notions that will be employed throughout the presentation of our approach, followed by a description of the proposed method for estimating the weights of the variables, and we close this section by a presentation of the similar cases retrieving approach.

Definitions and Notations

Definition 2. (Sensitivity distance). *The sensitivity distance* d^Q_{sen} *associated with a variable* Q *consists of a threshold defined by a domain expert and corresponding to the maximum distance above which two values of the variable* Q *are not considered similar.*

In practice, it is unusual to find feature values that match up perfectly, intuitively the Sensitivity distance notion allows to overcome this limitation. For instance, the change in temperature of the environment is perceptible by humans from the threshold of 1 °c. Another example of a sensitivity distance is the change in noise unnoticed by a human, which is 3 dB. More generally, we adopt the notion of sensitivity distance to compare the effect variables between two cases.

Definition 3. (Effect-based similarity). *Two cases are similarly based on their effects if the maximum distance between the values of each effect variable does not exceed the corresponding sensitivity distance:* $\forall (C1, C2) \in CB, \forall \mathcal{E}_i \in \mathcal{L}_E, \mathcal{E}_i^{C1} - \mathcal{E}_i^{C2} \leq d_{sen}^{\mathcal{E}_i}$. *With* $d_{sen}^{\mathcal{E}_i}$ *– the sensitivity distance for effect variable* $\mathcal{E}_i$.

Definition 4. (Performance). *The performance* $\mathcal{P}_{C_i}(\mathcal{C}_i, \mathcal{A}_i)$ *of a case* $C_i \triangleq (\mathcal{C}_i, \mathcal{A}_i, \mathcal{E}_i)$ *models the quality of the effects following the application of an action plan to a particular context, i.e. the user's (dis)satisfaction* $\mathcal{S}_{C_i}(\mathcal{E}_i)$ *of the effects generated by applying the actions* $\mathcal{A}_i$ *to the context* $\mathcal{C}_i$.

Note that thereafter no assumption is made regarding the used distance metrics. Any distance metric can be used so far as it can deal with the formalism introduced in the Sect. 2.1.

Lessons learned from literature [9] confirm that the integration of weight variables in the distance calculation is essential since it results in more accurate results. So, we propose to adopt a metric that integrates the variables' weights in its distance evaluation. Hereafter, we present a method based on a genetic algorithm (GA) and a clustering strategy to estimate the weights of the variables.

Features Weighting. The key idea behind our weighting approach is to look for the weights that group the most similar cases (with the closest context-action distances $\mathcal{D}_{CA}$, which stands for distance between Context and Action variables) and, therefore, with the closest effect distances $\mathcal{D}_E$. It is fundamental to associate the action and the context variables (through the $\mathcal{D}_{CA}$ distance) since the effects of the actions depend on the context in which they are applied and vice versa. For instance, for an EMS, the importance of sunshine to warm a room in winter depends on the opening/closing of the blinds. The main steps of the GA process are described shortly as follows:

1. *creation of the chromosome population*: The initial population of the GA is randomly generated. It is composed of a certain number of individuals which are composed of real values corresponding to the requested weights. The dimension of each individual corresponds to the number of context and action features in the case base.
2. *grouping similar cases*: Applying a K-means clustering on the cases of the case base based on the $\mathcal{D}_{CA}$ distance weighted with the chromosome values.
3. *performance evaluation*: Assess the fitness value of each individual according to the adopted fitness function. The latter consists of averaging the distance $\mathcal{D}_E$ between cases of the same cluster. The optimization criterion consists of two terms. The first one is to minimize the average of the distance $\mathcal{D}_E$, while the second is to minimize the number of the formed clusters.
4. *generation of new population*: Applying the evolution operators (selection, crossover, mutation) to produce a novel population.
5. reiteration of steps 2-4 until the satisfaction of the termination criteria. For the sake of computational efficiency, the optimization process is terminated when no improvement in the optimization criterion is observed after 10 consecutive iterations or the number of iterations reaches 500 iterations.

Similar Cases Retrieving. Unlike the traditional distance-based methods for extracting similar cases, which rely on a KNN-like approach, we introduce an approach that automatically adapts the number of similar source cases based on a similarity distance threshold. The similarity evaluation is based on the $\mathcal{D}_{CA}$ distance for the same motives explained in Sect. 2.2. The objective is to determine the similar source cases from a similarity metric, which assumes a threshold distance beyond which the cases are not considered similar.

Equation (1) introduces a similarity measure that answers this requirement. It provides a meaning to μ, it is the maximum $\mathcal{D}_{CA}$ that should not be exceeded between two cases considered as a similar. The learning of the threshold distance μ_* for the target case C_* involves two steps:

- *learn a distance threshold for each source case:* It is not possible to formulate a common approach to learning a threshold distance to identify the cases satisfying the similarity criterion, this distance is domain-dependent and subject to the context of each case. Given Hypothesis 1, we introduce Property 1, which defines the distance threshold μ_{C_i} corresponding to each source case C_i. We note $\Pi = \{\mu_{C_i}\}, \forall\, C_i \in CB$.
- *learn a distance threshold for target case:* The goal is to estimate a distance threshold μ_* for the target case C_* from the threshold distances computed in the previous step which generates, at worst, a number of distance thresholds equivalent to the number of source cases ($\max|\Pi| = |BC|$). Finding the distance threshold μ_* consists in looking, in the set Π, for the distance giving the best compromise between recall (the ratio of extracted relevant similar cases compared to the total number of relevant similar cases) and precision (the number of extracted relevant similar cases compared to the total number of extracted cases). This is achieved by using a statistically optimal decision method, which is the F1-score. So, the distance threshold μ_* corresponds to the distance $\mu \in \Pi$ maximizing the F1-score as expressed in Eq. (2).

$$\forall\, C_i, C_j \in CB,\ \mathcal{S}(C_i, C_j) = \begin{cases} \left(1 - \frac{\mathcal{D}_{CA}(C_i,C_j)}{\mu}\right), & \text{if } \mathcal{D}_{CA}(C_i, C_j) \leq \mu \\ 0, & \text{otherwise.} \end{cases} \tag{1}$$

Property 1. The context-action distance threshold μ_{C_i} for each case C_i should satisfy the condition that, $\forall C_j \in CB, \mathcal{D}_{CA}(C_i, C_j) \leq \mu_{C_i} \Rightarrow C$ and C_i are similarly based on their effects (see Definition 3).

$$\mu_* = \underset{\mu_{C_i}}{\operatorname{argmax}}(F1 - score(\mu_{C_i})) \tag{2}$$

However, it is not possible to apply the context-action threshold straightforwardly. Indeed, the actions of case C_* are unknown (that is precisely what we are looking for) and the solely information available for the new case C_* is the context data. Consequently, context variables are the only data used to assess the similarity of the case C_* to the ones in the case base using the context distance $\mathcal{D}_C$. So, the profile of the retrieval function is $F_{\mathcal{S}} : \mathcal{C}_* \mapsto (\mathcal{C}_i, \mathcal{A}_i, \mathcal{E}_i) \triangleq C_i \in CB$.

The challenge raised at this juncture is how to set the threshold $\overline{\mu}_*$ which represents the maximum context distance defining the similar neighbourhoods $\mathcal{S}_{C_*}$ of the case C_*, with $\forall\ C_i \in \mathcal{S}_{C_*}, \mathcal{D}_C(C_*, C_i) \leq \overline{\mu}_*$. One way to overcome this challenge is to deduce the threshold $\overline{\mu}_*$ for the context distance from the threshold μ_* of the context-action distance using a projection function ϕ that maps the feature space $\mathcal{L}_C^n \times \mathcal{L}_A^m$ to an n-dimensional features subspace $\phi(\mathcal{L}_C^n \times \mathcal{L}_A^m) \in \mathcal{L}_C^n$. Where n, the context features number and m, the action features number. The projection function ϕ seeks to conserve the data structure between the original data set and the data resulting from the transformation. The data structure is captured as distances between the cases. We do not present the projection process for lack of space, interested readers are referred to [6] for a broad review.

2.3 An Optimization Approach for Adaptation

The adaptation process aims at modifying one or several similar source cases' actions to produce the target actions that satisfy the constraints of the target context. Formally, the adaptation process is a couple $(F_\mathcal{S}, F_\mathcal{A})$, where $F_\mathcal{S}$ is a retrieval function between the source cases and the target case context as explained in the Sect. 2.2. The function $F_\mathcal{A}$ is a transformation function that produces a potential solution A_* by modifying the solutions of the source cases obtained thanks to the function $F_\mathcal{S}$. The profile of the $F_\mathcal{A}$ function is given by Formula (3).

$$\begin{aligned} &F_\mathcal{A} : (\mathcal{L}_C \times \mathcal{L}_A \times \mathcal{L}_E)^k \times \mathcal{L}_C \longrightarrow \mathcal{L}_A \\ &((\mathcal{C}_i, \mathcal{A}_i, \mathcal{E}_i)_{i \in [1,k]}, \mathcal{C}_*) \longmapsto \mathcal{A}_*, \forall\ (\mathcal{C}_i, \mathcal{A}_i, \mathcal{E}_i) \triangleq C_i \in \mathcal{S}_{C_*} \end{aligned} \tag{3}$$

where $k = |\mathcal{S}_{C_*}|$ – the number of similar source cases to case C_*.

Formula (3) does not constrain the number of similar cases required to perform the adaptation as long as it is not zero. The approach we propose in this paper is a compositional adaptation method, where the solutions of many source cases are efficiently mixed to produce a target solution. Indeed, previous studies [3] have pointed out the advantage of compositional adaptation compared to single-case adaptation which usually generates less accurate solutions. This is because often just a part of the source problem is relevant to the target problem, making the adaptation task difficult or even impossible.

The adaptation strategy proposed here assumes a two-phase approach to adaptation:

- *hybrid learning model*: An essential cause for the poor learning process in machine learning systems is the bias for either exclusively local learning or purely global learning approaches. In a fully local strategy, excessive importance is usually accorded to local irregularities in the input data and usually, the learning process fails to capture the structure of the data. In a strictly global strategy, the model complexity may prove to be inadequate on some parts of the input data set. It is obvious that both learning strategies are complementary.

To capitalize on the advantages of each strategy, a hybrid learning approach that conciliates the dichotomy between global and local learning should be suggested. We propose to learn a local model in the neighbourhood retrieved in the previous step using the weighted data. This model is not strictly local as it captures the global structure of the data thinks to the influence of the weights assigned to the features.
The literature provides several approaches commonly applied to model learning, including artificial neural networks, decision trees, support vector machines, etc. Nevertheless, there are no clear guidelines for the choice of a model, moreover, none of these techniques is distinguished by its effectiveness compared to the others. The adopted model is still the designer's choice. The study of the learning model algorithms lies outside the scope of this work. More information on learning models can be found in [15]. However, Learning a model on similar cases should be far simpler and more accurate than a global model of the system.
More formally, this step consists in learning a model $\mathcal{M}_{C_*}$ of the local behaviour of the effects according to the context and the actions:

$$\mathcal{M}_{C_*} : \mathcal{E}_i = g(\mathcal{C}_i, \mathcal{A}_i), \forall\ (\mathcal{C}_i, \mathcal{A}_i, \mathcal{E}_i) \triangleq C_i \in \mathcal{S}_{C_*} \tag{4}$$

With g – the modelling function, $\mathcal{S}_{C_*}$ – the set of similar cases.

- *model optimization*: The solution to be proposed (the actions) consists in solving an optimization problem with an objective function representing the performances $\mathcal{P}(\mathcal{C}_*, \mathcal{A}_*)$ of the proposed actions (cf. Definition 4). We outline below the mathematical foundations behind this optimisation:
  - the objective function, i.e., the criteria that one seeks to optimize consist of maximizing the performance $\mathcal{P}_{C_*}$ of the case C_* to solve.
  - the decision variables which represent the variables that can be changed to influence the objective function value; consist of actions variables.
  - the constraints, i.e. the conditions that the decision variables have to check so that the solution is accepted, are to satisfy the local model defined by the Formula (4).

  So, the general form of the optimization problem which consists in determining the actions $\mathcal{A}_*$ to be suggested to solve the case C_* should be as follows: $\mathcal{E}_* = \text{argmax}_{\mathcal{A}_*} \mathcal{P}(\mathcal{C}_*, \mathcal{A}_*)$, subject to the learned model $\mathcal{M}_{C_*} : \mathcal{E}_* = g(\mathcal{C}_*, \mathcal{A}_*)$.
  There are several methods to solve this optimization problem A comprehensive survey of existing techniques, including Population-based optimization algorithms such as Particle Swarm Optimization and Genetic Algorithm, as well as gradient-based methods, can be found in [17]. We do not impose any specific method for solving the optimization problem. The system designer has the freedom to choose the method that best suits their needs and preferences.

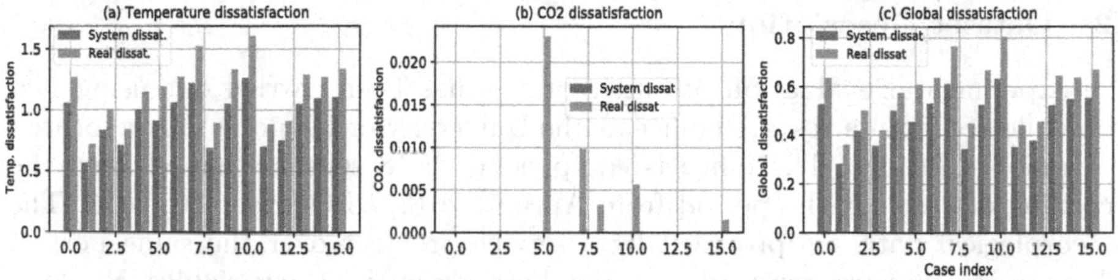

Fig. 1. Performance evaluation.

3 Experimental Evaluation

3.1 Case Study

A case study is carried out to evaluate the efficiency of the suggested approach. It consists in using the CBR paradigm in an EMS to improve the energy efficiency of buildings.

The goal of the EMS is to improve occupant comfort at the same or lower energy cost if possible. Most of the literature uses CBR to improve energy flexibility by implementing demand-response or demand-side management approaches in the form of load shifting, peak shaving strategies, etc. Unfortunately, these approaches are often rejected by occupants who are usually overlooked in the management process. Meanwhile, some studies have shown that occupant actions have a critical impact on the energy efficiency of buildings [16]. Thus, intelligently guiding the occupant in his actions (such as opening/closing windows, adjusting the setpoints of the air conditioner), can contribute to obtain better comfort in a building at the same energy rate if not less.

Based on our approach, we have developed an EMS whose objective is to make the occupant aware of the crucial impact of his actions on the energy efficiency of the building. The EMS guides the occupant by proposing an optimal action schedule to enhance his comfort (indoor temperature, air quality, etc.) without additional energy cost.

The performance function used in the adaptation process consists in evaluating the user dissatisfaction with the generated effects following the application of the actions proposed by the EMS. The user dissatisfaction with the thermal comfort $\mathcal{S}_T^h(T)$ and the air quality $\mathcal{S}_C^h(C)$ at the hth hour are modelled by the Formula (5). The global dissatisfaction $\mathcal{S}^h$ is defined as the average of $\mathcal{S}_T^h(T)$ and $\mathcal{S}_C^h(C)$.

$$\mathcal{S}_T^h(T) = \begin{cases} 0 & \text{if } T \in [21, 23] \\ \frac{T-23}{26-23} & \text{if } T > 23 \\ \frac{21-T}{21-18} & \text{if } T < 21 \end{cases}, \quad \mathcal{S}_C^h(C) = \begin{cases} 0 & \text{if } C \leq 500 \\ \frac{C-500}{1500-1000} & \text{if } C > 500 \end{cases} \tag{5}$$

3.2 Dataset Description

The experimental evaluation of our approach has been carried out using real data collected from a research office at the University of Grenoble Alpes located at Grenoble, France. The office is equipped with 18 sensors that allowed the recording of data for the period from April 1, 2015, to October 30, 2016. The meteorological data are provided by a service provider over the same period. All measurements are averaged over each hour except for the variables Windows opening (v_{13}) and Door opening (v_{14}) where the values correspond to the fraction of the hour during which the window (respectively the door) was open. For instance, if the window was open for 15 minutes in the kth hour of the day, then $v_{13} = 60/15 = 0.25$. The different variables that model the features considered in this experimentation are:

- *effect variables:* indoor temperature (v_1), indoor CO_2 concentration (v_2).
- *context variables:* corridor temperature (v_3), illuminance (v_4), solar radiation (v_5), wind speed (v_6), corridor CO_2 concentration (v_7), electricity power (v_8), heater temperature (v_9), occupancy (v_{10}), nebulosity (v_{11}), outdoor temperature (v_{12}).
- *action variables:* window opening (v_{13}), door opening (v_{14}).

Preprocessing. As with any data-driven approach, the performance and accuracy of our approach are heavily dependent on the quality of the available data used during the learning phase. To ensure optimal data quality, a preprocessing step is required. In this case study, the process carried out in the data preprocessing phase consists of three steps:

- *data cleaning*: The data cleaning process aims to complete the missing values, to eliminate the noise while identifying the outliers and correcting them. Each detected outlier or a missing value is replaced by the average value of the left and right neighbours in the values vector of the feature.
- *data normalization*: To reduce the effects of dominant features, we have opted for the MinMax method to rescale the features' values between 0 and 1.
- *data filtering*: The data are collected continuously. For efficiency reasons, we filter out the days on which there are no people in the office (e.g., closed days and weekends) since there are no actions recorded on these days.

Case Base Structure. The case base CB used hereafter consists in organizing the daily data obtained from the previously described database into cases. The vector representing each feature is coded as a 24-dimensional vector, whose elements are the feature values at each hour (from 0:00 to 23:00). As the studied building is a university building, we have restricted our analysis to the office hours only. So, we present to the office occupant an action plan for the time range between 08:00 and 20:00 (the office hours) since during the other hours, there are no people in the office and therefore there are no actions to propose.

3.3 Testbed Setup

Following the data processing phase, 98 d were retained for the experimentation. The case base is randomly split into disjoint training (74 d) and test (24 d) sets (i.e. 75% and 25% respectively). The training set corresponds to source cases, and the test set represents target cases.

To evaluate the accuracy of our approach, we need to reproduce the behaviour of the office following the application of the proposed action plan. The simulation is performed using the physical model of the office proposed in [1]. To be consistent with the results generated by the physical model, the values of the effect variables from the database presented in Sect. 3.2 are replaced by the values simulated by the physical model. The local behaviour of the data is predicted by a regressive approach using a Generalized Linear Model (GLM).

3.4 Empirical Evaluation

Features Weighting and Context-Distance Threshold. The features are weighted using the genetic algorithm presented in Sect. 2.2. The latter converged to the optimal weights in the 11th generation. The values depicted in Table 1 indicate that the variable v_3 (corridor temperature) has the highest weight which is in agreement with the findings of the physical model of the office, confirming the efficiency of our algorithm in estimating the weights of the features.

The proposed approach for evaluating the similarity allowed to define a threshold for the context distance of $\overline{\mu}_* = 1.2$. Unfortunately, due to the lack of data, this threshold allows obtaining neighbours for only 16 cases out of the 24 validation cases.

Table 1. Features weights.

Feature	v_3	v_4	v_5	v_6	v_7	v_8	v_9	v_{10}	v_{11}	v_{12}	v_{13}	v_{14}
Weight	0.211	0.102	0.026	0.042	0.040	0.088	0.108	0.039	0.004	0.029	0.016	0.094

Weighting Approach Accuracy. The accuracy of the proposed method to estimate weights was estimated using two commonly used statistical indicators, namely root mean square error (RMSE) and coefficient of determination (R^2) between the model output value (estimated temperature and estimated CO_2 concentration) and the real value from the case base. In particular, we study the impact of the number of nearest neighbours to be considered on the accuracy of the learned local model. We experiment the influence of several values for the number of neighbours K from the lowest value ($K = 2$) to the whole case base ($K = 74$) including the particular case $K = K(\overline{\mu}_*)$ corresponding to the threshold distance $\overline{\mu}_*$. The results are summarized in Table 2. The indicators values presented in Table 2 consider the average of the corresponding indicators values for the retained test cases (16 cases).

By examining the RMSE and R^2 values reported in Table 2, one can observe that a correlation can be established between the number of K and the RMSE and R^2 values. Indeed, it can be verified that whereas there is a net tendency of increasing R^2 when moving away from the value $\overline{\mu}_*$, a clear trend of decreasing RMSE can be observed. This emphasizes that considering fewer, but more similar cases is more suitable for the CBR process than considering a large number of less similar cases.

Table 2. Model accuracy.

K	Temperature		CO_2 concentration	
	RMSE	R^2(%)	RMSE	R^2(%)
all cases	0.659	76.21	18.85	74.09
60	0.619	76.84	15.14	76.28
50	0.602	78.02	14.35	81.92
40	0.559	78.51	12.63	83.78
30	0.553	79.02	9.04	86.97
20	0.531	82.17	8.76	88.64
10	0.315	85.42	5.28	90.96
2	0.273	89.67	4.44	91.66
$K(\overline{\mu}_*)$	0.103	91.85	4.02	92.97

Adaptation Performance. The efficiency of the adaptation approach is evaluated by comparing, for each test case, the performance (using Formulas (5)) of the proposed actions and the performance of the actions already recorded in the case base. The average of the performances of all effect variables is used to estimate the overall performance.

Figures 1.(a) and, 1.(b) show the performance of the system regarding variable v_1 and variable v_2. For both variables, the system (red bars) outperforms the recorded actions (grey bars) for all 16 test cases. Specifically, the system outperforms by 13% to 24% for v_1 and up to 100% for v_2. We observe that for the variable v_2, the actions recorded for the cases 0,1,2,3,4,6,9,11,12,13 and 14 are already optimal, the system is as good as these actions. For the five remaining cases, the results of the system are better. Figure 1.(c) depicts the overall performance of the system. The results show that the system improves the performance in 100% of the cases. The enhancement is between 13% and 31%.

4 Conclusion

We presented a novel approach to the case-based reasoning system conception with more generic methodologies. In particular, we proposed a refined case formalism. The latter is used to define a new method for estimating similarity using a genetic algorithm as a feature weighting mechanism. Furthermore, we introduced a domain-independent strategy for adaptation by combining machine learning and optimization methods. We illustrated the effectiveness of this approach through a detailed application example in the context of an energy management system. Future work will include the development of efficient methods for indexing and incorporating new cases into the case base (the retain stage) based on the knowledge learned from the retrieval and adaptation steps. Furthermore, we plan to investigate the feasibility of an adaptation approach based on optimization considering the context distance and the performance of each neighbouring case. Specifically, we will investigate the performance of an adaptation approach considering an operator which not only avoids poorly performing neighbors but also moves away from them.

References

1. Alyafi, A.A., Pal, M., Ploix, S., Reignier, P., Bandyopadhyay, S.: Differential explanations for energy management in buildings. In: 2017 Computing Conference, pp. 507–516 (2017)
2. Blanco, X., Rodríguez, S., Corchado, J.M., Zato, C.: Case-based reasoning applied to medical diagnosis and treatment. In: Omatu, S., Neves, J., Rodriguez, J.M.C., Paz Santana, J.F., Gonzalez, S.R. (eds.) Distributed Computing and Artificial Intelligence. AISC, vol. 217, pp. 137–146. Springer, Cham (2013). https://doi.org/10.1007/978-3-319-00551-5_17
3. Chedrawy, Z., Raza Abidi, S.S.: Case based reasoning for information personalization: using a context-sensitive compositional case adaptation approach. In: IEEE International Conference on Engineering of Intelligent Systems (2006)
4. Craw, S., Wiratunga, N., Rowe, R.C.: Learning adaptation knowledge to improve case-based reasoning. Artif. Intell. **170**(16), 1175–1192 (2006)
5. Cunningham, P.: A taxonomy of similarity mechanisms for case-based reasoning. IEEE Trans. Knowl. Data Eng. **21**(11), 1532–1543 (2009)
6. Espadoto, M., Martins, R.M., Kerren, A., Hirata, N.S., Telea, A.C.: Toward a quantitative survey of dimension reduction techniques. IEEE Trans. Vis. Comput. Graph. **27**(3), 2153–2173 (2021)
7. Fuchs, B., Lieber, J., Mille, A., Napoli, A.: An algorithm for adaptation in case-based reasoning. In: Horn, W. (ed.), ECAI 2000, pp. 45–49. IOS Press (2000)
8. Hanney, K., Keane, M.T.: Learning adaptation rules from a case-base. In: Smith, I., Faltings, B. (eds.) EWCBR 1996. LNCS, vol. 1168, pp. 179–192. Springer, Heidelberg (1996). https://doi.org/10.1007/BFb0020610
9. Iqbal, R.: Empirical learning aided by weak domain knowledge in the form of feature importance. In: Proceedings - 2011 International Conference on Multimedia and Signal Processing (2010)
10. Kolodner, J.: Case-Based Reasoning. Morgan Kaufmann Publishers, Burlington (1993)

11. Leake, D.B., Kinley, A., Wilson, D.: Acquiring case adaptation knowledge: A hybrid approach. In: Proceedings of the Thirteenth National Conference on Artificial Intelligence, vol. 1, pp. 684–689. AAAI Press (1996)
12. Leake, D.B., Ye, X.: Learning to improve efficiency for adaptation paths. In: International Conference on Case-Based Reasoning (2020)
13. Lees, B., Hamza, M., Irgens, C.: Chris: case-based reasoning support for engineering design. In: Intelligent Systems in Design and Manufacturing III, vol. 4192, pp. 394–402. International Society for Optics and Photonics, SPIE (2000)
14. Portinale, L., Leonardi, G., Artusio, P.: A smart financial advisory system exploiting case-based reasoning. In: CEUR Workshop (2016)
15. Shalev-Shwartz, S., Ben-David, S.: Understanding Machine Learning: From Theory to Algorithms. Cambridge University Press, USA (2014)
16. Uddin, M.N., Wei, H.H., Chi, H.L., Ni, M.: Influence of occupant behavior for building energy conservation: a systematic review study of diverse modeling and simulation approach. Buildings **11**(2), 41 (2021)
17. Wang, Y., Gao, W., Gong, M., Li, H., Xie, J.: A new two-stage based evolutionary algorithm for solving multi-objective optimization problems. Inf. Sci. **611**, 649–659 (2022)

Exploring Incompleteness in Case-Based Reasoning: A Strategy for Overcoming Challenge

Fateh Boulmaiz[1(✉)], Patrick Reignier[1], and Stephane Ploix[2]

[1] University Grenoble Alpes, INP Grenoble, LIG, 38000 Grenoble, France
{fateh.boulmaiz,patrick.reignier}@univ-genoble-alpes.fr
[2] University Grenoble Alpes, INP Grenoble, G-SCOP, 38000 Grenoble, France
stephane.ploix@univ-genoble-alpes.fr

Abstract. Data quality is a critical aspect of machine learning as the performance of a model is directly impacted by the quality of the data used for training and testing. Poor-quality data can result in biased models, overfitting, or suboptimal performance. A range of tools are proposed to evaluate the data quality regarding the most commonly used quality indicators. Unfortunately, current solutions are too generic to effectively deal with the specifics of each machine learning approach. In this study, a first investigation on data quality regarding the completeness dimension in the case-based reasoning paradigm was performed. We introduce an algorithm to check the completeness of data according to the open-world assumption leading to improving the performance of the reasoning process of the case-based reasoning approach.

Keywords: Case based reasoning · Data quality · Data completeness

1 Introduction

In the age of massive digitization and ubiquitous computing, assuring the quality of manipulated data becomes one of the major challenges for both companies and academic research, whatever discipline: database, artificial intelligence, image processing, information systems, etc. Several studies have confirmed the tremendous impact of data quality on the process in which such data are handled. For instance, it has been proven in [14] that the performance of machine learning algorithms is directly influenced by the quality of the data used in the learning process, while the survey conducted in [1] focused on the effect of poor data quality on the economy of countries, in particular, it estimates that poor data quality lost the USA's economy alone more than $3 trillion per year. This financial cost engendered by poor data quality is still increasing [2].

With organizations confronting more and more complex data issues which potentially influence the profitability of their business and with research proposing increasingly data quality-sensitive algorithms, the necessity for precise and

N. T. Nguyen et al. (Eds.): ACIIDS 2023, LNAI 13995, pp. 17–30, 2023.
https://doi.org/10.1007/978-981-99-5834-4_2

trusted data is more crucial than ever before. The different objectives and the multiple ways of using the data have led to different data quality dimensions (requirements). The later ones usually characterize quality properties such as accuracy, completeness, consistency, etc. Albeit the awareness of the need for acceptable data quality is reflected in the emergence of exhaustive literature devoted to this issue (see [7] for an overview), it is worth highlighting that: 1) Despite the ongoing research on data quality, there is no consensus on the properties that should be considered in defining a data quality norm [13]. For instance, the authors in [18] identify 179 dimensions for data quality, while in a more recent study [11] more than 300 properties to be considered for defining data quality are described; 2) Although some requirements have been unanimously identified to be important, there is no consensus on their precise definitions, i.e., the same requirement name is used with different semantics from one study to another; 3) Data quality assessment is a domain-specific process because of the diversity of data sources, the multiplicity of quality dimensions, and the specificity of the application domain. So, it is not possible to propose a generic data quality assessment approach applicable to all data-intensive applications.

While data quality is extensively studied by the database and data mining communities, it is entirely neglected in the machine learning domain in favor of developing learning algorithms and reasoning approaches, assuming high-quality data to feed learning algorithms. This paper addresses the issue of data quality in the context of machine learning to improve the robustness of learning algorithms in handling data of varying quality and to reduce the impact of poor quality data on the overall result of the machine learning process. However, since the scope of both the data quality and machine learning domains are broad, some limitations were imposed to make this work possible. The machine learning domain is too large to provide a single data quality assessment method that is valid for all machine learning approaches. We, therefore, restrict our scope of research to approaches based on the case-based reasoning paradigm. Concerning data quality, we only investigate the data completeness dimension, which is still widely unexplored, if not never investigated in the context of case-based reasoning. In this paper, we address the data incompleteness issue in the CBR approach using a heuristic based on the change point detection method.

The rest of the paper is structured as follows: Sect. 2 describes the background of the research. Section 3 outlines the problem statement through an motivating example followed by a formulation of the problem. Section 4 details the proposed approach to address the problem under consideration. Section 5 evaluates the proposed approach through a real case study and discusses the results. Section 6 concludes the paper and presents future work.

2 Background

2.1 Case-Based Reasoning and Data Completeness

Case-based reasoning (CBR) is a reasoning approach that relies on a collection of source cases, represented as a case base $\mathbb{CB}$. Each case $\mathbb{C}$ in the case base represents a problem-solving experience, defined as a pair $(\mathfrak{p}, \mathfrak{s})$, where $\mathfrak{p}$ denotes a problem in the specific application domain and $\mathfrak{s}$ represents its corresponding solution. In this discussion, we adopt a more detailed representation of a case $\mathbb{C}$, using a triplet $(\mathtt{C}^{\mathbb{C}}, \mathtt{A}^{\mathbb{C}}, \mathtt{E}^{\mathbb{C}})$ [5]. We define three sets: $\mathbb{C}^S$, $\mathbb{A}^S$, and $\mathbb{E}^S$. The context $\mathtt{C}^{\mathbb{C}}$ is an element of $\mathbb{C}^S$, representing the observed phenomena in the application domain. The actions $\mathtt{A}^{\mathbb{C}}$ are elements of $\mathbb{A}^S$, modeling controllable phenomena in the application domain. The effects $\mathtt{E}^{\mathbb{C}}$ are elements of $\mathbb{E}^S$, describing the consequences of applying the actions $\mathtt{A}^{\mathbb{C}}$ to the context $\mathtt{A}^{\mathbb{C}}$. The fundamental idea behind the CBR paradigm is expressed by Axiom 1. The process of solving a target case $\mathbb{C}_{tg}$, initially formed solely from the context, involves determining the relevant actions that, when applied, will produce effects and generate a new source case within the case base.

Axiom 1. (Consistency) *Similar actions applied to similar contexts yield similar effects.*

Specifically, the reasoning strategy begins by identifying a set $\mathtt{SIM}^{\mathbb{C}_{tg}}$ of source cases $\mathbb{C}_{sr}$ that are similar to the target case $\mathbb{C}_{tg}$ (retrieval stage). Next, the actions of the source cases $\mathbb{C}_{sr}$ are modified to align with the specific context of the target case $\mathbb{C}_{tg}$, resulting in the actions $\mathtt{A}^{\mathbb{C}_{\mathtt{tg}}}$ (adaptation stage). Depending on the chosen validation stage, the effects $\mathtt{E}^{\mathbb{C}_{\mathtt{tg}}}$ of applying $\mathtt{A}^{\mathbb{C}_{\mathtt{tg}}}$ to the context $\mathtt{C}^{\mathbb{C}_{\mathtt{tg}}}$ are generated, and if approved, the new target case $\mathbb{C}_{tg}(\mathtt{C}^{\mathbb{C}_{\mathtt{tg}}}, \mathtt{A}^{\mathbb{C}_{\mathtt{tg}}}, \mathtt{E}^{\mathbb{C}_{\mathtt{tg}}})$ is incorporated into the case base $\mathbb{CB}$ (memorization stage). This process can be formally expressed as follows:

$$\mathtt{CBR\ system} : \mathtt{Memorization} \circ \mathtt{Validation} \circ \mathtt{Adaptation} \circ \mathtt{Retrieval}$$

$$\mathtt{Retrieval\ function} : \mathtt{C}^{\mathbb{C}_{\mathtt{tg}}} \longmapsto \mathtt{SIM}^{\mathbb{C}_{tg}} = \{\mathbb{C}_{sr}\} \subseteq \mathbb{CB}$$

$$\mathtt{Adaptation\ function} : \mathtt{SIM}^{\mathbb{C}_{tg}} \cup \mathtt{C}^{\mathbb{C}_{\mathtt{tg}}} \longmapsto \mathtt{A}^{\mathbb{C}_{\mathtt{tg}}} \cup \{\mathtt{failure}\}$$

$$\mathtt{Validation\ function} : \mathtt{A}^{\mathbb{C}_{\mathtt{tg}}} \longmapsto \mathbb{C}_{tg}(\mathtt{C}^{\mathbb{C}_{\mathtt{tg}}}, \mathtt{A}^{\mathbb{C}_{\mathtt{tg}}}, \mathtt{E}^{\mathbb{C}_{\mathtt{tg}}})$$

$$\mathtt{Memorization\ function} : (\mathbb{CB}, \mathtt{C}^{\mathbb{C}_{\mathtt{tg}}}) \longmapsto \mathbb{CB} \cup \mathbb{C}_{tg}(\mathtt{C}^{\mathbb{C}_{\mathtt{tg}}}, \mathtt{A}^{\mathbb{C}_{\mathtt{tg}}}, \mathtt{E}^{\mathbb{C}_{\mathtt{tg}}})$$

To perform the various stages of the reasoning procedure, a CBR system relies on a collection of knowledge distributed across four distinct containers: domain knowledge, case knowledge, similarity knowledge, and adaptation knowledge [16]. Typically, multiple knowledge containers are involved at each stage of the reasoning process due to the interdependencies among them.

Completeness. Following established research in the field of knowledge bases [8] and databases [10], we adopt a perspective on completeness by introducing an ideal reference domain knowledge container $\mathtt{K}^{\mathtt{D}}_{\mathtt{R}}$. This container encompasses all real-world aspects pertaining to the application domain. The completeness of the domain knowledge $\mathtt{K}^{\mathtt{D}}$ within the CBR system is established when the application of any actions (defined within $\mathtt{K}^{\mathtt{D}}$) to any context (likewise defined within $\mathtt{K}^{\mathtt{D}}$) yields identical effects on $\mathtt{K}^{\mathtt{D}}$ as those observed on $\mathtt{K}^{\mathtt{D}}_{\mathtt{R}}$.

Definition 1. *(Completeness). Completeness relates to the capacity of the domain knowledge container within a CBR system to encompass all pertinent states of the application domain environment.*

The main obstacle in assessing and attaining completeness, as defined in Definition 1, arises from the Open World Assumption. This assumption posits that if a specific piece of real-world knowledge is not explicitly represented within the Knowledge domain $\mathtt{K}^{\mathtt{D}}$, it does not necessarily imply falsehood. Instead, it could be a valid piece of real-world knowledge that simply hasn't been incorporated into $\mathtt{K}^{\mathtt{D}}$.

Extensive research has been conducted in the field of data quality assessment, which remains a highly active area of investigation across various domains, including relational databases, big data, machine learning, data mining, and more. The verification of data quality poses significant challenges due to the following reasons:

- *Continuous data quality verification is an ongoing process.* This is necessary due to the nature of data, particularly their velocity, and the various data processing operations performed on them, such as data cleaning.
- *Data quality verification is highly contextual and domain-specific.* Metrics used to evaluate different aspects of data quality are tailored to the requirements of users/experts, taking into account the unique needs of the application domain. For instance, the aeronautics domain may have distinct data quality requirements compared to the education domain. Additionally, data quality requirements can vary depending on the specific task within a domain. In the healthcare domain, for example, there may be different quality requirements for the diagnostic phase and the treatment phase.

2.2 Change Point Analysis

Change points in a data set, which model a system, refer to sudden shifts in the data. These change points correspond to transitions that occur between different states of the modeled system, resulting from hidden changes in the data set's properties. The identification of these change points is the primary objective of change point analysis methods, which have garnered considerable attention in the field of statistics [17], as well as in various application domains including climate research [9], medical studies [19], and finance [6].

More precisely, let us consider a system characterized by non-stationary random phenomena, which is represented by a multivariate vector $\Psi = \{\psi_1, \ldots, \psi_m\}$ with values in $\mathbb{R}^{d \geq 1}$. This vector comprises m samples and it is assumed that the phenomena in the system exhibit piecewise stationarity, meaning that certain aspects of the system change abruptly at unknown instants $t_1, t_2, \ldots, t_m$. The task of change point detection involves addressing a model detection problem, wherein the objective is to determine the optimal segmentation S based on a quantitative criterion that needs to be minimized. Specifically, the goal is to identify the number m of changes and locate the indices $t_{i_{(1 \leq i \leq m)}}$.

3 The Completeness Challenge: A Problem Statement

3.1 Illustrative Scenario

To highlight the importance of ensuring data completeness in a CBR system, we present a compelling illustration. Let's envision an Energy Management System

(EMS) based on CBR that oversees a building equipped with an air-conditioning (AC) system. However, the EMS designer overlooked any means of detecting the functioning of the AC system. Consider two days with a similar context, such as identical weather conditions, and the same actions performed. If, on one day, the AC system was activated while on the other day it remained off (a phenomenon undetectable by the system), the divergent effects, such as distinct indoor temperatures, cast doubt on the fundamental assumption of the CBR paradigm.

3.2 Problem Formulation

Existing CBR systems utilize the case base directly to perform various stages of the CBR cycle, assuming that the domain knowledge remains consistent. Assumption 1 implicitly acknowledges the validity of the completeness hypothesis. However, in the context of modeling a complex domain with numerous interdependent variables, this assumption is arguably unwarranted. The violation of the completeness assumption raises significant concerns, which are as follows:

- There is no assurance that the CBR approach adheres to the principle stated in Assumption 1.
- The CBR system lacks the ability to recognize incomplete data, making it difficult to determine which data accurately represents reality for reasoning purposes.
- Consequently, the performance of the reasoning process may deteriorate when the case base contains cases that are incorrectly deemed similar.

The failure to adequately define one out of four knowledge containers can have a significant impact on the entire CBR system, potentially leading to overwhelming consequences. This occurs unless any of the remaining knowledge containers can compensate for the missing information. Consequently, the CBR system may either fail to provide a response or offer inaccurate solutions. Notably, research conducted in [3] has shown that incomplete domain knowledge plays a crucial role in generating such critical dysfunctions within a CBR system. When domain knowledge is incomplete in a CBR system, it is highly likely to result in the creation of incomplete cases. This incompleteness further burdens the retrieval process due to the absence of essential data, introducing bias into the similarity evaluation. Moreover, incomplete cases can also hinder the adaptation process, particularly when adaptation knowledge is automatically acquired from the case base.

The problem of verifying incompleteness can be effectively reframed as a problem of detecting hidden variables. To elaborate, when a case base exhibits an incompleteness situation, it means that a set of similar cases leads to disparate outcomes. This divergence is inherently caused by the presence of context and/or action variables that have not been taken into account during the evaluation of similarity. Therefore, it becomes evident that the issue of incompleteness verification can be approached by addressing the detection of these hidden variables.

Formally, consider a case base $\mathbb{CB} = \{\mathbb{C}_i\}_{1\leq i\leq n}$ consisting of a finite number n of cases $\mathbb{C}_i$. Each element of the latter is described by a set of features. The context $\mathtt{C}^{\mathbb{C}_i}$ of case $\mathbb{C}_i$ is specified by $\mathtt{C}^{\mathbb{C}_i} = \{O_{\mathtt{C}j}^{\mathbb{C}_i}\}_{1\leq j\leq n_1}$, where the observed features $O_{\mathtt{C}}$ are defined on the knowledge domain $\mathtt{K}_{\mathtt{C}}^{\mathtt{D}}$. The actions $\mathtt{A}^{\mathbb{C}_i}$ are modeled by the features $\{O_{\mathtt{A}j}^{\mathbb{C}_i}\}_{1\leq j\leq n_2}$ which are defined on the knowledge domain $\mathtt{K}_{\mathtt{A}}^{\mathtt{D}}$, and the effects are specified on the knowledge domain $\mathtt{K}_{\mathtt{E}}^{\mathtt{D}}$ by the features $\{O_{\mathtt{E}j}^{\mathbb{C}_i}\}_{1\leq j\leq n_3}$. The knowledge domain $\mathtt{K}^{\mathtt{D}}$ of the CBR system is defined by $\mathtt{K}^{\mathtt{D}} = \mathtt{K}_{\mathtt{C}}^{\mathtt{D}} \cup \mathtt{K}_{\mathtt{A}}^{\mathtt{D}} \cup \mathtt{K}_{\mathtt{E}}^{\mathtt{D}}$. Let's also assume, $\{H_{\mathtt{C}j}\}_{1\leq j\leq m_1}$, $\{H_{\mathtt{A}j}\}_{1\leq j\leq m_2}$, and $\{H_{\mathtt{E}j}\}_{1\leq j\leq m_3}$ are the hidden features of the context, action, and effect elements respectively. We denote the reference knowledge domain by $\mathtt{K}_{\mathtt{R}}^{\mathtt{D}} = \mathtt{K}^{\mathtt{D}} \cup \{H_{\mathtt{C}j}\}_{1\leq j\leq m_1} \cup \{H_{\mathtt{A}j}\}_{1\leq j\leq m_2} \cup \{H_{\mathtt{E}j}\}_{1\leq j\leq m_3}$.

The completeness evaluation problem of a CBR system against $\mathtt{K}_{\mathtt{R}}^{\mathtt{D}}$ consists in identifying eventual incompleteness situations in the case base. An incompleteness situation is formalized as:

$$\mathtt{CBR\ system} \Leftrightarrow \exists\ \mathbb{C}_1, \mathbb{C}_2 \in \mathbb{CB}/$$
$$(\{O_{\mathtt{C}j}^{\mathbb{C}_1}\} = \{O_{\mathtt{C}j}^{\mathbb{C}_2}\})_{1\leq j\leq n_1} \wedge (\{O_{\mathtt{A}j}^{\mathbb{C}_1}\} = \{O_{\mathtt{A}j}^{\mathbb{C}_2}\})_{1\leq j\leq n_2} \wedge (\{O_{\mathtt{E}j}^{\mathbb{C}_1}\} \neq \{O_{\mathtt{E}j}^{\mathbb{C}_2}\})_{1\leq j\leq n_3}$$
$$\Longrightarrow \exists f \in \{H_{\mathtt{C}j}\}_{1\leq j\leq m_1} \cup \{H_{\mathtt{A}j}\}_{1\leq j\leq m_2}$$

For effectiveness reasons, we argue that is a prerequisite to check the completeness of the data as early as possible in the problem-solving process, i.e., before starting the reasoning cycle. Furthermore, the incompleteness assessment process must be launched whenever the case base is updated.

4 Evaluating Data Incompleteness in CBR Systems

In this section, we provide a comprehensive explanation of the workflow employed by our I2CCBR (InCompleteness Checking CBR) algorithm, which serves to assess data incompleteness within a CBR system. To better organize the content, this section is divided into two parts, aligning with the overall architecture of the I2CCBR algorithm. In the following workflow, we begin by dividing the case base into optimal segments, achieved by grouping cases based on their respective effects. Subsequently, we leverage these partitions to identify potential instances of incompleteness using a highly effective approach that relies on context and action knowledge.

4.1 Segmentation of the Case Base

The goal of partitioning the case base is to find potential patterns in the effects of the cases. This involves identifying and estimating changes in the statistical properties of the effects, so that cases with similar effects can be grouped together in clusters. To achieve this goal, we propose a hybrid method in this section, which combines two techniques: the cumulative sum (CUMSUM) technique originally presented in [15] and the bootstrapping mechanism outlined in [12]. The process of detecting change points in the case effects model is an iterative one, consisting of the following two steps:

Step 1: Cumulative Sums. In light of the notation introduced in Sect. 2.1, we compute the cumulative sums $\mathtt{S}_i$ of the effect variables using the recursive formula outlined in Eq. (1). It is important to note that these cumulative sums do not directly correspond to the cumulative sums of the effect variables themselves. Instead, they represent the cumulative sums of the differences between the values and their average μ. As a result, the final cumulative sum $\mathtt{S}_n$ always equals zero.

By examining the chart of cumulative sums $\mathtt{S}_i$, potential change points in the effect variables can be detected by observing changes in the direction of the plot. However, it is important to note that the cumulative sums chart alone cannot definitively determine the presence of these change points or the corresponding indices of the cases. Addressing these two challenges becomes the focal point of the second step. In order to accomplish this, it becomes necessary to estimate the magnitude of change $\mathtt{S}_M$ in the cumulative sums $\mathtt{S}_i$. One approach to achieve this estimation is by employing Formula (2).

$$\forall\, i \leq n,\ \mathtt{S}_i = \begin{cases} 0, & \text{if } i = 0 \\ \mathtt{S}_{i-1} + (\mathtt{E}_{\mathbb{C}_i} - \mu), & \text{else} \end{cases} \tag{1}$$

Assuming n represents the total number of cases, and μ represents the average of the effect variable, it can be expressed as $\mu = \frac{1}{n}\sum_{i=1}^{n} \mathtt{E}_{\mathbb{C}_i}$, where $\mathtt{E}_{\mathbb{C}_i}$ represents the effect variable for each case $\mathbb{C}_i$.

$$\mathtt{S}_M = \max_{1 \leq i \leq n} \mathtt{S}_i - \min_{1 \leq i \leq n} \mathtt{S}_i \tag{2}$$

Step 2: Bootstrapping. To address the concern of establishing a confidence level for the identified change points, a bootstrapping approach is employed. The fundamental principle behind bootstrapping is to simulate the behavior of the cumulative sums $\mathtt{S}_i^{bs}$ under the assumption that there are no changes in the patterns of effects. By doing so, the resulting cumulative sums serve as a reference for comparing against the original order of the cumulative sums $\mathtt{S}_i$ calculated in step 1. The bootstrapping process involves applying the same procedure used in step 1 to a randomly rearranged case base. This generates the cumulative sums $\mathtt{S}_i^{bs}$ and the change magnitude $\mathtt{S}_M^{bs}$.

When the plot of the bootstrap cumulative sums $\mathtt{S}_i^{bs}$ tends to remain closer to zero compared to the plot of the original cumulative sums $\mathtt{S}_i$, it indicates the likelihood of a change occurring. To estimate the confidence in the presence of a change point, a significant number of bootstraps (denoted as z) are conducted. The number of situations (denoted as r) is then determined, where the magnitude of change $\mathtt{S}_M^{bs}$ is smaller than the change magnitude $\mathcal{S}_M$ observed in the original case base. The confidence index, denoted as $\mathtt{C}_{index}$, that signifies a shift in the pattern of effects can be calculated using Formula (3).

$$\mathtt{C}_{\mathtt{index}} = \frac{r}{z} \times 100 \tag{3}$$

Identifying Change Point Positions. When the confidence level reaches a significant threshold (typically around 90%), confirming the presence of a change point, one approach to estimate the case index corresponding to the change in the effect model is by utilizing the mean square error (MSE) metric. The case base is partitioned into two sections, one containing p cases and the other containing $n-p$ cases, where p represents the index of the last case prior to the change in the effect model. Estimating the index p involves solving an optimization problem that aims to minimize Function 4.

$$\mathtt{MSE} = \sum_{i=1}^{p} (\mathtt{E}_{\mathbb{C}_i} - \mu_1)^2 + \sum_{i=p+1}^{n} (\mathtt{E}_{\mathbb{C}_i} - \mu_2)^2 \tag{4}$$

With $\mu_1 = \frac{1}{p}\sum_{i=1}^{p} \mathtt{E}_{\mathbb{C}_i}$, $\mu_2 = \frac{1}{n-p}\sum_{i=p+1}^{n} \mathtt{E}_{\mathbb{C}_i}$.

After identifying a change point, the case base is split into two separate bases: the first base comprises cases from 1 to p, while the remaining cases form the second base. The iterative process described in steps 1 and 2 is then applied to each of these case bases, aiming to detect any additional change points. This iterative approach continues until no further change points are found within the case bases. Consequently, if there are additional changes present, they will be detected through this process.

4.2 Identification of Incompleteness Situations

Let $k \neq 0$ represent the number of identified change points in the case base $\mathbb{CB}$. The set of indices for cases whose effects indicate a change in the model is denoted as $\mathcal{K}$, with $|\mathcal{K}| = k$ and $\mathcal{K} = \{\mathtt{K}_j\}_{1 \leq i \leq k}$. Specifically, the index $\mathtt{K}_i$ corresponds to the case index preceding the ith change in the model of the effect variables. Consequently, the case base $\mathbb{CB}$ can be partitioned into $k+1$ groups denoted as $\mathbb{P}_{1 \leq i \leq k+1}$. These groups satisfy the constraints depicted in Eq. (5).

$$\begin{aligned} \mathbb{CB} &= \bigcup_{i=1}^{k+1} \mathbb{P}_i \\ \mathbb{P}_i &= \begin{cases} \{\mathbb{C}_t\}_{\mathtt{K}_{i-1} < t \leq \mathtt{K}_i}, & \text{if } 2 \leq i \\ \{\mathbb{C}_t\}_{1 \leq t \leq \mathtt{K}_i}, & \text{if } i = 1 \end{cases} \end{aligned} \tag{5}$$

The objective of the completeness evaluation is to identify scenarios in which two cases exhibit similar actions and contexts but yield different effects. To investigate potential incompleteness situations, the following approach is employed:

1. considering the set of groups $\{\mathbb{P}_i\}$, we proceed to compute the maximum context-action distance $\mathtt{D}_{\mathtt{CA},i}^{max}$ and the minimum context-action distance $\mathtt{D}_{\mathtt{CA},i}^{min}$ within each group $\mathbb{P}_i$, which represents a collection of cases with similar effects. We denote the resulting interval as $\mathcal{I}_i^{\mathtt{CA}}$, defined as $\left[\mathtt{D}_{\mathtt{CA},i}^{min}, \mathtt{D}_{\mathtt{CA},i}^{max}\right]$. Additionally, for group $\mathbb{P}_i$, we establish $\mathcal{I}_i^{\mathtt{E}}$ as the effect variable interval, represented by $\mathcal{I}_i^{\mathtt{E}} = \left[\mathtt{E}_i^{min}, \mathtt{E}_i^{max}\right]$.

2. the identification of an incompleteness situation is considered reliable if the following conditions are met: there exist two cases, $\mathbb{C}_i$ and $\mathbb{C}_j$, located in two distinct groups, $\mathbb{P}_i$ and $\mathbb{P}_j$, respectively, where the effect models of these cases differ. Additionally, the context-action distance between $\mathbb{C}_i$ and $\mathbb{C}_j$ falls within one of the intervals $\mathcal{I}_i^{\mathtt{CA}}$ or $\mathcal{I}_j^{\mathtt{CA}}$. This can be expressed formally as follows:

$$\exists\ \mathbb{C}_i \in \mathbb{P}_i, \mathbb{C}_j \in \mathbb{P}_j, k \in \{i,j\}/$$
$$\mathtt{D_{CA}}(\mathbb{C}_i, \mathbb{C}_j) \in \mathcal{I}_k^{\mathtt{CA}} \implies \mathcal{I}_i^{\mathtt{E}} \cap \mathcal{I}_j^{\mathtt{E}} = \emptyset\ \vee (\mathtt{E}^{\mathbb{C}_i} \notin \mathcal{I}_i^{\mathtt{E}} \cap \mathcal{I}_j^{\mathtt{E}} \wedge \mathtt{E}^{\mathbb{C}_j} \notin \mathcal{I}_i^{\mathtt{E}} \cap \mathcal{I}_j^{\mathtt{E}})$$
$$\vee (\mathtt{E}^{\mathbb{C}_m} \notin \mathcal{I}_i^{\mathtt{E}} \cap \mathcal{I}_j^{\mathtt{E}}, m \in \{i,j\} \wedge \mathbb{C}_m \notin \mathbb{P}_k)$$

5 Experimental Results

The aim of the experiment is to assess the reliability and efficiency of the proposed approach in identifying cases of incompleteness within a case base. Firstly, we provide a description of the dataset employed in the experiment. Subsequently, we outline the experimental setup utilized. Finally, we present and discuss the obtained results.

Data Collection. To assess the effectiveness of the I2CCBR approach, we conducted experiments using a real-world dataset. Specifically, we utilized the dataset obtained from the experiment presented in the motivation example (refer to Sect. 3.1). The dataset used in our experiments is sourced from [4], where the authors proposed a CBR-based approach aiming to enhance the energy efficiency of buildings while considering occupant comfort. The evaluation of their approach involved a case study conducted in an academic research office, where data were collected from multiple sensors deployed in the environment.

The data under consideration is categorized into three groups based on the case structure discussed in Sect. 2.1. The context data includes meteorological information and the number of occupants. The action data represents the opening/closing of doors and windows. The effect data encompasses the indoor temperature and the concentration of CO_2 in the office. Each case corresponds to measurements taken over a single day. For the evaluation conducted in this experiment, the focus is solely on assessing the incompleteness of the indoor temperature as the effect variable. The case base used in this evaluation comprises 98 cases arranged in chronological order according to their measurement dates.

Pre-processing of Data and Experimental Setup. To mitigate potential bias in the similarity assessment, the context and action data are normalized using the MinMax strategy, thereby transforming them into a range of 0 to 1 and reducing the influence of variables with large values.

In this experiment, the similarity between two cases based on context and action is evaluated using the weighted Euclidean distance as a similarity function.

However, the specific process of assigning weights to the context and action variables is not extensively discussed in this study. Instead, we employed the approach introduced in [5] to estimate these weights.

The experiments were conducted using a MacBook Pro laptop featuring an Intel® Core™ i7-8559U CPU running at 2.70 GHz, with 16 GB of RAM. The laptop was operating on the 64-bit version of Windows 10 Pro. The implementation of the proposed approach was done in Python 3.9. The code execution took place in JupyterLab 3.4.4.

Generation of Incompleteness Situations Within the Case Base. At present, it remains uncertain whether the case base is complete due to the challenges involved in accurately modeling the building environment. The complex interactions and numerous factors that influence the energy performance of a building make it difficult to achieve a complete representation. In order to assess the effectiveness of the I2CCBR algorithm, it is necessary to establish a baseline that undoubtedly includes situations of incompleteness.

We generated an incomplete case base $\mathbb{CB}_I$ based on the original case base $\mathbb{CB}$. The process of creating incomplete situations can be described as follows:

- We added incompleteness to the case base $\mathbb{CB}$ by randomly selecting and altering 5% of the cases, specifically modifying five cases.
- Due to the absence of an air-conditioning system in the experimental office, the selected cases were modified by incorporating a new action variable that represents the introduction of air conditioning. This variable serves to simulate the influence of an unobservable factor within the CBR system. The values assigned to this variable are randomly generated from a discrete uniform distribution spanning a range of 18 °C to 23 °C.
- The physical model of the office is utilized to generate the resulting effects that occur when the new actions, such as turning on the air conditioner, are applied within the selected cases. In order to maintain consistency, the physical model also simulates the actual effects of the other cases.

Empirical Results. Fig. 1 illustrates the average of the real effect variable (temperature) for the 98 cases (depicted by the green curve) alongside the corresponding simulated values (indicated by the red curve). It should be noted that the simulated effects of cases $\mathbb{C}6$, $\mathbb{C}19$, $\mathbb{C}25$, $\mathbb{C}59$, and $\mathbb{C}_{71}$ deviate significantly from their original counterparts. The substantial disparity between the original values and the simulated ones can be attributed to the impact of an undisclosed factor (representing air conditioning) on the effects of these instances. Essentially, these instances correspond to five cases that were randomly selected and modified.

Moreover, it is worth mentioning that the two curves largely coincide throughout the entire plot. The evaluation of Mean Absolute Percentage Error (MAPE), excluding the modified cases, demonstrates that the simulated data deviates from the real data by less than 2.50%, underscoring the robustness of the physical model employed for simulating the effect variables.

The outcomes of the initial stage of the C2CBR algorithm, aimed at identifying alterations in the effects model (temperature), are illustrated in Figs. 2 and 3. In Fig. 2, where the cumulative sums of the effect variable are plotted, every alteration in the background color signifies a sudden shift in the chart's direction, indicating the presence of a pattern modification. Specifically, each transition between the yellow and turquoise colors represents a change point. It becomes evident that there are a total of 6 variations in the background colors, corresponding to 6 change points.

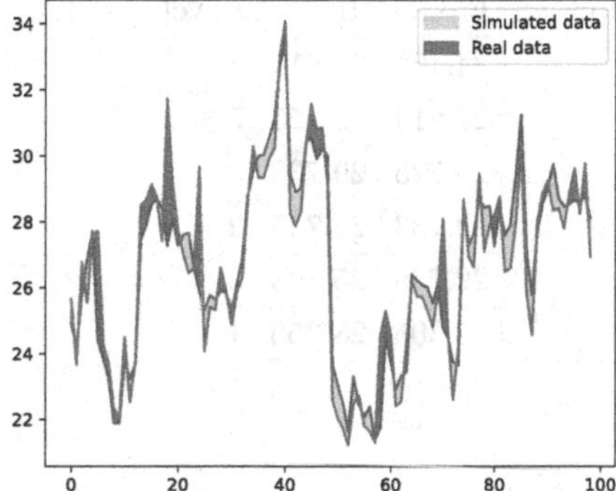

Fig. 1. Real and simulated effect.

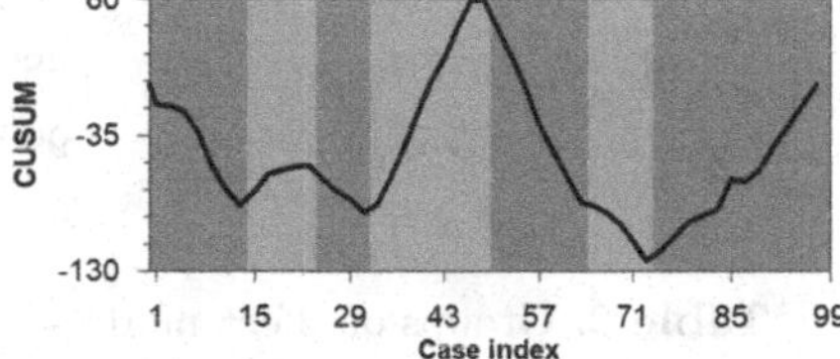

Fig. 2. Change point detection.

The six change points that have been identified are used to divide the case base into seven separate groups based on the chronological recording of case effects, as shown in Fig. 3. Each change point is represented by a shift in the background color to turquoise and corresponds to a specific case index in the case base. A segment shaded in turquoise represents a group that includes all cases based on the current effect variable model formed by two consecutive change points. Each row in Table 1 provides detailed information regarding each change point. A confidence interval, estimated with a 95% probability of accuracy, is assigned to the index of the change case where a model change is detected. For example, the 5th change point is estimated to be between cases 65 and 67 with a probability of 95%. Additionally, Formula 3 is used to report a confidence index for each detected change point, indicating the quality of the analysis. For instance, the system has 98% confidence that the 5th change point occurred. Further information is also provided in terms of average values for the effect variables of the groups before and after a change point. Table 2 presents the values of the $\mathcal{I}^{\mathtt{CA}}$ and $\mathcal{I}^{\mathtt{E}}$ metrics (defined in Sect. 4.2) for each of the seven groups.

After implementing the proposed heuristic described in Sect. 4.2, the five artificially generated incompleteness situations from the previous step have been accurately identified, as evidenced in Table 3. Every potential situation of incompleteness is described by two cases denoted as $\mathbb{C}_I^1$ and $\mathbb{C}_I^2$ that give rise to such a situation. It should be noted that the cases listed under column $\mathbb{C}_I^1$ in Table 3 correspond to the five modified cases. To illustrate, the incompleteness situation S3 occurs between cases $\mathbb{C}_{25}$ and $\mathbb{C}_{18}$ due to the fact that the distance $\mathtt{D_{CA}}(\mathbb{C}_{25}, \mathbb{C}_{18})$ is lower than the maximum distance $\mathtt{D}_{\mathtt{CA}}^{max}$ within the group $\mathbb{P}3$, to which case $\mathbb{C}_{25}$ belongs. However, it is worth noting that the effect variable of case $\mathbb{C}_{18}$ belongs to group $\mathbb{P}2$, indicated by $\mathtt{E}^{\mathbb{C}_{18}} = 27.36$, and $\mathtt{E}^{\mathbb{C}_{18}} = 25.52$.

Table 1. Detected change points.

Index	Confidence interval	Confidence index	From	To	Level
15	(11,15)	96%	24.49	27.814	4
25	(25,25)	94%	27.814	25.725	5
33	(33,37)	99%	25.725	29.733	3
51	(51,51)	100%	29.733	22.745	2
65	(65,67)	98%	22.745	25.305	3
75	(75,77)	99%	25.305	28.355	1

Table 2. Groups of effect models.

Group	$\mathcal{I}^{\mathtt{CA}}$	$\mathcal{I}^{\mathtt{E}}$
$\mathbb{P}_1$	$[0.138, 0.526]$	$[21.88, 27.25]$
$\mathbb{P}_2$	$[0.199, 0.550]$	$[26.61, 28.70]$
$\mathbb{P}_3$	$[0.156, 0.516]$	$[25.26, 26.14]$
$\mathbb{P}_4$	$[0.166, 0.390]$	$[23.65, 33.42]$
$\mathbb{P}_5$	$[0.188, 0.377]$	$[21.35, 24.54]$
$\mathbb{P}_6$	$[0.155, 0.440]$	$[23.64, 26.42]$
$\mathbb{P}_7$	$[0.126, 0.602]$	$[25.79, 31.23]$

Table 3. Detected incompleteness situations.

Situation	$\mathbb{C}_I^1$	$\mathbb{C}_I^2$	$\mathtt{D_{CA}}$
S1	$\mathbb{C}_6 \in \mathbb{G}_1$	$\mathbb{C}_{85} \in \mathbb{G}_7$	0.531
S2	$\mathbb{C}_{19} \in \mathbb{G}_2$	$\mathbb{C}_{39} \in \mathbb{G}_4$	0.184
S3	$\mathbb{C}_{25} \in \mathbb{G}_3$	$\mathbb{C}_{18} \in \mathbb{G}_2$	0.275
S4	$\mathbb{C}_{59} \in \mathbb{G}_5$	$\mathbb{C}_{69} \in \mathbb{G}_6$	0.369
S5	$\mathbb{C}_{71} \in \mathbb{G}_6$	$\mathbb{C}_{79} \in \mathbb{G}_7$	0.545

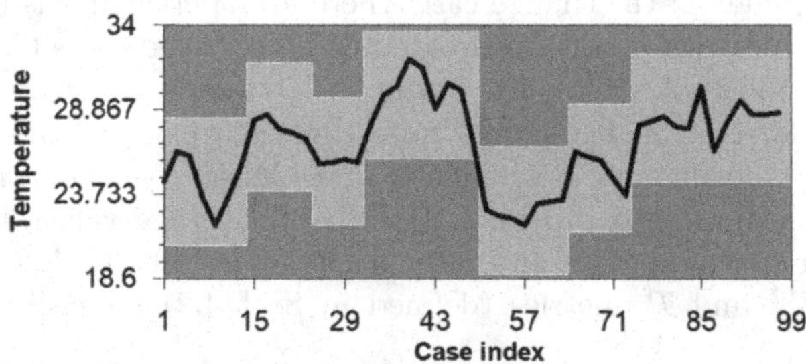

Fig. 3. Visualization of identified effect groups.

6 Conclusion

For the authors' knowledge, this work presents for the first time an attempt to address the problem of data incompleteness in a CBR system under the open-world assumption. The proposed approach I2CCBR combines a change point detection method and a heuristic strategy to identify potential incompleteness in the case base. A first experiment was performed on real-world data showing promising results on the applicability and efficiency of this approach.

The upcoming step of this work will be to extend the scope of our experimental evaluation by assessing the I2CCBR algorithm on larger datasets to confirm the obtained results.

References

1. Extracting business value from the 4 v's of big data. techreport, IBM, 2016. http://www.ibmbigdatahub.com/infographic/extracting-business-value-4-vs-big-data. Accessed 31 May 2022
2. 2019 Global data management research. Taking control in the digital age. Benchmarkreport, Experian UK&I, February 2019
3. Bergmann, R., Wilke, W., Vollrath, I.: Integrating general knowledge with object-oriented case representation and reasoning. In: 4th German Workshop: Case-Based Reasoning - System Development and Evaluation (1996)
4. Boulmaiz, F., Ploix, S., Reignier, P.: A data-driven approach for guiding the occupant's actions to achieve better comfort in buildings. In: IEEE 33rd International Conference on Tools with Artificial Intelligence (ICTAI) (2021)
5. Boulmaiz, F., Reignier, P., Ploix, S.: An occupant-centered approach to improve both his comfort and the energy efficiency of the building. Knowl.-Based Syst. **240**, 108970 (2022)
6. Charakopoulos, A., Karakasidis, T.: Backward degree a new index for online and offline change point detection based on complex network analysis. Phys. A Stat. Mech. Appl. **604**, 127929 (2022)
7. Cichy, C., Rass, S.: An overview of data quality frameworks. IEEE Access **7**, 24634–24648 (2019)
8. Galárraga, L., Razniewski, S., Amarilli, A., Suchanek, F.M.: Predicting completeness in knowledge bases. In: Proceedings of the Tenth ACM International Conference on Web Search and Data Mining. ACM, February 2017
9. Getahun, Y.S., Li, M.H., Pun, I.F.: Trend and change-point detection analyses of rainfall and temperature over the awash river basin of ethiopia. Heliyon **7**(9), e08024 (2021)
10. Grohe, M., Lindner, P.: Probabilistic databases with an infinite open-world assumption. In: Proceedings of the 38th ACM SIGMOD-SIGACT-SIGAI Symposium on Principles of Database Systems - PODS 2019. ACM Press (2019)
11. Haug, A.: Understanding the differences across data quality classifications: a literature review and guidelines for future research. Ind. Manage. Data Syst. **121**(12), 2651–2671 (2021)
12. Hinkley, D.V., Schechtman, E.: Conditional bootstrap methods in the mean-shift model. Biometrika **74**, 85–93 (1987)
13. Liaw, S.T., et al.: Towards an ontology for data quality in integrated chronic disease management: a realist review of the literature. Int. J. Med. Inf. **82**, 10–24 (2013)

14. Nguyen, P.T., Di Rocco, J., Iovino, L., Di Ruscio, D., Pierantonio, A.: Evaluation of a machine learning classifier for metamodels. Softw. Syst. Model. **20**(6), 1797–1821 (2021)
15. Pettitt, A.N.: A simple cumulative sum type statistic for the change-point problem with zero-one observations. Biometrika **67**(1), 79–84 (1980)
16. Richter, M.M.: The knowledge contained in similarity measures. In: International Conference on Case-Based Reasoning, ICCBR 1995, Sesimbra, Portugal (1995)
17. Truong, C., Oudre, L., Vayatis, N.: Selective review of offline change point detection methods. Sig. Process. **167**, 107299 (2020)
18. Wang, R.Y., Strong, D.M.: Beyond accuracy: What data quality means to data consumers. J. Manage. Inf. Syst. **12**(4), 5–33 (1996)
19. You, S.H., et al.: Change point analysis for detecting vaccine safety signals. Vaccines **9**(3), 2062021 (2021)

Leveraging both Successes and Failures in Case-Based Reasoning for Optimal Solutions

Fateh Boulmaiz(✉), Patrick Reignier, and Stephane Ploix

Université Grenoble Alpes, INP Grenoble, LIG, 38000 Grenoble, France
{fateh.boulmaiz,patrick.reignier,stephane.ploix}@univ-genoble-alpes.fr

Abstract. Usually, existing works on adaptation in case-based reasoning assume that the case base holds only successful cases, i.e., cases having solutions believed to be appropriate for the corresponding problems. However, in practice, the case base could hold failed cases, resulting from an earlier adaptation process but discarded by the revision process. Not considering failed cases would be missing an interesting opportunity to learn more knowledge for improving the adaptation process. This paper proposes a novel approach to the adaptation process in the case-based reasoning paradigm, based on an improved barycentric approach by considering the failed cases. The experiment performed on real data demonstrates the benefit of the method considering the failed cases in the adaptation process compared to the classical ones that ignore them, thus, improving the performance of the case-based reasoning system.

Keywords: Case-based reasoning · adaptation · successful case · failed case

1 Introduction

Case-based reasoning (CBR) is undoubtedly the most intuitive artificial intelligence approach for problem-solving, as it emulates human behavior. A CBR system searches through its memory, which is composed of a base of previously solved cases known as source cases, to find cases that exhibit similar problems to the target problem for which a solution is sought. It then adapts their solutions, if necessary, to solve the target problem. The target solution is thoroughly reviewed to ensure its suitability for resolving the target problem, and subsequently, the case base is updated with the new resolution experiment for the target case. Each step of the reasoning process is supported by a knowledge acquisition process required for that particular step.

Adaptation, one of the four key stages in the reasoning process, holds great significance as the quality of the solution heavily relies on its performance. The primary objective of adaptation is to tailor the solutions of similar source cases to meet the specific requirements of the target problem. This step is particularly

N. T. Nguyen et al. (Eds.): ACIIDS 2023, LNAI 13995, pp. 31–44, 2023.
https://doi.org/10.1007/978-981-99-5834-4_3

crucial because the source problems usually do not align perfectly with the target problem. Without successful adaptation, the CBR system cannot generate an appropriate solution for the target problem. The importance of adaptation has been recognized since the early days of CBR systems, leading to numerous studies that explore different approaches for acquiring adaptation knowledge to enhance its performance. According to [13], two distinct approaches to adaptation knowledge acquisition can be distinguished: knowledge-light approaches, which leverage existing knowledge within the system without requiring additional acquisition [11], and knowledge-intensive approaches, which rely on external knowledge sources, such as knowledge obtained from experts or users [5,6].

Existing adaptation approaches primarily concentrate on successful cases (referred to as $\mathbb{C}^+$) that provide relevant solutions to the corresponding problems. The definition of success is subjective and varies depending on the application domain. For example, in the context of a CBR application for an energy management system in a building, a successful case would involve achieving user comfort while minimizing energy expenditure. However, there are also cases that fail to meet the desired criteria. These failed cases (referred to as $\mathbb{C}^-$) have solutions that are deemed unsatisfactory and are typically rejected during the validation phase of the adaptation process. Additionally, the adaptation process often requires acquiring domain-specific knowledge to generate adaptation rules. This knowledge acquisition process is complex and challenging due to its strong dependence on the specific application domain, making it difficult to comprehend and grasp.

In spite of the abundance of research studies and the increased interest in the issue of adaptation, there are few works that specifically address the challenge of proposing a domain-independent adaptation approach. Moreover, there is limited research that considers adaptation from the perspective of solution quality, which encompasses both failed and successful cases. Surprisingly, these cases, which could potentially provide valuable knowledge, are rarely employed by CBR systems. This work introduces a fresh viewpoint on the adaptation process within the CBR paradigm, presenting a fully domain-independent approach that incorporates both successful and failed cases. The study proposes a novel method for acquiring adaptation knowledge, drawing inspiration from research on planning the path of a robot navigating through an unfamiliar and hazardous environment, including obstacles. The uniqueness of this approach lies in applying artificial forces to the proposed solution, aiming to distance itself from failed source solutions while gravitating towards successful ones.

The structure of this paper is organized as follows: Sect. 2 provides an overview of the motivation and background for this work. Section 3 elaborates on the contribution made towards leveraging failed and successful cases for a novel adaptation approach. The evaluation of the proposed approach is presented and discussed in Sect. 4. Finally, Sect. 5 concludes this work, highlights its key findings, and outlines future research directions.

2 Illustrative Example and Preliminary Concepts

A CBR-based energy management system (EMS) in a building serves as a representative case study within the scope of this research. The objective of an EMS is to meet user preferences for thermal comfort, air quality, and other factors, while minimizing energy consumption in the building. Undoubtedly, a building is a complex system influenced by various factors, including climate, building materials, geographical location, energy rates, and the occupants themselves, making it challenging to identify dependencies [3]. Earlier studies [9] have already highlighted the benefits of acquiring adaptation knowledge to enhance the performance of a CBR-based EMS. Additionally, the increasing awareness of environmental concerns has prompted numerous studies to explore the relationship between building energy consumption and occupant comfort, resulting in the establishment of standards [1,2,7] for evaluating user comfort. These standards provide a framework for assessing the quality of the target solution proposed by the adaptation process during the revision phase. Consequently, the solution can be classified as either a successful case ($\mathbb{C}^+$) or a failed case ($\mathbb{C}^-$).

In the CBR-based EMS described in [3], the primary goal is to raise the user's awareness of the impact of their actions on the energy efficiency of the building. To achieve this, the system assists the user by providing recommendations on a set of actions aimed at reducing energy wastage while taking their comfort into account. Each case within the system represents a specific energy management scenario for a building over a single day. The actions stored in the case base correspond to the actions actually performed by the building occupant; however, there is no guarantee that these actions yield satisfactory outcomes for the occupant. To address this, the system incorporates a function to evaluate the effectiveness of the actions stored in the case base, enabling the appropriate labeling ($\mathbb{C}^-$ or $\mathbb{C}^+$) of the corresponding cases.

2.1 Key Concepts and Notations Related to the CBR Paradigm

Each past experience of a CBR system, which forms the foundation for solving new problems, is stored in a structure called a source case ($\mathbb{C}^{sr}$), and the collection of source cases constitutes a case base ($\mathbb{CB}$). Below is a concise introduction to essential concepts within the CBR paradigm, necessary for comprehending our approach.

Case Organization. Consider three sets, $\mathbb{C}$, $\mathbb{A}$, and $\mathbb{E}$, which are mutually disjoint. A case is defined as a triplet $(\mathscr{C}, \mathcal{A}, \mathcal{E})$ where:

- $\mathscr{C}$ belongs to the context domain $\mathbb{C}$ and represents the fixed elements of the problem that cannot be controlled. For instance, in a CBR-based medical diagnostic system, $\mathscr{C}$ can represent physiological indicators of the patient such as heart rate, respiratory rate, etc.

- $\mathcal{A}$ belongs to the action domain $\mathbb{A}$, representing elements that can be controlled to achieve desired outcomes. It represents the suggested solution for the system. In a medical diagnostic system, this could entail the names of recommended medications and their corresponding administration protocols.
- $\mathcal{E}$ belongs to the effect domain $\mathbb{E}$, which characterizes the system's outcome resulting from action $\mathcal{A}$ in context $\mathscr{C}$. In a medical diagnostic system, $\mathcal{E}$ can denote the patient's post-treatment clinical observations or test results.

A target context, denoted as $\mathscr{C}^{tg}$, represents a specific context for which the CBR system aims to determine appropriate actions $\mathcal{A}^{tg}$ in order to produce desired effects $\mathcal{E}^{tg}$ and ultimately generate a target case $\mathbb{C}^{tg}$. The resolution of a problem within the CBR paradigm can be formally described by Eq. (1).

$$\begin{aligned} &\text{CBR system: } (\mathbb{CB}, \mathscr{C}^{tg}) \longmapsto \mathcal{A}^{tg} \\ &\mathbb{C}^{tg} \stackrel{\text{def}}{=} (\mathscr{C}^{tg}, \mathcal{A}^{tg}, \mathcal{E}^{tg}) \end{aligned} \tag{1}$$

The Retrieval and Adaptation Processes. While this paper does not delve into a comprehensive exposition of the reasoning process, it is crucial to acknowledge the inherent relationship between adaptation and knowledge retrieval. Consequently, it is often imperative to present the adaptation process alongside the retrieval process.

- *retrieval stage*: In the retrieval process, the goal is to find source cases that exhibit a context similar to the target context, using a threshold $\Theta_{\mathbb{C}^{tg}}$ to measure the distance between their context variables. Precisely, the process involves locating cases where the context distance from the target context is below $\Theta_{\mathbb{C}^{tg}}$. The retrieval function's profile is outlined by Eq. (2).

$$\text{Retrieval process:} \quad \mathscr{C}^{tg} \longmapsto \{\forall\ \mathbb{C}^{sr} \in \mathbb{CB}, D^{ct}(\mathbb{C}^{tg}, \mathbb{C}^{sr}) \leq \Theta_{\mathbb{C}^{tg}}\} = \mathbb{S}^{\mathbb{C}^{tg}} \tag{2}$$

  where D^{ct} represents the distance between the target context variables $\mathscr{C}^{tg}$ of the target case $\mathbb{C}^{tg}$ and the context variables $\mathscr{C}^{sr}$ of the source case $\mathbb{C}^{sr}$. There are no limitations on the choice of distance metric in order to accommodate the context variables. This flexibility allows for the utilization of various distance measures based on the nature of the context variables. For instance, in a CBR-based EMS, the Manhattan metric can be employed to calculate the contextual distance. This is particularly suitable when the context variables involved are real-valued.
- *adaptation stage*: Given that the source contexts often differ from the target context, it becomes necessary to establish a function that can modify the source actions in order to meet the requirements of the target context. The characteristics of this adaptation function can be described by Formula (3).

$$\begin{aligned} \text{Adaptation process: } &\forall\ \mathbb{C}^{sr} \stackrel{\text{def}}{=} (\mathscr{C}^{sr}, \mathcal{A}^{sr}, \mathcal{E}^{sr}) \in \mathbb{S}^{\mathbb{C}^{tg}}, \\ &(\{(\mathscr{C}^{sr}, \mathcal{A}^{sr}, \mathcal{E}^{sr})\}, \mathscr{C}^{tg}) \longmapsto \mathcal{A}^{tg} \end{aligned} \tag{3}$$

where the set $\mathbb{S}^{\mathbb{C}^{tg}}$ refers to the collection of source cases that are considered similar based on the definition provided by Eq. (2).
It is important to note that Eq. (3) does not place any limitations on the number of similar cases that can be considered during the adaptation process. As a result, we are dealing with a form of adaptation that involves combining solutions from multiple source cases to generate a target solution. This type of adaptation is known as compositional adaptation, with single case adaptation being a special case of it. Indeed, the experimental findings from [12] demonstrate that relying solely on a single case often produces less accurate outcomes. This can be attributed to the fact that, in many cases, only a portion of the problem exhibited in the similar source case is relevant to the target problem. Consequently, the process of adaptation becomes complex and, at times, even impossible.

2.2 Collision Avoidance Navigation

The primary objective of studying robot path planning is to address the movement of an autonomous robot within an unfamiliar environment. This involves guiding the robot from its starting point to a designated target position, while placing emphasis not only on finding the most efficient path but also on ensuring the utmost safety. The aim is to calculate a path that optimizes both efficiency and safety by effectively avoiding any obstacles that may arise along the trajectory leading to the target.

Numerous strategies have been proposed to address this challenge, with the Artificial Potential Field (APF) approach, originally introduced in [8], being widely utilized for robot guidance. The APF approach effectively handles the real-world environment in which a robot operates, taking into account both the desired objectives and the obstacles that need to be avoided during movement. The fundamental concept behind this approach is to treat the robot as a point moving within a two-dimensional space (in a basic scenario), influenced by a field created by the targets to be reached and the obstacles to be avoided. Consequently, the robot experiences two types of forces: an attractive force $\mathbb{F}^A$ generated by the targets, and a repulsive force $\mathbb{F}^R$ generated by the obstacles, which collectively determine the robot's movement.

While the repulsive forces exerted on the robot are stronger when it is closer to obstacles and weaker as the distance increases, the attractive forces acting on the robot are directly proportional to the distance between the robot and its target. By combining these forces, denoted as $\overrightarrow{\mathbb{F}} = \overrightarrow{\mathbb{F}^A} + \overrightarrow{\mathbb{F}^R}$, the robot's movement direction and speed can be determined while avoiding collisions with obstacles. Figure 1 illustrates the basic principle of this method, specifically designed for a robot moving in a two-dimensional environment, for the purpose of simplification.

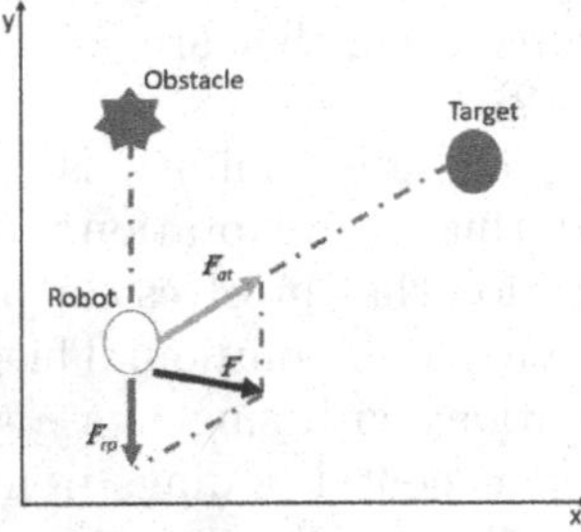

Fig. 1. Robotics's Artificial potential field.

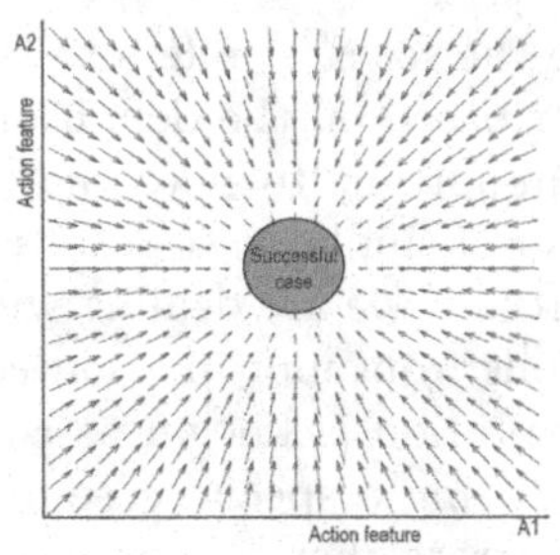

Fig. 2. CBR attractive force.

3 Adaptation Through Failed and Successful Cases

3.1 Problem Statement

The adaptation problem, taking into account both failed and successful cases, can be expressed through the following problem statement. Considering the following observations:

- the case base $\mathbb{CB}$ is partitioned into two subsets: failed cases denoted as $\mathbb{CB}^-$, and successful cases denoted as $\mathbb{CB}^+$. Therefore, $\mathbb{CB} = \mathbb{CB}^- \cup \mathbb{CB}^+$.
- through language misuse, we use the term "target case" to refer to the elements within a specific context for which we are seeking a solution. However, the structure of this case is not fully defined, particularly regarding the elements representing the actions and their subsequent effects, which remain unknown.

The objective of finding a solution for a target case that is currently under construction involves inferring a set of target actions from source cases that share a similar context. These target actions are aimed at satisfying the specific context of the target case. This process leads to the identification and definition of the target effects. Ultimately, the goal is to construct a comprehensive case that encompasses the three parties: context, actions, and effects.

When dealing with similar source cases, which encompass both successful (member of $\mathbb{CB}^+$) and failed cases (member of $\mathbb{CB}^-$), it is important to handle them differently based on their outcome (failure or success) and their level of similarity to the target case. To address this, the proposed method should incorporate mechanisms that guide the approach towards successful solutions for similar source cases while moving away from failed ones. It should also consider that the closer a source case is to the target case, the greater impact its solution will have on the desired target solution.

3.2 Principle

Our approach to addressing failed cases in the adaptation process draws inspiration from navigation algorithms employed in the programming of autonomous robots. Specifically, we adopt the concept of artificial potential field discussed in Sect. 2.2.

Prior to delving into the specifics of our approach in the following section, we have established a set of assumptions to guarantee the effective integration of an artificial potential field-like concept within the framework of this study:

- although the study does not cover the labeling process, it is assumed that prior experiences (referred to as source cases) have already been labeled as either successful or failed cases. Additionally, it is presumed that the CBR system is equipped with a quality function QF, which evaluates the effectiveness of the actions taken within a given context. Higher scores indicate better performance. Consequently, this implicitly establishes a threshold value $\mathcal{TS}^{\mathcal{E}_i}$ for each effect feature $\mathcal{E}_i$, as defined by Eq. (4).

$$\forall\, \mathbb{C} \stackrel{\text{def}}{=} (\mathscr{C}, \mathcal{A}, \mathcal{E}) \in \mathbb{CB},\ \ \text{QF} : \mathcal{E} \longmapsto \mathbb{R}$$
$$\text{LF}(\mathbb{C}) = \begin{cases} \mathbb{C}^+ & \text{if } \text{QF}(\mathcal{E}) \geq \mathcal{TS}_s^{\mathcal{E}}\ ,\ \forall\, \mathcal{E} \in \mathbb{E} \\ \mathbb{C}^- & \text{otherwise.} \end{cases} \tag{4}$$

  With LF – the labeling function, $\mathcal{E}$ – an effect variable of the case $\mathbb{C}$.
- classical CBR methods typically retrieve a fixed number of neighboring cases from the case base $\mathbb{CB}$, without considering the optimal number of similar cases specific to the target case. This approach, resembling KNN, gives rise to certain issues. Not all target cases necessarily possess the same number of similar neighbors; some may have more while others may have fewer. Additionally, handling situations where there are significantly more source cases equidistant from a target case than the predefined number becomes challenging. In this study, we assume the presence of a retrieval approach that adjusts the number of source cases based on their similarity to the target case $\mathbb{C}^{tg}$. This adjustment is achieved by dynamically defining a similarity threshold $\varphi^{\mathbb{C}^{tg}}$ for the context distance between $\mathbb{C}^{tg}$ and its neighboring source cases. For example, a method proposed in [3] offers a technique to determine this threshold by combining statistical analysis and a genetic algorithm.

The main concept behind the approach proposed in this work involves associating the types of source cases available in the case base, namely successful and failed cases, with the types of objects involved in the domain of robot movement, specifically targets and obstacles. As a result, failed cases are interpreted as obstacles, while successful cases are treated as targets. In this framework, cases $\mathbb{C}^+ \in \mathbb{S}^{\mathbb{C}^{tg}}$ that exhibit positive outcomes generate an attractive force $\mathbb{F}^A$, which draws the target solution towards them. On the other hand, failed cases $\mathbb{C}^- \in \mathbb{S}^{\mathbb{C}^{tg}}$ produce a repulsive force $\mathbb{F}^R$ pushing the solution away from them.

The source, both successful and failed, cases are utilized to create a CBR potential field that represents the characteristics of the desired solution. Similar to the approach in the robotic potential field method, the CBR potential field consists of two fields. For exemple, in the case of the attractive potential field, a force of attraction is generated from the target solution towards the source solutions of the successful cases. This is achieved by configuring the latter in a way that enables pulling the target solution closer to the solutions of these cases.

To help explain this idea, let's imagine a system that incorporates domain knowledge with just two action variables. In this system, we can visualize the attractive potential field created by any successful case as shown in Fig. 2. In this figure, at every point in the context space representing the target context, the force vectors point towards the successful source case. On the other hand, the repulsive potential field generates a pushback force from the failed case towards the target solution. This force helps move the target solution away from the solutions associated with these failed cases. Figure 3 provides a visual representation of the repulsive force in a configuration similar to the example that demonstrates the CBR attractive force.

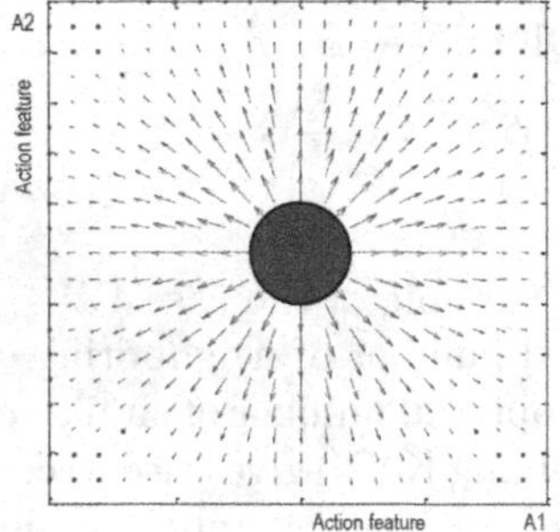

Fig. 3. Repulsive force in CBR context.

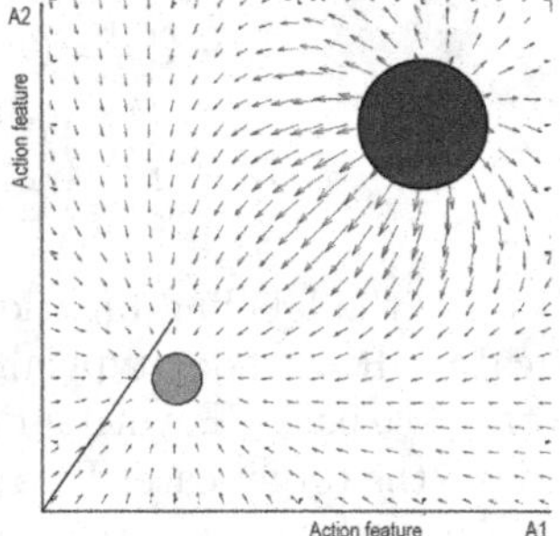

Fig. 4. Total potential force in CBR.

In the end, the positioning of the target solution within the solution space (actions) is established by combining the attractive and repulsive forces exerted by neighboring successful and failed cases, respectively. In the case where there are only two similar cases, one being a successful case and the other a failed case, the overall potential field takes the form depicted in Fig. 4.

3.3 Local Prediction of the Target Solution

While we draw inspiration from the potential artificial field method, applying it in the context of this work, as it is typically done in the robotics field, is not suitable for determining the target solution due to several reasons:

- in the realm of robotics, the overall force that a robot can exert is determined solely by the distance between the robot and the goal or obstacles it encounters. However, in the context of case-based reasoning, the strength of the attractive and repulsive forces is not solely dependent on the distance between the target context and the surrounding similar cases. It also takes into account the performance or quality of the neighboring similar cases.

- within the realm of robotics, the repulsive force in proximity to an obstacle is strongest, gradually diminishing as the distance from the obstacle increases, unlike the attractive force. In the context of case-based reasoning applications, the magnitude of both forces should be directly proportional to the performance of the source solutions, but inversely proportional to the distance between the source similar contexts and the target context.
- typically, robotic applications involve a single goal to be reached. However, in the scenario of a multi-goal environment, the objective is to find a path that sequentially traverses all these goals while optimizing specific criteria. In the context of CBR systems, the goal is to leverage the knowledge from neighboring source cases to infer the desired solution for the target case.
- while the objective of the potential artificial field in robotics is to identify a secure path towards the goal, its purpose within the context of CBR is to acquire fresh knowledge that facilitate the adaptation process in constructing the target solution. In other words, its role is to guide the reasoning process towards the most valuable solutions, which are typically the closest and best-performing cases, while steering away from unfavorable cases that are either farthest away or exhibit poor performance.

Table 1. Summery of results on synthetic dataset.

Approach	Test Set																	
	P_1			P_2			P_3			P_4			P_5			GLOBAL		
	Metrics			Metrics			Metrics			Metrics			Metrics			Metrics		
	IRP (%)	*EII(%)*	*EER(%)*	*IRP*	*EII*	*EER*	*IRP*	*EII*	*EER*	*IRP*	*EII*	*EER*	*IRP*	*EII*	*EER*	*IRP*	*EII*	*EER*
CBR_S	16.73	59.13	59.13	17.85	48.57	48.57	19.53	60.12	60.12	20.48	56.07	56.07	18.79	64.48	64.48	18.68	57.67	57.67
CBR_B	18.27	57.51	57.51	15.36	63.90	63.90	22.85	59.69	59.60	24.23	65.52	65.52	21.10	662.71	62.71	20.36	61.87	61.87
CBR_P	22.62	42.26	57.10	18.54	48.85	63.71	20.14	50.21	60.10	22.48	52.92	70.19	23.47	39.86	60.09	21.45	46;82	62,24
CBR_R	−2.56	32.18	49.75	9.12	29.80	51.19	14.71	43.07	64.24	17.45	39.52	57.74	12.04	41.26	62.84	10.15	37.18	57.15
$APF-CBR$	**34.68**	**100**	**100**	**28.85**	**99.76**	**99.76**	**33.91**	**100**	**100**	**31.27**	**100**	**100**	**38.73**	**99.88**	**99.88**	**33.49**	**99.92**	**99.92**

To effectively incorporate the specificities of the CBR adaptation process, it becomes necessary to modify the artificial potential field approach. In our proposed approach, the target solution (actions), denoted as $\mathcal{A}^{tg}$, is determined by the vector sum of all attractive forces ($\mathbb{F}^A_{\mathbb{C}+}, \forall\, \mathbb{C}^+ \in \mathbb{S}^{\mathbb{C}^{tg}}$) and all repulsive forces ($\mathbb{F}^R_{\mathbb{C}-}, \forall\, \mathbb{C}^- \in \mathbb{S}^{\mathbb{C}^{tg}}$), as defined in Eq. (5).

$$\forall\, \mathbb{C}^+, \mathbb{C}^- \in \mathbb{S}^{\mathbb{C}^{tg}}, \sum_{\mathbb{C}+} \mathbb{F}^A_{\mathbb{C}+} \overrightarrow{\mathcal{A}_{tg} A_{C_i+}} + \sum_{\mathbb{C}-} \mathbb{F}^R_{\mathbb{C}-} \overrightarrow{\mathcal{A}_{tg} A_{\mathbb{C}^-}} = 0 \tag{5}$$

As previously stated, the strength of the repulsion and attraction forces is influenced by both the distance between the target context and the context of the similar source case, as well as the performance of the source case. Eq. (5) provides a metric $\mathbb{F}_{\mathbb{C}}$, which determines the magnitude and direction of the corresponding force associated with case $\mathbb{C}$. To estimate the value of this force, we introduce Eq. (6).

$$\forall\, \mathbb{C} \in \mathbb{S}^{\mathbb{C}^{tg}}, \mathbb{F}_{\mathbb{C}} = \begin{cases} \left(1 - \frac{D^{ct}(\mathbb{C}^{tg},\mathbb{C}^{sr})}{\Theta_{\mathbb{C}^{tg}}}\right) \times (\mathrm{QF}_{\mathbb{C}} - \mathcal{TS}) & \text{if } \mathrm{QF}_{\mathbb{C}} \neq \mathcal{TS} \\ 1 - \frac{D^{ct}(\mathbb{C}^{tg},\mathbb{C}^{sr})}{\Theta_{\mathbb{C}^{tg}}} & \text{else} \end{cases} \tag{6}$$

where $D^{ct}(\mathbb{C}^{tg}, \mathbb{C}^{sr})$ indicate the context distance between $\mathbb{C}^{tg}$ and its neighboring case $\mathbb{C}^{sr}$, $\Theta_{\mathbb{C}^{tg}}$ represent the context distance threshold, $\mathcal{TS}$ represent the performance threshold, and $\mathrm{QF}_{\mathbb{C}}$ denote the performance of $\mathbb{C}$.

The Eq. (6) demonstrates that regardless of the force's nature, its strength gradually diminishes as the contextual distance increases, until it reaches zero when the contextual distance reaches the similarity threshold $\Theta_{\mathbb{C}^{tg}}$. In addition to determining the force's strength, the term $\mathrm{QF}_{\mathbb{C}} - \mathcal{TS}$ specifies the type of force. If $\mathrm{QF}_{\mathbb{C}} \geq \mathcal{TS}$, then $\mathbb{F}_{\mathbb{C}} \geq 0$, indicating an attractive force. Conversely, if $\mathrm{QF}_{\mathbb{C}} < \mathcal{TS}$, the force is repulsive. Therefore, it is necessary for the proposed actions $\mathcal{A}^{tg}$ to adhere to the following conditions:

$$\forall\, \mathbb{C} \stackrel{\text{def}}{=} (\mathscr{C}, \mathcal{A}, \mathcal{E}) \in \mathbb{S}^{\mathbb{C}^{tg}},\ \mathcal{A}^{tg} = \frac{1}{\sum_{\mathbb{C}} \mathbb{F}_{\mathbb{C}}} \sum_{\mathbb{C}} \mathbb{F}_{\mathbb{C}} \mathcal{A} \tag{7}$$

4 Evaluation

In this section, we provide a practical evaluation of the proposed approach, referred to as APF-CBR hereafter. The evaluation aims to accomplish two main objectives. Firstly, investigate the influence of considering both failed and successful cases on enhancing the effectiveness of the CBR system. Secondly, evaluate the efficacy of the APF-CBR approach in comparison to existing adaptation approaches.

4.1 Experimental Design

As indicated in Sect. 2, the APF-CBR approach is applied within an EMS, aiming to raise the user's consciousness about the consequences of their actions on energy consumption in a building. More specifically, the EMS suggests a set of measures to the occupant that enhance comfort while simultaneously reducing energy consumption.

To assess the effectiveness of the APF-CBR approach, we carried out an experiment utilizing semi-synthetic data derived from actual data presented in [4]. The data base used for this experiment consisted of a total of 15,948 cases, with each case comprising three types of variables: effect variables, action variables, and context variables. The effect variables in our cases represent the temperature and air quality within the building. The action variables are used to model the opening of the door and window, while the context variables capture the weather conditions. We utilized a 24-value vector to represent each variable, corresponding to a single day. We utilized a 5-fold cross-validation approach to assess the variables in our study. To evaluate our model's performance, we employed a 5-fold cross-validation approach. Initially, we randomly divided the original case base into five subsets of equal size, namely P_1, P_2, P_3, P_4, and P_5. During each iteration of the cross-validation process, one subset was selected as the test set, denoted as $\mathbb{CB}_T$, consisting of target cases. The remaining four subsets served as the learning set, denoted as $\mathbb{CB}_L$, which comprised the source cases. This process was repeated five times, with each subset being used once as the test set. The final values of the metrics adopted in the evaluation correspond to the average of the values obtained in the five iterations

To assess the effectiveness of the proposed actions, we employed functions that measured the level of user dissatisfaction with temperature ($\text{QF}_{\text{C}}^{\mathcal{E}_T}$) and air quality ($\text{QF}_{\text{C}}^{\mathcal{E}_{CO_2}}$), as depicted by Formula (8). To simulate the consequences resulting from the implementation of the suggested actions, we constructed a physical model of the building that was utilized in the experiment.

$$\text{QF}_{\text{C}}^{\mathcal{E}_T}(h) = \begin{cases} 0 & \text{if } \mathcal{E}_T(h) \in [21, 23] \\ \frac{\mathcal{E}_T(h)-23}{26-23} & \text{if } \mathcal{E}_T(h) > 23 \\ \frac{21-\mathcal{E}_T(h)}{21-18} & \text{if } \mathcal{E}_T(h) < 21 \end{cases}, \quad \text{QF}_{\text{C}}^{\mathcal{E}_{CO_2}}(h) = \begin{cases} 0 & \text{if } \mathcal{E}_{CO_2}(h) \leq 500 \\ \frac{\mathcal{E}_{CO_2}(h)-500}{1500-1000} & \text{if } \mathcal{E}_{CO_2}(h) > 500 \end{cases} \tag{8}$$

4.2 Baselines and Metrics

The evaluation process includes several baselines:

1. the CBR_S approach, which is discussed in [4], utilizes both failed and successful cases. However, it lacks an adaptation process, as it simply involves taking a vote among the solutions of similar cases and selecting the solution with the best performance (maximizing the quality function) to be directly applied to the target case. This baseline choice aims to assess the importance of incorporating multiple source cases in establishing an adaptation process.
2. one approach, referred to as CBR_B, employs a standard barycentric method to merge solutions from both successful and failed similar cases, without the use of artificial forces. The primary objective of this approach is to assess the effectiveness of artificial forces in enhancing the reasoning process.
3. to illustrate the advantages of considering both negative and positive cases over only positive cases, we tested a modified variant of our approach called CBR_P. This variant exclusively considers positive cases, relying solely on attractive forces. The objective behind this modification was to highlight the benefits of incorporating both negative and positive cases in comparison to exclusively focusing on positive cases.
4. as an additional baseline, the method presented in [10] (denoted as CBR_R) is employed. CBR_R utilizes a K-Nearest Neighbors (KNN) algorithm to identify source cases that are similar to the target case. From these similar cases, a generalized case is created. Moreover, the similar cases are utilized to train a linear regression model. This regression model is then employed to predict the solution for the target case based on the generalized case.

It's important to note that in the experiment, the comparison of all the approaches being tested is based on the actions performed by the user without any assistance. This evaluation is conducted using three specific metrics:

- *Improvement Ratio of Performance (IRP)*: The IRP metric evaluates the performance enhancement achieved by each tested approach. It is determined by comparing the average of the global satisfaction $\mathrm{QF}^P_{\mathbb{C}}$, of the proposed actions with the corresponding value $\mathrm{QF}^U_{\mathbb{C}}$, of the actions performed by the user without assistance for each test case $\mathbb{C}^t$.

$$IRP_{\mathbb{C}^t} = \frac{\mathrm{QF}^P_{\mathbb{C}} - \mathrm{QF}^U_{\mathbb{C}}}{\mathrm{QF}^U_{\mathbb{C}}} \tag{9}$$

- *Effectiveness Improvement Index (EII)* is calculated as the average ratio of the number of test cases that show performance improvement when the actions recommended by this approach are applied, to the total number of test cases.

$$EII = \frac{\beta^+}{\beta} \tag{10}$$

  With $\beta = |\mathbb{CB}_T|$ – the set of test cases, $\beta^+ = \{\mathbb{C} \in \mathbb{CB}_T \ / \ IRP_{\mathbb{C}^t} > 0\}$
- *Effective Enhancement Ratio (EER)* refers to the average ratio between the number of test cases that show improved performance when the recommended actions from the approach are applied, and the total number of test cases for which the approach successfully suggests a solution (whether it improves or degrades performance compared to the user's actions).

4.3 Results

Regardless of the adaptation approach used in a CBR system, the retrieval process plays a significant role in determining its performance. Although this paper does not delve into analyzing the retrieval process, we adopt the methodology proposed in [4] to assess similarity and identify similar source cases within the training set. After applying this approach, it is observed that each target case from the test set has at least one similar source case from the training set.

Table 1 provides a summary of the results obtained through 5-fold cross-validation, comparing our approach to the four baselines. Notable observations from this experiment are:

- while the EER metric aligns with the EII value for the $\mathrm{CBR_S}$, $\mathrm{CBR_B}$, and APF-CBR approaches, the $\mathrm{CBR_P}$ and $\mathrm{CBR_R}$ approaches exhibit a lower EII value compared to EER. This disparity can be attributed to the fact that the first three approaches are capable of producing a solution even when provided with a collection of exclusively failed cases. On the other hand, the $\mathrm{CBR_P}$ and $\mathrm{CBR_R}$ approaches do not possess this capability.
- irrespective of the test set employed, our APF-CBR approach consistently outperforms all other baseline methods, demonstrating superior performance in terms of EII and EER.

- the quality of the adaptation process is significantly influenced by the number of similar source cases. When employing a compositional adaptation approach, the IRP tends to be superior compared to using a single similar case, as exemplified by the comparison between APF-CBR (compositional approach) and CBR_S (single similar case).
- the inclusion of attraction and repulsion forces significantly impacts the results of the adaptation process. Utilizing these forces, our APF-CBR approach surpasses the CBR_B baseline, which does not incorporate them, even when considering an equal number of similar cases. APF-CBR demonstrates a superior performance, being 1.64 times more effective than CBR_B in terms of enhancing case performance (global IRP = 33.49% compared to 20.36%). Furthermore, APF-CBR is 1.61 times more efficient in terms of the number of cases for which it successfully finds a solution. It enhances the performance of user-proposed solutions without assistance in 99.92% of cases, as opposed to 61.87% for CBR_B.
- the performance of a Case-Based Reasoning (CBR) system is significantly influenced by the utilization of failed cases. By incorporating both successful and failed cases, the system enhances the outcomes of the reasoning process. When comparing the performance of three different approaches-APF-CBR, CBR_P, and CBR_R, the EER results demonstrate that the APF-CBR approach surpasses the other baselines. The APF-CBR approach exhibits more than three times greater efficiency than CBR_R and more than 1.5 times greater efficiency than CBR_P in enhancing the performance (PER).

5 Conclusion

This paper introduces a novel method for improving the adaptation process in the Case-Based Reasoning paradigm. Instead of relying solely on successful source cases, we consider both failed and successful cases. We draw inspiration from studies on planning safe paths for robots in unknown environments. Our approach involves generating attraction and repulsion forces from successful and failed cases, respectively, to guide reasoning towards the best solutions and away from the failed ones. Experimental results in an EMS context demonstrate a significant ienhancement in system performance by considering both successful and failed cases. We have developed and evaluated an approach that incorporates the entire set of successful and failed similar cases. Further evaluation could explore the impact of the number of neighboring successful and failed cases, focusing on the top-performing cases and the worst-performing cases. Additionally, future research could investigate the potential influence of failed cases on the domain ontology. Understanding this impact could provide insights into how the domain ontology could be refined or modified to prevent the recurrence of negative cases in the future.

References

1. Ashrae, (ed.). ASHRAE Standard Thermal Environmental Conditions for Human Occupancy. American Society of Heating, Refrigerating and Air-Conditioning Engineers., Atlanta, USA (1992)
2. Ashrae, (ed.). Indoor air quality guide: best practices for design, construction, and commissioning. American Society of Heating, Refrigerating and Air-Conditioning Engineers., Atlanta, USA (2009)
3. Boulmaiz, F., Alyafi, A.A., Ploix, S., Reignier, P.: Optimizing occupant actions to enhance his comfort while reducing energy demand in buildings. In: 11th IEEE IDAACS (2021)
4. Boulmaiz, F., Reignier, P., Ploix, S.: An occupant-centered approach to improve both his comfort and the energy efficiency of the building. Knowl.-Based Syst. **249**, 108970 (2022)
5. Díaz-Agudo, B., González-Calero, P.A.: An architecture for knowledge intensive CBR systems. In: Blanzieri, E., Portinale, L. (eds.) EWCBR 2000. LNCS, vol. 1898, pp. 37–48. Springer, Heidelberg (2000). https://doi.org/10.1007/3-540-44527-7_5
6. Govedarova, N., Stoyanov, S. and Popchev, I.: An ontology based CBR architecture for knowledge management in bulchino catalogue. In: CompSysTech (2008)
7. CSA Group. Z412–17 Office ergonomics - an application standard for workplace ergonomics (2017)
8. Khatib, O.: Real-time obstacle avoidance for manipulators and mobile robots. In: Proceedings of IEEE International Conference on Robotics and Automation (1985)
9. Minor, M., Marx, L.: Case-based reasoning for inert systems in building energy management. In: Aha, D.W., Lieber, J. (eds.) ICCBR 2017. LNCS (LNAI), vol. 10339, pp. 200–211. Springer, Cham (2017). https://doi.org/10.1007/978-3-319-61030-6_14
10. Patterson, D., Rooney, N., Galushka, M.: A regression based adaptation strategy for case-based reasoning, In: AAAI/IAAI (2002)
11. Petrovic, S., Khussainova, G., Jagannathan, R.: Knowledge-light adaptation approaches in case-based reasoning for radiotherapy treatment planning. Artif. Intell. Med. **68**, 17–28 (2016)
12. Sizov, G., Öztürk, P., Marsi, E.: Compositional adaptation of explanations in textual case-based reasoning. In: Goel, A., Díaz-Agudo, M.B., Roth-Berghofer, T. (eds.) ICCBR 2016. LNCS (LNAI), vol. 9969, pp. 387–401. Springer, Cham (2016). https://doi.org/10.1007/978-3-319-47096-2_26
13. Wilke, W., Vollrath, I., Althoff, K.D., Bergmann, R.: A framework for learning adaptation knowledge based on knowledge light approaches. In: Fifth German Workshop on Case-BasedReasoning, pp. 235–242 (1997)

Transfer Learning for Abnormal Behaviors Identification in Examination Room from Surveillance Videos: A Case Study in Vietnam

Pham Thi-Ngoc-Diem[1(✉)], Lan Ngoc Ha[1,2], and Hai Thanh Nguyen[1]

[1] College of Information and Communication Technology, Can Tho University, Can Tho, Vietnam
{ptndiem,nthai.cit}@ctu.edu.vn
[2] Thieu Van Choi High School, Soc Trang, Vietnam
hangoclan.c3tvc@soctrang.edu.vn

Abstract. The examination evaluates the learners' ability to achieve a specific goal, representing a crucial assessment of the knowledge acquired during the learning process. To attain success in the examination, it is imperative to prevent cheating in the examination hall. Despite this, cheating still occurs due to the limited availability of human resources. Therefore, we have collected surveillance videos of examination halls from a high school in Vietnam to analyze and implement deep learning architectures such as You Only Look Once (YOLO) and Single Shot Detector (SSD) MobileNet V2 to detect anomalous behavior among students during the examination. Our study focuses on detecting five common abnormal behaviors, including looking around, bending over the desk, putting one or two hands under the table, waving, and standing up. YOLO achieved the best results, with a performance of 83.55%, 99.65%, 97%, 99.2%, and 98.0866% in Intersection over Union (IoU), Mean Average Precision (mAP), Precision, Recall, and F1-Score, respectively, across 2639 images. This approach is expected to assist educators and teachers in detecting and preventing cheating activities in examination rooms.

Keywords: Abnormal behavior · Deep learning · Neural networks · Object detection · YOLO V4 · SDD MobileNet V2

1 Introduction

The phenomenon of students "sitting in the wrong class" is no longer an unfamiliar problem. This situation is widespread at all school levels. It is not uncommon for teachers to witness students using cheat sheets or sophisticated electronic devices like headphones, pocket calculators, etc., during an exam. Some students even collaborate to access textbooks and documents or exchange answers. Such situations stem from various reasons, such as students' lack of motivation, overdependence on others, and laziness. Furthermore, schools face competitive

N. T. Nguyen et al. (Eds.): ACIIDS 2023, LNAI 13995, pp. 45–57, 2023.
https://doi.org/10.1007/978-981-99-5834-4_4

pressure, and some families fail to prioritize their children's education. This situation could lead to unexpected consequences for future society if not prevented in time.

Meanwhile, advances in technology and research in artificial intelligence have enabled the creation of many automatic systems that can help solve problems in education, such as the attendance checking system [1,2] and detecting systems for students' abnormal behaviors during exams [3,4].

In this study, we collected real videos from a high school in Vietnam that captured five actions marked as abnormal behaviors, including looking around, bending over the desk, putting one or two hands under the table, waving, and standing up in the examination hall and collected samples for normal actions. We utilized deep learning architectures, including YOLO V4 [5] and SSD MobileNet V2 [6], to detect abnormal actions in surveillance videos from examination halls. We conducted experiments to compare the performance of these two models. The results on the dataset, including 2639 images extracted from videos, revealed that YOLO V4 outperformed SSD MobiletNet V2. However, YOLO V4 took longer to train the models.

This study is organized as follows. Section 2 introduces related works. Section 3 describes the dataset collection, labeling samples for behavior types, and presents the overall proposed method and deep learning architecture hyperparameters. Section 4 presents and explains the experimental results. Finally, we provide some closing remarks in Sect. 5.

2 Related Work

Numerous recent studies have been detecting human behavior, such as studies in [7–14].

In the study by Cao et al. [7], the authors compared solutions to estimate the 2D posture of multiple people in real time. Yu et al. [8] developed a behavioral detection system based on Convolutional Neural Networks (CNNs). They segmented the video into images, extracted the moving image of the body by separating the image background, trained the data using a CNN built by gradient descent, and identified and classified behaviors. Compared to other behavioral recognition methods, their results demonstrate that CNN can study and identify human behavior without manual training and labeling. Xu et al. [9] developed a motion behavior recognition system based on optical flow to detect moving objects. They used Convolutional Neural Networks to select features and reduce image size and a support vector classifier to train, classify, and recognize actions. The experimental results showed that this method effectively distinguished human actions and significantly improved action recognition accuracy under different lens situations, such as close-up, far-away, or slight camera movement. In Song et al. [10], a goal-based behavior detection solution was proposed by combining YOLO v3 algorithms to detect objects, using KCF for target tracking, TSN algorithm for action recognition, and HMOF to extract features of each target. After testing on the UMN dataset, the suggested solution's accuracy was the greatest at 99.8%, while the UCSD Ped24 dataset's EER error rate was the lowest at 5.5% (pixel-level) and 14.5% (frame-level). Lina et al. [12] proposed a model using a combination of OpenPose and YOLO for detecting human behaviors, including walking, standing, sitting, and falling, on a dataset of 400 images. The experimental results show that the training accuracy of the model reaches 95%. Lu et al. [13] investigated the ability to recognize human behavior using deep learning based on the YOLO v3 model. After some tests, the accuracy reached 80.20%. Schuldt et al. [14] used a support vector classifier to classify six human behaviors. They experimented on a dataset containing 2391 behavioral sequences of 25 people in four different situations, and their results showed that the proposed method has advantages over other methods.

In recent years, several studies have been conducted to investigate abnormal/violent human behavior detection, such as those by Qian et al. [15], Wu et al. [16], and Nasaruddin et al. [17]. These studies propose various models for identifying and detecting abnormal behavior in human activities. Qian et al. [15] proposed an improved ResNet model to recognize abnormal behavior. The study used the UTI dataset to evaluate the performance of the proposed model, which showed an accuracy improvement of 2.8% compared to the original ResNet-50 model. The highest improved model achieved an accuracy of 89.2%. Wu et al. [16] proposed combining the Gaussian model and fuzzy C-means clustering (FCM) to detect anomalous behavior in house environments. The method segmented the background image of each frame in the video and used FCM to detect outliers in the data sample. Fang et al. [17] proposed a combination of YOLO v3, K-Means, GIoUloss, focal loss, and Darknet 32 algorithms to detect abnormal behavior of candidates. The study used a dual frame instead of the stream method to optimize the detection process. The results showed that improving the YOLO v3 algorithm improved detection accuracy and speed, achieving an average accuracy (mAP) of 88.53% on the test suite and a detection rate of 42 frames per second. Ullah et al. [11] proposed a solution to detect violent behavior in videos based on 3D CNN. This study could be applied in public places such as airports, streets, hospitals, and schools. The test results obtained are very high on the Violence in Movies5 (99.9%), Violent Crowd6 (98%), and Hockey Fight5 (96%) datasets.

In particular, many studies related to the detection of abnormal behavior in the exam room have also been carried out [18–21]. It is worth noting that the system proposed by [21] using Neural Networks and Gaussian Distribution achieved accuracy of 97% and the F1 Score of 88%. While Soman et al. [19] proposed a method for detecting five anomaly behaviors (normal, turning around, passing paper, peeping, and signaling) in videos based on HOG and KNN. The overall accuracy of this method is 88.2%.

These studies provide valuable insights and techniques for developing efficient and accurate abnormal human behavior detection systems.

3 Methods

The process of detecting abnormal behavior in the exam room is described as follows. Identifying and detecting abnormal behavior in the exam room is done in two stages. The first stage is the model training using two architectures, YOLO V4, and SSD Mobilenet V2. In this stage, collected videos from exam rooms in a high school were preprocessed by splitting them into frames (images) using the OpenCV library. The unclear, similar, or images that do not include students are removed. The dataset for training abnormal behavior detection models contains the rest of the images. Each dataset image is labeled to determine the student's position in the image. The second stage is to conduct the student behavior classification. In this phase, frames of videos or images are used as the experimental set tested on the behavior detection models generated in stage 1. As a result, the models produce the bounding box annotation along with one of six labels (looking around, bending over the desk, putting one or two hands under the table, waving, standing up, and normal action) that can locate the object in the input images or video frames and students' behavior.

3.1 Abnormal Behaviors in the Examination Hall and Data Collection

The videos used to build and evaluate models are collected in Thieu Van Choi high school in Soc Trang province, Vietnam. The five videos gathered should be approximately 40 min in total length. Videos numbered 1,2,3 and 4 are used as the training set. The videos are captured from a surveillance system equipped in the corner of the classroom, where one can observe students' faces. The dataset for training the models consists of 2639 images extracted from these videos after removing the inappropriate frames. 5th video is used for testing models. The images in the training dataset are manually labeled in the YOLO labeling format using the LabelImg[1] tool. Each label contains the object class (the first column), the object center coordinates (the second and the third), and the width and height of the bounding box as shown from left to right, respectively, in Fig. 1. Based on high school graduation examination regulations of the Vietnamese Ministry of Education and Training in 2020 [22], we discriminated against six classes of actions. Therefore, the considered abnormal behaviors include looking around, bending over the desk, putting a hand under the table, waving, standing up, and normal action. After labeling, we obtain 15,788 labels corresponding to

```
5 0.727083 0.612963 0.158333 0.298148
5 0.852865 0.385185 0.093229 0.198148
2 0.249219 0.423148 0.026562 0.109259
2 0.282292 0.429630 0.036458 0.112963
3 0.402604 0.385185 0.048958 0.074074
```

Fig. 1. An illustration of labels of an image.

[1] https://github.com/heartexlabs/labelImg.

6 of these classes from the dataset of 2639 images. Table 1 describes in detail the number of labels corresponding to each class in the dataset. For example, class 0 describes the objects with the action of looking around, while class 1 describes bending over the desk. Other classes, including 2, 3, 4, 5, denote putting one or two hands under the table, waving, standing up, and normal actions, respectively.

Table 1. Classes of the dataset.

Description	Class	Number of objects
Looking around	0	3186
Bending over the desk	1	2164
Putting the hand under the table	2	2914
Waving	3	1426
Standing up	4	1186
Normal action	5	4914

3.2 Deep Learning Architectures for Abnormal Behavior Detection

In this work, YOLO V4 and SSD on MobileNet V2 are used to detect abnormal behaviors of students in the exam room. SSD MobileNet V2 [6] is an object detection approach that receives an image as input and returns bounding boxes for the objects in the image. SSD MobileNet V2 extracts feature from the image, then processed through SSD prediction layers that reduce the image size to recognize objects at different scales. YOLO [5] is a deep learning algorithm that uses a single-stage CNN for fast and accurate object detection. The input of the YOLO algorithm is the image pixels. Then, the bounding box for each object after detection is drawn according to the coordinates and probability of the object's occurrence. Many studies have improved the architecture through each version in terms of accuracy and speed.

4 Experimental Results

4.1 Environmental Settings

All experiments and evaluations were performed on the Colab environment, including Tesla T4 GPU with 15109MB memory. Camera with 2.0 Megapixel resolution is used to collect data. The camera is located at the top of the exam room and it ensures that all students are covered. The exam room has a length of 7.5 m and a width of 7.5 m with an area of 56.25 m^2.

We compared two architectures YOLO V4 and SSD MobileNet V2 for abnormal behavior detection in the examination hall. Each architecture is trained on the same training and testing sets and is evaluated by mAP, IoU, F1-Score, and precision metrics. The dataset for training is split into an 80:20 ratio, with 80% for training and 20% for validation.

The number of iterations in both SSD MobileNet V2 and YOLO V4 is 12000, calculated by the number of classes × 2000. Filters are equal to 33, performed according to the formula (the number of classes + 5) × 3 as proposed by[2]. We use default values for other hyper-parameters.

4.2 Abnormal Behavior Detection Using YOLO V4

As presented in Table 2, the results are reported across several evaluation metrics. The IoU metric achieved a value of 83.55%, while the mAP metric demonstrated the highest performance across all metrics, with a score of 99.65%. Furthermore, the remaining metrics all achieved values above 95%.

Table 2. Performance of the abnormal behavior detection model using YOLO V4.

mAP (%)	IoU (%)	Precision (%)	Recall (%)	F1-Score (%)
99.65	97.00	99.20	83.55	98.09

Figure 2a shows that the loss function change gradually decreases and tends to converge when approaching round 12000. At the same time, the model's accuracy also increases through specific training rounds, as shown in Fig. 2b. This shows that the model is learning and learning effectively. Figure 4a reveals the performance in detecting the action of Standing up, normal actions, and the behavior of putting one or two hands under the table reached 100% while the other obtained 99%.

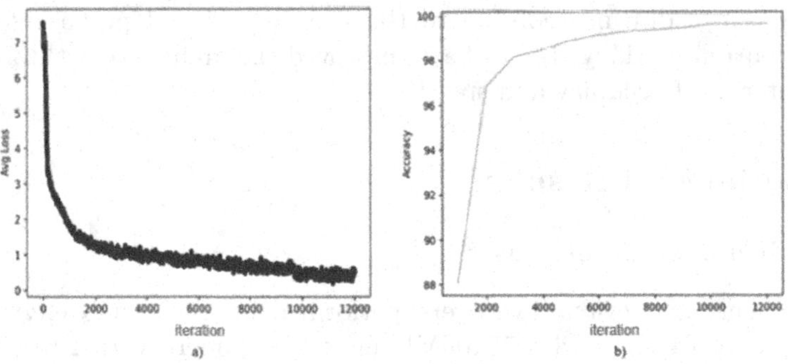

Fig. 2. YOLO V4 performance in Loss (a) and Accuracy (b).

[2] https://www.ccoderun.ca/programming/2019-08-18_Training_with_Darknet/.

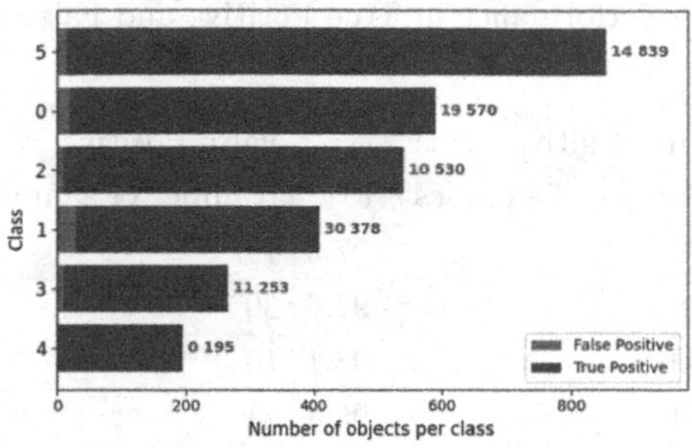

Fig. 3. Action detection performance of YOLO V4.

The object detection result on the validation set of images was presented in Fig. 3. This graph indicates the total number of predictions with IoU > 0.5 (True Positive) and IoU < 0.5 (False Positive) per label. Table 3 describes the correct recognition rate as True Positive and False Positive of 6 labels in the dataset. The rate of models correctly recognizing behavior is high, from 92.6% or more. The action of standing up is easy to detect, while bending over desk action is more challenging. These results prove that the YOLO V4 model performs well in recognizing the trained behaviors. In addition, Fig. 4b also illustrates that the object miss rate is low during the object detection process. The omission rate of labels 0, 1, and 2 is not more than 1%. Articularly for label 5, the images of behaviors are pretty clear, independent, and easier to distinguish than other behaviors. So during the object detection process, the model does not miss this behavior.

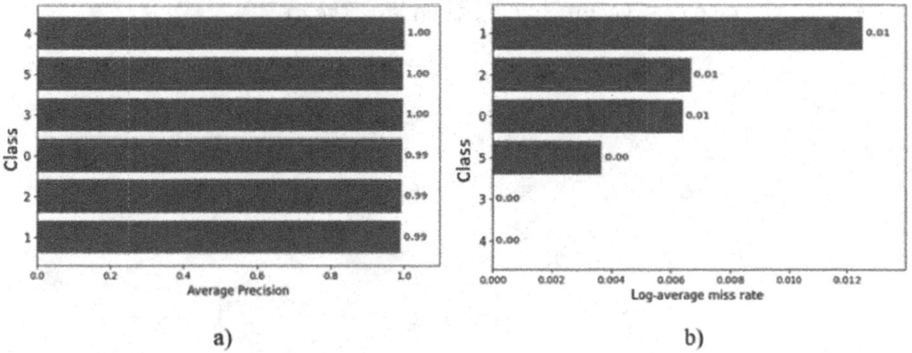

Fig. 4. Detection performance in Average Precision (a) and Log-average miss rate (b).

Table 3. The detection performance in True Positive and False Positive with YOLO V4.

Class	True Positive		False Positive	
	Number of samples	%	Number of samples	%
0	570	97.3	19	2.7
1	378	92.6	30	7.4
2	530	98.1	10	1.9
3	253	95.8	11	4.2
4	195	100.0	0	0.0
5	839	98.4	14	1.6

4.3 Abnormal Behavior Detection with SSD MobileNet V2

The labels in YOLO format are automatically converted to the SSD format before training. After the training, the values of the model evaluation measures are presented in Table 4. Specifically, it shows that the Precision, Recall, and F1-Score of the model in the training dataset reached the average value, respectively, 86.42%, 68.8%, and 76.61%. In addition, Fig. 5 indicates that the Average Loss value on the training set tends to reduce and converges from the 8,000th iteration. The orange line represents the Average Loss value of the training set, while the blue color represents the Average Loss value of the test set.

Table 4. Metrics obtained with the SSD MobileNet V2 model.

IoU (%)	Precision (%)	Recall (%)	F1-Score (%)
69.92	86.42	68.80	76.61

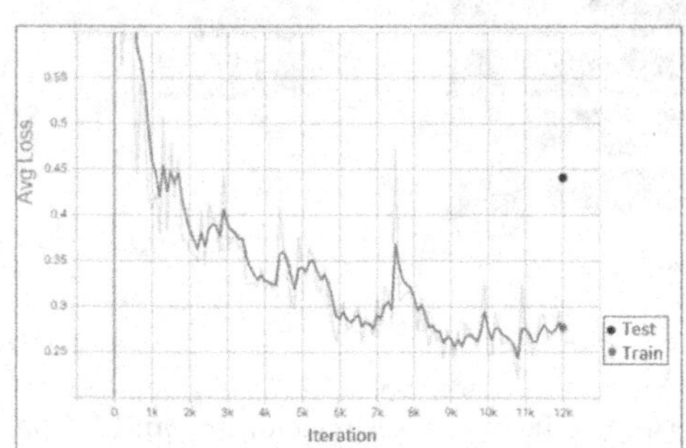

Fig. 5. Avg loss during training SSD MobileNet V2 model.

Figure 6a reveals that the Average Precision (AP) value of class 0 (Looking around) is the highest at 36% and class 2 (putting one or two hands under the table) has the lowest AP of 21%. From the AP results, the mAP value of the dataset reached 29.61%.

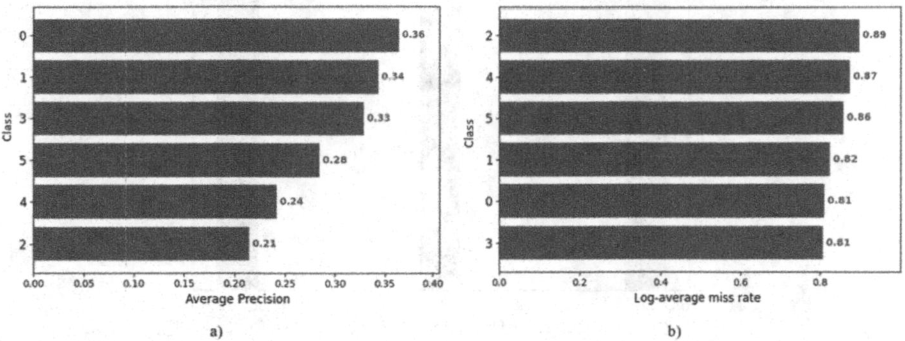

Fig. 6. Each action type recognition performance of SSD MobileNet V2 in Average Precision (a), Log-average miss rate object detection (b).

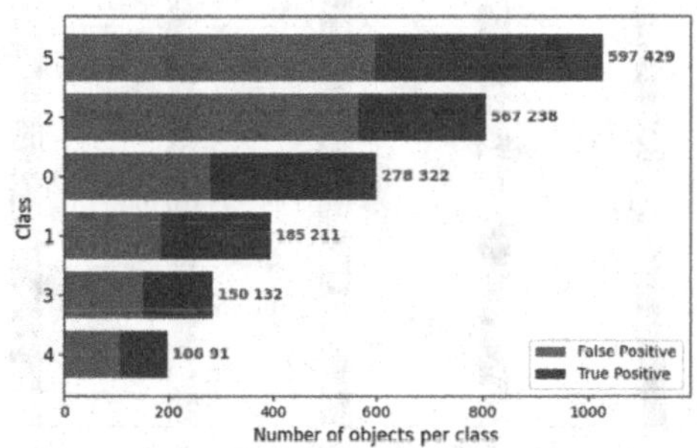

Fig. 7. Action detection performance of SSD MobileNet V2.

Figure 6b indicates that SSD MobileNet V2 performs less than YOLO V4. Specifically, the omission rate of classes 0, 1, 2, 3, 4, and 5, respectively 0.81, 0.82, 0.89, 0.81, 0.87, 0.86. The highest omission rate is class 2 (89%), and the class with the lowest rate is class 0 and class 3 (81%). Figure 7 shows the total number of predictions with IoU $>$ 0.5 (True Positive) and IoU $<$ 0.5 (False Positive) in each class. Table 5 exhibits that the rate of models correctly recognizing the behavior is very low; in all classes, the rate does not exceed 53.7%.

4.4 Discussion

As illustrated in Table 6, the YOLO V4 outperformed SSD MobileNet V2 on all measures. Figures 8 and 9 show that YOLO V4 detects most objects while many

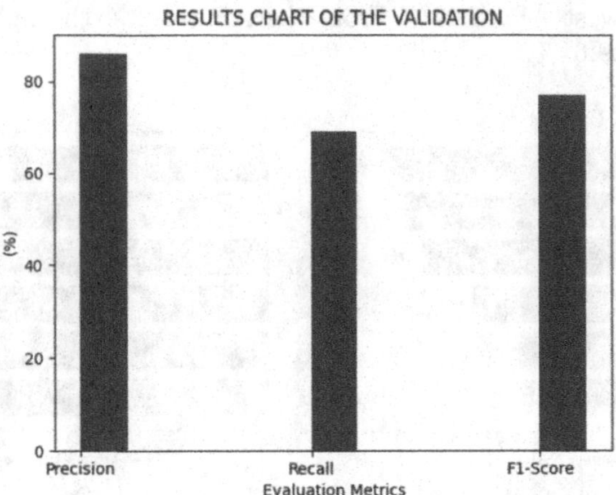

Fig. 8. Object detection performance of SSD MobileNet V2.

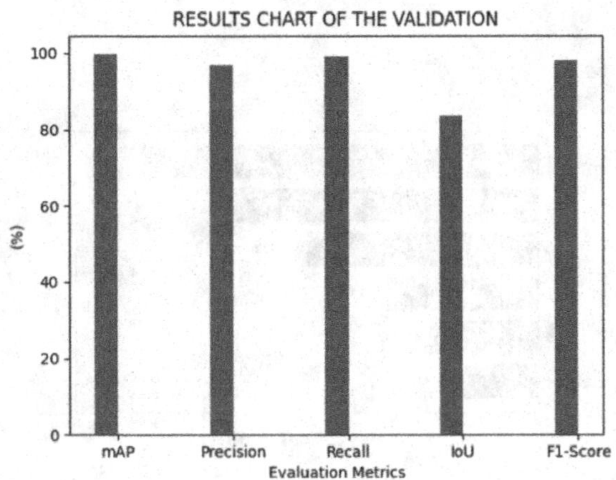

Fig. 9. Object detection performance of YOLO V4.

Table 5. True Positive and False Positive of the SSD MobileNet V2 model

Class	True Positive		False Positive	
	Number	%	Number	%
0	322	53.7	278	46.3
1	211	53.3	185	46.7
2	238	29.6	567	70.4
3	132	46.8	150	53.2
4	91	46.2	106	53.8
5	429	41.8	597	58.2

Table 6. Performance measures comparison of YOLO V4 and SSD MobileNet V2.

Model	IoU (%)	Precision (%)	Recall (%)	F1-Score (%)	Training time
YOLO V4	83.6	97.0	99.2	98.1	19 h 40 min
SSD MobileNet V2	69.9	86.4	68.8	76.6	3 h 15 min

objects are ignored by SSD MobileNet V2 architecture. However, the training time of YOLO was five times more than that of MobileNet. This can be explained because the number of parameters of YOLO (60 million parameters [23]) is many times larger than that of SSD MobileNet V2 (3.4 million).

5 Conclusion

This study presents an approach to detect abnormal actions in examinations via videos, including looking around, bending over the desk, putting the hand under the table, waving, standing up, and normal actions. More specifically, a dataset was collected, including 2639 images containing these behaviors. Each image of this dataset is manually labeled using Labelimg tool. This research used and compared two object detection approaches, YOLO V4 and SSD MobileNet V2. As a result, the YOLO V4 model outperforms SSD MobileNet V2 and is used to build the abnormal behavior detection model in the exam rooms. Besides the obtained results, the work still has some limitations. The dataset used to train the models is small. The confusion between bowing to write articles and bowing under the table of students near the camera still occurs. In addition, the concealed positions will not be detected, other behaviors can be mistakenly detected, and a candidate's sequence of abnormal behaviors has not been identified.

For building a complete system, more data should be collected with various actions, such as various types of bowing. In addition, future work can evaluate different factors such as lighting conditions, camera angles, and the inability to detect concealed positions. Furthermore, further research should study object-tracking methods and face-recognition methods to identify candidates with abnormal behaviors. Besides, the research on the problem of determining a sequence of a candidate's behavior is also a problem that needs to be solved to integrate into the abnormal behavior detection system. Finally, it combines detection methods, face recognition, object tracking methods, and behavior recognition to build a complete system that can be applied in practice.

References

1. Rohini, V., Sobhana, M., Chowdary, C.S.: Attendance monitoring system design based on face segmentation and recognition. Recent Patents Eng. **17**(2), 82–89 (2023). https://doi.org/10.2174/1872212116666220401154639
2. Budiman, A., Yaputera, R.A., Achmad, S., Kurniawan, A.: Student attendance with face recognition (LBPH or CNN) systematic literature review. Proc. Comput. Sci. **216**, 31–38 (2023). https://doi.org/10.1016/j.procs.2022.12.108

3. Li, D., Qu, P., Jin, T., Chen, C., Bai, Y.: An anomaly detection of learning behaviour data based on discrete Markov chain. Int. J. Continuing Eng. Educ. Life-Long Learn. **33**(1), 69 (2023). https://doi.org/10.1504/ijceell.2023.127872
4. Wei, F.: Study on behaviour anomaly detection method of English online learning based on feature extraction. Int. J. Reason. Based Intell. Syst. **15**(1), 41 (2023). https://doi.org/10.1504/ijris.2023.128372
5. Redmon, J., Divvala, S., Girshick, R., Farhadi, A.: You only look once: unified, real-time object detection (2015). https://arxiv.org/abs/1506.02640
6. Sandler, M., Howard, A., Zhu, M., Zhmoginov, A., Chen, L.C.: Mobilenetv 2: inverted residuals and linear bottlenecks (2018). https://arxiv.org/abs/1801.04381
7. Cao, Z., Simon, T., Wei, S.E., Sheikh, Y.: Realtime multi-person 2d pose estimation using part affinity fields. In: Proceedings of the IEEE Conference on Computer Vision and Pattern Recognition, pp. 7291–7299 (2017)
8. Yu, B.: Design and implementation of behavior recognition system based on convolutional neural network. In: ITM Web of Conferences, vol. 12, p. 01025. EDP Sciences (2017)
9. Xu, H., Li, L., Fang, M., Zhang, F.: Movement human actions recognition based on machine learning. Int. J. Online Eng. **14**(4), 193–210 (2018)
10. Song, L., Liu, B., Zhu, H., Chu, Q., Yu, N.: Abnormal behavior detection based on target analysis. arXiv preprint arXiv:2107.13706 (2021)
11. Ullah, F.U.M., Ullah, A., Muhammad, K., Haq, I.U., Baik, S.W.: Violence detection using spatiotemporal features with 3d convolutional neural network. Sensors **19**(11), 2472 (2019)
12. Lina, W., Ding, J.: Behavior detection method of openpose combined with yolo network. In: 2020 International Conference on Communications, Information System and Computer Engineering (CISCE), pp. 326–330. IEEE (2020)
13. Lu, J., Yan, W.Q., Nguyen, M.: Human behaviour recognition using deep learning. In: 2018 15th IEEE International Conference on Advanced Video and Signal Based Surveillance (AVSS), pp. 1–6. IEEE (2018)
14. Schuldt, C., Laptev, I., Caputo, B.: Recognizing human actions: a local SVM approach. In: Proceedings of the 17th International Conference on Pattern Recognition, 2004. ICPR 2004, vol. 3, pp. 32–36. IEEE (2004)
15. Qian, H., Zhou, X., Zheng, M.: Abnormal behavior detection and recognition method based on improved Resnet model. CMC-Comput. Mater. Continua **65**(3), 2153–2167 (2020)
16. Wu, C., Cheng, Z.: A novel detection framework for detecting abnormal human behavior. Math. Probl. Eng. **2020**, 1–9 (2020)
17. Fang, M.t., Chen, Z., Przystupa, K., Li, T., Majka, M., Kochan, O.: Examination of abnormal behavior detection based on improved Yolov3. Electronics. **10**(2), 197 (2021)
18. Alairaji, R.M., Aljazaery, I.A., ALRikabi, H.T.S.: Abnormal behavior detection of students in the examination hall from surveillance videos. In: Gandhi, T.K., Konar, D., Sen, B., Sharma, K. (eds.) Advanced Computational Paradigms and Hybrid Intelligent Computing, pp. 113–125. Springer Singapore, Singapore (2022)
19. Soman, N., Devi, R., Srinivasa, G.: Detection of anomalous behavior in an examination hall towards automated proctoring, pp. 1–6 (2017)
20. Yong, L., Dongjian, H.: Video-based detection of abnormal behavior in the examination room. vol. 3, pp. 295–298 (2010)
21. Al Ibrahim, A., Abo Samra, G.A.A., Dahab, M.: Real-time anomalous behavior detection of students in examination rooms using neural networks and Gaussian distribution. Int. J. Sci. Eng. Res. **9**, 1716–1724 (2018)

22. The Ministry of Education and Training of Vietnam: Circular No. 15/2020/TT-BGDDT dated May 26, 2020 (20). https://moet.gov.vn/van-ban/vanban/Pages/chi-tiet-van-ban.aspx?ItemID=1351
23. Liu, H., Fan, K., Ouyang, Q., Li, N.: Real-time small drones detection based on pruned YOLOv4. Sensors **21**(10), 3374 (2021). https://doi.org/10.3390/s21103374

A Novel Question-Context Interaction Method for Machine Reading Comprehension

Tuan-Anh Phan, Hoang Ngo, and Khac-Hoai Nam Bui(✉)

Viettel Cyberspace Center, Viettel Group, Hanoi, Vietnam
{anhpt161,hoangnv74,nambkh}@viettel.com.vn

Abstract. Machine reading comprehension (MRC) is a challenging NLP task that requires machines to model the complex interactions between questions and specific contexts. In Question-Answering (QA) tasks, most existing works rely on the powerful encoder of pre-trained language models (PrLM) in order to represent word/subword embeddings for extracting the answer. In this study, we present a novel method for enriching the context representation by exploiting the question-context interaction at the sentence level. In particular, we introduce the sentence-based question-context interaction (S-QCI) block, which combines two main layers such as the question-aware layer and the cross-sentence layer, to represent the sentence embedding of the context. The sentence information is then used to enrich question information for the context representation at the word level. The main idea is that the word units in the sentence, which have a high attention score of question-sentence interaction, can be enriched with more question information for the final output of the extractive-span MRC task. The experiment on NewsQA, a benchmark dataset in this research field, indicates that the proposed method has significant improvements compared with the baselines using PrLM and achieves new state-of-the-art results.

Keywords: Machine Reading Comprehension · Sentence-based Context Representation · Question-Context Interaction

1 Introduction

Machine reading comprehension (MRC) is teaching machines to read a text that can be applied in various natural language processing (NLP) tasks such as text summarization, dialog state tracking, and question answering. Specifically, MRC is a core component of the modern question-answering (QA) systems, which is regarded as *Reader* component, as shown in Fig. 1 (a) [2]. Accordingly, given a query/question, *Retriever* aims to retrieve relevant contexts and *Reader* is to extract the final answer from the selected context using MRC-based models. In this study, we focus on improving the performance of the span-extraction

N. T. Nguyen et al. (Eds.): ACIIDS 2023, LNAI 13995, pp. 58–69, 2023.
https://doi.org/10.1007/978-981-99-5834-4_5

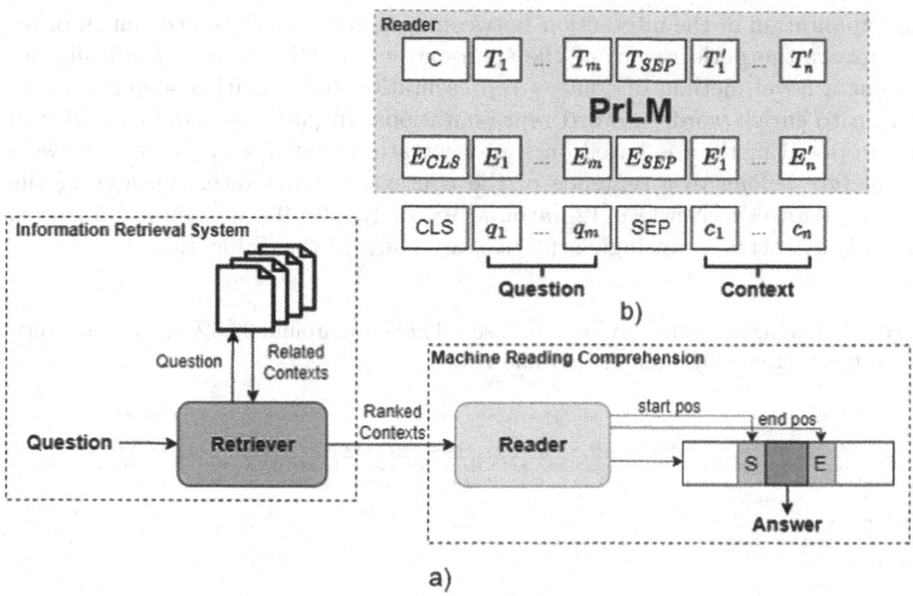

Fig. 1. a) The pipeline of extractive MRC-based modern QA with two main components: Retriever and Reader; b) The general Reader framework using PrLMs.

MRC for the *Reader*. Specifically, a typical span extraction MRC task can be formulated as follows:

$$a = f(q, p) \tag{1}$$

where q and p denote the given question and selected context, respectively. f is a predictor, which is learned to extract the answer a. The state-of-the-art models in this research field utilize the capability of pre-trained language models (PrLM) (e.g., BERT [4] and RoBERTa [9]) as the encoder. Specifically, the general framework of span extraction MRC using PrLMs is illustrated in Fig. 1 (b), which is regarded as the baseline model of this study. Accordingly, the task is solved based on the capacity of PrLMs in terms of capturing the contextualized embedding in word-level representations. Consequently, the decoder mechanisms have become a bottleneck due to the powerful PrLM encoder [17]. Furthermore, most previous MRC models are designed with the assumption that all questions can be answered. Sequentially, the recent approach focuses on proposing models which are able to distinguish unanswerable questions. In particular, the models, following the *read-then-verify* strategy, are technically decomposed into two subtasks: *reading comprehension* and *verification* in order to improve the accuracy of the final prediction. Despite the recent significant improvement in this research field, we identify the major limitation of previous work is that most models try to build a robust language model as the encoder for the context representation at the word level without exploiting the information from other semantic units (e.g., sentence, passage, latent topic and so on). In this regard, this study takes the exploitation of the interaction between semantic units into account in order to improve the performance of the span-extraction MRC task. Specifically, we present a novel method of context representation by exploiting sentence information to enrich word/subword representations. In particular, the core idea of the proposed method is based on the observation that most extracted answers completely belong to a sentence of the context. For instance, considering the training dataset of NewQA [12], around 98,5% (91120/92550 samples) answers, which is extracted by a single sentence, as shown in the Table. 1.

Table 1. Examples of the QANews dataset. There are around 98,5% extract answers, which belong to a single sentence (italic texts).

Context-1: The wife of Robert F. Kennedy Jr. was charged with drunken driving after a police officer saw her run over a curb outside a school, authorities said Tuesday night. *Mary Richardson Kennedy was arrested Saturday night in Bedford, New York, a sergeant with the town 's police department said.* Kennedy 's blood alcohol level was 0.11, said Sgt. Matthew Dunn. The legal limit is 0.08. He confirmed published reports that an officer saw Kennedy drive over a curb outside a school in her station wagon. A message left for Robert Kennedy 's spokesperson was not returned. Bedford, in Westchester County, is about an hour north of New York City
Question:
When did the arrest take place?
Gold Answer:
Bedford, New York

Generally, the main contributions of this study are two folds as follows:

- We present a novel method of span-extraction MRC task by exploiting sentence-level information for enriching the context representation. To the best of our knowledge, this paper is the first study that uses different levels of semantic units (i.e., word and sentence) for improving the performance of the task.
- We execute the proposed method on NewQA, a benchmark dataset in this research field. The evaluation indicates promising results of the proposed method. Specifically, the reported results not only outperform baseline models using PrLM but also achieve the new state-of-the-art on the NewQA dataset. Our source code is available for further investigation on Github[1].

The rest of this paper is organized as follows: Sect. 2 takes a brief review of MRC models and recent approaches. The proposed method is presented in Sect. 3. Section 4 provides the evaluation results on the NewQA dataset. Section 5 is the conclusion of this study.

[1] https://github.com/tuananhphan97vn/SentenceLevelMRC.

2 Related Work

Recently, MRC has received extensive attention, which has been applied to various NLP tasks, especially for QA systems [2,16]. Technically, a standard solution for MRC generally consists of two steps [6]: i) utilizing the power of the PrLMs for encoding the question-context in the word/subword level; ii) the outputs of the encoder are then processed by designing simple mechanism as the decoder following the characteristics of the MRC task. The main reason is that using PrLMs as an encoder has a significant impact on MRC performance [4]. Accordingly, several well-known PrLMs such as BERT [4], RoBERTa [9], XLNet [15], and ELECTRA [3] have been used as encoders and achieved remarkable results. Recent progress on the MRC task requires the models to deal with unanswerable questions, which is closer to real-world applications [5]. Specifically, these models follow the *read-then-verify* strategy, which contains two stages: i) the first stage uses a neural reader (i.e., using PrLMs) to exploit attention matching between question and context for extracting candidate answers and detecting unanswerable questions; ii) the second stage verifies the answers and produces final predictions. Accordingly, this approach currently achieves the best performance of MRC tasks.

In this study, we propose a new approach for the MRC models. Specifically, instead of executing verify layer, our second stage tries to enrich the semantic unit embedding, which is extracted in the first stage. In particular, the proposed S-QCI block is able to exploit the context-question interaction at the sentence level. The sentence information is then used to enrich the word representation for the final prediction. For unanswerable questions, a simple solution is to follow the work in [8] by using an empty word token to the context with a classification layer in the reader block. More details of the proposed method are described in the following section.

3 Methodology

Compared with baseline models, the proposed method add an S-QCI block in order to exploit the sentence information in terms of enriching word representation for the final prediction. The proposed architecture is illustrated in Fig. 2 (a). Accordingly, the input of the block includes a set of tokens (T_W), which are the output of the pre-trained language model, the question embedding (T_q), and the sentence embedding of the context (T_S).

3.1 Sentence Embedding

For the sentence embedding process, we first concatenate the question and context as the input of the PrLM (i.e., RoBERTa). The output is subword/word tokens (T_W) as shown in Fig. 2 (a). For the bounding of a sentence, we use the external sentence tokenizer (i.e., NLTK sentence tokenizer) tool to split the context into sentences and use one dictionary object to map each subword to

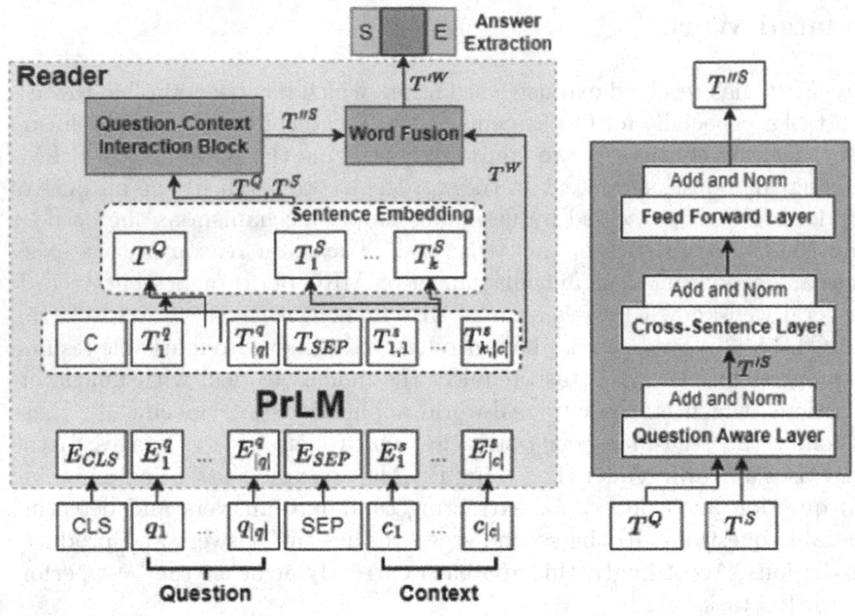

Fig. 2. Overview of the proposed model.

appropriate sentences. In particular, the sentence embedding process can be sequentially formulated as follows:

$$\begin{aligned}(T_1^q, ..., T_{|q|}^q), (T_{1,1}^s, ..., T_{1,|s_1|}^s, ..., T_{k,1}^s, ..., T_{k,|s_k|}^s) \\ = PrLM(E_1^q, ..., E_{|q|}^q, E_1^c, ..., E_{|c|}^c)\end{aligned} \tag{2}$$

$$\begin{aligned}T^Q &= \frac{1}{|q|}\sum_{i}^{|q|} T_{q_i} \\ T_i^S &= \frac{1}{|S_i|}\sum_{j=1}^{|S_i|} T_{S_{ij}}\end{aligned} \tag{3}$$

Accordingly, given the token embeddings that are generated by the PrLM, the sentence representations are calculated based on the sum of token embedding, position embedding, and token-type embedding. For producing sentence embeddings, we use the mean-pooling layer through all of the tokens for both question and sentence in the context.

3.2 S-QCI Block

S-QCI block is the main focus of the proposed model, which includes two main layers such as *Question Aware Layer* and *Cross-Sentence Layer* as shown in Fig. 2 (b).

Question Aware Layer: The main idea of this layer is to enrich question information for each sentence of the context using an attention mechanism. Specifically, the scaled dot-product mechanism [13] is adopted for computing the correction between vector query and matrix keys, where query and keys are question and sentence embedding, respectively. Moreover, the value of the attention score expresses the compatibility of the question with each sentence. The output of sentence embedding is computed as a linear combination of questions with its information. Therefore, if a sentence has a high attention score, its embedding is enriched by more question information. Consequentially, the output of this layer is the new representation of sentence embedding (T'^S), which is formulated as follows:

$$\begin{aligned}\alpha_i &= \frac{e^{z_i}}{\sum_j e^{z_j}} \\ z_i &= \frac{(\mathcal{W}_1 T^Q)^\top (\mathcal{W}_1 T_i^S)}{\sqrt{d_s}} \\ T_i'^S &= tanh(\alpha_i * \mathcal{W}_2 T^Q + (1-\alpha_i) * \mathcal{W}_2 T_i^S)\end{aligned} \tag{4}$$

where $\mathcal{W}_1$ and $\mathcal{W}_2$ are learnable parameters and updated in the training process.

Cross-Sentence Layer: The new sentence representations T'^S of the context, which obtains the question information, are then put into the cross-sentence layer. Specifically, this layer is built to exploit cross-sentence relations of the context. In this regard, similar to *Question Aware* layer, scale dot product attention is adopted to update sentence embedding. Generally, the main technical difference between the two layers is that the query is the sentence embedding matrix instead of the question embedding. Consequentially, the output of this layer is the new sentence embedding (T''^S), which obtains the information of other sentences in the context. The cross-sentence layer is formulated as follows:

$$\begin{aligned}\beta_{i,j} &= \frac{e^{z_{i,j}}}{\sum_m e^{z_{i,m}}} \\ z_{i,j} &= \frac{(\mathcal{W}_3 T_i'^S)^\top (\mathcal{W}_3 T_j'^S)}{\sqrt{d_s}} \\ T_i''^S &= \sum_{j=1}^{k} \beta_{i,j} \mathcal{W}_4 T_j'^S\end{aligned} \tag{5}$$

where $\mathcal{W}_3$, $\mathcal{W}_4$ are the learnable parameters.

3.3 Word Fusion:

The output T''^S (Eq. 5) of sentence representations are then used for updating the word representation, which can be formulated as follows:

$$T'^W_i = RELU(\mathcal{W}_5(T_i^W || T_{W_i}''^S)) \tag{6}$$

where $\mathcal{W}_5$ is the learnable parameter and $||$ denotes the concatenation operation.

3.4 Training Process

The final outputs of word representations T'^{W} are used to extract a span in the context, which is regarded as the answer. Specifically, following prior works in this research field, we employ a linear layer using the Softmax layer to predict start (s) and end (s) point probabilities:

$$\begin{aligned} p^s &= Softmax(\mathcal{W}_6 T'^{W}) \\ p^e &= Softmax(\mathcal{W}_7 T'^{W}) \end{aligned} \tag{7}$$

where $\mathcal{W}_6$ and $\mathcal{W}_7$ represent the learnable parameters. Sequentially, the cross-entropy loss is used to minimize the sum of negative log probabilities as follows:

$$\mathcal{L}^{span} = -\frac{1}{|c|}\sum_{i}^{|c|} log(p^s_{y^s_i}) + log(p^e_{y^e_i}) \tag{8}$$

where y^s_i and y^e_i denote the ground-truth start and end points of the sample i.

4 Experiment

4.1 Experimental Setup

We use RoBERTa-large [9], which has 24 layers, as the backbone of the proposed model. Regarding the decoder, we remain the post-processing for extracting answers from Transformers, a well-known Pytorch implementation library[2]. Specifically, for the fine-tuning task, the initial learning rate is set in {1e-5, 2e-5, 3e-5} and adopted decayed linearly after each epoch. The warm-up step is 1000 (batch size equals 32). The maximum number of epochs for training is set to 5. Question and answer lengths are set to 64 and 30 tokens, respectively. For the proposed S-QCI block, we use the NLTK library[3] for tokenizing sentences. Regarding sentence embedding, the hidden size is set to 1024. The number of S-QCI blocks is selected in {1,2,3,4} and determined during the validation process. All models are trained with an NVIDIA A100 40GB.

4.2 Benchmark Dataset and Baseline Models

We use NewsQA [12], which is a well-known question-answering dataset, for the evaluation. The dataset is conducted with more than 100,000 human-generated question-answer pairs, which are collected from over 10,000 CNN news articles. Following the work of the respective paper[4], the dataset is split into 97313, 5456, and 5412 samples for training, validation, and testing, respectively. More detail of the dataset is shown in Table. 2. Accordingly, the training dataset includes

[2] https://github.com/huggingface/transformers.
[3] https://www.nltk.org/api/nltk.tokenize.html.
[4] https://github.com/Maluuba/newsqa.

Table 2. Statistics of the evaluated dataset.

	Train	Dev	Test
Answer	76.560	4.341	4.292
No-Answer	20.753	1.115	1.120
Total	97.313	5.456	5.412

around 20k unanswerable questions among 97k questions. Regarding the comparison, we report the results of recent models, which mainly belong to three approaches such as the *Deep Learning-based* (i.e., mLSTM [14], AMANDA [7], DecaProp [11]), *PrLMs-based*(i.e., BERT [4]), and *Read-then-Verify-based* (i.e., NeurQuRI [1], Retro-Reader [17]) approaches. Specifically, the modes are sequentially described as follows:

- **mLSTM**: A well-known benchmark deep learning model for MRC tasks based on the LSTM network. First, an LSTM model is used to encode the document and question (i.e., using Glove embeddings), and the mLSTM network is then used to compare the document encodings with the question encoding.
- **AMANDA**: The model focuses on identifying the essential words of the question to extract the answer in terms of deeper understanding question requires tasks such as multi-sentence reasoning and co-reference resolution.
- **DecaProp**: The model exploits the interaction between question and context with densely hierarchical levels.
- **BERT**: A benchmark PrLM for MRC task, which achieved remarkable results in this research field. Specifically, BERT is adopted as an encoder for capturing the contextualized information between question and context for extracting the answer.
- **NeurQuRI**: The model takes the unanswerable question into account by developing an attention-based satisfaction score between question and candidate answering embedding.
- **Retro-Reader**: The model is a state-of-the-art model in this research field by exploiting a better verify layer with unanswerable questions, which include two reading stages: the first stage encoder question and context using PrLMs for extracting candidate answers; the second stage verifies the extracted answers by using various loss functions for the final prediction.

4.3 Main Results

Regarding the span-extraction MRC task, two common metrics are used for the evaluation: Exact Match (EM) and F1 score (F1) [16]. *EM* represents the percentage of questions that the extracted answer exactly matches the correct answer and *F1 score* measures the average overlap between the prediction and correct answer at the token level. Furthermore, the significance test is also adopted for measuring the difference. Specifically, following the previous work [17], we use McNemar's test, which uses EM as the metric measurement. Table 3 shows the reported results.

Table 3. Experiment results for the NewsQA dataset. Report results with * are from [17]. Other results are obtained from respective papers. Our results are calculated by averaging values of three runs.

Model	Dev		Test	
	EM	F1	EM	F1
mLSTM*	34.4	49.6	34.9	50.0
AMANDA	48.4	63.3	48.4	63.7
DecaProp	48.4	65.7	53.31	66.3
BERT*	-	-	46.5	56.7
NeurQuRI	-	-	48.2	59.5
Retro-Reader	58.5	68.6	55.9	66.8
Proposed Model	**64.1**	**71.7**	**62.9**	**71.2**

Based on the evaluation results, several hypotheses for extracting the MRC task can be expressed as follows:

- Using pre-trained models (e.g., BERT) to exploit the contextualized information between question and context can achieve new benchmark results in this research field compared with deep learning models (e.g., LSTM)
- AMANDA and DecaProb models achieve promising results without using PrLMs indicating that exploiting the interaction between question-context is able to significantly improve the performance.
- Regarding the reading-then-verify approach, the model with PrLMs (Retro-Reader model) outperforms the model without PrLMs (NeurQuRI) and achieves remarkable results.
- Our model exploits the interaction between question and context by utilizing the sentence information can achieve new state-of-the-art results in this research field.

Our model achieves the best results with three S-QCI blocks. Particularly, our method reaches the highest scores with 62.9% for EM and 71.2% for F1. Compared with the result of the state-of-the-art model as Retro-Reader [17], our results are higher than approximately 7% and 4,4% for EM and F1, respectively. In addition, for the significant test, the results of the proposed model significantly outperform the baseline with p-value < 0.01, proving the remarkable improvement of the proposal.

In general, the reported results indicate that exploiting the information at the sentence level in our method is able to boost performance compared with strong baseline models in this research field. On the other hand, a limitation of this study is that the proposed method can not cover long documents since the PrLMs are limited by 512 tokens. An appropriate solution is to divide the long documents into different chunks, and updated global information at the sentence level by using graph neural network (GNN) [10]. We leave this issue for future work of this study.

4.4 Ablation Study

In order to the impact of each module in the proposed S-QCI block, we evaluate two ablated variants such as i) **w/o Question-Aware** removes the question aware layer; ii) **w/o Cross-Sentence** removes cross-sentence layer. Furthermore, we also design **S-W-QCI**, a new variant of the S-QIC block by integrating a sentence-based word representation module into the S-QCI block. Table 4 shows the results of different variants of the proposed model. As reported results, the proposed architecture outperforms all variants, which proves that combining both modules can achieve the best results.

Table 4. Ablation Study.

Model	Dev		Test	
	EM	F1	EM	F1
w/o Question-Aware	62.7	71.1	62.4	71.1
w/o Cross-Sentence	63.1	71.1	62.3	70.4
S-W-QCI	63.1	70.6	62.6	70.4
Proposed Model	**64.1**	**71.7**	**62.9**	**71.2**

5 Conclusion

MRC has become an advantageous approach for various NLP tasks, especially extractive span MRC for QA systems. Most previous works utilize the powerful PrLMs as the encoder to extract the answers. In this study, taking the observation that most of the extracted answers belong to a single sentence, we exploit the attention between the question and each sentence of the context to enrich the final output of word-level representation. The experiment on NewsQA, a well-known dataset of question-answering systems, indicates the promising results of the proposed method. Regarding the future works of this study, we try to extend S-QIC blocks across different chunks in terms of global interaction to deal with the long-text issue of the MRC task.

References

1. Back, S., Chinthakindi, S.C., Kedia, A., Lee, H., Choo, J.: NeurQuRI: neural question requirement inspector for answerability prediction in machine reading comprehension. In: 8th International Conference on Learning Representations, ICLR 2020, Addis Ababa, Ethiopia, April 26–30, 2020. OpenReview.net (2020). https://openreview.net/forum?id=ryxgsCVYPr
2. Chen, D., Fisch, A., Weston, J., Bordes, A.: Reading Wikipedia to answer open-domain questions. In: Barzilay, R., Kan, M. (eds.) Proceedings of the 55th Annual Meeting of the Association for Computational Linguistics, ACL 2017, Vancouver, Canada, July 30 - August 4, Volume 1: Long Papers. pp. 1870–1879. Association for Computational Linguistics (2017). https://doi.org/10.18653/v1/P17-1171
3. Clark, K., Luong, M., Le, Q.V., Manning, C.D.: ELECTRA: pre-training text encoders as discriminators rather than generators. In: 8th International Conference on Learning Representations, ICLR 2020, Addis Ababa, Ethiopia, April 26–30, 2020. OpenReview.net (2020). https://openreview.net/forum?id=r1xMH1BtvB
4. Devlin, J., Chang, M., Lee, K., Toutanova, K.: BERT: pre-training of deep bidirectional transformers for language understanding. In: Burstein, J., Doran, C., Solorio, T. (eds.) Proceedings of the 2019 Conference of the North American Chapter of the Association for Computational Linguistics: Human Language Technologies, NAACL-HLT 2019, Minneapolis, MN, USA, June 2–7, 2019, Volume 1 (Long and Short Papers), pp. 4171–4186. Association for Computational Linguistics (2019). https://doi.org/10.18653/v1/n19-1423
5. Hu, M., Wei, F., Peng, Y., Huang, Z., Yang, N., Li, D.: Read + verify: machine reading comprehension with unanswerable questions. In: The Thirty-Third AAAI Conference on Artificial Intelligence, AAAI 2019, The Thirty-First Innovative Applications of Artificial Intelligence Conference, IAAI 2019, The Ninth AAAI Symposium on Educational Advances in Artificial Intelligence, EAAI 2019, Honolulu, Hawaii, USA, January 27 - February 1, 2019, pp. 6529–6537. AAAI Press (2019). https://doi.org/10.1609/aaai.v33i01.33016529
6. Huang, Z., et al.: Recent trends in deep learning based open-domain textual question answering systems. IEEE Access **8**, 94341–94356 (2020). https://doi.org/10.1109/ACCESS.2020.2988903
7. Kundu, S., Ng, H.T.: A question-focused multi-factor attention network for question answering. In: McIlraith, S.A., Weinberger, K.Q. (eds.) Proceedings of the Thirty-Second AAAI Conference on Artificial Intelligence, (AAAI 2018), the 30th innovative Applications of Artificial Intelligence (IAAI 2018), and the 8th AAAI Symposium on Educational Advances in Artificial Intelligence (EAAI 2018), New Orleans, Louisiana, USA, February 2–7, 2018, pp. 5828–5835. AAAI Press (2018). https://www.aaai.org/ocs/index.php/AAAI/AAAI18/paper/view/17226
8. Liu, X., Shen, Y., Duh, K., Gao, J.: Stochastic answer networks for machine reading comprehension. In: Gurevych, I., Miyao, Y. (eds.) Proceedings of the 56th Annual Meeting of the Association for Computational Linguistics, ACL 2018, Melbourne, Australia, July 15–20, 2018, Volume 1: Long Papers, pp. 1694–1704. Association for Computational Linguistics (2018). https://doi.org/10.18653/v1/P18-1157, https://aclanthology.org/P18-1157/
9. Liu, Y., et al.: Roberta: a robustly optimized BERT pretraining approach. CoRR abs/1907.11692 (2019). http://arxiv.org/abs/1907.11692

10. Phan, T., Nguyen, N.N., Bui, K.N.: HeterGraphLongSum: heterogeneous graph neural network with passage aggregation for extractive long document summarization. In: Proceedings of the 29th International Conference on Computational Linguistics, COLING 2022, Gyeongju, Republic of Korea, October 12–17, 2022, pp. 6248–6258. International Committee on Computational Linguistics (2022). https://aclanthology.org/2022.coling-1.545
11. Tay, Y., Luu, A.T., Hui, S.C., Su, J.: Densely connected attention propagation for reading comprehension. In: Bengio, S., Wallach, H.M., Larochelle, H., Grauman, K., Cesa-Bianchi, N., Garnett, R. (eds.) Advances in Neural Information Processing Systems 31: Annual Conference on Neural Information Processing Systems 2018, NeurIPS 2018(December), pp. 3–8, 2018. Montréal, Canada, pp. 4911–4922 (2018). https://proceedings.neurips.cc/paper/2018/hash/7b66b4fd401a271a1c7224027ce111bc-Abstract.html
12. Trischler, A., et al.: NewsQA: a machine comprehension dataset. In: Proceedings of the 2nd Workshop on Representation Learning for NLP, pp. 191–200. Association for Computational Linguistics, Vancouver, Canada, August 2017. https://doi.org/10.18653/v1/W17-2623, https://aclanthology.org/W17-2623
13. Vaswani, A., et al.: Attention is all you need. In: Guyon, I., von Luxburg, U., Bengio, S., Wallach, H.M., Fergus, R., Vishwanathan, S.V.N., Garnett, R. (eds.) Advances in Neural Information Processing Systems 30: Annual Conference on Neural Information Processing Systems 2017(December), pp. 4–9, 2017. Long Beach, CA, USA, pp. 5998–6008 (2017). https://proceedings.neurips.cc/paper/2017/hash/3f5ee243547dee91fbd053c1c4a845aa-Abstract.html
14. Wang, W., Yang, N., Wei, F., Chang, B., Zhou, M.: Gated self-matching networks for reading comprehension and question answering. In: Barzilay, R., Kan, M. (eds.) Proceedings of the 55th Annual Meeting of the Association for Computational Linguistics, ACL 2017, Vancouver, Canada, July 30–August 4, Volume 1: Long Papers, pp. 189–198. Association for Computational Linguistics (2017). https://doi.org/10.18653/v1/P17-1018, https://doi.org/10.18653/v1/P17-1018
15. Yang, Z., Dai, Z., Yang, Y., Carbonell, J.G., Salakhutdinov, R., Le, Q.V.: XLNet: generalized autoregressive pretraining for language understanding. In: Wallach, H.M., Larochelle, H., Beygelzimer, A., d'Alché-Buc, F., Fox, E.B., Garnett, R. (eds.) Advances in Neural Information Processing Systems 32: Annual Conference on Neural Information Processing Systems 2019, NeurIPS 2019(December), pp. 8–14, 2019. Vancouver, BC, Canada, pp. 5754–5764 (2019). https://proceedings.neurips.cc/paper/2019/hash/dc6a7e655d7e5840e66733e9ee67cc69-Abstract.html
16. Zeng, C., Li, S., Li, Q., Hu, J., Hu, J.: A survey on machine reading comprehension-tasks, evaluation metrics and benchmark datasets. Appl. Sci. **10**(21), 7640 (2020). https://doi.org/10.3390/app10217640, https://www.mdpi.com/2076-3417/10/21/7640
17. Zhang, Z., Yang, J., Zhao, H.: Retrospective reader for machine reading comprehension. In: Thirty-Fifth AAAI Conference on Artificial Intelligence, AAAI 2021, Thirty-Third Conference on Innovative Applications of Artificial Intelligence, IAAI 2021, The Eleventh Symposium on Educational Advances in Artificial Intelligence, EAAI 2021, Virtual Event, February 2–9, 2021, pp. 14506–14514. AAAI Press (2021). https://ojs.aaai.org/index.php/AAAI/article/view/17705

Granular Computing to Forecast Alzheimer's Disease Distinctive Individual Development

Andrzej W. Przybyszewski[1,2(✉)], Jerzy P. Nowacki[1], Aldona Drabik[1], and the BIOCARD Study Team

[1] Polish-Japanese Academy of Information Technology, 02-008 Warszawa, Poland
{przy,jerzy.nowacki,drabik}@pjwstk.edu.pl

[2] Department of Neurology, UMass Medical School, Worcester, MA 01655, USA

Abstract. Our study aimed to test if clinically tested normal subjects might have patterns in the results of their cognitive tests that have some similarities to suits obtained from patients in different AD stages.

Due to the aging population, the prevalence of Alzheimer's Disease (AD) related dementia is fast increasing. That is already a worldwide problem. AD-related neurodegeneration starts several decades before the first symptoms and AD biomarkers were identified in recent years. The purpose of our study was to find AD-related biomarkers in healthy subjects. We have estimated such changes based on the Model consisting of subjects in different AD stages from normal to patients with dementia from Biocard data. By using the granular computing method, we found reducts (sets of attributes) related to different stages of the disease. By applying this classification to psychophysical test results of normal subjects, we have demonstrated if some of them might show some similarities to the first symptoms related to AD. As such psychophysical tests can be easily implemented in computers connected to the internet, our method has the potential to be used as a new AD preventive method. We have analyzed Biocard data by comparing a group of 150 subjects in different AD stages with normal subjects. we have found granules that classify cognitive attributes with disease stages (CDRSUM). By applying these rules to normal (CDRSUM = 0) 21 subjects we have predicted that one subject might get mild dementia (CDRSUM > 4.5), one very mild dementia (CDRSUM > 2.25), and five others might get questionable impairment (CDRSUM > 0.75). AI methods can find, invisible for neuropsychologists, patterns in cognitive attributes of normal subjects that might indicate their pre-dementia stage.

Keywords: Neurodegeneration · multi-granular computing · feedbacks

N. T. Nguyen et al. (Eds.): ACIIDS 2023, LNAI 13995, pp. 70–81, 2023.
https://doi.org/10.1007/978-981-99-5834-4_6

1 Introduction

The prevalence of Alzheimer's Disease (AD) related dementia is fast increasing due to our aging population. The prevalence worldwide is estimated to be as high as 24 million, by 2050, AD number could potentially rise to 139 million worldwide. There is no cure for AD, as during the first clinical symptoms and neurological diagnosis many parts of the brain are already affected without the possibility to recovery. As the neurodegenerations begin two to three decades before observed symptoms, the best chance to fight AD is to estimate the beginning period of the AD-related brain changes.

The BIOCARD* study was initiated in 1995 by NIH with 354 normal individuals interrupted in 2005 and continued from 2009 as Johns Hopkins (JHU) study. At JHU, patients have yearly cognitive and clinical visits that measured total of over 500 attributes with 96 cognitive parameters [1, 2]. Albert [2] has successfully predicted conversion from normal to MCI (Mild Cognitive Impairment) due to AD, 5 years after baseline, for 224 subjects by using the following parameters: beta-amyloid (b-Symbol format), MRI hippocampal and entorhinal cortex volumes, cognitive tests scores, and APOE genotype. Even if their predictions were for all tested patients, they found important biomarkers that we are using in our study.

But our approach is different, as we have performed classification by using granular computing (GrC) [3] connecting cognitive test results with genetic data (related to the apolipoprotein E ApoE genotype) and AD-related clinical symptoms in the group of different subjects from normal to AD. We took this group as our Model for the supervised training of different granules related to various stages of the disease, from normal subjects to MCI (Mild Cognitive Impairment) and subjects with AD-related dementia. In the next step, we applied these granules to individual, normal subjects to predict in every individual tested subject, a possibility of the beginning of the neurodegeneration (AD-related brain changes). To confirm our method, we have also applied it to the early stages in patients that were diagnosed with AD if we can predict their future Alzheimer's disease stage.

As AD-related dementia is so fast increasing, we need an easy and accessible internet method for the preclinical classification of all potential patients, otherwise, we will soon live in a different world.

Our granular computing method and its implementation with a rough set are well established in neurodegeneration disease, especially used in many publications related to Parkinson's disease patients [4].

• Data used in preparation of this article were derived from BIOCARD study, supported by grant U19 – AG033655 from the National Institute on Aging. The BIOCARD study team did not participate in the analysis or writing of this report, however, they contributed to the design and implementation of the study. A listing of BIOCARD investigators can be found on the BIOCARD website (on the 'BIOCARD Data Access Procedures' page, 'Acknowledgement Agreement' document)

2 Methods

It is a continuation of our previous study [5] therefore methods are similar, but in this part, we are looking into longitudinal changes in our subjects. We have analyzed predominantly cognitive of several different groups of subjects. The first group consists of 150 subjects with 40 normal subjects, 70 MCI (Mild Cognitive Impairment), and 40 subjects with dementias (AD). It was chosen this way as in the whole population of 354 normal subjects followed from 1995, only 40 subjects became demented. Therefore, we have added 40 normal subjects and 70 MCI as they are in between AD and normal subjects. The second group was 40 AD subjects, and the last group was 21 subjects, clinically classified as normal.

In all subjects with recorded their age, had the following neuropsychological tests formed every year:

1. Logical Memory Immediate (LOGMEM1A) - participants are read a logically organized story and asked to recall the story immediately after its presentation
2. Logical Memory Delayed (LOGMEM2A) - approximately 20 min later, the participants are again asked to recall the story from memory
3. Trail Making, Part A (TrailA - connecting time in a sec of randomly placed numbers),
4. Trail Making Part B (TrailB - connecting time in a sec of randomly placed numbers and letters),
5. Digit Symbol Substitution Test (DSST) from the Wechsler Adult Intelligence Scale - requires a subject to match symbols to numbers according to a key located on the top of the page (associative learning).
6. Verbal Fluency Letter F (FCORR) requires the generation of words from the initial letters F under 60 s of time constraints.
7. Rey Figure Recall (REYRECAL), in which examinees are asked to reproduce a complicated line drawing from memory (visual memory test).
8. Paired Associate Immediate (PAIRED1), from the Wechsler Memory Scale – subjects are asked to learn unrelated word pairs (e.g., stove-letter).
9. Paired Associate Delayed (PAIRED2), from the Wechsler Memory Scale – as above but delayed matching-to-sample tasks (testing hippocampus function).
10. Boston Naming Test (BOSTON) determining confrontational picture-naming abilities in patients.
11. CVLT (California Verbal Learning Test) - an examinee listens to series of words and is then asked to recall the terms and the category to which they belong (test of verbal learning and memory).

In addition, we have registered APOE genotype; individuals who are *ApoE4* carriers vs. non-carriers (digitized as 1 vs. 0).

The decision attributes were CDRSUM (sum of boxes) as precise and quantitative general index of the Clinical Dementia Rating [6]. There are the following CDRSUM values related to different AD stages:

1. (0.0) – normal; 2. (0.5–4.0) – questionable cognitive impairment: 2a. (0.5–2.5) – questionable impairment; 2b. (3.0–4.0) – very mild dementia,

3. (4.5–9.0) – mild dementia [6].

2.1 Rough Set Implementation of GrC

Our data mining granular computing (GrC) analysis was implemented by rough set theory (RST) discovered by Zdzislaw Pawlak [7] whose solutions of the vague concept of boundaries were approximated by sharp sets of the upper and lower approximations [7]. In the first step, our data were inserted in the decision table with rows stand out the actual attributes' values for the different or for the same subject and columns were linked to diverse attributes with the decision attribute CDRSUM. Following Pawlak [7] an *information system* is a pair $S = (U, A)$, where U, A are nonempty finite sets. The set U is the universe of objects, and A is the set of attributes. If $a \in A$ and $u \in U$, the value $a(u)$ is a unique element of V (where V is a value set).

Based on our classification, we have estimated CDRSUM, compared with CDRSUM obtained by neurologists and determined the predicted stage of individual patient.

The *indiscernibility relation IND(B)* of any subset B of A is defined after Pawlak (1991): $(x, y) \in IND(B)$ *iff* $a(x) = a(y)$ for every $a \in B$ where the value $a(x) \in V$. This relation divides U into *elementary granules* and it is the basis of the rough set theory. The most important rough set properties are the *lower approximation* of set $X \subseteq U$ in relation to an attribute B defined as: $\underline{B}X = \{u \in U : [u]_B \subseteq X\}$, and the *upper approximation* of set X defined as $\overline{B}X = \{u \in U : [u]_B \cap X \neq \phi\}$.

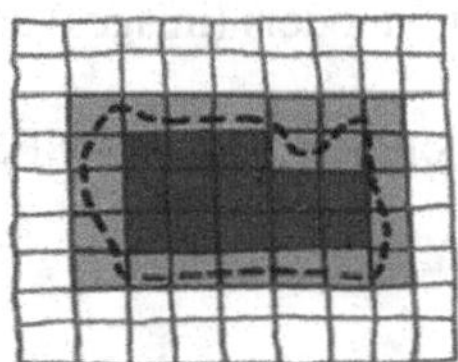

Fig. 1. Interrupted curve symbolizes properties of the set X. Squares represent elementary granules $[u]_B$ those in the black are associated with the *lower approximation* of X, grey and black squares are associated with the *upper approximation* of X, and white squares are outside of the set X.

The difference between the upper and lower approximation sets is the boundary region (BR). As it is visible on Fig. 1, squares marked in grey (boundary region) are related to imprecision of our classification. If BR is empty then set X is classified in an exact way, if not is rough classified with respect to $[u]_B$.

Therefore, we can divide U into three parts: the positive region *POS(X)* (black in Fig. 1), the boundary region *BR(X)* (gray in Fig. 1), and the negative region *NEG(X)* (white in Fig. 1). These three regions generate, on the RST basis, different type of decision rules: the certain rules are generated by the positive region, the uncertain rules by the boundary region, and the incorrect rules by the negative region.

We can also extend the decision table to a triplet: $S = (U, C, D)$ where the set of attributes A is divided into C *as* condition and D as decision attributes [8, 9]. In a single row there are many conditions and only one decision attribute, all related to a specific test of the individual subject. Names and values of conditions attributes related to the value of the decision attribute give a unique rule. One hard problem, especially in the medical field, is related to the inconsistent results, leading to contradictory rules. In this situation, doctors often are using averaging techniques. In the case of the inconsistent (conflicting) data, it means that the same attributes values are classified as different concepts. In general, it means that in the data set some attributes are missing. We will show an important example in our data when we wrongly have classified symptoms of one subject because missing values of attributes, or because missing attributes. LERS (Learning from example based on Rough Sets) [10] computes lower and upper approximation of all concepts. Conflicting data are not removed from the data set. Instead, concepts are approximated by new sets of the lower and upper approximations. Rough Set Theory generalizes individual measurements (rules) to universal principles (knowledge) but rules have different confidence. There are always true rules related to the *lower approximation set* and rules that are only partly true associated with the *upper approximation set*. If the border set (Fig. 1) is nonempty it is related to the uncertain rules.

In this work we have used different intelligent algorithms implemented in RSES 2.2 software such as: exhaustive algorithm, genetic algorithm [9, 11] covering algorithm, or LEM2 algorithm [10].

We have based our approach on the mechanisms in the visual brain related to advanced processes of the complex objects' recognition [12]. The processes in the higher visual brain areas that are related to different objects classification are using GrC to find upper and lower approximations of the retinal image [12]. These approximations are compared with the different objects' models (images) saved in the visual cortex (as the Model).

We have used Rough Set Exploration System RSES 2.2 as a toolset for analyzing data with rough set methods [13].

3 Results

3.1 Statistics

In Table 1 * means stat.sig.difference of means, if in all three columns all combinations are significant, if in two columns * or # mean values in these columns are significantly different.

Table 1 presents the statistical calculations for Group1 (40 normal subjects, 70 MCI subjects, and 40 (AD patients), Group2 (40 AD patients same as in Group1), and GroupN (40 normal subjects different to Group1 normal subjects) as mean ± SD. The age of subjects in different groups is similar, but other parameters show differences: Lgm1A (LOGMEM1A) is smallest for AD patients and largest for N, Lgm2A (LOGMEM2A) has similar changes as Lgm1A, execution functions: TrailA and TrailB are growing from N to AD, DSST is decreasing from N to AD in a similar way as Fcorr (FCORR). We did not show other parameters because of lack of the space. The changes of the CDRSUM are obvious as in normal subjects its values are 0 and significantly larger for AD patients. There are large differences between the values of individual subjects, so the mean values (except CDRSUM) were not all statistically significant, but we are not looking for means here, but for similarities in patterns.

Table 1. Statistic calculations of tested data

	Group1 n = 150	Group2 n = 40	GroupN n = 21
age	76 ±9.1	78.5 ±12	76.6 ± 8.4
LOGMEM1A	16.1 ±4.6	12.2 ±4.4	18.0 ± 3.0
LOGMEM2A	15.2 ±5.3	10.1 ±5.2	17.5 ± 3.9
TrailA	39.6 ±20	50.6 ±30	30.7 ± 12
TrailB	99.9 ±59	151 ±78	64.2 ± 21
DSST	46.5 ±14	37.3 ±13	57 ±13
FCORR	15.4 ±5*	12.6 ±6*	18.2 ±5 *
REYRECAL	19.7 ±8*	14.6 ±7*	24.2 ±6*
PAIRED1	18.4 ±8*	15.5 ±4*#	19.9 ±3#
PAIRED2	6.9 ±1*	6.2 ±1.5*#	7.2 ±1#
BOSTON	27.7 ±3*	25.3 ±4*	29.1 ±1*
CVLT	53 ±14*	41 ±13*	64.4 ±10*
APOE	0.35 ±0.5	0.3 ±0.5	0.3 ±0.5
CDRSUM	1.3 ±1.8	3.5 ±2.5#	0 ±0#

3.2 Granular Computing for Reference of Group1 Group

Previously [5], we have used all **14 attributes** to find with help of RSES which attributes are significant, and after discretization which rules can describe patients from Group1.

Comparing with all 14 condition attributes [5], there were the following 3 ranges of the decision attribute *CDRSUM: "(-Inf, 0.75)", "(0.75, 2.25)", "(2.25, Inf)"* for patients in Group1.

From 324 rules we have found *several rules* with a small number of attributes (used in actual study):

$$(TRAILB = "(153.0, Inf)")\&(REYRECAL = "(-Inf, 15.75)")\&(APOE = 1) \\ => (CDRSUM = "(2.25, Inf)"[6])6 \quad (1)$$

$$(TRAILB = "(74.5, 153.0)")\&(FCORR = "(16.5, Inf)")\&(REYRECAL = "(15.75, 25.25)") \\ \&(age = "(-Inf, 76.5)") => (CDRSUM = "(0.75, 2.25)"[2])2 \quad (2)$$

There is the following interpretation of above equations: Eq. 1 claims for 6 cases that if *TrailB* has bad time above 153s, and *REYRECALL* (visual memory) is also bad then *CDRSUM* is also above 2.25 that means the questionable impairment [6]. In Eq. 2, *TrailB* is better than in Eq. 1 but still has a relatively long time, and *FCORR* (speech fluency) is good (above 15), *REYRECALL* is about medium then *CDRSUM* is below 2.25 which suggests questionable cognitive impairment.

We have used 324 general rules from Group1 to predict CDRSUM of the normal patients (GroupN), we found that three patients might have a potential impairment:

$$\begin{aligned}&(Pat = 164087)\&(LOGMEM1A = "(-Inf, 15.5)"))\&(LOGMEM2A = "(-Inf, 16.5)")\&\\&(TRAILA = "(35.5, Inf)")\&(TRAILB = "(74.5, 153.0)")\&(FCORR = "(-Inf, 16.5)")\&\\&(REYRECAL = "(15.75, 25.25)")\&(PAIRD2 = "(-Inf, 6.5)")\&(age = "(-Inf, 76.5)")\&\\&(APOE = 1) => (CDRSUM = "(2.25, Inf)")\end{aligned} \tag{3}$$

$$\begin{aligned}&(Pat = 401297)\&(LOGMEM1A = "(-Inf, 15.5)"))\&(LOGMEM2A = "(-Inf, 16.5)")\&\\&(TRAILA = "(-Inf, 23.5)")\&(TRAILB = "(-Inf, 74.5)")\&(FCORR = "(-Inf, 16.5)")\&\\&(REYRECAL = "(-Inf, 15.75)")\&(PAIRD2 = "(6.5, Inf)")\&(age = "(-Inf, 76.5)")\&\\&(APOE = 1) => (CDRSUM = "(2.25, Inf)")\end{aligned} \tag{4}$$

$$\begin{aligned}&(Pat = 808698)\&(LOGMEM1A = "(15.5, 20.5)"))\&(LOGMEM2A = "(16.5, Inf)")\&\\&(TRAILA = "(-Inf, 23.5)")\&(TRAILB = "(-Inf, 74.5)")\&(FCORR = "(-Inf, 16.5)")\&\\&(REYRECAL = "(15.75, 25.25)")\&(PAIRD2 = "(6.5, Inf)")\&(age = "(-Inf, 76.5)")\&\\&(APOE = 1) => (CDRSUM = "(0.75, 2.25)")\end{aligned} \tag{5}$$

The first patient *(Pat = 164087)* Eq. 3 has affected logical memories: immediate and delayed, executive functions (*TrialA&B*), (especially *TrialB* longer execution time) executive functions, bad *FCORR*, low *REYRECALL,* and in addition bad APOE genotype (*ApoE4* carrier) that determines that *CDRSUM* is larger than 2.25 (questionable impairment to very mild dementia [6]). The second patient *(Pat = 401297)* Eq. 4 is like the first one, affected logical memories: immediate and delayed, but good executive functions (*TrialA&B*), and bad *FCORR* and *REYRECALL* and in addition bad APOE genotype that determines *CDRSUM* above 2.25. The third patient is *(Pat = 808698)* in better shape with good logical memories, and executive functions, but bad *FCORR* and *APOE* genotype then with *CDRSUM* below 2.25 (questionable impairment [6]).

Therefore, in this study, at first, we reduced number of attributes from 14 to the following **five**: *APOE, FCORR, DDST, TrailB,* and as the decision attribute was *CDRSUM*. These condition attributes are significant as the early predictors of AD.

The APOE genotype; individuals who are *ApoE4* is important genetic factor, which influence probability of AD. One of the early predictors of AD is the poor language performance that is quantify by FCORR test. Another early indication are difficulties in reasoning that may be estimated by DDST test. Slowing processing speed is also observed as an early AD indicator that can be quantified by Trail B tests.

We put all data in the decision table (as described in the Methods section), and with RSES help, after discretization we found that because large data set and small number of parameters, the decision attribute has 7 ranges: "(-Inf,0.25)", "(0.25,0.75)", "(0.75,1.25)", "(1.25,2.25)", "(2.25,3.25)", "(3.25,4.25)", "(4.25,Inf)". After generalization, there were 82 rules, below are some examples:

$$(APOE = 0)\&(DSST = "(66.5, Inf)") => (CDRSUM = "(-Inf, 0.25)"[4])\ 4 \tag{6}$$

$$(APOE = 1)\&(DSST = "(43.5, 45.5)") => (CDRSUM = "(0.25, 0.75)"[4])4 \tag{7}$$

$$(APOE = 1)\&(DSST = "(46.5, 49.5)")\&(TRAILB = "(72.5, 128.5)") => (CDSUM = "(2.25, 3.25)"[2])2 \tag{8}$$

$$(APOE = 1)\&(TRAILB = "(128.5, Inf)")\&(FCORR = "(6.5, 10.5)") => (CDSUM = "(4.25, Inf)"[2])2 \tag{9}$$

As one may notice above, a small number of the condition attributes are used in the above rules. One significant attribute in the genetic *APOE* genotype, and in these approximate rules lack of *ApoE4* carriers (*APOE* = *0*) is related to health (very low *CDRSUM*). Another attribute *DSST* - digit symbol substitution test is related to associative learning, and higher numbers are better. In Eq. 6 is the best result, in Eq. 7 lower value of *DSST* and *APOE* = *1* (higher genetic chance to get AD) gives a small increase of *CDRSUM*. In Eq. 8 *DSST* looks better, but the execution function attribute *TrailB* is on the slower side with *APOE* = *1* influence *CDRSUM*. In Eq. 9 execution (*TrailB*) is very slow and, *APOE* = *1* and language fluency problems (low value of *FCORR*) are the main factors that such patients have indications of mild dementia (*CDRSUM* is larger than 4.5).

By applying all 82 rules to the healthy patients (GroupN) with clinically confirmed *CDRSUM* = *0,* we found one patient with *CDRSUM* significantly larger than 0 in the following classification:

$$(Pat = 164087)\&(APOE = 1))\&(DSST ='' (46.5, 49.5)'')\&(FCORR ='' (10.5, 13.5)'')\& \\ (TRAILB ='' (72.5, 128.5)'' => (CDRSUM ='' (2.25, 3.25)'' \tag{10}$$

The Eq. 10 indicates based on 4 condition attributes that patient *164087* might have *CDRSUM* = *"(2.25,3.25)"* that suggests very mild dementia [6].

In the next step, we have increased number of attributes to **seven**: *APOE, BOSTON, FCORR, DDST, TrailB, REYRECAL,* and *CDRSUM* as the decision attribute. These additional attributes are also related to the early AD symptoms, like forgetting names of objects (BOSTON), or problems with remember place and environment where you are (as visual memory of the complex figure).

We have obtained 104 rules from Group1 patients and applied them to GroupN normal subjects, and got the following classifications (two examples):

$$(Pat = 558865)\&(APOE = 1)\&(FCORR = ''(10.5, 13.5)''))\&(REYRECAL = ''(15.75, 25.25 \\ )'')\&(TRAILB = ''(75.0, 114.5)'')\&(DSST = ''(33.5, 41.5)'')\&(BOSTON = ''(26.5, 27.5)'') \\ => (CDRSUM = ''(0.75, 1.25)) \tag{11}$$

$$(Pat = 164087)\&(APOE = 1))\&(FCORR = ''(10.5, 13.5)''))\&(REYRCAL = ''(15.75, 25.25) \\ '')\&(TRAILB = ''(75.0, 114.5)'')\&(DSST = ''(47.5, 53.5)'')\&(BOSTON = ''(25.5, 26.5)'') => \\ (CDRSUM = ''(2.0, 3.25)) \tag{12}$$

It is interesting that for *Pat* = *164087* more condition attributes are in Eq. 12 in comparison to Eq. 10 give almost identical results for *CDRSUM*, as the main factors are related to bad speech fluency (*FCORR*) and the *APOE* genome. These differences are subtle and are probably difficult to be noticed by clinical researchers.

3.3 Granular Computing for Reference of Group2 Patients

In this part, we have reduced the number of attributes from 14 to the following, same as above, **five** condition attributes: *APOE, DDST, FCORR, TrailB,* and as the decision attribute was *CDRSUM*. We put all data in the decision table (as described in the Methods section), and with RSES help, after discretization, we found that the decision attribute has 3 following ranges: *"(-Inf,0.75)"*, *"(0.75,3.25)"*, and *"(3.25, Inf)"*. After generalization, there were 4 rules, and 2 rules were shown below:

$$(APOE = 1)\&(FCORR = ''(-Inf, 15.5)'') => (CDRSUM = ''(3.25, Inf)''[6])6 \quad (13)$$

$$(DSST ='' (37.5, 45.0)'')\&(APOE = 0)\&(FCORR ='' (-Inf, 15.5)'') => (CDSUM = ''(0.75, 3.25)''[5])5 \quad (14)$$

What is unusual is that there is there no rule with decision attribute *CDRSUM = "(-Inf,0.75)"*. It is the consequence of a small number of condition attributes for the group of only 40 patients with Alzheimer's disease. We applied the above rules obtained from Group1 to predict the CDRSUM of GroupN.

Table 2. Confusion matrix for CDRSUM of GroupN by rules obtained from Group1 Predicted

Actual	"(3.25, Inf)"	"(0.75, 3.25)"	"(-Inf, 0.75)"	ACC
"(3.25, Inf)"	0.0	0.0	0.0	0.0
"(0.75,3.25)"	0.0	0.0	0.0	0.0
"(-Inf, 0.75)"	4.0	1.0	0.0	0.0
TPR	0.0	0.0	0.0	

TPR: True positive rates for decision classes; ACC: Accuracy for decision classes: the global coverage was 0.24 and the global accuracy was 0.0, the coverage for decision classes was 0.0, 0.0, 0.24.

In Table 2 for healthy patients, we got on our rule's basis, *CDRSUM > 0*. In one patient *Pat = 377216 CDRSUM* was between 0.75, and 3.25, but in four patients *Pat = 164087, Pat = 242874, Pat = 40129, Pat = 808698* above 3.25. Below are two examples, on this basis we have identified subjects with the possible impairments:

$$(Pat = 164087)\&(APOE = 1))\&(DSST ='' (45.0, 56.0)'')\&(FCORR ='' (-Inf, 15.5)'')\& (TRAILB ='' (-Inf, 128.5)'' => (CDRSUM ='' (3.25, Inf)'') \quad (15)$$

$$(Pat = 242874)\&(APOE = 1))\&(DSST ='' (45.0, 56.0)'')\&(FCORR ='' (-Inf, 15.5)'')\& (TRAILB ='' (-Inf, 128.5)'' => (CDRSUM ='' (3.25, Inf)'') \quad (16)$$

In the two above equations, speech fluency (*FCORR*) is affected. The executive function *TrialB* is below 128.5s, but this classification is rather crude and does not give us the lower value (from other equations (e.g., Eq. 11), we know that for *Pat = 164087* is above 72s which is a good value, but almost one minute more is not so good). In two last Eqs. 15 and 16 DSST is very good, but it seems that it does not have a strong influence on our results.

In our previous study [5] with 14 attributes, we obtained 58 rules for Group2 subjects, and as an example, we present below two rules with only 3 attributes each:

$$(DSST = ''(39.5, Inf)'')\&(FCORR = ''(20.5, Inf)'') => (CDRSUM = ''(-Inf, 4.5)''[2])2 \quad (17)$$

$$(LOGMEM1A = ''(13.5, 15.5)'')\&(FCORR = ''(-Inf, 12.5)'') => (CDRSUM = ''(4.5, 6.5)''[2])2 \quad (18)$$

Interpretation: Eq. 17, if *DSST* is above 39.5 and *FCORR* above 20.5 then *CDRSUM* is below 4.5; Eq. 18, if Logical Memory Immediate (LOGMEM1A) is above between 13.5 and 15.1 and the Verbal Fluency (Letter F -FCORR), is poor (below 12.5) then CDRSUM is between 4.5 and 6.5, which means mild dementia [6]. However, notice that this rule was fulfilled in principle in only one case in our 40 AD patients.

$$(Pat = 164087)\&(LOGMEM1A = ''(13.5, 15.5)'')\&(DSST = ''(39.5, Inf)'')\&(FCOR = ''(-Inf, 12.5)'')\&(BOSTON = ''(24.5, Inf)'')\&(APOE = 1) => (CDRSUM = ''(4.5, 6.5)'') \tag{19}$$

$$(Pat = 164087)\&(TRAILA = ''(-Inf, 73.5)'')\&(TRAILB = ''(52.5, Inf)''))\&(FCOR = ''(-Inf, 17.0)'')\&(PAIRD1 = ''(14.5, Inf)'')\&(PAIRD2 = ''(-Inf, 5.5)''))\&(BOSTON = ''(-Inf, 27.5)'')\&(CVLT = ''(33.5, Inf)'')\&(APOE = 1) => (CDRSUM = ''(5.75, 6.5)'') \tag{20}$$

However, two subjects had *CDRSUM* between 4.5 and 6, which means that they might have mild dementia [6]. The one our patient *Pat* = *164087*, with his/her results described in Eq. 19 seems to be easier to interpret with poor logical memory intermate and poor verbal fluency also with sensitive genetics with APOE = 1. To confirm this result, we have repeated the same subject classification using an additional attribute *CVLT* as a more universal test of verbal learning and memory. The result is in Eq. 20 not only confirms our previous results (Eqs. 15, 19), but also gives a narrower *CDRSUM* range between 5.75 and 6.5 which means mild dementia.

4 Discussion

Alzheimer's disease has about 20–30 years long prodromal phase, with brain changes before the first symptoms onset. This makes it a challenge to find biomarkers related to this period as when the first symptoms are clinically confirmed, there is no cure. Brain plasticity and adaptation processes may explain why patients do not notice several decades of extensive neurodegeneration. We proposed novel AI-related tools to increase the precision, sensitivity, and accuracy in monitoring ongoing neurodegenerations by looking not only into individual results in a clinical way but into patterns of many different cognitive test results and by comparing them with results of more advanced AD patients

We have studied these patterns with a granular computing approach and by comparing different sets of attributes (granules) to find possible patterns in normal subjects (n = 21) that might have similarities to granules observed in AD patients. We have used data from the BIOCARD study and analyzed all their AD patients (n = 40), part of their MCI patients (n = 70), and verified their normal, healthy patients (n = 20), as the reference groups. We were looking for different granules in all reference groups together as Group1, and in only the AD group Group2. In this study, we have increased the number of different granules by changing the number of used attributes. As in the previous study [5] we have always used 14 attributes, in this part we have changed from 5 to 7 attributes and compared results with 14 attributes. The other new and important part was the interpretability of obtained rules.

We have tried to estimate what is the meaning of individual granules, as we have performed similar estimations for Parkinson's disease patients [14]. We have used two groups: Group1 has granules related to normal subjects, MCI, and AD patients. On this basis, we have obtained a large set of rules that have represented subjects' different stages of the disease from normal to dementia. We have tested several of such models mostly changing normal subjects and getting different rules, which we have applied to other normal subjects to get in the reference group 'pure' normal. Also, rules can be created with different granularity and algorithms that might give different classifications. Therefore, we were looking for classifications that are complete e.g., they give similar results with different sets of rules. Groupl1 has given us rules that are subtle and determine the beginning of possible symptoms. In addition to our previous classifications of 14 attributes [5], we have added classifications with 5 or 7 attributes. These new granules gave us rules supporting our previous classifications like Eq. 10 or gave new rules Eqs. 11–12.

In the next step, we used a more advanced model – Group2 that gave rules based on AD patients. We got higher values of the CRDSUM that gave us only classifications of the possible subjects with mild dementia.

Using only 5 attributes, we obtained 5 new rules that partly confirmed our previous classifications and were easier to interpret. We have also performed classifications with 15 attributes that gave us conformation and better precision of the patient with mild dementia (Eq. 19, 20). As it is the first, to our knowledge, work that estimates a singular complex pattern of the individual patient's symptoms, our rules are taken from one population and applied to different subjects, so they are not certain. Therefore, the next step is to find different methods for their conformation.

References

1. Albert, M., et al.: The BIOCARD research team, cognitive changes preceding clinical symptom onset of mild cognitive impairment and relationship to *ApoE* genotype. Curr Alzheimer Res. **11**(8), 773–784 (2014)
2. Albert, M., Zhu, Y., Moghekar, A., Mori, S., Miller, M.I., Soldan, A. et al.: Predicting Progression from Normal Cognition to Mild Cognitive Impairment for Individuals at 5years Brain. **141**(3): 877–887 (2018)
3. Pedrycz, W. (ed.): Granular Computing: an Emerging Paradigm. Physica Verlag, Heidelberg, New York (2001)
4. Przybyszewski, A.W., Kon, M., Szlufik, S., Szymanski, A., Koziorowski, D.M.: Multimodal learning and intelligent prediction of symptom development in individual Parkinson's patients. Sensors **16**(9), 1498 (2016). https://doi.org/10.3390/s16091498
5. Przybyszewski A.W., Bojakowska, K., Nowacki, J.P., Drabik, A.: Rough set rules (RSR) predominantly based on cognitive tests can predict Alzheimer's related dementia. Nguyen, N.T., et al. (eds.): ACIIDS 2022, LNAI 13757, pp. 129–141 (2022)
6. O'Bryant, S.E., Waring, S.C., Cullum, C.M., et al.: Staging dementia using clinical dementia rating scale sum of boxes scores: a Texas Alzheimer's research consortium study. Arch Neurol. **65**(8), 1091–1095 (2008)
7. Pawlak, Z.: Rough Sets: Theoretical Aspects of Reasoning About Data. Kluwer, Dordrecht (1991)
8. Bazan, J., Nguyen, S.H., et al.: Desion rules synthesis for object classification. In: Orłowska, E. (ed.), Incomplete Information: Rough Set Analysis, Physica – Verlag, Heidelberg, pp. 23–57 (1998)

9. Bazan, J., Nguyen, H.S., Nguyen, et al.: Rough set algorithms in classification problem. In: Polkowski, L., Tsumoto, S., Lin, T. (eds.), Rough Set Methods and Applications, Physica-Verlag, Heidelberg New York, pp. 49–88 (2000)
10. Grzymała-Busse, J.: A New Version of the Rule Induction System LERS Fundamenta Informaticae, **31**(1), 27–39 (1997)
11. Bazan, J., Szczuka, M.: The Rough Set Exploration System. Peters, J.F., Skowron, A. (Eds.): Transactions on Rough Sets III, LNCS 3400, pp. 37–56 (2005)
12. Przybyszewski, A.W.: The neurophysiological bases of cognitive computation using rough set theory. Peters, J.F., et al. (eds.): *Transactions on Rough Sets IX*, LNCS 5390, pp. 287–317 (2008)
13. Bazan, J., Szczuka, M.: RSES and RSESlib - a collection of tools for rough set computations. Ziarko, W., Yao, Y. (eds.): RSCTC 2000, LNAI 2005, pp. 106–113 (2001)
14. Przybyszewski, A.W.: Parkinson's disease development prediction by C-granule computing. Nguyen, N.T., et al. (eds.): ICCCI 2019, LNAI 11683, pp. 1–11 (2019)

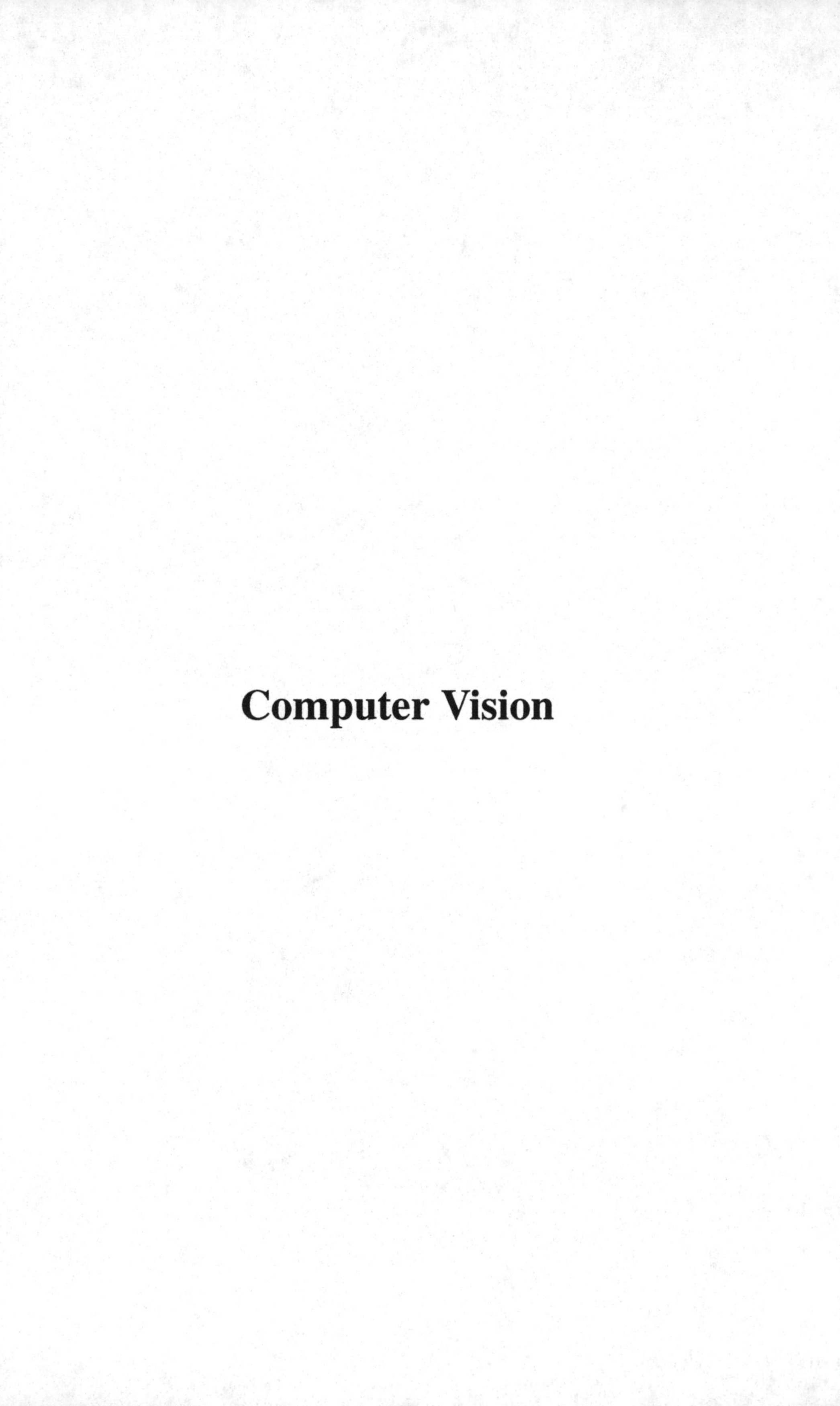

Computer Vision

AdVLO: Region Selection via Attention-Driven for Visual LiDAR Odometry

Han Lam(✉), Khoa Pho, and Atsuo Yoshitaka

Graduate School of Advanced Science and Technology, Japan Advanced Institute of Science and Technology, Nomi, Ishikawa, Japan
{hanlam,khoapho,ayoshi}@jaist.ac.jp

Abstract. Simultaneous Localization and Mapping (SLAM) aims to estimate the position and reconstruct the map of mobile robotics. Odometry is an essential component that tries to calculate the translations and rotations between frames of the sensors attached to the vehicle on the fly. Visual-LiDAR Odometry (VLO) is a prominent approach that has advantages in the sensor costs of cameras and robustness to environmental changes of LiDAR sensors. In general, one of the critical tasks in Odometry is selecting the important features between frames. In this paper, we proposed an end-to-end visual LiDAR odometry method named AdVLO that selects the important regions between frames via an attention-driven mechanism. A mask of essential regions of the input frame is generated via the attention mechanism. We then fuse the attention mask with the corresponding frame to maintain the essential regions. Instead of concatenating like previous works in VLO, we fuse the visual features and LiDAR using the Guided attention technique. The translation and rotation of the camera are calculated via the sequential computation of the LSTM. Experimental results on the KITTI dataset show that our proposed method achieves promising results compared to other odometry methods.

Keywords: SLAM · Odometry · Robot Localization

1 Introduction

Simultaneous Localization and Mapping (SLAM) aims to localize the position and draw a map of the unknown environment. SLAM has many applications, including 3D reconstruction, mobile robotics, self-driving cars, and autonomous underwater vehicles. SLAM can help to discover environments that are unsafe for humans. One of the essential components in SLAM is Odometry, which continuously estimates the sensor's movement by comparing the differences in the information from the consecutive input sequence while the vehicle is moving. The vehicle's trajectory is then calculated by accumulating the movements estimated by Odometry. Figure 1 demonstrates Odometry for SLAM. In this figure,

N. T. Nguyen et al. (Eds.): ACIIDS 2023, LNAI 13995, pp. 85–96, 2023.
https://doi.org/10.1007/978-981-99-5834-4_7

the triangular represents the vehicle's pose at the time step $t, t+1, \ldots, t+n$, and the red dots are the landmarks. The transitions of the camera's poses are estimated based on the changes in landmarks' positions in the sensors' captured views between two consecutive time steps.

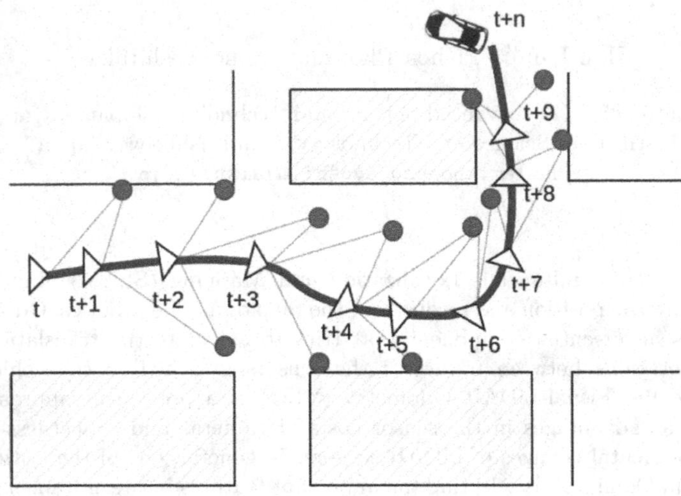

Fig. 1. Demonstration of Odometry in SLAM

Visual Odometry (VO), which uses cameras, is commonly used since they are inexpensive and convenient. The trajectory is estimated through a sequence of images while the vehicle moves. Hand-crafted features such as SIFT [13] and SURF [2] are extracted as the key points for each frame in the sequence. The depth information is then estimated for the key points based on the triangulate algorithm [15] for multiple cameras or RGB-D cameras. The key points arc then translated into the same 3D coordinate. The movement is estimated by calculating the differences of the corresponding key points in the same coordinate between frames. However, VOs are sensitive to illumination changes and texture deficiency; therefore insufficient for outdoor environments.

An alternative is to use LiDAR sensors to capture depth information in outdoor environments. LiDAR-Odometry (LO) extracts point features from 3D point cloud sequences and applies the ICP algorithm [19] for matching. LiDAR is a sensor that emits laser beams to measure the distance to objects in the surrounding environment. LiDAR provides better accuracy, field-of-views, and robustness to the environmental changes for odometry tasks. However, they may fail easily in structureless scenarios. Moreover, the raw 3D point clouds are too large to process directly and have noises.

Visual-LiDAR Odometry is a reasonable solution by combining the advantages of both VO and LO. The main challenge in Odometry is the key point selection for matching. In this paper, we focus on VLO for outdoor environments. We proposed Attention-driven Visual LiDAR Odometry (AdVLO), an end-to-end method that selects the important regions between frames via an attention-driven mechanism for VLO. This method generates a mask of essential regions of the input frame via the attention mechanism. The import regions of the input frame are maintained by fusing the input frame with the mask of attention. To reduce the size of the 3D point clouds from LiDAR, we produce the 2D range images from the point clouds. The visual and LiDAR features are fused by applying the Guided attention technique. The movement, which is represented as translation and rotation, is estimated via sequential computation of the Long-short term memory network (LSTM). Experimental results on the KITTI dataset show that our proposed method achieves promising results compared to other odometry methods for outdoor environments.

The rest of this paper is structured as follows: Sect. 2 surveys Related Works, Sect. 3 presents the proposed method, Sect. 4 shows the Experimental Result, and Sect. 5 is the conclusion.

2 Related Works

In this section, we briefly review the previous works related to our work. In principle, Odometry aims to match the key points between two-time steps (Fig. 2). The vehicle's movement in 3D space is calculated based on the change in position of each key point in the captured viewpoints. There are Visual Odometry, LiDAR Odometry, and Visual-LiDAR Odometry, which are different in capturing sensors.

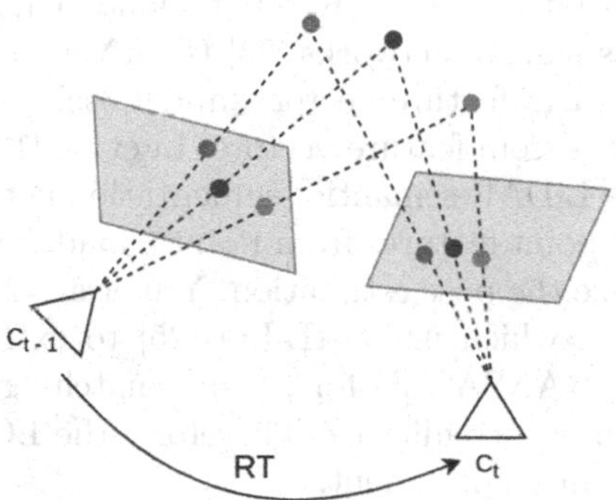

Fig. 2. Visual Odometry Principle

2.1 Visual Odometry

Davison et al. [4] propose the MonoSLAM, the first monocular SLAM to extract Shi and Tomasi features [20] from images to estimate the camera poses. Other methods use some other features, such as feature from scale-invariant (SIFT) [18], speeded up robust features (SURF) [2], or oriented fast and rotated brief (ORB) [18] in feature extraction. Mur-Artal et al. introduced ORB-SLAM2 [14], which expands the monocular camera to stereo and RGB-D cameras.

The deep learning approach gets attention from researchers because it outperforms traditional methods. In Visual SLAM and Visual Odometry, many researchers propose solutions based on deep learning. These methods can apply in many stages, such as feature extraction, feature matching, or pose estimation. Kendall et al. [8] propose PoseNet, which uses CNN to estimate the transition between frame with an image sequence as an input. Wang et al. [21] extract features from a CNN network and use an RNN network to match the feature. DeepSLAM [11] uses stereo images in the training phase and single images in the testing phase. However, all these methods only use single image sequences.

Generally, Odometry requires key points in 3D coordinates to calculate the movement. However, the VO approach only estimates the depth of information for key points using 2D images. Therefore, the accuracies of VO are limited.

2.2 LiDAR Odometry

LiDAR Odometry performs better than Visual Odometry, which uses LiDAR to scan surrounding objects in 3D coordinates. LiDAR Odometry has higher accuracy when compared with Visual Odometry. Zhang et al. [28] proposed LiDAR odometry and mapping (LOAM). LOAM extracts feature from the point cloud and match them in Cartesian space. Li et al. [9] proposed a learning-based LiDAR Odometry method (LO-Net). The LO-Net has a similar accuracy when compared with the LOAM. LO-Net uses a deep learning model to learn feature representation from the LiDAR point cloud. Cho et al. [3] introduce DeepLO, which includes feature networks [23] (FeatNet) and a pose network [8] (PoseNet). FeatNet extracts feature vectors, and PoseNet estimates the transformation between frames from feature vectors. Li et al. [10] propose a semantic SLAM, which includes a LiDAR semantic segmentation network. L3-Net [22] use PointNet [17] to extract point features from Point Cloud. This method requires a pre-built map to optimize the pose estimation. Yin et al. [25] proposed a LiDAR Odometry method (LO), which use CAE-LO [25] to extract feature from the 3D point cloud and use RANSAC [5] for feature matching. However, LiDAR is highly affected by luminance conditions. Therefore, the LO approach is limited in applying to the outdoor environment.

2.3 Visual LiDAR Odometry

Visual-LiDAR Odometry (VLO) takes advantage of LiDAR and camera information, which can help the VO to have precise depth information and LO overcome luminance problems. This paper follows the VLO approach to solve the Odometry problem. Zhang et al. [28] proposed V-LOAM, a fusion of data from the camera and LiDAR. This method estimates motion from Visual Odometry. LiDAR Odometry is then applied to refine motion estimates and incrementally build maps. Yi An et al. [1] convert Point Cloud to depth images and fusion with images to extract features. Loose Coupling Visual-LiDAR Odometry [24] combines the VISO2 method [7] and LOAM [27]. VISO2 is used for motion estimation between frames and applies this result to correct distortion of LiDAR data.

The main bottlenecks of the VLO approach are the keypoint selection and the visual and LiDAR information fusion. Previous works used calibration information to translate the 3D point cloud in LiDAR coordinate to the Camera coordinate or vice versa. The image feature points or 3D key points are extracted by adding depth information to 2D features. Instead of extracting feature points, our method selects important regions via the self-attention mechanism. A region on an image consists of many adjacent points, which is more robust than individual key points. Moreover, we only used frontal view depth images instead of raw 3D point clouds. Instead of simply applying concatenation layers, we apply a guided attention layer to fuse LiDAR and the Camera information to emphasize the interaction between objects in images and the point clouds.

3 Multimodel SLAM with Attention

This section describes the Odometry problem and our proposed method in detail.

3.1 Problem Description

Given a sequence of frames and the corresponding LiDAR point clouds, Odometry aims to estimate the transition between continuous frames. The transition P between frames is characterized by the translation T and the rotation matrix R of the camera's pose in the 3D dimension. R is a matrix representing the rotation around the axes Ox, Oy, Oz by the angles ω, ϕ, κ (roll, pitch, yaw) and $T = (T_x, T_y, T_z)$ is the robot translation vector at time $t-1$ to time t.

3.2 Feature Embedding

Image feature extraction aims to derive higher information from raw pixels. In recent approaches, Convolution Neural Networks are dominating in image embedding. In this paper, we apply Resnet50 as feature embedding for input image. Since we perform Odometry for outdoor environments, we applied a pre-trained weight of Resnet50 on ImageNet-1K, which contains objects in the wild.

The 3D point cloud captured from LiDAR is massive to process directly. Therefore, the 3D point cloud should be reduced for more efficiency. We first convert the 3D point cloud into a 2D range image to reduce the size of LiDAR. We only keep the points in the camera's view to reduce as much as possible. Since we only need the depth information of the selected regions in the images, the 3D points of LiDAR that are out of view can be considered redundant. We also apply Resnet50 to extract the feature from the 2D range image. The output of this module is the feature vector I.

3.3 Attention-Driven Region Selection

We propose AdVLO, which focuses on region selection for Visual LiDAR Odometry via an Attention-driven mechanism. Figure 3 shows the architecture of our proposed AdVLO. The embedding components described above feed the input image and point cloud. The attention-driven mechanism is applied for the image embedding to find the essential region in the image. Guided attention is then applied to attach the depth information from LiDAR to the extracted essential regions. Finally, an estimation module based on LSTM is applied to calculate the vehicle's transition through the continuous frames.

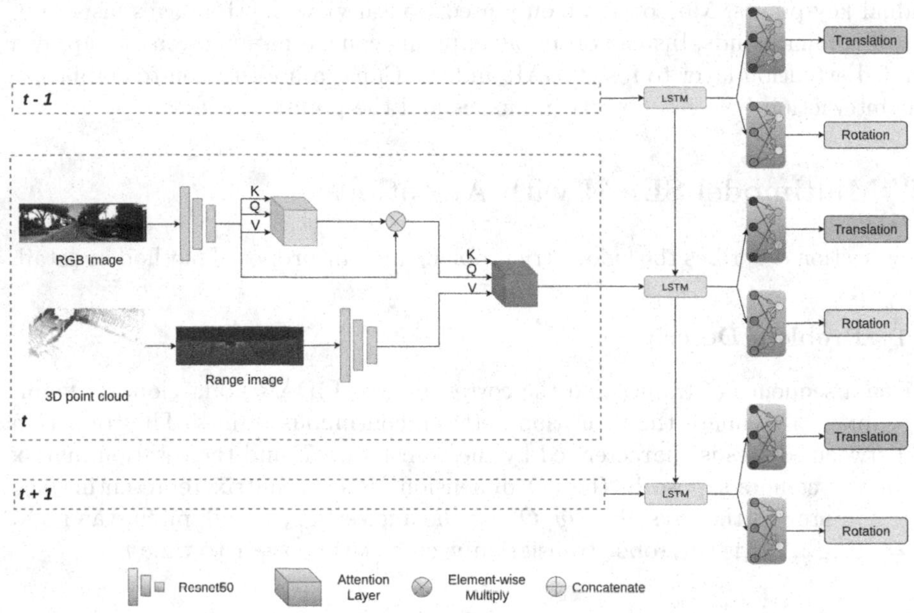

Fig. 3. The proposed network architecture

The attention mechanisms are to intensify the relationship between entities. Moreover, Multi-head Attention considers entities in many parallel viewpoints, which is helpful for practical applications. The Multi-head Attention calculates the refined signal as (1).

$$\text{MultiheadAttention}(q, K, V) = [head_1, head_2, \ldots, head_h]W^O \tag{1}$$

Each head in Multi-head Attention is pre-defined by users and reflects various considerations. The detail of each head is pointed out in (2).

$$head_j = softmax(\frac{QW_j^Q(KW_j^K)^T}{\sqrt{d_k}})VW_j^V \tag{2}$$

Each query q is accumulated in each specific viewpoint by the attentive score from the key K and latent information from value V. W^O is the trainable weight for all the heads, and d_k is the dimension's number we choose. In most approaches utilizing this attention, the values of K and V are often similar. It allows query features to be considered and intensified in the simultaneously meaningful space. In this problem, we apply attention mechanisms to find the essential regions in the image and their relationship with the depth information from LiDAR. The essential regions in the image embedding I are determined by a self-attention module in (3).

$$a = \text{SelfAttention}(I) = \text{MultiheadAttention}(I, I, I) \tag{3}$$

Unlike the previous works, which often apply entire attention maps with high dimensions to update the embedding, we produce an attention mask with one dimension to filter the essential regions via a convolutional layer with one kernel as in (4).

$$\hat{a} = \text{conv2D}_{\text{num\_kernel}=1}(a) \tag{4}$$

This attention mask with one channel plays the role of segmentation of the essential regions in the input image. The attention-driven mechanism is applied to keep only the extracted essential regions via an element-wise multiply operator as in (5).

$$s = I \otimes \hat{a} \tag{5}$$

The goal of the attention-driven mechanism is to focus on the key points from the static objects. An explanation is that the static objects' key points help produce a more precise transition estimation than moving objects. However, preparing the labeled segmentation of the essential regions requires massive human efforts. The attention is applied to automatically learn to determine the essential regions without labeled ground truths.

A combination module is required to fuse the information of the image and LiDAR. We apply guided attention to update the information of the image embedding by its interaction with the LiDAR embedding L in (6).

$$g = \text{GuidedAttention}(L, s) = \text{MultiheadAttention}(L, s, s) \tag{6}$$

We then build an estimation module to calculate the rotation and translation from the combined information g. An LSTM is applied to calculate the accumulation of information throughout a sequence of input information. In this paper, we choose the sequence length of 3 for the input. Rotation and translation sub-networks are built over the LSTM to calculate the rotation and translation. Each sub-networks has three layers whose layer size is 128, and a classification layer with three output nodes corresponds to the three components of rotation or translation.

3.4 Loss Function

To train the proposed network, we optimize the joint loss function, which is as follows:

$$\mathcal{L} = w_{\mathcal{L}_{trans}} \mathcal{L}_{trans} + w_{\mathcal{L}_{rot}} \mathcal{L}_{rot} \tag{7}$$

where $\mathcal{L}_{trans}$ and $\mathcal{L}_{rot}$ are the translation and rotation loss, and $w_{\mathcal{L}_{trans}}$ and $w_{\mathcal{L}_{rot}}$ are their weights, respectively. This work applies to mean squared error (MSE) as the loss function to calculate the difference between the estimated values and the ground truths. For each estimated translation between frame $t-1$ and t, the translation loss function:

$$\mathcal{L}_{trans} = \frac{1}{N}\sum_{i}^{N}(t_i - \widehat{t_i})^2 \tag{8}$$

Similarly, the rotation loss function is defined as follow:

$$\mathcal{L}_{rot} = \frac{1}{N}\sum_{i}^{N}(r_i - \widehat{r_i})^2 \tag{9}$$

In this paper, we set $w_{\mathcal{L}_{trans}}$ and $w_{\mathcal{L}_{rot}}$ as 1 and 100, respectively, to normalize the scale between those two.

4 Experiments

4.1 Experiments Setting

Dataset and Evaluation Metrics. We evaluate our network on the KITTI odometry dataset [6], which contains traffic with pre-collected image sequences, 3D points cloud from LiDAR, and vehicle GPS. The image sequences and the 3D point cloud are used as the input. The training translation and rotation ground truth are calculated from the GPS ground truth. Similar to previous works, we take sequences 00 to 08 for the training and sequences 09 and 10 for testing. Following the official KITTI benchmark, we compute the mean squared error for translation error (measured in percent) and rotation error (degrees per meter) on sequences of length $100, 200, \ldots, 800$ meters for evaluation.

We augment the training data for better accuracy by flipping the horizontal RGB images and the corresponding 2D range images of sequences 00 to 08. We denote the original image sequences are $\langle I_t, I_{t+1}\rangle$ which have the pose values $E = [t_x, t_y, t_z, r_x, r_y, r_z]$. The flip sequences $\left\langle I_t^f, I_{t+1}^f\right\rangle$ thus have the pose values $E^f = [t_x^f, t_y^f, t_z^f, r_x^f, r_y^f, r_z^f] = [-t_x, t_y, t_z, r_x, -r_y, -r_z]$.

Experimental Environment. The proposed network is trained and tested on a 3.9 GHz CPU, 24GB RAM, and a single NVIDIA GeForce RTX 3090 with Ubuntu 18.04 Operating system. The proposed network is implemented based on the Pytorch framework [16]. We use Resnet50 with a pre-trained weight on ImageNet-1K for feature embedding in the training phase. The batch size is set as 8, and the number of iterations is 5000 per epoch. The initial learning rate is 0.00001 and decreases a half after every ten epochs. Adam optimizer is applied with $\beta = 0.9$ to optimize the loss function in training.

Table 1. Comparision of the proposed method with another learning-based method. t_{error} is the translational error and r_{error} is the rotational error. The lower is the better.

Method	Sequence 09		Sequence 10	
	t_{error}	r_{error}	t_{error}	r_{error}
Deep-VO [21]			**8.11**	8.83
UnDeepVO [12]	**7.01**	3.16	10.63	4.65
SfMLearner [29]	18.77	3.21	14.33	**3.30**
Zhan et al. [26]	11.92	3.60	12.62	3.43
AdVLO (Unidirectional LSTM)	12.79	3.12	11.80	5.18
AdVLO (Bidirectional LSTM)	10.04	**2.41**	10.13	4.33

4.2 Experimental Results

Table 1 shows the experimental results of the proposed method. We experiment with Unidirectional LSTM and Bidirectional LSTM models. Experimental results show that the error is reduced in the Bidirectional LSTM model. The reason is that the hidden state of Bidirectional LSTM at each time step is influenced by past and future information. It can be seen that the performance of the proposed method is almost better than other recently published works. More specifically, in sequence 09, the proposed method can reduce the translation error (t_{error}) and rotation error (r_{error}) down to 10.04 and 2.41 in MSE, respectively, which outperform those produced by Zhan and SfMLearner. The translation error produced by AdVLO is higher than that by UnDeepVO; however, the rotation error by AdVLO is lower than that by UnDeepVO. In sequence 09, the proposed method's rotation error is the lowest compared to previous methods. The interesting point is that the higher the speed, the smaller the rotation error; the opposite is the translation error. At higher speeds, the object moves more quickly, and the camera captures more frames per second, improving the pose estimates' accuracy. Because of the higher vehicle speed, the rotation error is decreased in sequence 09.

In sequence 10, the translation error is cut down to 10.13, which is lower than those produced by Zhan, SfMLearner, and UnDeepVO. Our translation error in sequence 10 is higher than that by Deep-VO; however, our rotation error is lower than that by Deep-VO more than twofold. AdVLO can reduce rotation error to 4.33 in sequence 10, which is lower than those by UnDeepVO and Deep-VO. Although the rotation error is slightly higher than those by Zhan and SfMLearner, our translation error is higher than those by Zhan and SfMLearner.

Figure 4 visualizes the result of the feature selection module. We projected the attention mask as the green region onto corresponding RGB images. In Fig. 4, a moving car appears in a sequence of consecutive frames. The attention-driven mechanism focuses on static objects (landmarks), as the green masks highlight. The attention-driven mechanism focuses on static regions and does not focus on this moving car. This visualization demonstrates that the proposed method's self-attention mechanism has determined static regions in images. However, not all static regions are essential regions. The essential region must robustness and provide more information for the system. Some static objects, such as traffic signs or roads, yield essential information. However, the proposed method also focuses on static regions with less information, such as trees or road surfaces. Based on this observation, essential region segmentation can improve the proposed method.

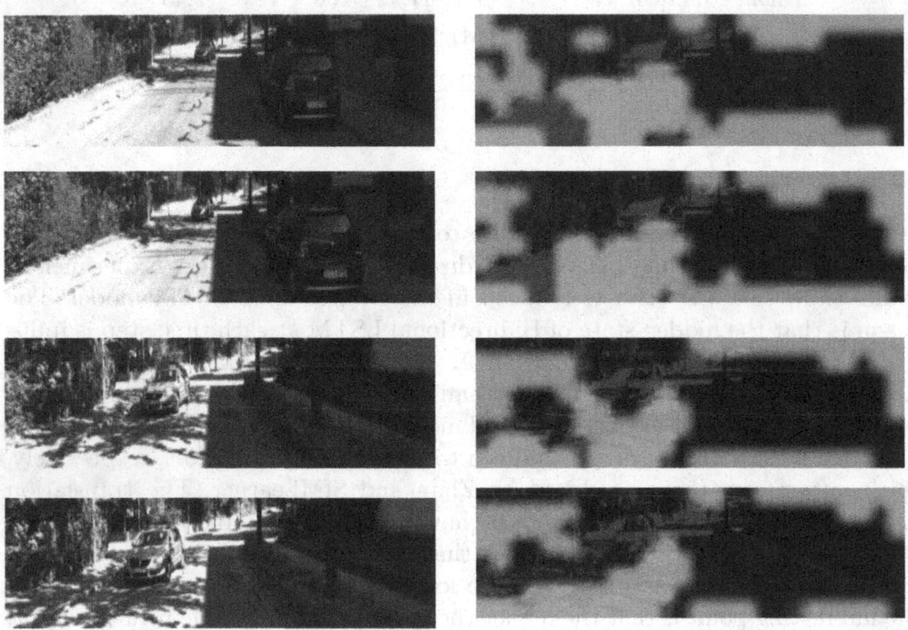

Fig. 4. The result of attention mask in region selection module.

5 Conclusion

Selecting essential regions is a bottleneck in the Odometry problem. The key points for estimating the vehicle's transition should be extracted from static objects. However, preparing annotations for static objects in video sequences and the corresponding LiDAR point clouds requires massive human effort. We propose AdVLO, which extracts the essential regions via an attention-driven mechanism, for the Odometry problem. Guided attention is applied to fuse the information of the image and LiDAR. Experiments show that our proposed method is almost better than other related works. The advantages of our proposed method come from the attention to the regions of static objects in the input images. The accuracy of our method can be improved further by discriminating the static objects from the moving objects in the images.

References

1. An, Y., Shi, J., Gu, D., Liu, Q.: Visual-lidar slam based on unsupervised multi-channel deep neural networks. Springer Cogn. Comput. **14**, 1496–1508 (2022)
2. Bay, H., Tuytelaars, T., Van Gool, L.: SURF: speeded up robust features. In: Leonardis, A., Bischof, H., Pinz, A. (eds.) ECCV 2006. LNCS, vol. 3951, pp. 404–417. Springer, Heidelberg (2006). https://doi.org/10.1007/11744023_32
3. Cho, Y., Kim, G., Kim, A.: DeepLO: geometry-aware deep lidar odometry. In: IEEE International Conference on Robotics and Automation (ICRA), pp. 2145–2152 (2022)
4. Davison, A.J., Reid, I., Molton, N., Stasse, O.: MonoSLAM: real-time single camera slam. IEEE Trans. Pattern Anal. Mach. Intell. **26**(6), 1052–1067 (2007). https://doi.org/10.1109/tpami.2007.1049
5. Fischler, M.A., Bolles, R.C.: Random sample consensus: a paradigm for model fitting with applications to image analysis and automated cartography. Commun. ACM **24**, 381–395 (1981)
6. Geiger, A., Lenz, P., Stiller, C., Urtasun, R.: Vision meets robotics: the kitti dataset. Int. J. Robot. Res. **32**, 1231–1237 (2013). https://doi.org/10.1177/0278364913491297
7. Geiger, A., Ziegler, J., Stiller, C.: StereoScan: dense 3d reconstruction in real-time. In: IEEE Intelligent Vehicles Symposium (IV), pp. 963–968 (2011)
8. Kendall, A., Grimes, M., Cipolla, R.: PoseNet: a convolutional network for real-time 6-DOF camera relocalization. In: International Conference on Computer Vision (ICCV), pp. 2938–2946 (2015)
9. Li, Q., Chen, S., Wang, C., Li, X., Wen, C., Cheng, M., Li, J.: Lo-net: deep real-time lidar odometry. In: IEEE/CVF Conference on Computer Vision and Pattern Recognition (CVPR) (2019). https://doi.org/10.1109/cvpr.2019.00867
10. Li, R., Gu, D., Liu, Q., Long, Z., Hu, H.: Semantic scene mapping with spatio-temporal deep neural network for robotic applications. Cogn. Comput. (2018). https://doi.org/10.1007/s12559-017-9526-9
11. Li, R., Wang, S., Gu, D.: DeepSLAM: a robust monocular slam system with unsupervised deep learning. IEEE Trans. Industr. Electron. (2020). https://doi.org/10.1109/tie.2020.2982096
12. Li, R., Wang, S., Long, Z., Gu, D.: UnDeepVO: monocular visual odometry through unsupervised deep learning. ArXiv e-prints (2017). https://doi.org/10.48550/ARXIV.1709.06841, arXiv:1709.06841
13. Lowe, D.: Object recognition from local scale-invariant features. In: Proceedings of the Seventh IEEE International Conference on Computer Vision, vol. 2, pp. 1150–1157 (1999). https://doi.org/10.1109/ICCV.1999.790410
14. Mur-Artal, R., Tardós, J.D.: Orb-SLAM2: an open-source slam system for monocular, stereo and RGB-D cameras. IEEE Trans. Rob. (2017). https://doi.org/10.1109/tro.2017.2705103
15. Nguyen, X.D., You, B.J., Oh, S.R.: A simple framework for indoor monocular slam. Int. J. Control. Autom. Syst. **6**, 62–75 (2008)
16. Paszke, A., et al.: PyTorch: an imperative style, high-performance deep learning library. In: Wallach, H., Larochelle, H., Beygelzimer, A., d'Alché-Buc, F., Fox, E., Garnett, R. (eds.) Advances in Neural Information Processing Systems (NIPS), vol. 32 (2019). https://proceedings.neurips.cc/paper/2019/file/bdbca288fee7f92f2bfa9f7012727740-Paper.pdf

17. Qi, C., Su, H., Mo, K., Guibas, L.J.: PointNet: deep learning on point sets for 3d classification and segmentation. In: IEEE/CVF Conference on Computer Vision and Pattern Recognition (CVPR), pp. 77–85 (2017)
18. Rublee, E., Rabaud, V., Konolige, K., Bradski, G.: Orb: an efficient alternative to sift or surf. In: International Conference on Computer Vision, pp. 2564–2571 (2011). https://doi.org/10.1109/ICCV.2011.6126544
19. Rusinkiewicz, S., Levoy, M.: Efficient variants of the ICP algorithm. In: Proceedings Third International Conference on 3-D Digital Imaging and Modeling, pp. 145–152 (2001). https://doi.org/10.1109/IM.2001.924423
20. Shi, J., Tomasi: Good features to track. In: Proceedings of IEEE Conference on Computer Vision and Pattern Recognition, pp. 593–600 (1994). https://doi.org/10.1109/CVPR.1994.323794
21. Wang, S., Clark, R., Wen, H., Trigoni, N.: DeepVO: towards end-to-end visual odometry with deep recurrent convolutional neural networks. In: 2017 IEEE International Conference on Robotics and Automation (ICRA) (2017). https://doi.org/10.1109/icra.2017.7989236
22. Weixin, L., Lu, W., Zhou, Y., Wan, G., Hou, S., Song, S.: L3-net: towards learning based lidar localization for autonomous driving. In: IEEE/CVF Conference on Computer Vision and Pattern Recognition (CVPR), pp. 6389–6398 (2019). https://doi.org/10.1109/cvpr.2019.00655
23. Xu, C., Feng, Z., Chen, Y., Wang, M., Wei, T.: FeatNet: large-scale fraud device detection by network representation learning with rich features. In: Proceedings of the 11th ACM Workshop on Artificial Intelligence and Security, pp. 57–63 (2018)
24. Yan, M., Wang, J., Li, J., Zhang, C.: Loose coupling visual-lidar odometry by combining VISO2 and LOAM. In: 36th Chinese Control Conference (CCC), pp. 6841–6846 (2017)
25. Yin, D., et al.: CAE-LO: lidar odometry leveraging fully unsupervised convolutional auto-encoder for interest point detection and feature description. arXiv: Computer Vision and Pattern Recognition (2020)
26. Zhan, H., Garg, R., Weerasekera, C.S., Li, K., Agarwal, H., Reid, I.: Unsupervised learning of monocular depth estimation and visual odometry with deep feature reconstruction. In: Proceedings of the IEEE/CVF Conference on Computer Vision and Pattern Recognition (CVPR), pp. 340–349 (2018)
27. Zhang, J., Singh, S.: LOAM: lidar odometry and mapping in real-time. In: Proceedings of Robotics: Science and Systems (RSS 2014) (2014)
28. Zhang, J., Singh, S.: Visual-lidar odometry and mapping: low-drift, robust, and fast. In: 2015 IEEE International Conference on Robotics and Automation (ICRA), pp. 2174–2181 (2015)
29. Zhou, T., Brown, M., Snavely, N., Lowe, D.G.: Unsupervised learning of depth and ego-motion from video. In: IEEE/CVF Conference on Computer Vision and Pattern Recognition (CVPR), pp. 6612–6619 (2017). https://doi.org/10.1109/CVPR.2017.700

Intelligent Retrieval System on Legal Information

Hoang H. Le[1,3,4], Cong-Thanh Nguyen[2,3], Thinh P. Ngo[1,3], Phu V. Vinh[1,3], Binh T. Nguyen[1,3,4], Anh T. Huynh[2,3], and Hien D. Nguyen[2,3](✉)

[1] University of Science, Ho Chi Minh City, Vietnam
[2] University of Information Technology, Ho Chi Minh City, Vietnam
hiennd@uit.edu.vn
[3] Vietnam National University, Ho Chi Minh City, Vietnam
[4] AISIA Research Lab, Ho Chi Minh City, Vietnam

Abstract. Nowadays, intelligent retrieval systems in law are vital in facilitating legal research and providing access to vast legal information. These systems allow users to search for legal information more efficiently and accurately. This paper investigates retrieval systems, their technological advancements, and their impact on legal research. The experimental results show that the proposed method is emerging to apply for analysis queries of practical law cases and extract suitable information from legal documents. It also discusses the challenges associated with law retrieval systems and explores future research directions to improve them.

Keywords: Intelligent system · Information Retrieval · BERT · Natural language processing · Ontology

1 Introduction

After the Covid-19 pandemic, many workers find policies to support their benefits based on legal regulations [9,21]. Vietnamese labor law, which has many related legal documents, plays a crucial role in safeguarding the rights and benefits of employees. It includes access to healthcare insurance [1], social insurance [2], and unemployment insurance [7,8]. In this context, healthcare insurance refers to medical coverage provided by an employer, while social insurance encompasses benefits such as retirement and occupational accident insurance. Unemployment insurance provides temporary financial assistance to employees who have lost their jobs.

The law information retrieval system is designed to assist users in ding relevant legal information quickly and efficiently [22]. The system allows users to search a vast collection of legal documents and databases to retrieve relevant information. The system employs algorithms and techniques to provide accurate and relevant search results [25]. Modern legal retrieval systems use sophisticated technologies such as natural language processing to provide accurate and relevant results [4,25].

N. T. Nguyen et al. (Eds.): ACIIDS 2023, LNAI 13995, pp. 97–108, 2023.
https://doi.org/10.1007/978-981-99-5834-4_8

This paper studies the techniques for legal text retrieval using two popular text retrieval models, BERT and TF-IDF/BM25. BERT (Bidirectional Encoder Representations from Transformers) is a state-of-the-art language model that can capture the context and semantics of words in a sentence [5]. TF-IDF (Term Frequency-Inverse Document Frequency) and BM25 are traditional text retrieval model that assigns weights to words based on their frequency in a document, and inverse frequency in a corpus [18]. Those techniques are designed to extract legal information for inputted queries based on legal documents about healthcare insurance [1], social insurance [2], and unemployment insurance [7,8]. The experiments show that combining both models can further improve the accuracy and efficiency of legal text retrieval, especially for long and complex legal texts.

The following section presents related work for designing information retrieval on law documents. Section 3 proposes extracting information from multiple legal documents using BERT and TF-IDF/BM25. It also gives the metric to evaluate the effectiveness of methods. Section 4 shows the experimental results of the proposed methods and the combination method. The last section concludes this study and gives future works.

2 Related Work

Numerous labor law restrictions have an impact on workers. So, it is essential to process legal paperwork correctly. Presently, several techniques have made substantial progress in the intelligent processing of legal documents. Those techniques include extracting, classifying, and question-answering [10,24].

Ontology is an effective method to organize the knowledge base of legal documents [6,16]. Legal-Onto is an ontology for organizing legal documents [15,17]. That ontology was built based on the foundation of the relational, intellectual model, Rela-model [14]. Legal-Onto was applied to represent the Land Law [15], and road traffic law [17] in Vietnam. Another ontology for a legal informatics document is also presented in [6]. This ontology can be used to represent the structure of a legal resource, legal temporal events, legal activities that have an impact on the document, and the semantic organization of the legal document. However, those methods have yet to be mentioned as the solution to analyze queries as law cases in practice.

An ontology that specifies domain knowledge and a database of document repositories can be combined using a technique created by the study in [13]. This method applies a model of domain knowledge called the categorized key phrase-based ontology to a number of information retrieval tasks. However, this model's graph-based measure has yet to be used to evaluate the semantic relevance of legal documents.

The authors of [11] created a system for analyzing Vietnamese legal text by fusing the benefits of standard information retrieval methods, pretrained masked language models (BERT), and legal domain knowledge. It was also advised to use a novel data augmentation technique based on knowledge of the legal field and legal textual entailment. Nevertheless, that method fails to convey the meaning of the legal instrument accurately.

3 Methodology

This section presents the main problem of information retrieval systems for legal documents, the data collection process, data processing, and evaluation metrics.

3.1 Problem Formulation

In this research, the general idea is to identify a list of legal documents that includes the intended meaning of the questions. Specifically, the goal is to develop a broad approach to generate a set of legal documents that can automatically extract the relevant information from a given question. The resulting list of legal documents will provide a comprehensive framework for accurately extracting the questions' intended meaning and improving the system's overall performance.

The model employed various evaluation metrics and techniques to evaluate the proposed approach's effectiveness. Using multiple evaluation approaches, the system aims to provide a comprehensive assessment of the proposed methodology and its ability to extract the intended meaning of questions accurately.

3.2 Data Collection

Questions Data. The data are published and belong to the Vietnam Social Insurance Portal[1]. There is a section on the Portal for users to submit questions about social insurance, view answers, and view other questions and answers.

Data are crawled from people's questions and official answers from the source: including question id, sender, submission date, field, question name, question content, and response content. For each question, data have been extracted from the law contained in the response. First, the method has been used semi-automatic to withdraw, in the answer, the sentences that start with an article, clause, or point. Then, we use the manual method to make the extracted sentence more accurate. After that, we separate that sentence into the corresponding four components: Article, Clause, Point, and Law document (Code, Resolution, Joint Circular, Circular, Decree, Decision, etc.).

Currently, the data crawled has about 2000 records. In it, there are questions and answers about VSSID applications, administrative procedures, and legal documents. However, with the VSSID application and some administrative procedures that are not related to legal documents, they are not extracted from legal documents.

Legal Documents. The legal documents were collected from Thuvienphapluat[2]. We use a web scraping tool that automates retrieving legal documents from

[1] baohiemxahoi.gov.vn - This website provides information on social insurance policies, social insurance duties and procedures related to social insurance.

[2] thuvienphapluat.vn - Thuvienphapluat is a Vietnamese website that provides online legal documents of Vietnam and related legal documents.

the website. The collected data were evaluated for their quality and completeness, they were also compared with question data obtained from other sources to assess the reliability of the Law Library as a data source for legal research.

After collecting the legal documents from that webpage, we further process the data by splitting the documents into smaller segments based on the hierarchical structure of the legal text. Specifically, we will identify and extract articles within each document and assign a unique ID to each segment. This process will enable us to organize the legal data in a structured format that can be easily queried and analyzed (Table 1).

3.3 Data Processing

Due to the legal text data and questions collected from the website, many unwanted components exist. We eliminated irrelevant parts and special characters (e.g., new-line and extra white space) to improve the model's accuracy. But there are specific difficulties in processing raw data. As a result, we have to manually handle and verify the data to ensure its accuracy. This extra step is necessary to avoid errors arising from incomplete or inaccurate data, which could impact our analysis and decision-making process.

After collecting data and filtering information, we obtained a reliable dataset for further analysis and research. The dataset comprises 23 legal documents appropriate to healthcare, social, and unemployment insurance, with 1094 articles and 618 questions.

3.4 Vocabulary Frequency

To gain insights into the field of Vietnamese labor law, WordClouds are used to represent the most commonly occurring words within our dataset visually (Fig. 1 and Fig. 3). Additionally, Fig. 2 and Fig. 4 present the top 10 words with the highest frequencies, all relevant to the legal domain of Vietnamese labor law.

3.5 Feature Extraction and Modeling

Prior to the training phase, we examine **TF-IDF** and **BM25** [20] as the initial methods to tackle the problem. They are statistical methods for estimating the relevance of a given query matching to documents based on the frequency of word occurrences.

Then, we used **word segmentation** from **VnCoreNLP** [23] to improve the performance of **TF-IDF** and **BM25** [20] by segmenting Vietnamese words into meaningful phrases.

Table 1. Different examples for legal texts.

ID	Articles
law1_1	"Điều 1. Phạm vi điều chỉnh Luật này quy định chế độ, chính sách bảo hiểm xã hội; quyền và trách nhiệm của người lao động, người sử dụng lao động; cơ quan, tổ chức, cá nhân có liên quan đến bảo hiểm xã hội, tổ chức đại diện tập thể lao động, tổ chức đại diện người sử dụng lao động; cơ quan bảo hiểm xã hội; quỹ bảo hiểm xã hội; thủ tục thực hiện bảo hiểm xã hội và quản lý nhà nước về bảo hiểm xã hội." "Article 1. Scope of regulation This Law provides for the social insurance regime and policies, the rights and responsibilities of employees, employers, relevant agencies, organizations, and individuals involved in social insurance, representative organizations of labor collectives and employers, social insurance agencies, social insurance funds, social insurance implementation procedures, and state management of social insurance."
law1_2	"Điều 2. Đối tượng áp dụng 1. Người lao động là công dân Việt Nam thuộc đối tượng tham gia bảo hiểm xã hội bắt buộc, bao gồm: a) Người làm việc theo hợp đồng lao động không xác định thời hạn, hợp đồng lao động xác định thời hạn, hợp đồng lao động theo mùa vụ hoặc theo một công việc nhất định có thời hạn từ đủ 03 tháng đến dưới 12 tháng, kể cả hợp đồng lao động được ký kết giữa người sử dụng lao động với người đại diện theo pháp luật của người dưới 15 tuổi theo quy định của pháp luật về lao động..." "Article 2. Applicable subjects 1. Employees who are Vietnamese citizens are mandatory participants in social insurance, including a) Workers under an indefinite-term labor contract, a definite-term labor contract, a seasonal labor contract, or a specific job with a duration of 03 months or more, but less than 12 months, including labor contracts signed between the employer and a legal representative of a person under 15 years of age, as regulated by labor law;..."
law23_8	Điều 8. Trách nhiệm của Tổ kiểm soát các cấp 1. Tổ kiểm soát cấp huyện 1.1. Kiểm soát việc tuân thủ quy trình thực hiện, hồ Sơ đính kèm, thông tin của người tham gia và thành viên HGĐ được cập nhật vào phần mềm HGĐ đảm bảo đầy đủ, chính xác từ đề nghị của cán bộ sổ, thẻ... "Article 8. Responsibilities of Control Boards at all levels 1. Responsibilities of Control Boards at district level 1.1. Control the compliance with the implementation procedures, the attached documents, information of the participants and members of the Household Management Board updated in the Household Management software, ensuring their completeness and accuracy based on the proposal of the bookkeepers, card issuers,..."

As previously stated, each article belonging to a certain law document is granted a unique ID. When we feed a query into the pipeline, it is analyzed, and our objective is to guess which article is linked to the query. Each one may be related to more than one article or be assigned multiple labels simultaneously. This part will introduce fundamental approaches for solving our task. Examples are shown in Table 2. At outperform modeling, our core architecture implements SBERT [19], a sentence embedding model that employs PhoBERT [12], a large-scale monolingual language pre-trained model for Vietnamese. The model is

Fig. 1. Wordcloud Legal corpus.

Word	English	Frequency
Lao	Labor (Part-of)	6131
ng	Labor (Part-of)	5540
nh	Regulations (Part-of)	4681
ca	Of	4632
ngi	People	4349
him	insurance (Part-of)	3840
vị	And	3822
bo	insurance (Part-of)	3740
theo	follow	3234
quy	Regulations (Part-of)	3217

Fig. 2. Word Frequency Legal corpus.

Fig. 3. Wordcloud User Questions.

Word	English	Frequency
tũi	I	1234
cú	Have	1111
c	Get	994
em	I	803
BHXH	Social insurance (acronym)	792
vị	And	754
bo	Insurance (Part- of)	703
thòng	Month	653
him	Insurance (Part- of)	628
lị	is	620

Fig. 4. Word Frequency User Questions.

trained throughout two stages, as follows: (Fig. 5 depicts the flow of our model in further detail.)

(a) **Stage 1:** Queries and article documents are transformed into vectors through TF-IDF [18] or BM25 [20] algorithm to retrieve relevant and irrelevant article documents to each query which is computed by cosine similarity between vectors. It can be understood that negative samples for each question are generated by selecting the top-n article documents from **Article Retriever TF-IDF/BM25** layer that did not contain the correct answer. The following phase is the SBERT method. First, at **Word Embedding** layer, those samples are embedded by sentence transformer based on PhoBERT [12]. The goal of the next mean **Pooling** layer is to produce fixed-size sentence embeddings. Finally, the **Cosine Similarity Loss** is used to determine how similar two embeddings are. It enables our architecture to be fine-tuned to recognize the similarity between the negative and positive samples (The positive sample includes query and article documents target). This technique, namely contrastive learning, is commonly utilized when there is a shortage of data and the need to heighten the contrast between the positive and negative ones. After training, our system can make reasonably accurate predictions.

Table 2. A query sample and its corresponding ground-truth labels as well as prediction of our outperforming system.

Query	**Ground-truth labels**	**Prediction**
"Bố em làm việc cơ quan và tham gia bảo hiểm 35 năm.Nay được quyết định nghỉ hưu. Vậy khi nào bố em được lãnh tiền bảo hiểm và lãnh thế nào.có được lãnh BHXH 1 lần không?" **English translation:** ***"My father worked in a company and participated in social insurance for 35 years. Now he has decided to retire. So when will my father receive the insurance money, and how to receive it? Can he get one-time social insurance?"***	**law1_59**	**law1_59** **law1_13**
"Mình đã nghỉ làm ở công ty cũ từ đầu năm 2021, đến hiện tại công ty cũ chưa chốt sổ bảo hiểm cho mình. Hiện tại mình đã đi làm và đóng bảo hiểm tại công ty mới, nhưng trên VSSID chỉ hiện quá trình đóng bảo hiểm tại công ty mới mà không hiện quá trình ở công ty cũ. Cho mình hỏi làm thế nào để lấy lại quá trình trên vssid được ạ. Trong trường hợp công ty cũ làm sai không chốt sổ bảo hiểm cho mình thì quá trình cũ của mình có được tính hay không?" **English translation:** ***"I quit my job at my old company in 2021, and until now, the old company has not closed my social insurance. Currently, I have worked and paid insurance at the new company, but on VSSID, only the insurance payment process at the new company is displayed, but not the process at the old company. Please tell me how to get the process back on VSSID. Also, in case the old company makes a mistake and does not close the insurance book for me, will my old process be counted or not?"***	**law13_46.96** **law1_21** **law3_48**	**law13_46.96** **law1_21** **law3_48**

(b) **Stage 2:** Negative samples for each question are generated once again by selecting the top-n documents, which are embedded from **SBERT** [19] trained in **stage 1.** This phase quickly and effectively provides our model more embedded negative data samples, which are of high quality as well. Furthermore, our model gets strengthened, reinforced, and fine-tuned due to learning the contrastive information in those data.

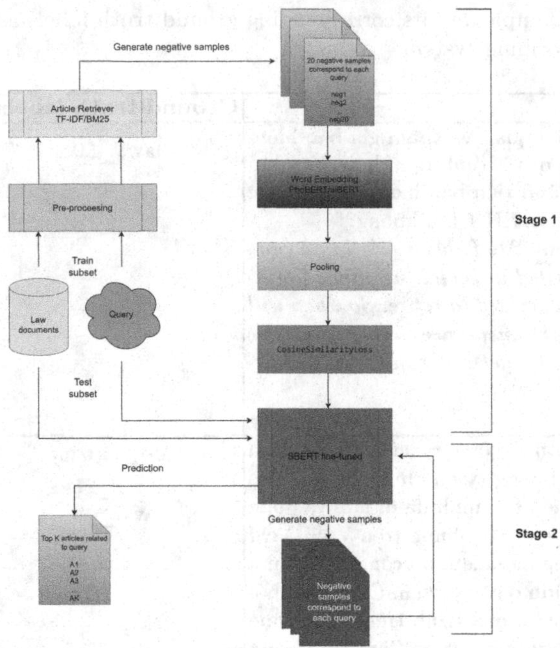

Fig. 5. TF-IDF/BM25 + SBERT system (Explained at **Feature extraction and modeling** subsection of Sect. 3.).

3.6 Metric Evaluation

We evaluate the performance of different approaches using $Top_K@acc$ as the metric. Accuracy is calculated as the proportion of questions with all correct labels in the **Top** K documents returned by our methods. L_K is a collection containing **k** labels, or IDs of Article, which our system predicts are most related to the query, l_q is the query's actual collection of labels.

$$Top_K@acc = \frac{1}{n}\sum_{1}^{n}\begin{cases} 1, & l_q \subseteq L_K \\ 0, & \text{otherwise} \end{cases} \tag{1}$$

4 Experiments

4.1 Experimental Design

The following experimental is designed to compare the performance of different methods. In the word embedding phase, the proposed model utilizes **vinai/phobert-base** [12], **fptai/vibert** [3] with a maximum sequence length of 256. SBERT is fine-tuned using three epochs and a batch size of 16. On the local machine, we run our tests on a single NVIDIA RTX 3060.

TF-IDF [18] is a numerical statistic that indicates how important a lexical unit is in a document or text in a dataset. Specifically, the TF-IDF weight is composed of two terms: the first term computes the normalized term frequency (TF) is defined as the number of times a word appears in a document divided by the total number of words in that document; the second the term is the Inverse Document Frequency (IDF). Practically, TF-IDF can be computed as follows:

$$tfidf(t,d,D) = tf(t,d) \times idf(t,D)$$
$$tf(t,D) = log(1 + freq(t,d))$$
$$idf(t,D) = log(\frac{|D|}{\{d \in D : t \in d\}}),$$

where t is a unigram or bigram term in a document d from a collection of documents D. $freq(t,d)$ measures how many times t a term appears in d.

BM25 [20] is an improved variant of TF-IDF. BM25 has two new parameters: k, which helps balance the relevance of term frequency and IDF, and b, which modifies the weight of document length normalization. Suggested values are $k = [1.2, 2.0]$ and $b = 0.75$.

$$BM25(D,Q) = \sum_{i=1}^{n} log(\frac{N - n(q_i) + 0.5}{n(q_i) + 0.5} + 1) \times \frac{f(q_i,D) \cdot (k+1)}{f(q_i,D) + k \cdot (1 - b + b \cdot \frac{|D|}{avgdl})}$$

where $f(q_i, D)$ is the frequency with which q_i appears in document D. $|D|$ is the number of words in D, and $avgdl$ is the average document length in the text collection from which documents are drawn; N is the total number of documents in the collection, and $n(q_i)$ is the number of documents containing q_i.

SBERT [19] is a framework for computing sentence embeddings using BERT [5] models. Sentence-BERT modifies the original BERT model using a Siamese or triplet network structure. These are networks that share weights and can encode multiple inputs at once. **PhoBERT** [12] and **viBERT** [3], pretrained BERT embeddings for the Vietnamese language, has been developed by VinAI and FPTAI, respectively.

Cosine similarity can be computed by the following equation:

$$similarity(q, d_i) = \frac{q \cdot d_i}{\|q\| \|d_i\|}, \tag{2}$$

where q is the vector form of the query and d_i is vector form of the article i from a collection of law documents D ($d \in D$).

CosineSimilarityLoss

$$similarityLoss = \frac{1}{N} * \sum (label - \tanh(similarity(e_i, e_j)))^2 \tag{3}$$

where N is the number of pairs, the *label* is the target similarity (-1 or 1), *tanh* is the hyperbolic tangent function, *similarity* is the cosine similarity function, and e_i is the vector form of the article i.

4.2 Results

We conduct different experiments to compare these approaches in the collected legislation dataset. This dataset mainly focuses on a particular subject, the Labor Code of Viet Nam that we have recently formed. Table 3 displays the results of eight different methods, where each notation stands for every distinct approach as follows. **TIWS** employs TF-IDF to retrieve the top K most relevant article documents to the query, where each sentence can be tokenized as tokens by Word Segmentation from Undersea library. Similarly, **BMWS** is the approach where BM25 replaces TF-IDF and combines with Word Segmentation.

For **TPS1**, TF-IDF/BM25 is used to generate negative samples, and then they are embedded by PhoBERT and fed into the model for training. After the training process at Stage 1 is completed, one can get PhoBERT fine-tuned. This fine-tuned one is used to embed article documents and queries in the testing dataset and to predict which documents are most relevant to the question using cosine similarity. By the inheritance of PhoBERT fine-tuned from the previous **TPS1**, **TPS2** continuously re-embeds all of the article documents and queries from the training subset, and negative samples are generated based on the similarity between article and query embedding vectors. After that, they are fed into the model for training the second time, and PhoBERT fine-tuned-v2 is created. Finally, this fine-tuned-v2 can embed and predict the relevant article documents and queries efficiently.

TvS1 is similar to the **TPS1**. The only change between this model and **TPS1** is that viBERT has replaced PhoBERT. Finally, **TvS2** is analogous to the **TPS2**. Again, the only change between this model and **TPS2** is that viBERT has replaced PhoBERT.

As illustrated in Table 3, the proposed model combining TF-IDF/BM25 and SBERT achieves outperformance on the dataset. Given that TF-IDF + SBERT and BM25+ SBERT have virtually identical scores, we only display one representative on the table. The score increases when k in $Top_K@acc$ increases.

Table 3. Evaluation results on our dataset using several techniques with varying $Top_K@acc$.

Methods	$Top_5@acc$	$Top_{10}@acc$	$Top_{20}@acc$	$Top_{50}@acc$	$Top_{100}@acc$
TF-IDF	0.0356	0.0777	0.1472	0.3867	0.4919
BM25	0.0307	0.0728	0.1634	0.3155	0.4660
TIWS	0.0518	0.0728	0.1408	0.3608	0.4854
BMWS	0.0388	0.0841	0.2023	0.3414	0.4693
TPS1	0.4345	0.4935	0.5575	**0.6812**	0.7540
TPS2	**0.5518**	**0.5939**	**0.6359**	0.6618	0.6942
TvS1	0.2335	0.2875	0.3620	0.5475	**0.7691**
TvS2	0.2573	0.3020	0.3921	0.5021	0.5523

5 Conclusion and Future Work

This study proposed a method to design intelligent retrieval systems on law documents based on NLP approaches, TF-IDF, BM25 and BERT. Although each technique did not work well on legal domain, the performance of model is enhanced when combining TF-IDF/BM25 and BERT. Thus, the combination model is emerging to serve as a foundation for future text retrieval models geared at legal inquiries. Ultimately, the efforts will result in a more robust and efficient legal retrieval system.

In the future work, the proposed model will continue to improve and experiment with new techniques to increase its performance. It can be incorporated a knowledge graph [15] to assist in the search for relevant documents and improve the overall efficiency of the system.

References

1. National Assembly: Labor on Employment 2013, No. 38/2013/QH13 (2013)
2. National Assembly: Labor Code 2019, No. 45/2019/QH14 (2019)
3. Bui, T.V., Tran, O.T., Le-Hong, P.: Improving sequence tagging for Vietnamese text using transformer-based neural models. CoRR abs/2006.15994 (2020). https://arxiv.org/abs/2006.15994
4. Dale, R.: Law and word order: NLP in legal tech. Nat. Lang. Eng. **25**(1), 211–217 (2019)
5. Devlin, J., Chang, M.W., Lee, K., Toutanova, K.: BERT: pre-training of deep bidirectional transformers for language understanding. In: Proceedings of the 2019 Conference of the North American Chapter of the Association for Computational Linguistics: Human Language Technologies, Volume 1 (Long and Short Papers). Association for Computational Linguistics, Minneapolis, Minnesota, June 2019
6. Fernández-Barrera, M., Sartor, G.: The legal theory perspective: doctrinal conceptual systems vs. computational ontologies. In: Sartor, G., Casanovas, P., Biasiotti, M., Fernández-Barrera, M. (eds.) Approaches to Legal Ontologies. Law, Governance and Technology Series, vol. 1, pp. 15–47. Springer, Dordrecht (2011). https://doi.org/10.1007/978-94-007-0120-5_2
7. Vietnam Government: Decree on Detailing Unemployment Insurance of the Law on employment - No. 28/2015/ND-CP (2015)
8. Vietnam Government: Decree on detailing and guiding the implementation of a number of articles of the labour code regarding working conditions and labour relations, No. 145/2020/ND-CP (2020)
9. Le, T.A.T., Vodden, K., Wu, J., Atiwesh, G.: Policy responses to the covid-19 pandemic in Vietnam. Int. J. Environ. Res. Public Health **18**(2), 559 (2021)
10. Mirończuk, M.M.: The BigGrams: the semi-supervised information extraction system from html: an improvement in the wrapper induction. Knowl. Inf. Syst. **54**(3), 711–776 (2018)
11. Ngo, H., Nguyen, T., Nguyen, D., et al.: AimeLaw at ALQAC 2021: enriching neural network models with legal-domain knowledge. In: 2021 13th International Conference on Knowledge and Systems Engineering (KSE). IEEE (2021)
12. Nguyen, D.Q., Nguyen, A.T.: PhoBERT: pre-trained language models for Vietnamese. In: Findings of the Association for Computational Linguistics: EMNLP 2020, pp. 1037–1042 (2020)

13. Nguyen, H., Tran, T.V., Pham, X.T., Huynh, A.: Ontology-based integration of knowledge base for building an intelligent searching chatbot. Sens. Mater. **33**(9), 3101–3123 (2021)
14. Nguyen, H.D., Pham, V.T., Le, T.T., Tran, D.H.: A mathematical approach for representing knowledge about relations and its application. In: Proceedings of 7th International Conference on Knowledge and Systems Engineering (KSE 2015), pp. 324–327. IEEE (2015)
15. Nguyen, T.H., Nguyen, H.D., Pham, V.T., Tran, D.A., Selamat, A.: Legal-Onto: an ontology-based model for representing the knowledge of a legal document. In: Proceedings of 17th Evaluation of Novel Approaches to Software Engineering (ENASE 2022), Online streaming, pp. 426–434 (2022)
16. de Oliveira Rodrigues, C.M., de Freitas, F.L.G., Barreiros, E.F.S., de Azevedo, R.R., de Almeida Filho, A.T.: Legal ontologies over time: a systematic mapping study. Exp. Syst. Appl. **130**, 12–30 (2019)
17. Pham, V.T., Nguyen, H.D., Le, T., et al.: Ontology-based solution for building an intelligent searching system on traffic law documents. In: Proceedings of 15th International Conference on Agents and Artificial Intelligence (ICAART 2023), Lisbon, Portugal, pp. 217–224 (2023)
18. Qaiser, S., Ali, R.: Text mining: use of TF-IDF to examine the relevance of words to documents. Int. J. Comput. Appl. **181**(1), 25–29 (2018)
19. Reimers, N., Gurevych, I.: Sentence-BERT: sentence embeddings using Siamese BERT-networks. In: Proceedings of the 2019 Conference on Empirical Methods in Natural Language Processing. Association for Computational Linguistics (2019). https://arxiv.org/abs/1908.10084
20. Robertson, S., Zaragoza, H.: The probabilistic relevance framework: BM25 and beyond. Foundations Trends Inf. Retrieval **3**, 333–389 (2009). https://doi.org/10.1561/1500000019
21. Trinh, N.T.T.: Impact of the Covid-19 on the labor market in Vietnam. Int. J. Health Sci. **6**, 6355–6367 (2022)
22. Villata, S., Araszkiewicz, M., Ashley, K., et al.: Thirty years of artificial intelligence and law: the third decade. Artif. Intell. Law **30**, 561–591 (2022)
23. Vu, T., Nguyen, D.Q., Nguyen, D.Q., Dras, M., Johnson, M.: VnCoreNLP: a Vietnamese natural language processing toolkit. In: Proceedings of the 2018 Conference of the North American Chapter of the Association for Computational Linguistics: Demonstrations, pp. 56–60. Association for Computational Linguistics, New Orleans, Louisiana, June 2018. https://doi.org/10.18653/v1/N18-5012. https://aclanthology.org/N18-5012
24. Zhao, G., Liu, Y., Erdun, E.: Review on intelligent processing technologies of legal documents. In: Sun, X., Zhang, X., Xia, Z., Bertino, E. (eds.) Artificial Intelligence and Security: 8th International Conference, ICAIS 2022, Qinghai, China, 15–20 July 2022, Proceedings, Part I, pp. 684–695. Springer, Cham (2022). https://doi.org/10.1007/978-3-031-06794-5_55
25. Zhong, H., Xiao, C., Tu, C., et al.: How does NLP benefit legal system: a summary of legal artificial intelligence. In: Proceedings of the 58th Annual Meeting of the Association for Computational Linguistics (COLING 2020), pp. 5218–5230. Association for Computational Linguistics (2020)

VSNet: Vehicle State Classification for Drone Image with Mosaic Augmentation and Soft-Label Assignment

Youlkyeong Lee, Jehwan Choi(✉), and Kanghyun Jo(✉)

Department of Electrical, Electronic and Computer Engineering, University of Ulsan, Ulsan 44610, South Korea
choijh1897@gmail.com, acejo@ulsan.ac.kr

Abstract. Numerous architectures are under development to comprehend object information and background in images by analyzing features extracted through Convolutional Neural Networks (CNNs). Autonomous driving requires understanding diverse information and collecting data from heterogeneous environments to generalize classification models. However, the patterns of feature maps extracted through convolution layers in drone image data, which encompass assorted types of vehicles and road shapes, tend to be simple, leading to overfitting during model training. To prevent overfitting, this study applies Mosaic Augmentation to increase data diversity and brings generalization to the data. This data augmentation method randomly combines four selected images to create a new mosaic image. Soft-label Assignment is used to determine the labels of the mosaic images. The dataset is collected using a drone flying along roads, and approximately 4,000 images are used for training. In the experiment, the classification performance of vehicle status is listed based on the weight of the loss function of the soft label and hard label. Having achieved an accuracy of 83.41%, the effectiveness of the proposed method is compared with dilated residual networks in terms of improving model performance.

Keywords: Drone image · Transportation system · Vehicle state · Classification

1 Introduction

Generating desired information through algorithms using various CNN models is crucial for collecting and analyzing image-based traffic information. Utilizing drones for data analysis provides a different perspective on the images compared to the black box or installed cameras in conventional vehicles. Therefore, a drone capturing a wider area at once is essential for comprehending comprehensive

This work was supported by the National Research Foundation of Korea (NRF) grant funded by the government (MSIT) (No. 2020R1A2C200897212).

N. T. Nguyen et al. (Eds.): ACIIDS 2023, LNAI 13995, pp. 109–120, 2023.
https://doi.org/10.1007/978-981-99-5834-4_9

vehicle movement and flow. Understanding vehicle movement is necessary and can be categorized into three basic states: normal state, where the vehicle moves along the lane; lane-changing state, where the vehicle changes its lane; and stop state, where the vehicle comes to a halt.

This research aims to design a CNN model that classifies three vehicle movement states through drone images and applies mosaic data augmentation [1] and soft label assignment [2]. The dataset is collected data using a drone, which captured images with a bird's-eye view of the road. This work utilizes mosaic augmentation to increase data diversity and prevent overfitting during model training. This data augmentation technique randomly combines four selected images to generate a new mosaic image. Soft-label Assignment is used to determine the labels of the mosaic images. These techniques demonstrate the potential of drones for traffic information analysis and the effectiveness of the proposed methodology in improving classification model performance for autonomous driving systems.

2 Related Work

2.1 Autonomous Vehicle Dataset for Object Detection

Images and annotation data collected in the past environment on various roads have been continuously accumulated. With the advancement of object classification and detection technology, autonomous driving technology is being developed rapidly. The Cityscapes [3] and KITTI [4] datasets were created as datasets for autonomous vehicle research [5–10]. The dataset generated image collection and annotation data for traffic conditions on the road through cameras installed in the vehicle. The KITTI dataset also includes 3D bounding box location and camera calibration information through a 3D laser scanner.

2.2 Drone-Based Dataset for Object Detection

Stanford drone dataset [11] is the first public aerial image dataset using drones. This dataset contains ten kinds of tracking information (Track ID, (xmin, ymin), (xmax, ymax), frame, lost, occluded, generated, label) about objects on the road in the video image. Images taken at eight locations on the Stanford campus were collected. The targets are six classes (Bicyclist, Pedestrian, Skateboarder, Cart, Car, and Bus). However, the annotation quality of the bounding box of the object is roughly expressed, which has a problem with the performance of the object detection algorithm. The VisDrone [12] dataset is a large-scale drone image produced by AISKYEYE team at Lab of Machine Learning and Data Mining, Tianjin University, China. The dataset aims to develop applications that can be used for computer vision through drones. Through cameras installed in drones, 288 video images were collected from 14 urban areas in China. It produced 2.6 million bounding boxes, including ten classes (pedestrian, person, bus, car, van, truck, bicycle, awning tricycles, motorcycles, and tricycles). Data validation is tested through VisDrone challenge [13–15] and various kinds of research [16–18] are utilized. The Institut für Kraftfahrwesen Aachen research team had built a drone-based road user trajectory dataset for various situations. The test for vehicles related to autonomous driving is conducted based on the scenario. Therefore, we present reliable and high-quality data criteria. The highD [19] is a large-scale vehicle trajectory dataset for German high roads. It includes six locations, 16.5 h, and 110,000 trajectory information. In inD [20], automated vehicles require data-based analysis methods to understand complex environments. By collecting road images using drones, it was proposed to collect road trajectories and natural road conditions through vehicle movement. Finally, the dataset provides a dataset including road conditions and vehicles, bicycles, and pedestrians over four kinds of German intersections. The roundD [21] includes the movement trajectories of cars, vans, trucks, buses, pedestrians, bicycles, and motorcycles in three traffic circles in Germany. In addition, positions, headings, speeds, accelerations, and classes of objects were extracted from the video and provided as data.

3 Proposed Algorithm

Figure 1 illustrates the overall process for classifying vehicle status. The process consists of four components: 1) Vehicle detection with YOLOv5, 2) Mosaic data augmentation, 3) Soft-label Assignment, and 4) a network for vehicle state classification. These four components work together to determine the movement status of vehicles. This chapter provides an explanation of the proposed methods in detail.

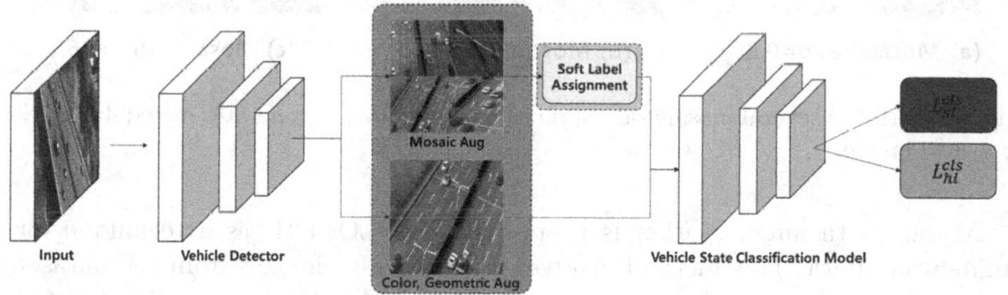

Fig. 1. An overview process of vehicle state classification that contains object detection, data augmentation (mosaic augmentation), soft label assignment, and VSNet (vehicle state network).

3.1 Vehicle Detection

This study first presents an approach for detecting vehicles using YOLOv5 [22], an advanced object detection algorithm that has achieved state-of-the-art performance on a variety of visual recognition tasks. YOLOv5 is an abbreviation for "You Only Look Once version 5", and is an extension of the original YOLO algorithm with improvements in speed and accuracy. It is based on a deep neural network architecture that efficiently extracts features from images and predicts object bounding boxes and class probabilities in a single forward pass.

The YOLOv5 algorithm comprises two main components: a feature extraction backbone and a detection head. The backbone network is built on efficient architecture, which has been shown to be highly efficient and effective in a wide range of vision tasks. The detection head employs anchor boxes and grid cells to predict object locations and classes at multiple scales. To adapt YOLOv5 for vehicle detection task, the train fine-tunes the model on a custom dataset of drone flight images using transfer learning. Vehicle types are limited to car, truck, and bus. Specifically, this work initializes the network with pre-trained weights on the COCO dataset [23] with advanced data augmentation and optimization techniques.

3.2 Mosaic Data Augmentation

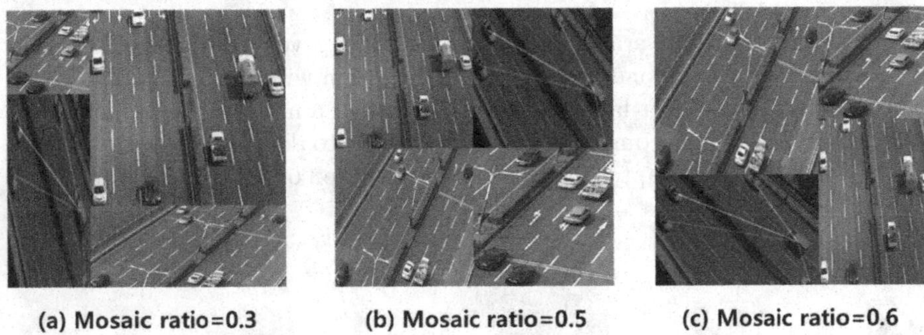

(a) Mosaic ratio=0.3 **(b) Mosaic ratio=0.5** **(c) Mosaic ratio=0.6**

Fig. 2. With the several mosaic ratios, mosaic augmentation generates mixed 4 images to a new image.

Mosaic data augmentation is proposed in YOLOv4 [1] as a technique for augmenting data. This method involves selecting four images from the dataset and arranging them in a manner that is determined by the mosaic ratio, denoted by $\mathcal{M}_r$ so that they are represented as a single image in Fig. 2. $\mathcal{M}_r$ is chosen randomly from the range [0.3, 0.7]. $I(i)$ is image among dataset at index i. Index, i is randomly selected. Every $I(i)$ is resized to 512×512. w_i and h_i denote the width and height of $I(i)$. $I(n)$ contains 4 images that n is an order, $n = 0, \ldots, 3$. Based on $\mathcal{M}_r$, the width and height sizes of each of the four images are determined as follows:

$$I(n) = \begin{cases} \text{new w}_0 = \mathcal{M}_r \times 512,\ \text{new h}_0 = \mathcal{M}_r \times 512 & \text{if } n \text{ is } 0 \\ \text{new w}_1 = 512 - \text{new w}_0,\ \text{new h}_1 = 512 - \text{new h}_3 & \text{if } n \text{ is } 1 \\ \text{new w}_2 = \mathcal{M}_r \times 512,\ \text{new h}_2 = 512 - \text{new h}_0 & \text{if } n \text{ is } 2 \\ \text{new w}_3 = 512 - \text{new w}_2,\ \text{new h}_3 = \mathcal{M}_r \times 512 & \text{if } n \text{ is } 3 \end{cases} \tag{1}$$

This approach enables the model to learn from multiple images simultaneously, improving generalization by incorporating diverse contextual information into a single image. Mosaic is applied by selecting random numbers in quantity equal to the batch size for each iteration.

3.3 Soft Label Assignment

After applying mosaic augmentation in this study, a method of soft label assignment for label allocation is proposed. In the mosaic image, four original images correspond to four labels. A soft label is created by referring to label smoothing [2]. The soft label, $\mathcal{S}(x)$ is shown in Eq. (2). In training sample x, $h_i(k|x)$ represents the hard label distribution of the four images at classes, $k \in 0, 1, 2$ and index of distribution i. The hyperparameter α is assigned a weight value between 0 and 1. The value of K, which denotes the number of images, is 4. Equation (2) are defined as the following:

$$\mathcal{S}(x) = \frac{1}{N}\sum_{i=1}^{N}\{(1-\alpha)h_i(k|x) + \alpha/K\} \tag{2}$$

The ground truth label distribution multiples the weight α and interpolates the hard label through the α/K. The label of the mosaic image generates a soft label by calculating the average value from $(1-\alpha)h_i(k|x) + \alpha/K$. Equation (2) represents the ground truth soft label, which adjusts the ground truth label distribution by applying label smoothing to mosaic images for classification models. In conclusion, these methods adapt the use of mosaic augmentation and soft label assignment resulting in improved classification model performance.

3.4 Vehicle State Classification

This classification model is adapted [24] as the previous work. The proposed model comprises the Wide Area Feature Extraction (WAFE) module and Deformable Residual (DR) module. These modules play critical roles in extracting and focusing on feature information. The following section provides a detailed layer-by-layer explanation of these modules.

Wide Area Feature Extraction Module (WAFE module). To classify the state of the target vehicle, the input image considers the position and state of the surrounding vehicles. Figure 5 shows that the vehicles in the image are mostly separated. To exclude unnecessary information like background, the first convolutional layer passes a 5×5 kernel size with 64 filters, a stride of 4, and a dilated ratio of 3. Next 1×1 convolutional layer extends the number of channels, 64 to 128. To stabilize the learning process on the feature map, batch normalization (BN) [25] is performed after all convolutional layers, and the proposed network employs Gaussian Error Linear Units (GELU) [26] as the activation function. The feature map is further processed by dividing the 32 channels into

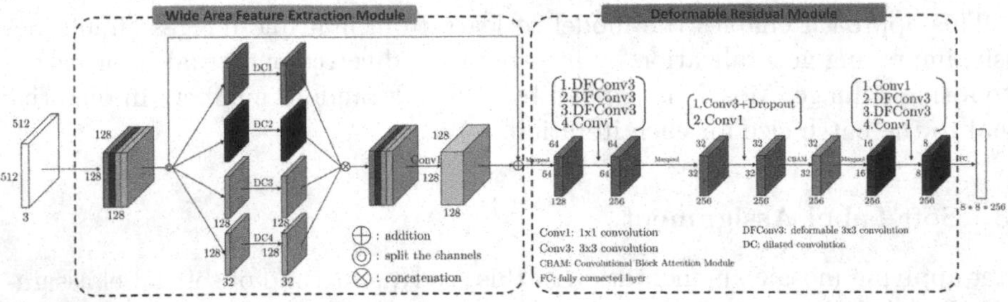

Fig. 3. The left module in the illustration is the WAFE module. It uses a dilated convolutional layer to appropriately extract features while reducing computation when objects in the image are far apart. The right module is the DR module, which is designed to extract meaningful features for vehicle status judgment by applying a variety of receptive fields using a deformable convolutional layer.

four groups, and each group is passed through four kinds of dilated ratio, [1, 3, 5, 7] of 3×3 convolutional layer. The four groups of outputs are concatenated, and a 1×1 kernel is applied. Additionally, a residual block is used to incorporate previous information before the maxpooling operation into the feature map.

Deformable Residual Module (DR Module). Deformable residual is modified from deformable convolutional layer [27] to extract flexible spatial information through output feature from WAFE module. As illustrated in Fig. 4(a), the traditional 3×3 convolutional layer has a fixed receptive field in the image area, represented by the red and blue dots. However, in the image data used for vehicle detection, the vehicles are often separated from each other. Therefore, using a fixed receptive field would extract feature information that includes unnecessary background information. To address this issue and perform more effective convolutional operations, deformable convolution is employed. Figure 4(b) shows how deformable convolution generates an offset as the convolutional layer and performs convolution operations through the offset information.

$$\mathbf{y}(\mathbf{p}_0) = \sum_{\mathbf{p}_n \in \mathcal{R}} \mathbf{w}(\mathbf{p}_n) \cdot \mathbf{x}(\mathbf{p}_0 + \mathbf{p}_n + \triangle\mathbf{p}_n) \tag{3}$$

Equation (3) represents the offset that determines the kernel position in deformable convolution. $\mathbf{y}$ represents the output feature map. The kernel grid, $\mathcal{R}$, is defined as the receptive field, where $\mathcal{R} = (-1,-1), (-1,0), \ldots, (0,1), (1,1)$. The convolution occurs at the pixel position of the input image $\mathbf{x}$, which is $\mathbf{p}_0$, and at the individual positions in $\mathcal{R}$, which are $\mathbf{p}_n$, along with the offset, $\triangle\mathbf{p}_n$. In particular, the offset $\triangle\mathbf{p}_n$ value is generated based on the convolution layer value and is trained in each iteration. Thus, $\mathbf{p}_0+\mathbf{p}_n+\triangle\mathbf{p}_n$ ultimately determines the position of the input value, and the convolution operation is performed by multiplying it with the convolution kernel weight, $\mathbf{w}(\mathbf{p}_n)$, at that position.

The Convolutional Block Attention Module (CBAM) [28] allows for complementary attention of both channel-wise and spatial-wise information and is

Fig. 4. (a) 3 × 3 conventional convolution, (b) 3 × 3 deformable convolution, In deformable convolution, deep red and dark blue dots are focused on the vehicle in the input image. (Color figure online)

applied through the output of three deformable convolutional layer operations. The fully connected layer receives the feature map that has been calculated by two deformable convolutional layers.

Loss Function. During training, hard and soft labels are utilized to adjust the loss function. For hard labels, the original loss function uses in this study is Focal Loss [29], which helps to balance the training process and prevent bias towards one class when dealing with data imbalance issues. For soft labels, the loss function is the mean squared error (MSE). Since the value of the soft labels is a float number, it has been computed average value. The proposed total loss function, $\mathcal{L}$ is as follows:

$$\mathcal{L} = \alpha_l \mathcal{L}_{hl}^{cls} + (1 - \alpha_l)\mathcal{L}_{sl}^{cls} \tag{4}$$

$\mathcal{L}_{hl}^{cls}$ is the loss calculated for the hard label, and $\mathcal{L}_{sl}^{cls}$ represents the loss result for the soft label. The parameter, α_l assigns weights to the hard and soft losses. Since soft labels are selected less frequently than hard labels, the weight of hard labels is higher. α_l is 0.9.

4 Experiment

Drone Image Dataset: The dataset for drone images is captured from a top-down perspective in Fig. 5, and vehicle detection is performed using YOLOv5 large model on the collected images. After detection, the image is cropped based on the five vehicles surrounding the target vehicle. The dataset consists of three classes: lane_change, safe, and stop, and the total number of training and test data is shown in Table 1.

Configuration Details. In this study, the Adam optimizer [30] is employed and the learning rate is set to 0.001. Epoch is 200. Four NVIDIA RTX 3090 GPU, each with 24GB of memory, are used, with a batch size of 16.

Table 1. Information of train and test dataset for vehicle state classification.

Class	train	test	Total
lane_change	**860**	**214**	**1,074**
safe	**1,241**	**310**	**1,551**
stop	**1,222**	**305**	**1,525**
Total	**3,323**	**829**	**4,152**

Fig. 5. Illustration for drone image dataset. The view in the picture is bird's-eye view.

Object Detection. In this study, YOLOv5 [22] is adopted as the object detection algorithm, and car and truck are the two classes considered for train and test. Table 3 presents the object detection performance for the train and test datasets. The training on 9,776 images a performance of 95.75 mAP(AP_{50}) and 83.8 mAP($AP_{50:95}$), while testing on 2,200 images a performance of 91.8 mAP(AP_{50}) and 80.3 mAP($AP_{50:95}$). Utilizing this detector, other traffic videos are analyzed to identify and extract vehicle information, including their position and class (Table 2).

VSNet Performance. The performance of the network for classifying the final vehicle state in Table 3 is compared to the Dilated Residual Network (DRN) [31]. DRN is a classification model derived from ResNet [32] and is a network that replaces the convolutional layers with dilated convolutional layers. Proposed model utilizes dilated and deformable convolutional layers to extract features from a wide area, and therefore, its network is compared with DRN, which is composed of dilated convolutional layers. DRN has four types, A, B, C, and D,

Table 2. The mAP performance of YOLOv5 on drone train and test dataset.

Class	Images	Instance	mAP@50	mAP@50:95
all_train	**9,776**	**309,470**	**95.75**	**83.8**
car_vehicle	**9,776**	**277,263**	**97.2**	**86.0**
truck_vehicle	**9,776**	**32,207**	**94.3**	**81.6**
all_test	**2,200**	**85,398**	**91.8**	**80.3**
car_vehicle	**2,200**	**78,765**	**96.1**	**85.3**
truck_vehicle	**2,200**	**6,633**	**87.5**	**75.4**

with additional dilated blocks and skip-connections. In this paper, types C and D are used, and type D is a simplified version of type C.

Compared to DRN_D_22, the first proposed model shows a 16.9% difference in accuracy results, but it reduces the number of parameters by 92.2%. In addition, the proposed model presents the results of applying mosaic and color data augmentation. When both augmentations are applied, it shows the best performance among the results presented, with an accuracy of 83.41%. Furthermore, compared to the DRN_C_42 model, it achieves a 1.63% higher accuracy and saves 96% of the parameters.

Table 4 presents the accuracy performance of the proposed model according to the soft label values. The highest accuracy performance of 83.41% is achieved when α is set to 0.7. As α gradually decreases, the performance decreases as well. This is because the soft label values differ from the original hard label values, leading to differences in learning performance.

Table 3. Comparison result with dilated residual networks (DRN) and vehicle state network (VSNet) for data augmentation.

Method	Data augmentation		#para	Acc(%)
	Mosaic	Color		
DRN_C_26	-	-	**21,126,584**	**89.62**
DRN_C_42	-	-	**31,234,744**	**81.78**
DRN_D_22	-	-	**16,393,752**	**87.69**
DRN_D_38	-	-	**26,501,912**	**86.49**
DRN_D_54	-	-	**35,809,176**	**89.26**
Proposed	-	-	**1,273,504**	**72.85**
Proposed	-	O	**1,273,504**	**79.73**
Proposed	O	-	**1,273,504**	**81.25**
Proposed	O	O	**1,273,504**	**83.41**

Table 4. According to Soft label, α, accuracy of vehicle state classification.

Method	Soft label = α	Acc (%)
Proposed	**0.3**	**79.62**
Proposed	**0.4**	**81.73**
Proposed	**0.5**	**83.12**
Proposed	**0.6**	**82.57**
Proposed	**0.7**	**83.41**

5 Conclusion

This study applies mosaic augmentation and soft-label assignment techniques to classify vehicle states using drone images. Mosaic augmentation combines existing images to create a new image, increasing the amount of data and improving generalization for a limited dataset. Additionally, soft-label assignment is used to generate labels for the mosaic images in vehicle state classification. These two techniques contribute to smooth training and enhance the accuracy performance of the proposed classification model.

Acknowledgment. This work was supported by the National Research Foundation of Korea(NRF) grant funded by the government(MSIT).(No.2020R1A2C200897212).

References

1. Bochkovskiy, A., Wang, C.-Y., Liao, H.-Y.M.: YOLOv4: optimal speed and accuracy of object detection. ArXiv, abs/2004.10934 (2020)
2. Szegedy, C., Vanhoucke, V., Ioffe, S., Shlens, J., Wojna, Z.: Rethinking the inception architecture for computer vision. In: 2016 IEEE Conference on Computer Vision and Pattern Recognition (CVPR), pp. 2818–2826 (2015)
3. Cordts, M.: The cityscapes dataset for semantic urban scene understanding. In: 2016 IEEE Conference on Computer Vision and Pattern Recognition (CVPR), pp. 3213–3223 (2016)
4. Geiger, A., Lenz, P., Stiller, C., Urtasun, R.: Vision meets robotics: The KITTI dataset. Int. J. Rob. Res. **32**, 1231–1237 (2013)
5. Chen, L.-C., Wang, H., Qiao, S.: Scaling wide residual networks for panoptic segmentation. ArXiv, abs/2011.11675 (2020)
6. Chen, Z.: Vision transformer adapter for dense predictions. ArXiv, abs/2205.08534 (2022)
7. Xu, J., Xiong, Z., Bhattacharyya, S.: PIDNet: a real-time semantic segmentation network inspired from PID controller. ArXiv, abs/2206.02066 (2022)
8. Li, S., Yan, Z., Li, H., Cheng, K.-T.: Exploring intermediate representation for monocular vehicle pose estimation. In: 2021 IEEE/CVF Conference on Computer Vision and Pattern Recognition (CVPR), pp. 1873–1883 (2020)
9. Zhang, Y., Zhang, Q., Zhu, Z., Hou, J., Yuan, Y.: GLENet: boosting 3D object detectors with generative label uncertainty estimation. ArXiv, abs/2207.02466 (2022)

10. Hong, Y., Dai, H., Ding, Y.: Cross-modality knowledge distillation network for monocular 3D object detection. ArXiv, abs/2211.07171 (2022)
11. Robicquet, A., Sadeghian, A., Alahi, A., Savarese, S.: Learning social etiquette: human trajectory understanding in crowded scenes. In: Leibe, B., Matas, J., Sebe, N., Welling, M. (eds.) ECCV 2016. LNCS, vol. 9912, pp. 549–565. Springer, Cham (2016). https://doi.org/10.1007/978-3-319-46484-8_33
12. Zhu, P.: Detection and tracking meet drones challenge. IEEE Trans. Pattern Anal. Mach. Intell., 1 (2020)
13. Cao, Y.: VisDrone-DET2021: the vision meets drone object detection challenge results. In: 2021 IEEE/CVF International Conference on Computer Vision Workshops (ICCVW), pp. 2847–2854 (2021)
14. Du, D., et al.: VisDrone-CC2021: the vision meets drone crowd counting challenge results, pp. 2830–2838 (2021)
15. Fan, H.: VisDrone-MOT2021: the vision meets drone multiple object tracking challenge results. In: 2021 IEEE/CVF International Conference on Computer Vision Workshops (ICCVW), pp. 2839–2846 (2021)
16. Wang, J., Xu, C., Yang, W., Yu, L.: A normalized Gaussian Wasserstein distance for tiny object detection. ArXiv, abs/2110.13389 (2021)
17. Lee, Y., Tang, Q., Choi, J.-W., Jo, K.: Low computational vehicle re-identification for unlabeled drone flight images. In: IECON 2022–48th Annual Conference of the IEEE Industrial Electronics Society, pp. 1–6 (2022)
18. Lee, Y., Tang, Q., Choi, J.-W., Jo, K.: Low computational vehicle lane changing prediction using drone traffic dataset. In: 2022 International Workshop on Intelligent Systems (IWIS), pp. 1–4 (2022)
19. Krajewski, R., Bock, J., Kloeker, L., Eckstein, L.: The highD dataset: a drone dataset of naturalistic vehicle trajectories on German highways for validation of highly automated driving systems. In: 2018 21st International Conference on Intelligent Transportation Systems (ITSC), pp. 2118–2125 (2018)
20. Bock, J., Krajewski, R., Moers, T., Runde, S., Vater, L., Eckstein, L.: The inD dataset: a drone dataset of naturalistic road user trajectories at German intersections. In: 2020 IEEE Intelligent Vehicles Symposium (IV), pp. 1929–1934 (2019)
21. Krajewski, R., Moers, T., Bock, J., Vater, L., Eckstein, L.: The round dataset: a drone dataset of road user trajectories at roundabouts in Germany. In: 2020 IEEE 23rd International Conference on Intelligent Transportation Systems (ITSC), pp. 1–6 (2020)
22. Jocher, G.R.: ultralytics/yolov5: v5.0 - YOLOv5-P6 1280 models, AWS, supervise.ly and YouTube integrations (2021)
23. Lin, T.-Y., et al.: Microsoft COCO: common objects in context. In: Fleet, D., Pajdla, T., Schiele, B., Tuytelaars, T. (eds.) ECCV 2014. LNCS, vol. 8693, pp. 740–755. Springer, Cham (2014). https://doi.org/10.1007/978-3-319-10602-1_48
24. Lee, Y., Kim, S., Choi, J., Jo, K.: Vehicle state classification from drone image. In: IEEE International Conference on Industrial Technology (ICIT), pp. 1–5 (2023)
25. Ioffe, S., Szegedy, C.: Batch normalization: accelerating deep network training by reducing internal covariate shift. In: International Conference on Machine Learning (2015)
26. Hendrycks, D., Gimpel, K.: Gaussian error linear units (GELUs). arXiv: Learning (2016)
27. Dai, J.: Deformable convolutional networks. In: 2017 IEEE International Conference on Computer Vision (ICCV), pp. 764–773 (2017)

28. Woo, S., Park, J., Lee, J.-Y., Kweon, I.S.: CBAM: convolutional block attention module. In: Ferrari, V., Hebert, M., Sminchisescu, C., Weiss, Y. (eds.) ECCV 2018. LNCS, vol. 11211, pp. 3–19. Springer, Cham (2018). https://doi.org/10.1007/978-3-030-01234-2_1
29. Lin, T.-Y., Goyal, P., Girshick, R.B., He, K., Dollár, P.: Focal loss for dense object detection. IEEE Trans. Pattern Anal. Mach. Intell. **42**, 318–327 (2017)
30. Kingma, D.P., Ba, J.: Adam: a method for stochastic optimization. CoRR, abs/1412.6980 (2014)
31. Yu, F., Koltun, V., Funkhouser, T.: Dilated residual networks. In: Computer Vision and Pattern Recognition (CVPR) (2017)
32. He, K., Zhang, X., Ren, S., Sun, J.: Deep residual learning for image recognition. In: 2016 IEEE Conference on Computer Vision and Pattern Recognition (CVPR), pp. 770–778 (2016)

Creating High-Resolution Adversarial Images Against Convolutional Neural Networks with the Noise Blowing-Up Method

Franck Leprévost, Ali Osman Topal, and Enea Mancellari(✉)

University of Luxembourg, House of Numbers, 6, avenue de la Fonte, 4364 Esch-sur-Alzette, Grand Duchy of Luxembourg
{Franck.Leprevost,Aliosman.Topal,Enea.Mancellari}@uni.lu

Abstract. Convolutional Neural Networks (CNNs) are widely used for image recognition tasks but are vulnerable to attacks. Most existing attacks create adversarial images of a size equal to the CNN's input size; mainly because creating adversarial images in the high-resolution domain leads to substantial speed, adversity, and visual quality challenges. In a previous work, we developed a method that lifts any existing attack working efficiently in the CNN's input size domain to the high-resolution domain. This method successfully addressed the first two challenges but only partially addressed the third one. The present article provides a crucial refinement of this strategy that, while keeping all its other features, substantially increases the visual quality of the obtained high-resolution adversarial images. The refinement amounts to a *blowing-up* to the high-resolution domain of the adversarial noise created in the low-resolution domain. Adding this blown-up noise to the clean original high-resolution image leads to an almost indistinguishable high-resolution adversarial image. The noise blowing-up strategy is successfully tested on an evolutionary-based black-box targeted attack against VGG-16 trained on ImageNet, with 10 high-resolution clean images.

Keywords: Black-box attack · Convolutional Neural Network · Evolutionary Algorithm · High resolution adversarial image · Noise Blowing-Up

1 Introduction

The profusion of images in today's society and the need to efficiently assess the information they contain for a large series of applications (self-driving cars, face recognition and security controls, satellite images, medical images, etc.) have led to the development of tools to automatically process and sort this type of data. Trained CNNs are among the most powerful and reliable tools available. Nevertheless, specifically designed adversarial images may lead CNNs to erroneous classifications, potentially resulting in catastrophic consequences.

N. T. Nguyen et al. (Eds.): ACIIDS 2023, LNAI 13995, pp. 121–134, 2023.
https://doi.org/10.1007/978-981-99-5834-4_10

Vice versa, efficient attacks reveal CNNs weaknesses, which in turn may lead to more robust CNNs.

Attacks depend on the scenario considered. For instance, starting with an original image classified by a CNN in a given category, the target scenario essentially consists of choosing a target category, different from the original one, and in creating a variant of the original image that the CNN will classify in the target category, although a human would classify this adversarial image still in the original category, or would be unable to notice any difference between the original and adversarial images. Attacks also depend on the level of knowledge of the CNN at the disposal of the attacker. While White-box attacks (see e.g., [2,16]) have full knowledge of the architecture of the CNN to attack (number and type of layers, weights, etc.), Black-box attacks [9,10] have no access to the CNN to attack and are therefore more challenging.

Our objective is to create adversarial images that closely resemble the original ones. Since original digital images in the real world are often in high-resolution, we focus on generating high-resolution adversarial images. Therefore, we aim at creating images, that can replace the original ones without losing visual quality while being able to deceive classification tools. Such achievements have significant potential in the context of privacy preservation, e.g. on social media where images are naturally of high resolution.

1.1 Standard Methodology

CNNs assess images by initially resizing them to fit their input size. In particular, high-resolution images are down-scaled, such as to 224×224 for most ImageNet-trained CNNs. So far, all attacks - black box or otherwise - have involved images of moderate size, or resized to values that CNNs handle natively, what we call here the "low-resolution" $\mathcal{R}$ domain. The construction of adversarial images is then achieved by adding some carefully designed adversarial noise to the potentially resized original image. In particular, the adversarial noise created by all these attacks is in the "low resolution" domain handled natively by the CNNs so that the obtained adversarial images are as large as the CNN's input size. In particular, these attacks explore a search space of size that does not depend on the size of the original image, but that coincides with the size of the CNN input.

1.2 Three Challenges

Creating adversarial images of large size (with any type of attack) leads to three challenges regarding speed, adversity, and visual quality. Firstly, the complexity of the problem increases quadratically with the size of the images, which of course impacts the speed of the attacks. For instance, we showed in [12] that an EA-based attack, that succeeded in creating adversarial images in the 224×224 domain, did not even indicate any convincing sign of potential success after 40 hours for any of the high-resolution image in Table 1. Secondly, the adversarial noise, introduced in the "high resolution" $\mathcal{H}$ domain, should "survive" the downsizing process from $\mathcal{H}$ to $\mathcal{R}$ to fit the CNN. Thirdly, the noise introduced in the "high resolution" domain should be indiscernible to the human eye when viewing the images at their original size, not only when they are scaled down to fit in the "low resolution" domain.

1.3 Our Contribution

Our previous works [11,12] provided the design of the first effective strategy that lifts to the high-resolution domain any existing attack working efficiently in the CNN's input size domain. This was achieved by lifting an adversarial image obtained in the $\mathcal{R}$ domain to an adversarial image in the $\mathcal{H}$ domain. This approach successfully addressed the first two challenges of speed and adversity. However, it only partially addressed the third challenge of visual quality in the $\mathcal{H}$ domain.

Our contributions to the present article are twofold. Firstly, we provide a substantial refinement of the strategy given in [11,12] that, while keeping all its other features – in particular it continues to lift to the high-resolution domain any attack working in the CNN's input size domain –, substantially increases the visual quality of the high-resolution adversarial images, as well as the speed and efficiency in creating them. The refinement amounts to a "blowing-up" to the high-resolution domain of the adversarial noise – only of the adversarial noise, and not of the full adversarial image—created in the low-resolution domain. Adding this high-resolution noise to the original high-resolution image leads to a tentative high-resolution adversarial image.

Secondly, we apply this adversarial noise blowing-up strategy to one black-box attack for the target scenario against VGG-16 trained on ImageNet. We use the same 10 high-resolution clean images as in [11,12], and run the attack 10 times for each clean image. We then show that the obtained tentative high-resolution adversarial images are indeed adversarial.

To illustrate the visual quality of adversarial images obtained by this refined approach, we consider a challenging example of a high-resolution image. We compare this clean image with the HR adversarial image obtained by the method of [11,12] on the one hand, and with the HR adversarial image obtained by the new method on the other hand. We demonstrate that our new method creates high-resolution adversarial images of enhanced visual quality.

1.4 Organisation of the Paper

Section 2 briefly recalls some standard attack scenarios in $\mathcal{R}$, clarifies what are their lifted version to $\mathcal{H}$, and fixes some notations. Section 3 formalizes the noise blowing-up method and provides the scheme of our attack $atk^{scenario}_{\mathcal{H},\mathcal{C}}$ that lifts to $\mathcal{H}$ any attack $atk^{scenario}_{\mathcal{R},\mathcal{C}}$ against a CNN $\mathcal{C}$ that works in the $\mathcal{R}$ domain, and that takes advantage of lifting the adversarial noise only. It sets the main indicators used to assess the quality of the obtained tentative adversarial images.

Section 4 presents a case study. The noise blowing-up strategy is applied for the target scenario to the evolutionary algorithm-based attack presented already in [3,5]. To illustrate the gain in visual quality provided by our new approach, one sample is detailed in Sect. 5. Section 6 summarizes our findings and indicates directions for future research.

All algorithms and experiments were implemented using Python 3.8 [18] with NumPy 1.17 [13], TensorFlow 2.4 [1], Keras 2.2 [6], and Scikit 0.24 [19] libraries. Computations were performed on nodes with Nvidia Tesla V100 GPGPUs of the IRIS HPC Cluster at the University of Luxembourg.

2 CNNs and Attack Scenarios

CNNs used for image classification undergo training on a large dataset, denoted as $\mathcal{S}$, to categorize images into predetermined categories $c_1, \cdots, c_\ell$. The categories, along with their index number ℓ, are specifically associated with dataset $\mathcal{S}$ and remain consistent across all CNN models trained on $\mathcal{S}$. One denotes by $\mathcal{R}$ the set of images of size $r_1 \times r_2$ (where r_1 is the height and r_2 is the width of the image) natively adapted to such CNNs.

Once trained, a CNN can be exposed to images (typically) in the same domain $\mathcal{R}$ as those on which it was trained. Given an input image $\mathcal{I} \in \mathcal{R}$, the trained CNN produces a classification output vector

$$\mathbf{o}_{\mathcal{I}} = (\mathbf{o}_{\mathcal{I}}[1], \cdots, \mathbf{o}_{\mathcal{I}}[\ell]), \tag{1}$$

where $0 \leq \mathbf{o}_{\mathcal{I}}[i] \leq 1$ for $1 \leq i \leq \ell$, and $\sum_{i=1}^{\ell} \mathbf{o}_{\mathcal{I}}[i] = 1$. Each c_i-label value $\mathbf{o}_{\mathcal{I}}[i]$ measures the plausibility that the image $\mathcal{I}$ belongs to the category c_i.

Consequently, the CNN classifies the image $\mathcal{I}$ as belonging to the category c_k if $k = \arg\max_{1 \leq i \leq \ell}(\mathbf{o}_{\mathcal{I}}[i])$. If there is no ambiguity on the dominating category, one denotes $(c_k, \mathbf{o}_{\mathcal{I}}[k])$ the pair specifying the dominating category and the corresponding label value. In this case, we consider that $\mathcal{C}$'s classification of $\mathcal{I}$ is

$$\mathcal{C}(\mathcal{I}) \in \mathcal{V} = \{(c_i, v_i), \text{ where } v_i \in]0, 1] \text{ for } 1 \leq i \leq \ell\}. \tag{2}$$

The higher the c_k-label value $\mathbf{o}_{\mathcal{I}}[k]$, the higher the confidence that $\mathcal{I}$ represents an object of the category c_k.

Remark. The dominant category is without ambiguity for most images used in practice. Still, the situation differs when there are different categories for which their corresponding label values while being larger than the remaining ones, are almost equal between themselves. This occurs *a fortiori* when all ℓ label values are almost equi-distributed, like for instance for adversarial images created in the context of the *flat scenario* (see Subsect. 2.2). If $\mathcal{I}$ is such an image, then one considers instead that:

$$\mathcal{C}(\mathcal{I}) \in \mathcal{V} = \{((c_1, v_1), \cdots, (c_\ell, v_\ell)), \text{ where } v_i \in]0, 1] \text{ for } 1 \leq i \leq \ell\}. \tag{3}$$

2.1 Assessment of the Human Perception of Distinct Images

Given two images $\mathcal{A}$ and $\mathcal{D}$ of the same size (belonging or not to the $\mathcal{R}$ domain), there are different methods to numerically assess the human perception of the difference between them. In the present study, this assessment is performed mainly by computing the values of $L_p(\mathcal{A},\mathcal{D})$ for $p = 1, 2$, or ∞. In a nutshell, the L_p-distance measures the difference between the pixel values of $\mathcal{A}$ and $\mathcal{D}$ as follows, where $\mathcal{I}(r)$ represents the value of the rth-pixel of the image $\mathcal{I}$:

$$\begin{cases} L_p(\mathcal{A},\mathcal{D}) &= (\sum_r |\mathcal{A}(r) - \mathcal{D}(r)|^p)^{1/p} \text{ for } p = 1, 2. \\ L_\infty(\mathcal{A},\mathcal{D}) &= \text{Max}_r|\mathcal{A}(r) - \mathcal{D}(r)|. \end{cases} \tag{4}$$

2.2 Attack Scenarios in the $\mathcal{R}$ Domain

Let $\mathcal{C}$ be a trained CNN, c_a be a category among the ℓ possible categories, and $\mathcal{A}$ be a clean image in the $\mathcal{R}$ domain classified by $\mathcal{C}$ as belonging to c_a. Let τ_a be its c_a-label value. Based on these initial conditions, we describe three attack scenarios aiming at creating an adversarial image $\mathcal{D} \in \mathcal{R}$ accordingly.

Whatever the scenario, one requires that $\mathcal{D}$ remains so close to $\mathcal{A}$, that a human would not notice any difference between $\mathcal{A}$ and $\mathcal{D}$. This is performed in practice by fixing the value of the parameter ϵ, that controls (or restricts) the global maximum amplitude allowed for the value modifications of each individual pixel of $\mathcal{A}$ to obtain the adversarial image $\mathcal{D}$. For a given attack scenario, note that the value set to ϵ usually depends on the concrete performed attack. It depends more specifically on the L_p distance used in the attack to assess the human perception between the original image and the adversarial image.

The (c_a, c_t) *target scenario* performed on $\mathcal{A}$ requires first to select a category $c_t \neq c_a$. The attack then aims at constructing an image $\mathcal{D}$ that is either a *good enough adversarial image* or a *τ-strong adversarial image.*

A *good enough adversarial image* is an image that, when subjected to classification by $\mathcal{C}$, is classified as belonging to the target category c_t, without any strict requirement on the specific label value of c_t, as long as it is dominant compared to all other label values. An adversarial image is considered a τ-strong adversarial image if it is classified by classifier $\mathcal{C}$ as belonging to the target category c_t and its label value for the c_t label, denoted as τ_t, is equal to or greater than a predetermined threshold value τ. Here, τ is a fixed value between 0 and 1 (exclusive) that is determined beforehand.

In the *untarget scenario* performed on $\mathcal{A}$, the attack aims at constructing an image $\mathcal{D}$ that $\mathcal{C}$ classifies in any category $c \neq c_a$.

In the *flat scenario* performed on $\mathcal{A}$, the attacks aim at constructing an image $\mathcal{D}$ that $\mathcal{C}$ is unable to classify in any category with sufficient confidence. In other words, for $\mathcal{D}$, all categories are likely possible. Put otherwise, one has $\mathbf{o}_{\mathcal{D}}[i] \simeq \frac{1}{\ell}$ for all $1 \leq i \leq \ell$.

One writes $atk^{scenario}_{\mathcal{R},\mathcal{C}}$ the specific attack performed to deceive $\mathcal{C}$ in the $\mathcal{R}$ domain according to the selected scenario, and $atk^{scenario}_{\mathcal{R},\mathcal{C}}(\mathcal{A})$ the adversarial image obtained by running successfully this attack on the clean image $\mathcal{A}$.

2.3 Attack Scenarios Expressed in the $\mathcal{H}$ Domain

In the context of high resolution (HR) images, let us denote by $\mathcal{H}$ the set of images that are larger than those of $\mathcal{R}$. In other words, an image of size $h \times w$

belongs to $\mathcal{H}$ if $h \geq r_1$ and $w \geq r_2$. One assumes given a fixed *degradation function*

$$\rho : \mathcal{H} \longrightarrow \mathcal{R}, \tag{5}$$

that transforms any image $\mathcal{I} \in \mathcal{H}$ into a "degraded" image $\rho(\mathcal{I}) \in \mathcal{R}$. Then there is a well-defined composition of maps $\mathcal{C} \circ \rho$.

Given $\mathcal{A}_a^{\text{hr}} \in \mathcal{H}$, one obtains that way the classification of the reduced image $\mathcal{A}_a = \rho(\mathcal{A}_a^{\text{hr}}) \in \mathcal{R}$ as $\mathcal{C}(\mathcal{A}_a) \in \mathcal{V}$. Although not mandatory, we shall assume, for the sake of simplicity, that the dominating category of the reduced image $\mathcal{A}_a$ is without ambiguity. Therefore, let $\mathcal{C}(\mathcal{A}_a) = (c_a, \tau_a) \in \mathcal{V}$ be the outcome of $\mathcal{C}$'s classification of $\mathcal{A}_a$.

An adversarial HR image against $\mathcal{C}$ for the (c_a, c_t) *target scenario* performed by an attack $atk_{\mathcal{H},\mathcal{C}}^{target}$ on $\mathcal{A}_a^{\text{hr}} \in \mathcal{H}$ is an image $\mathcal{D}_t^{\text{hr},\mathcal{C}}(\mathcal{A}_a^{\text{hr}}) = atk_{\mathcal{H},\mathcal{C}}^{target}(\mathcal{A}_a^{\text{hr}}) \in \mathcal{H}$, that satisfies two conditions.

On the one hand, a human should not be able to notice any visual difference between the original $\mathcal{A}_a^{\text{hr}}$ and the adversarial $\mathcal{D}_t^{\text{hr},\mathcal{C}}(\mathcal{A}_a^{\text{hr}})$ HR images. On the other hand, $\mathcal{C}$ should classify the degraded image $\mathcal{D}_t^{\mathcal{C}}(\mathcal{A}_a^{\text{hr}}) = \rho(\mathcal{D}_t^{\text{hr},\mathcal{C}}(\mathcal{A}_a^{\text{hr}}))$ in the category c_t for a sufficiently convincing c_t-label value. The (c_a, c_t) *target scenario* performed on the HR image $\mathcal{A}_a^{\text{hr}}$ can be visualized by the following scheme.

$$\begin{array}{ccc}
\mathcal{A}_a^{\text{hr}} \in \mathcal{H} & \xrightarrow{atk_{\mathcal{H},\mathcal{C}}^{target}} & \mathcal{D}_t^{\text{hr},\mathcal{C}}(\mathcal{A}_a^{\text{hr}}) \in \mathcal{H} \\
\rho \downarrow & & \downarrow \rho \\
\mathcal{A}_a \in \mathcal{R} & & \mathcal{D}_t^{\mathcal{C}}(\mathcal{A}_a^{\text{hr}}) \in \mathcal{R} \\
\mathcal{C} \downarrow & & \downarrow \mathcal{C} \\
(c_a, \tau_a) \in \mathcal{V} & & (c_t, \tau_t) \in \mathcal{V}
\end{array}$$

The image $\mathcal{D}_t^{\text{hr},\mathcal{C}}(\mathcal{A}_a^{\text{hr}}) \in \mathcal{H}$ is then a *good enough adversarial image* or a τ-*strong adversarial image* if its reduced version $\mathcal{D}_t^{\mathcal{C}}(\mathcal{A}_a^{\text{hr}}) = \rho(\mathcal{D}_t^{\text{hr},\mathcal{C}}(\mathcal{A}_a^{\text{hr}}))$ is.

Thanks to the degradation function ρ, one can express in a similar way in the $\mathcal{H}$ domain any attack scenario that makes sense in the $\mathcal{R}$ domain. This holds in particular for the *untarget scenario* and for the *flat scenario*. One denotes by $\mathcal{D}_{\text{untarget}}^{\text{hr},\mathcal{C}}(\mathcal{A}_a^{\text{hr}}) = atk_{\mathcal{H},\mathcal{C}}^{untarget}(\mathcal{A}^{\text{hr}})$ the HR adversarial images obtained by an attack $atk_{\mathcal{H},\mathcal{C}}^{untarget}$ for the untarget scenario performed on $\mathcal{A}_a^{\text{hr}} \in \mathcal{H}$, and by $\mathcal{D}_{\text{untarget}}^{\mathcal{C}}(\mathcal{A}_a^{\text{hr}}) \in \mathcal{R}$ its degraded version. *Mutatis mutandis*, one denotes by $\mathcal{D}_{\text{flat}}^{\text{hr},\mathcal{C}}(\mathcal{A}_a^{\text{hr}}) = atk_{\mathcal{H},\mathcal{C}}^{flat}(\mathcal{A}^{\text{hr}})$ the HR adversarial images obtained by an attack $atk_{\mathcal{H},\mathcal{C}}^{flat}$ for the flat scenario performed on $\mathcal{A}_a^{\text{hr}} \in \mathcal{H}$, and by $\mathcal{D}_{\text{flat}}^{\mathcal{C}}(\mathcal{A}_a^{\text{hr}}) \in \mathcal{R}$ its degraded version.

3 The Noise Blowing-Up Strategy

We present here a method that attempts to circumvent the speed, adversity, and visual quality challenges, that are encountered when one intends to create HR adversarial images. While speed and adversity were successfully addressed in [11, 12] *via* a strategy similar to some extent to the present one, the visual quality challenge remained partly unsatisfying. The refinement provided by the noise-blowing up method presented here addresses this issue, simplifies and generalises the attack scheme described in [11,12], and lifts to the $\mathcal{H}$ domain any attack working in the $\mathcal{R}$ domain.

The design of the noise blowing-up strategy, that aims at creating, in seven steps, an efficient attack in the $\mathcal{H}$ domain once given an efficient attack in the $\mathcal{R}$ domain, is given in Subsect. 3.1. The description of the process is detailed here for the challenging *target scenario* (any other scenario can easily be derived from the presented scheme). Subsection 3.2 gives a series of indicators. The assessment of these indicators depends on the choice of the degrading and enlarging functions used to move from $\mathcal{H}$ to $\mathcal{R}$, and *vice versa.* These choices are made in the experiments performed in Sect. 4.

3.1 Construction of Adversarial Images in $\mathcal{H}$ for the Target Scenario

Given a CNN $\mathcal{C}$, the starting point is a large-size clean image $\mathcal{A}_a^{\text{hr}} \in \mathcal{H}$.
In Step 1, one constructs its degraded image $\mathcal{A}_a = \rho(\mathcal{A}_a^{\text{hr}}) \in \mathcal{R}$.
In Step 2, one runs $\mathcal{C}$ on $\mathcal{A}_a$ to get its classification in a category c_a. More precisely, one gets $\mathcal{C}(\mathcal{A}_a) = (c_a, \tau_a)$.
In Step 3, one assumes given an image $\widetilde{\mathcal{D}}^{\mathcal{C}}_{t,\tilde{\tau}_t}(\mathcal{A}_a) \in \mathcal{R}$, that is adversarial for the (c_a, c_t) target scenario performed on $\mathcal{A}_a = \rho(\mathcal{A}_a^{\text{hr}})$ for a c_t-label value $\tilde{\tau}_t$ exceeding a threshold $\tilde{\tau}$. As already stated, it does not matter how such an adversarial image is obtained.
Step 4 consists in getting the adversarial noise $\mathcal{N}^{\mathcal{C}}(\mathcal{A}_a) \in \mathcal{R}$ as the difference

$$\mathcal{N}^{\mathcal{C}}(\mathcal{A}_a) = \widetilde{\mathcal{D}}^{\mathcal{C}}_{t,\tilde{\tau}_t}(\mathcal{A}_a) - \mathcal{A}_a \in \mathcal{R} \tag{6}$$

of images living in $\mathcal{R}$, one being the adversarial image of the clean other.
To perform Step 5, one needs a fixed *enlarging function*

$$\lambda: \ \mathcal{R} \longrightarrow \mathcal{H} \tag{7}$$

that transforms any image of $\mathcal{R}$ into an image in $\mathcal{H}$ (see Sect. 4.1 for the specific used λ function). Anticipating Step 4, it is worth noting that, although the *reduction function* ρ and the *enlarging function* λ have opposite purposes, these functions are not necessarily inverse one from the other. In other words, $\rho \circ \lambda$ and $\lambda \circ \rho$ may differ from the identity maps $id_{\mathcal{R}}$ and $id_{\mathcal{H}}$ respectively (usually they do).

One applies the enlarging function λ to the low-resolution adversarial noise $\mathcal{N}^{\mathcal{C}}(\mathcal{A}_a)$, what leads to the blown-up noise $\mathcal{N}^{hr,\mathcal{C}}(\mathcal{A}_a^{\text{hr}}) = \lambda(\mathcal{N}^{\mathcal{C}}(\mathcal{A}_a)) \in \mathcal{H}$. Then one creates the HR tentative adversarial image by adding this blown-up noise to the original high-resolution image as follows:

$$\mathcal{D}^{\text{hr},\mathcal{C}}_{t,\tau_t}(\mathcal{A}_a^{\text{hr}}) = \mathcal{A}_a^{\text{hr}} + \mathcal{N}^{hr,\mathcal{C}}(\mathcal{A}_a^{\text{hr}}) \in \mathcal{H}. \tag{8}$$

In Step 6, the application of the reduction function ρ to this HD tentative adversarial image creates an image $\mathcal{D}^{\mathcal{C}}_{t,\tau_t}(\mathcal{A}^{\mathrm{hr}}_a) = \rho(\mathcal{D}^{\mathrm{hr},\mathcal{C}}_{t,\tau_t}(\mathcal{A}^{\mathrm{hr}}_a))$ in the $\mathcal{R}$ domain. In Step 7, one runs $\mathcal{C}$ on $\mathcal{D}^{\mathcal{C}}_{t,\tau_t}(\mathcal{A}^{\mathrm{hr}}_a)$ to get its classification.

The attack succeeds if $\mathcal{C}$ classifies this image in c_t, potentially for a c_t-label value τ_t exceeding the threshold value τ fixed in advance, and if a human is unable to notice any difference between the images $\mathcal{A}^{\mathrm{hr}}_a$ and $\mathcal{D}^{\mathrm{hr},\mathcal{C}}_{t,\tau_t}(\mathcal{A}^{\mathrm{hr}}_a)$ in the $\mathcal{H}$ domain. The key point is to set the value of $\tilde{\tau}_t$ so that this occurs.

The following scheme, summarizing the seven steps, shows how to create, from a targeted attack $atk^{target}_{\mathcal{R},\mathcal{C}}$ efficient against $\mathcal{C}$ in the $\mathcal{R}$ domain, the attack $atk^{target}_{\mathcal{H},\mathcal{C}}$ in the $\mathcal{H}$ domain obtained by the noise blowing-up method:

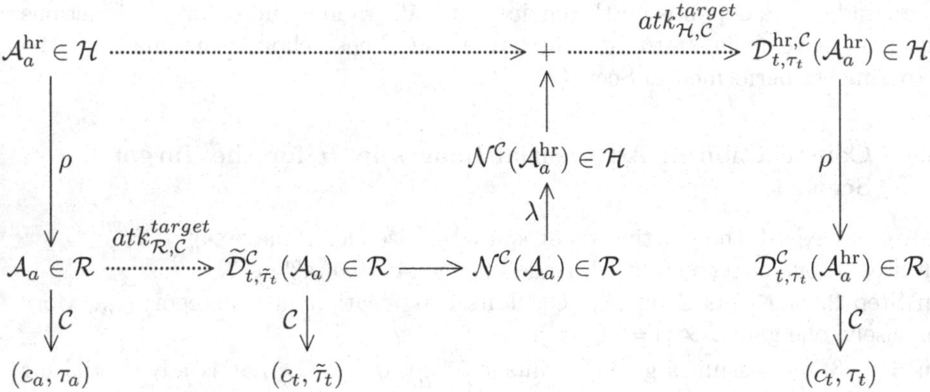

$$
\begin{array}{ccccccc}
\mathcal{A}^{\mathrm{hr}}_a \in \mathcal{H} & \dashrightarrow & & & + & \overset{atk^{target}_{\mathcal{H},\mathcal{C}}}{\dashrightarrow} & \mathcal{D}^{\mathrm{hr},\mathcal{C}}_{t,\tau_t}(\mathcal{A}^{\mathrm{hr}}_a) \in \mathcal{H} \\
 & & & & \uparrow & & \\
\downarrow \rho & & & & \mathcal{N}^{\mathcal{C}}(\mathcal{A}^{\mathrm{hr}}_a) \in \mathcal{H} & & \downarrow \rho \\
 & & & & \uparrow \lambda & & \\
\mathcal{A}_a \in \mathcal{R} & \overset{atk^{target}_{\mathcal{R},\mathcal{C}}}{\dashrightarrow} & \widetilde{\mathcal{D}}^{\mathcal{C}}_{t,\tilde{\tau}_t}(\mathcal{A}_a) \in \mathcal{R} & \longrightarrow & \mathcal{N}^{\mathcal{C}}(\mathcal{A}_a) \in \mathcal{R} & & \mathcal{D}^{\mathcal{C}}_{t,\tau_t}(\mathcal{A}^{\mathrm{hr}}_a) \in \mathcal{R} \\
\downarrow \mathcal{C} & & \downarrow \mathcal{C} & & & & \downarrow \mathcal{C} \\
(c_a, \tau_a) & & (c_t, \tilde{\tau}_t) & & & & (c_t, \tau_t)
\end{array}
$$

3.2 Indicators

Although both $\widetilde{\mathcal{D}}^{\mathcal{C}}_{t,\tilde{\tau}_t}(\mathcal{A}_a)$ and $\mathcal{D}^{\mathcal{C}}_{t,\tau_t}(\mathcal{A}^{\mathrm{hr}}_a)$ stem from $\mathcal{A}^{\mathrm{hr}}_a$, and belong to the same set $\mathcal{R}$ of low-resolution images, these images nevertheless differ in general, since $\rho \circ \lambda \neq id_{\mathcal{R}}$. Therefore, the verification process performed in Step 7 on the HR tentative adversarial image, which checks whether its reduction belongs to c_t, is mandatory. Moreover, should it be the case, $\tilde{\tau}_t$ and τ_t are likely to differ. The real-valued *loss function* $\mathcal{L}$ defined for $\mathcal{A}^{\mathrm{hr}}_a \in \mathcal{H}$ gives the difference:

$$\mathcal{L}_{\mathcal{C}}(\mathcal{A}^{\mathrm{hr}}_a) = \tilde{\tau}_t - \tau_t. \tag{9}$$

Our attack is effective if one can set accurately the value of $\tilde{\tau}_t$ to match the inequality $\tau_t \geq \tau$ for the threshold value τ, or to make sure that $\mathcal{D}^{\mathcal{C}}_{t,\tau_t}(\mathcal{A}^{\mathrm{hr}}_a)$ is a good enough adversarial image in the $\mathcal{R}$ domain, while controlling the distance variations between $\mathcal{A}^{\mathrm{hr}}_a$ and the adversarial $\mathcal{D}^{\mathrm{hr},\mathcal{C}}_{t,\tau_t}(\mathcal{A}^{\mathrm{hr}}_a)$.

Additionally, the visual proximity between images for a human eye is assessed by L_p distances (see Subsect. 2.1). There are two pairs of images that one wants to compare. On the one hand, there is the pair $(\mathcal{A}_a, \mathcal{D}^{\mathcal{C}}_{t,\tau_t}(\mathcal{A}^{\mathrm{hr}}_a))$ of images in the $\mathcal{R}$ domain, for which one uses the same L_p distance as in the attack $atk^{target}_{\mathcal{R},\mathcal{C}}$. On the other hand, there is the pair $(\mathcal{A}^{\mathrm{hr}}_a, \mathcal{D}^{\mathrm{hr},\mathcal{C}}_{t,\tau_t}(\mathcal{A}^{\mathrm{hr}}_a))$ of images in the $\mathcal{H}$ domain, for which one uses the L_2 distance systematically. In this case, the most important of both actually, one writes more simply $L_2^{hr} = L_2(\mathcal{A}_a^{\text{hr}}, \mathcal{D}_{t,\tau_t}^{\text{hr},\mathcal{C}}(\mathcal{A}_a^{\text{hr}}))$ when there is no ambiguity.

Note that the present approach, unlike the approach introduced in [11,12], does not require frequently scale up and down via λ, ρ the adversarial images. In particular, if one knows how the loss function behaves (in the worst case, or on average) for a given attack, then one can adjust *a priori* the value of $\tilde{\tau}_t$ accordingly, and be satisfied with one scaling up and down.

4 Case Study

This section provides a (first) proof of concept of our noise blowing-up strategy with one CNN, one scenario, one attack and 10 HR images.

4.1 The CNN, the Scenario, the Images

We consider $\mathcal{C}$ = VGG-16 trained on ImageNet [7], and the 10 clean HR images $\mathcal{A}_1^{\text{hr}}, \cdots, \mathcal{A}_{10}^{\text{hr}}$ pictured in Table 1. These images, including the two images $\mathcal{A}_9^{\text{hr}}$ and $\mathcal{A}_{10}^{\text{hr}}$ graciously provided by the French artist Speedy Graphito [15], are those considered in [11,12]. More precisely, Table 1 gives 10 categories $c_1, \cdots, c_{10}$, and, for each c_a, it gives a HR image $\mathcal{A}_a^{\text{hr}}$, whose degraded version is classified by $\mathcal{C}$ = VGG-16 in c_a. Taking advantage of the outcomes of [11,12] for the choice of most parameters used in the case study, we use (ρ, λ) = (Lanczos,Lanczos) (see [8,14] for the Lanczos method). Table 1 gives the original size of $\mathcal{A}_a^{\text{hr}}$, the classification (c_a, τ_a) by VGG-16 of $\rho(\mathcal{A}_a^{\text{hr}})$, and the category c_t used for the (c_a, c_t) target scenario (identical to those used in [11,12], picked at random among the categories of ImageNet).

Table 1. For $1 \leq a \leq 10$, the image $\mathcal{A}_a^{\text{hr}}$ classified by VGG-16 in the category c_a, and their respective target categories c_t.

a	1	2	3	4	5	6	7	8	9	10
c_a	Cheetah	Eskimo Dog	Koala	Lamp Shade	Toucan	Screen	Comic Book	Sports Car	Binder	Coffee Mug
$h \times w$	604×910	640×960	607×910	2913×2462	607×910	600×641	800×1280	800×1280	2011×1954	1710×1740
$\mathcal{A}_a^{hr}$										
τ_a	0.9527	0.3434	0.9974	0.5359	0.4553	0.7064	0.4916	0.4802	0.2825	0.0844
c_t	poncho	goblet	Weimaraner	weevil	wombat	swing	altar	beagle	triceratops	hamper

4.2 The Attack

We apply the noise blowing-up strategy with the black-box evolutionary algorithm (EA) based attack developed in [4,17]. For this EA attack, we keep the same parameters as those of [4,17]: $\alpha = 1$, $\epsilon = 16$, and $X = 20.000$. The pseudocode of the EA-based attack, expressed in the $\mathcal{R}$ domain, is given as follows:

Algorithm 1. EA attack pseudocode [4,17]

```
Input: CNN $\mathcal{C}$, ancestor $\mathcal{A}$, perturbation magnitude $\alpha$, maximum perturbation $\epsilon$,
  ancestor class $c_a$, ordinal $t$ of target class $c_t$, $g$ current and $X$ maximum generation;
Initialize population as 40 copies of $\mathcal{A}$, with $I_0$ as first individual;
Compute fitness for each individual;
while $(o_{I_0}[t] < \tau)$ & $x < X$ do
    Rank individuals in descending fitness order and segregate: elite 10, middle class
      20, lower class 10;
    Select random number of pixels to mutate and perturb them with $\pm\alpha$. Clip all
      mutations to $[-\epsilon, \epsilon]$. The elite is not mutated. The lower class is replaced with
      mutated individuals from the elite and middle class;
    Cross-over individuals to form a new population;
    Evaluate fitness of each individual;
```

We set $\tilde{\tau} = 0.55$ to ensure that the $\tilde{\tau}_t$-strong adversarial images, obtained by this attack in the $\mathcal{R}$ domain, are clearly in the c_t target category, with a convincing margin ≥ 0.10 with respect to the second best category. Since different seed values for the EA lead to different adversarial images, to ensure the reliability of our results, we performed, for each clean HR image $\mathcal{A}_a^{\text{hr}}$, and each (c_a, c_t) pair, 10 independent runs with random seed values. The EA succeeded in all cases, creating a total of 100 adversarial images, 10 for each clean image $\mathcal{A}_a^{\text{hr}}$.

4.3 Experimental Results

Referring to the steps specified in Subsect. 3.1, for each ancestor image $\mathcal{A}_a^{\text{hr}}$ specified in their 1st column, Table 2 and Table 3 summarize the results of the case study, computed as averages over the 10 independent runs. Note that Step 3, which corresponds to the concrete attack performed in the $\mathcal{R}$ domain, should be considered essentially as "outside" our strategy, in the sense that it is an input on which we have no influence *a priori*. Therefore the computational efforts performed in this Step 3 do not impact the performance of our scheme.

In Table 2, the 2nd column, which corresponds to Step 3 of the noise blowing-up strategy, gives the average number of generations required by the EA to create a 0.55-strong adversarial image for the (c_a, c_t) target scenario (note that the two artistic images are the most challenging of all). The 3rd column gives the average value of $\tilde{\tau}_t$, which of course exceeds $\tilde{\tau} = 0.55$ as expected. The 4th column provides the average c_t-label value for the degraded adversarial images. The 5th column gives the average loss (Eq. 9). This difference between the adversarial images in $\mathcal{R}$ varies between 0.0132 and 0.1950. Still, in all cases, the degraded adversarial image remained classified in the target category c_t. The last column assesses the visual quality difference between the HR clean and adversarial images.

Table 2. Average $\tilde{\tau}_t$ and τ_t with corresponding loss values and average L_2 distances between the ancestor and adversarial images in the HR domain.

	$avgGens^{0.55}$	$avg\_\tilde{\tau}_t$	avg_τ_t	$avg\_\mathcal{L}$	$avg\_L_2^{hr}$
$\mathcal{A}_1^{\text{hr}}$	9994	0.5505	0.3929	0.1576	9803
$\mathcal{A}_2^{\text{hr}}$	3985	0.5502	0.5233	0.0270	10476
$\mathcal{A}_3^{\text{hr}}$	3529	0.5510	0.4930	0.0581	10052
$\mathcal{A}_4^{\text{hr}}$	3212	0.5510	0.4815	0.0695	31833
$\mathcal{A}_5^{\text{hr}}$	2845	0.5512	0.4957	0.0556	9532
$\mathcal{A}_6^{\text{hr}}$	5188	0.5505	0.5373	0.0132	8405
$\mathcal{A}_7^{\text{hr}}$	3000	0.5506	0.4177	0.1329	27091
$\mathcal{A}_8^{\text{hr}}$	3377	0.5503	0.4968	0.0535	26237
$\mathcal{A}_9^{\text{hr}}$	15603	0.5504	0.3553	0.1950	12136
$\mathcal{A}_{10}^{\text{hr}}$	11770	0.5501	0.5246	0.0255	12819
Avg	**6250**	**0.5506**	**0.4718**	**0.0788**	**15838**

Table 3 lists the average execution time spent on each step of the noise blowing-up method. Out of those, recall again that Step 3 is used in, but is independent from the noise blowing-up strategy. The time overhead required by the noise blowing-up strategy is the sum of the time of all steps except Step 3. Its value, given in the last column, amounts to 0.14571 seconds on average, which is negligible both in absolute terms as well as compared to the circa one hour required by the EA attack referred to in Step 3: the noise blowing-up time overhead amounts to 0.004% for this specific attack.

Table 3. Average time (in seconds) spent on the main steps of the noise blowing-up technique, and noise blowing-up time overhead.

	Step1	Step2	Step 3	Step 4	Step 5	Step 6	Step 7	Overhead
$\mathcal{A}_1^{\text{hr}}$	0.00727	0.03362	5048	0.00018	0.00857	0.00700	0.03569	0.09233
$\mathcal{A}_2^{\text{hr}}$	0.00976	0.03484	2299	0.00019	0.00968	0.00789	0.03679	0.09914
$\mathcal{A}_3^{\text{hr}}$	0.00827	0.03631	1848	0.00020	0.00856	0.00718	0.03689	0.09740
$\mathcal{A}_4^{\text{hr}}$	0.07146	0.03663	2199	0.00020	0.11523	0.06922	0.03831	0.33104
$\mathcal{A}_5^{\text{hr}}$	0.00980	0.03573	1660	0.00020	0.00875	0.00721	0.03716	0.09885
$\mathcal{A}_6^{\text{hr}}$	0.00831	0.03726	2920	0.00019	0.00611	0.00576	0.03764	0.09528
$\mathcal{A}_7^{\text{hr}}$	0.01424	0.03484	1773	0.00021	0.01556	0.01221	0.03736	0.11441
$\mathcal{A}_8^{\text{hr}}$	0.01478	0.03576	1716	0.00020	0.01480	0.01217	0.03627	0.11398
$\mathcal{A}_9^{\text{hr}}$	0.04199	0.03558	9072	0.00024	0.06720	0.04020	0.03731	0.22252
$\mathcal{A}_{10}^{\text{hr}}$	0.03445	0.03637	6564	0.00020	0.05190	0.03186	0.03740	0.19218
Average	**0.02203**	**0.03569**	**3510**	**0.00020**	**0.03064**	**0.02007**	**0.03708**	**0.14571**

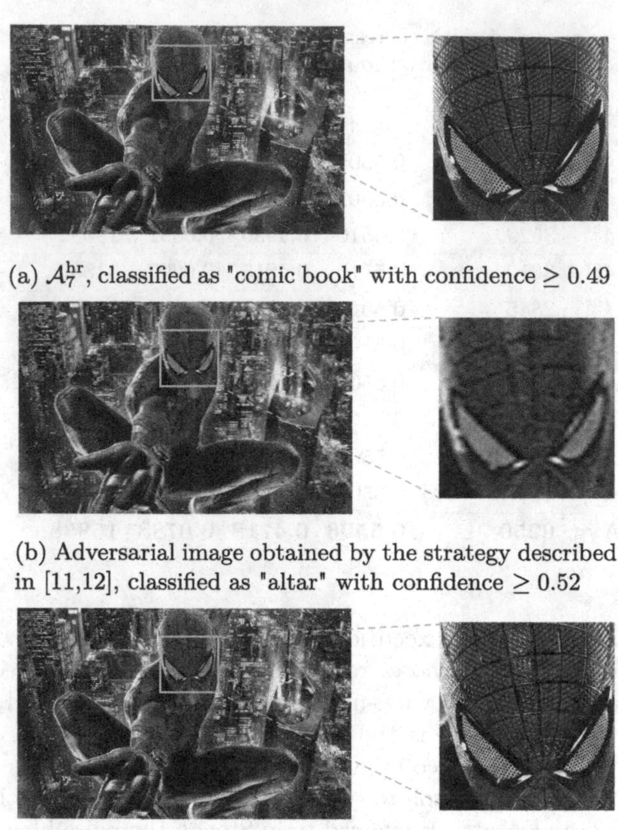

(a) $\mathcal{A}_7^{\text{hr}}$, classified as "comic book" with confidence ≥ 0.49

(b) Adversarial image obtained by the strategy described in [11,12], classified as "altar" with confidence ≥ 0.52

(c) $\mathcal{D}_{altar,0.41}^{hr,VGG-16}(\mathcal{A}_7^{\text{hr}})$, classified as "altar" with confidence ≥ 0.41

Fig. 1. Visual comparison of the clean HR image $\mathcal{A}_7^{\text{hr}}$ with the adversarial HR images obtained by $\text{EA}^{\text{target},\mathcal{C}}$ for $\mathcal{C}$ = VGG-16 with $\tilde{\tau}_t \geq 0.55$, and c_t = altar. The clean $\mathcal{A}_7^{\text{hr}}$ (a), the HR adversarial image obtained from [11,12] (b), and the HR adversarial image obtained from the noise blowing-up strategy (c).

5 One Detailed Example

The "true" visual quality for a human eye is assessed by looking at some representative examples either from some distance or by zooming in on some areas. This section highlights on the clean HR image $\mathcal{A}_7^{\text{hr}}$ the visual quality enhancements that benefit the HR adversarial images obtained by the noise blowing-up strategy, as compared with the HR adversarial images constructed in [11,12]. Especially, one considers areas that remained visually problematic with the method used in these latter papers.

Figure 1a represents this clean HR image $\mathcal{A}_7^{\text{hr}}$, and a zoom of that picture in some areas. Figure 1b shows the HR adversarial image obtained by the method described in [11,12]. Figure 1c shows the HR adversarial image obtained by the noise blowing-up method. For both methods, $\tilde{\tau}$ was set to 0.55.

At some distance, both HR adversarial images present a good visual quality. However, the zoomed areas show differences between the HR adversarial images. Details from the HR adversarial image shown in Fig. 1b become blurry for a human eye. Therefore, a human is able to distinguish between the clean image shown in Fig. 1a and the adversarial image shown in Fig. 1b. The situation differs significantly with the adversarial image obtained from the noise blowing-up method. Zooming into the same area does not exhibit any visible blurriness anymore. It becomes much more challenging for a human to distinguish between the clean HR image in Fig. 1a and the HR adversarial image in Fig. 1c.

6 Conclusion

This paper describes the noise blowing-up strategy that constructs high-resolution adversarial images against CNNs at their image recognition task. This strategy applies to any scenario and any effective attack in the low-resolution domain. We presented a convincing proof of concept for this strategy, thanks to one CNN, one scenario, one attack, and a few high-resolution images. This strategy successfully addressed the speed and adversity challenges raised by the construction of HR adversarial images. Foremost, our method substantially enhanced the visual quality of the obtained adversarial images, as compared to previous methods. Finally, our experiments showed that the noise blowing-up strategy overhead is extremely modest compared to the time required by the concrete attack at hand.

This paper will be extended in many ways. Firstly, we intend to apply the strategy to at least 10 diverse and state-of-the-art CNNs, to different attacks (black-box, white-box, GANs) performed on more scenarios, and on many more clean HR images, and explore the deep reasons for the enhanced visual quality provided by our strategy. Secondly, we intend to study variants of this strategy. For instance, instead of blowing up one layer of some strong adversarial noise, one can blow up several layers of lighter adversarial noise. Implementing this variant in parallel may accelerate the overall process. Thirdly, we intend to compare (or combine) this strategy with another one, of a completely different nature, that would involve some pre-processing to select the areas of interest on which to focus the construction of adversarial noise.

Acknowledgements. We thank Bernard Utudjian and Speedy Graphito for providing artistic images used in Sect. 4, and Uli Sorger for fruitful discussions.

References

1. Abadi, M., et al.: TensorFlow: large-scale machine learning on heterogeneous distributed systems. arXiv preprint arXiv:1603.04467 (2016)
2. Carlini, N., Wagner, D.: Towards evaluating the robustness of neural networks. In: 2017 IEEE Symposium on Security and Privacy (SP), pp. 39–57. IEEE (2017)

3. Chitic, R., Bernard, N., Leprévost, F.: A proof of concept to deceive humans and machines at image classification with evolutionary algorithms. In: Nguyen, N.T., Jearanaitanakij, K., Selamat, A., Trawiński, B., Chittayasothorn, S. (eds.) ACIIDS 2020. LNCS (LNAI), vol. 12034, pp. 467–480. Springer, Cham (2020). https://doi.org/10.1007/978-3-030-42058-1_39
4. Chitic, R., Leprévost, F., Bernard, N.: Evolutionary algorithms deceive humans and machines at image classification: an extended proof of concept on two scenarios. J. Inf. Telecommun., 1–23 (2020)
5. Chitic, R., Topal, A.O., Leprévost, F.: Evolutionary algorithm-based images, humanly indistinguishable and adversarial against convolutional neural networks: efficiency and filter robustness. IEEE Access **9**, 160758–160778 (2021)
6. Chollet, F., et al.: Keras. https://keras.io (2015)
7. Deng, J., Dong, W., Socher, R., Li, L.J., Li, K., Fei-Fei, L.: The ImageNet image database (2009). http://image-net.org
8. Duchon, C.E.: Lanczos filtering in one and two dimensions. J. Appl. Meteorol. Climatol. **18**(8), 1016–1022 (1979)
9. Guo, C., Gardner, J., You, Y., Wilson, A.G., Weinberger, K.: Simple black-box adversarial attacks. In: International Conference on Machine Learning, pp. 2484–2493. PMLR (2019)
10. Hu, W., Tan, Y.: Generating adversarial malware examples for black-box attacks based on GAN. In: Tan, Y., Shi, Y. (eds.) Data Mining and Big Data: 7th International Conference, DMBD 2022, Beijing, China, 21–24 November 2022, Proceedings, Part II, pp. 409–423. Springer, Singapore (2023). https://doi.org/10.1007/978-981-19-8991-9_29
11. Leprévost, F., Topal, A.O., Avdusinovic, E., Chitic, R.: Strategy and feasibility study for the construction of high resolution images adversarial against convolutional neural networks. In: Nguyen, N.T., Tran, T.K., Tukayev, U., Hong, TP., Trawiński, B., Szczerbicki, E. (eds.) Intelligent Information and Database Systems. 14th Asian Conference, ACIIDS 2022, Ho-Chi-Minh-City, Vietnam, 28–30 November 2022, pp. 467–480. Springer, Heidelberg (2022). https://doi.org/10.1007/978-3-031-21743-2_23
12. Leprévost, F., Topal, A.O., Avdusinovic, E., Chitic, R.: A strategy creating high-resolution adversarial images against convolutional neural networks and a feasibility study on 10 CNNs. J. Inf. Telecommun., 1–31 (2022)
13. Oliphant, T.E.: A guide to NumPy. Trelgol Publishing USA (2006)
14. Parsania, P.S., Virparia, P.V.: A comparative analysis of image interpolation algorithms. Int. J. Adv. Res. Comput. Commun. Eng. **5**(1), 29–34 (2016)
15. SpeedyGraphito: Mes 400 Coups. Panoramart (2020)
16. Szegedy, C., et al.: Intriguing properties of neural networks. arXiv preprint arXiv:1312.6199 (2013)
17. Topal, A.O., Chitic, R., Leprévost, F.: One evolutionary algorithm deceives humans and ten convolutional neural networks trained on ImageNet at image recognition. Appl. Soft Comput. **143**, 110397 (2023). https://doi.org/10.1016/j.asoc.2023.110397. https://www.sciencedirect.com/science/article/pii/S1568494623004155
18. Van Rossum, G., Drake, F.L.: Python 3 Reference Manual. CreateSpace, Scotts Valley (2009)
19. Van der Walt, S., et al.: The scikit-image contributors: scikit-image: image processing in Python. PeerJ **2**, e453 (2014). https://doi.org/10.7717/peerj.453

Faster Imputation Using Singular Value Decomposition for Sparse Data

Phuc Nguyen[2,3], Linh G. H. Tran[2,3], Bao H. Le[1,2,3], Thuong H. T. Nguyen[2,3], Thu Nguyen[4], Hien D. Nguyen[3,5], and Binh T. Nguyen[1,2,3](✉)

[1] AISIA Research Lab, Ho Chi Minh City, Vietnam
ngtbinh@hcmus.edu.vn
[2] University of Science, Ho Chi Minh City, Vietnam
[3] Vietnam National University, Ho Chi Minh City, Vietnam
[4] Simula Metropolitan, Oslo, Norway
[5] University of Information Technology, Ho Chi Minh City, Vietnam

Abstract. With the emergence of many knowledge-based systems worldwide, there have been more and more applications using different kinds of data and solving significant daily problems. Among that, the issues of missing data in such systems have become more popular, especially in data-driven areas. Other research on the imputation problem has dealt with partial and missing data. This study aims to investigate the imputation techniques for sparse data using the Singular Value Decomposition technique, namely SVDI. We explore the application of the SVDI framework for image classification and text classification tasks that involve sparse data. The experimental results show that the proposed SVDI method improves the speed and accuracy of the imputation process when compared to the PCAI method. We aim to publish our codes related to the SVDI later for the relevant research community.

Keywords: Sparse data · Data imputation · Singular Value decomposition

1 Introduction

The issue of missing data is a significant one that regularly emerges in many data-driven areas. Partial or missing data can occur due to several circumstances, including data entry mistakes, measurement flaws, or simply the inability to obtain specific information. It can result in skewed or incomplete studies and other problems, such as diminished statistical power, increased uncertainty, and poor interpretability. Imputation procedures, which fill in missing values in a data set to generate a complete data matrix, are frequently used to overcome this problem. On the other hand, sparse data refers to rows of data that include a significant percentage of zeroes as values. For example, it is frequently the case

Supported by Vietnam National University Ho Chi Minh City under the grant number DS2023-18-01.

N. T. Nguyen et al. (Eds.): ACIIDS 2023, LNAI 13995, pp. 135–146, 2023.
https://doi.org/10.1007/978-981-99-5834-4_11

in some issue areas, such as recommender systems, when a user has 0 ratings for all but a small number of movies or music in the database. Another typical illustration is a "bag of words" model of a text document, where most words have a value of 0, and other words in the document have a count or frequency. Examples of sparse data suitable for dimensionality reduction using Singular value decomposition (SVD) include Text Classification, One Hot Encoding, Bag of Words Counts, Recommender Systems, Customer-Product Purchases, User-Song Listen Counts, and User-Movie Ratings.

Singular value decomposition (SVD) is a widely used tool for data analysis with applications throughout science and engineering. In general, SVD operates by disassembling the initial matrix. SVD aims to approximate a dataset with many dimensions using fewer dimensions. The data are arranged in decreasing order of variation upon exposure of the substructure. This makes it easier to identify the area with the most variance, which may be reduced using SVD. By extracting the initial few singular vectors or eigenvectors, it may be used for dimension reduction, data visualization, data compression, and information extraction; for examples, see Alter et al. [1], Prasantha et al. [29], and Nguyen et al. [23,24]. On the one hand, SVD can solve several fundamental data analysis methods, such as the Principal component analysis (PCA) [8], the Canonical correlation analysis (CCA) [29], and the Singular Value Thresholding (SVT) [18]. On the other hand, SVD is also connected to several potent tools in different fields, such as the Latent matrix factorization (LMF) [33] and the Latent semantic analysis (LSA) [3]. Thus, when data is sparse, SVD may be the method with the highest level of popularity for dimensionality reduction.

In summary, the contribution of this paper can be listed as follows:

(a) We focus on evaluating the effectiveness of SVDI in parsing sparse data in two specific application domains: image and text classification.
(b) We compare the performance of SVD Imputation with Principal Component Analysis Imputation (PCAI) measures regarding their ability to handle missing data in these two application domains.

The rest of the paper is structured as follows. Section 2 presents a survey about imputation methods and their application in practical problems. Section 3 describes the process combining Singular Value Decomposition and imputation techniques. The results and discussion are performed in Sect. 4. The paper ends with conclusions and future works in the last section.

2 Related Works

Instead of removing or ignoring the unknown data, a large amount of research has been tackling this problem by using imputation methods [25]. While Suthar et al. [30] surveys classifying the imputation methods of missing data in data mining, Musil et al. [22] and Lüdtke et al. [19] compare imputation strategies in different designs. There are also advanced imputation methods, such as K-nearest Neighbor (KNN) imputation [20] and Machine Learning-based imputation [10,12]. Besides that, Lakshminarayan et al. [12] experiment with two Machine Learning (ML) systems: Autoclass and C4.5, for the problem.

In addition to various methods available for handling missing data, one noteworthy example is the Generative Adversarial Imputation Nets (GAIN) [34], which is a modified version of the Generative Adversarial Nets (GAN) framework. GAIN uses a generator and discriminator network to impute missing data and is trained using additional information as a hint vector to focus on imputation quality. Another technique is the Missing GP (MGP) [9], which uses sparse Gaussian processes to predict missing values at each dimension using all the variables from other dimensions. MGP outputs a predictive distribution for each missing value and can be trained simultaneously to impute all observed missing values. Finally, Khan et al. [11] suggests a hybrid technique of single and multiple imputation techniques, which extends the Multivariate Imputation by Chained Equation (MICE) algorithm to impute categorical and numeric data. Additionally, Awan et al. [2] presents the Conditional Generative Adversarial Imputation Network (CGAIN), which imputes missing data using class-specific distributions based on class-specific characteristics of the data. Moreover, DPER algorithm [26] directly computes maximum likelihood estimates (MLEs) for randomly missing data sets, eliminating the need for separate imputation steps. It provides computational efficiency and superior estimation performance compared to existing methods.

Many studies have applied to impute methods to address missing data in real-world problems. Firstly, Jerez et al. [10] use statistical and machine learning methods to impute missing data in an actual breast cancer problem. Furthermore, Liu et al. [16] have a systematic review of deep learning-based imputation techniques for handling missing values in healthcare data. The study aims to evaluate the use of these techniques, with a particular focus on data types, to assist healthcare researchers in dealing with missing values. Hassan et al. [7] propose a missing data imputation method based on the salp swarm algorithm for diabetes disease. The study aims to impute missing values in the Pima Indian diabetes disease dataset using a proposed algorithm, namely ISSA.

Singular value decomposition (SVD) has recently been widely used in different fields, including multi-environment trials and transforming genome-wide expression data. However, in the research of Alter et al., [1], imputing missing values using standard SVD can lead to low-quality results when affected by outliers. Still, the Yan method proposed four robust SVD extensions to address this issue. Singular value decomposition can also be used in transforming genome-wide expression data [5], enabling meaningful comparisons of the expression of different genes across different arrays and experiments. Moreover, to improve the issue of handling missing data, a proposed Bayesian model that is based on the SVD components of a continuous data matrix is shown by Zhai et al. [35] to be the most accurate and precise method compared to the current imputation methods in simulated and real datasets.

3 Methodology

3.1 Sparse Data

Sparse data refers to datasets characterized by a significant proportion of zero or missing entries, indicating that the vast majority of the data points possess values of zero. Unlike missing data, where values are unknown or undefined, sparse data values are generally known but non-existent or specifically set to zero. This terminology finds frequent application in fields such as machine learning, data science, and information retrieval, where dealing with sparse data poses a common challenge.

Consider a movie recommendation system that suggests movies to users based on their viewing history. The system has a comprehensive database with information about movies, users, and ratings. However, not all users have watched or rated every movie, resulting in a sparse dataset with many missing entries. For example, in a dataset with 100,000 users and 1,000 movies, only 1 million non-zero entries exist, representing less than 1% of the total dataset. This high sparsity means that most entries in the dataset are zero values.

3.2 Mechanisms of Missing Data

The impact of missing data depends on the method used to generate the missing data. Rubin and his colleagues [13–15,28] established the foundations of missing data theory. Central to missing data theory is his classification of missing data problems into three categories: (1) missing completely at random (MCAR); (2) missing at random (MAR); and (3) missing not at random (MNAR). These three classes of missing data are referred to as missing data mechanisms (for a slightly different classification, see [6]). Despite the name, they are not reasons for missing data that are causative. Instead, the statistical link between observations (variables) and the risk of missing data is represented by missing data mechanisms. Another word that is sometimes confused with missing data mechanisms is missing data patterns; these are descriptions of which values in a dataset are missing.

MCAR occurs when the missing data is independent of the observed or unobserved data. For example, participants flip a coin in a survey to decide whether to answer questions. MAR occurs when the missingness can be explained using observed data. For instance, survey participants that live in specific postal codes may refuse to fill in the questionnaire. MNAR occurs when the missingness depends on an unobserved or missing attribute. For example, people who own six-bedroom houses may refuse to participate in a survey, as owning a bigger house may indicate greater wealth and a better-paying job. A researcher's choice of approach is made more accessible when the data are MCAR or MAR since they allow them to overlook the causes of missing data. Every approach is viable in this situation. The data may be MCAR or MAR, but it is challenging to provide actual proof of this. It is a sound method to compare findings from many studies to see their sensitivity to the MCAR and MAR assumptions. The outcomes of the various analyses differ from one another, and this reveals which assumptions are the most important.

3.3 Singular Value Decomposition (SVD)

The concept of singular value decomposition (SVD) is a fundamental tool in linear algebra. Given an $n \times d$ matrix A, one can express it as the product of three matrices:

$$A = USV^T, \tag{1}$$

where U is an $n \times n$ orthogonal matrix, V is a $d \times d$ orthogonal matrix, and S is an $n \times d$ diagonal matrix with nonnegative entries. Notably, the diagonal entries of S are sorted from highest to lowest, progressing from the "northwest" to the "southeast" of the matrix. Assuming we use r eigenvalues, the projection matrix can be defined as $V = W_r$, where W_r is formed by selecting the first r columns of matrix V. Consequently, the reduced dimension version of matrix A is given by the product AV as well.

The following formula, as depicted below, is helpful in providing a more detailed illustration of the singular value decomposition (SVD) method. SVD is a powerful mathematical tool used for matrix factorization, wherein each singular value in S is accompanied by an associated left singular vector in U and a right singular vector in V.

$$\mathrm{A} = \mathrm{USV^T} = \underbrace{\begin{bmatrix} | & | & \&| \\ u_1 & u_2 \ldots & u_n \\ | & | & \&| \end{bmatrix}}_{n \times n} \times \underbrace{\begin{bmatrix} \sigma_1 & 0 & \ldots & 0 \\ 0 & \sigma_2 & \ldots & 0 \\ \vdots & \vdots & \ddots & \vdots \\ 0 & 0 & \ldots & \sigma_d \\ & & \vdots & \\ 0 & 0 & \vdots & 0 \end{bmatrix}}_{n \times d} \times \underbrace{\begin{bmatrix} | & v_1^T & | \\ | & v_2^T & | \\ \&\vdots & & \\ | & v_d^T & | \end{bmatrix}}_{d \times d}$$

It is important to note that the orthogonal matrices U and V that are part of the singular value decomposition (SVD) of matrix A are not necessarily the same. It is because A may not be a square matrix, resulting in U and V having different dimensions. The columns of U represent the left singular vectors of A, while the columns of V, or the rows of V^T, represent the right singular vectors of A. The singular values of matrix A are represented by the entries of S, with each singular value being associated with a singular vector. The singular vectors are ordered such that the first or top singular vector corresponds to the largest singular value, which is illustrated in the figure above. It is worth noting that every matrix A can be decomposed into its SVD, a remarkable fact with a straightforward proof that is better suited for a linear algebra course. Geometrically, this means that no matter how peculiar a matrix may be, it can always be decomposed into a rotation (multiplication by V^T), scaling plus dimension addition or removal (multiplication by S), and a rotation within the range (multiplication by U). The SVD is "more or less unique," with the singular values of a matrix being unique. If a singular value is repeated, the subspaces created by the corresponding left and right singular vectors have a distinct definition. However, there is flexibility in selecting orthonormal bases for each of these subspaces.

3.4 SVD Imputation (SVDI)

The below algorithms describe the approach of "SVD Imputation" (SVDI). For example, suppose there exists a dataset $\mathcal{D} = [\mathcal{F}, \mathcal{M}]$, which can be decomposed into a partition of fully observed features denoted by $\mathcal{F}$, and another partition $\mathcal{M}$ containing features with missing values. To facilitate the imputation process on this incomplete dataset, one may adopt the approach of SVD Imputation (SVDI), which involves reducing the dimensionality of the fully observed partition $\mathcal{F}$ via *svd*(A), generating a new reduced feature matrix $\mathcal{R}_{\mathcal{F}}$. Subsequently, the imputation process can be carried out on the union of $\mathcal{R}_{\mathcal{F}}$ and the partition with missing values, $\mathcal{M}$, instead of the original full dataset $[\mathcal{F}, \mathcal{M}]$.

The rationale for this approach is twofold. First, by reducing the dimensionality of $\mathcal{F}$, one can accelerate the computational efficiency of the imputation method. This is particularly beneficial in scenarios where the size of the covariance matrix, a key component of SVD, is smaller than that of the full dataset $\mathcal{F}$ due to $\mathcal{F}$ having more samples than features. In such cases, implementing SVD based on the covariance matrix is expected to be faster. Conversely, when the number of features in $\mathcal{F}$ exceeds the sample size, the covariance matrix of $\mathcal{F}$ is larger than that of $\mathcal{F}$ itself. In this scenario, SVD formulation based on the data is a more favorable approach. By considering these factors, researchers and practitioners can optimize the SVDI methodology to suit their particular dataset and computational resources best. The variations in the mean squared error of the imputed version and the ground truth for various procedures are only marginally different, as demonstrated in the studies. SVDI appears to perform somewhat better on several occasions. That is feasible because SVD keeps just the essential information from the data while eliminating some noise, improving imputation quality.

Algorithm 1. SVD imputation framework

Require:
$\mathcal{D} \leftarrow [\mathcal{F}, \mathcal{M}]$
Imputer I
SVD algorithm *svd*
Procedure:
$(\mathcal{R}, V) \leftarrow svd(\mathcal{F})$
$\mathcal{M}' \leftarrow I([\mathcal{R}, \mathcal{M}])$
Return Imputed version $\mathcal{M}'$ of $\mathcal{M}$

4 Experiments

In this section, we validate the performance of SVDI using multiple real-world datasets with various settings (such as on datasets with different missing rates), and we compare SVDI and PCAI when the objective is to perform classification on the imputed dataset. We report RMSE, running time, and average accuracy as the standard performance metric. Unless specified, missingness is applied to the datasets with a missing rate is 20%, and the default missing mechanism is MCAR.

4.1 Datasets

We experiment on three datasets:

1. **IMDB**[1] The IMDB dataset contains 50000 movies and TV shows divided into 25000 training and 25000 test samples. Each review is labeled as positive or negative based on sentiment. The reviews are preprocessed. Each review sentence can be represented by TF-IDF 5000 features.
2. **Fashion MNIST**[2] includes clothing images is also selected in our experiments. The dataset consists of 60000 training images, 10000 testing images of size 28 $\times$ 28 (784 features), and ten labels corresponding to 10 different types of fashion.
3. **MNIST**[3] is a large collection of handwritten digits. It also has a training set of 60000 images, a test set of 10000 images of size 28$\times$28, and ten labels corresponding to 10 digits.

The detail of each dataset can be listed below (TAble 1).

Table 1. The description of datasets used in our experiments.

Dataset	# classes	# features	Samples	Sparsity
Fashion MNIST	10	784	70000	50.1
MNIST	10	784	70000	80.4
IMDB	2	5000	50000	98.05

4.2 Experimental Design

We compare the running time, average RMSE, and average accuracy of PCAI with our SVD Imputation (SVDI) methods. We calculate the running time by the sum of dimensional reduction and imputation time. The imputation methods we use in our experiments are:

1. **SoftImpute** [21]: Matrix completion by iterative soft thresholding of SVD decompositions. The algorithm fills in the missing values with the current guess and then solves the optimization problem on the complete matrix using a soft-thresholded SVD.

[1] https://www.kaggle.com/datasets/lakshmi25npathi/imdb-dataset-of-50k-movie-reviews.

[2] https://github.com/zalandoresearch/fashion-mnist.

[3] http://yann.lecun.com/exdb/mnist/.

2. **Multiple Imputation by Chained Equation (MICE)** [4]: models each feature with missing values as a function of other features and uses that estimate for imputation in an iterated round-robin fashion.
3. **kNN Imputation (KNNI)** [32]: Nearest neighbor imputations which weight samples using the mean squared difference on features for which two rows both have observed data.
4. **GAIN** [34]: A deep learning approach for imputing missing data by utilizing Generative Adversarial Network (GAN).

All methods are implemented with default configurations in their original papers. For all PCA computations, the number of eigenvectors is chosen so that the minimum amount of variance explained is 95%. We utilize logistic regression as the classifier.

It is worth noting that any dataset can be rearranged so that the first q features are not missing while the remaining features have missing values. Therefore, we assume that each dataset's first q features are not missing, while the remaining ones contain missing values. The default value for q is half of the total number of attributes in each dataset. Then, we randomly simulate missing data in the missing partition M at default missing rates of 20%, 40%, and 60%. Here, a missing rate of x% refers to the percentage of missing entries in the missing partition M. To introduce a fixed missing rate, we use two different missing mechanisms inspired by [17].

(a) **MCAR**: Set all features in missing partition M to have missing values when $v_i \leq, t, i \in (1:n)$ rate with t is the missing rate.
(b) **MNAR**: Randomly sample 2 features x_1 and x_2 from the missing partition M, calculate their median m_1 and m_2. Then we set all features to the missing value where $v_i \leq t, i \in (1:n)$ and ($x_1 \leq m_1$ or $x_2 \leq m_2$) and t is the missing rate.

Unless otherwise stated, missingness is applied to the datasets by randomly removing 20% of all missing partition M, with MCAR as the default missing mechanism. We conduct all experiments using the Kaggle notebook with a default Intel Xeon CPU and 30 GB RAM. If no results are produced after 20000 s of running, or if a memory allocation issue arises, we terminate the experiment and denote it as **NA** in the result tables.

4.3 Results and Discussion

We perform experiments comparing the performance between SVDI and PCAI and present the results in the tables below. From now on, the bold values on the tables indicate better performance for each metric on each dataset. According to Table 2, SVDI outperforms PCAI by having lower average RMSE values for most methods and datasets. Specifically, SVDI proves more effective than PCAI with SoftImpute, Mice, and KNNI methods on Fashion MNIST and MNIST datasets. In the IMDB datasets, there is an insignificant difference between PCA and SVD, and we could not retrieve the results of MICE due to a memory issue.

Table 2. The average RMSE of Imputation methods.

Methods	Strategy	Fashion MNIST	MNIST	IMDB
SoftImpute	PCAI	0.20675	0.3618	**0.448**
	SVDI	**0.155**	**0.1995**	0.458
GAIN	PCAI	**0.156**	1.2557	0.452
	SVDI	0.261	**0.1754**	**0.443**
Mice	PCAI	0.244	2.712	**NA**
	SVDI	**0.09**	**0.11**	
KNNI	PCAI	0.162	1.2495	**0.6204**
	SVDI	**0.122**	**0.166**	0.6211

Table 3. The average accuracy (%) of Imputation methods.

Methods	Strategy	Fashion MNIST	MNIST	IMDB
SoftImpute	PCAI	**84.19**	92.05	85.94
	SVDI	83.64	**92.38**	**88.51**
GAIN	PCAI	**84.06**	**92.5**	85.952
	SVDI	83.9	92.3	**88.548**
Mice	PCAI	80.1	92.43	**NA**
	SVDI	**83.8**	**92.46**	
KNNI	PCAI	**84.35**	**92.85**	85.9
	SVDI	83.84	92.83	**88.54**

But overall, SVDI is a better strategy for imputing missing values than PCAI, especially when combined with SoftImpute, Mice, and KNNI methods.

Based on the results presented in Table 3, the average accuracy table, it is evident that the SVDI imputation strategy generally surpasses the PCAI strategy. GAIN and KNNI with PCAI achieve slightly higher average accuracy on the Fashion and MNIST datasets than their SVDI counterparts. However, on the IMDB dataset, PCAI performs notably worse, with an average accuracy of nearly 3% lower than that of SVDI. Mice method with PCAI tends to yield lower average accuracies on the Fashion MNIST and MNIST datasets, with values of 80.1% and 92.43%, respectively. For the SoftImpute technique, while PCAI outperforms SVDI on the Fashion MNIST, the MNIST dataset shows that PCAI is 0.33% less effective than SVDI. Notably, all SVDI strategies yielded results approximately 3% better than PCAI ones on the IMDB dataset.

Table 4 displays each method's average running time values on different datasets. PCAI demonstrates faster running time in the Fashion MNIST dataset with all four methods compared to SVDI. On the contrary, in the IMDB and MNIST datasets, SVDI is consistently more effective than PCAI, with lower running time values. Overall, it is evident that SVDI exhibits faster perfor-

Table 4. The average running time (s) of Imputation methods.

Methods	Strategy	Fashion MNIST	MNIST	IMDB
SoftImpute	PCAI	**17.92**	18.97	644.24
	SVDI	18.38	**17.81**	**411.47**
GAIN	PCAI	**383.9**	**417.63**	14070.26
	SVDI	422.9	423.58	**10813.95**
Mice	PCAI	**1984.63**	**4664.094**	**NA**
	SVDI	2664.87	6211.2	
KNNI	PCAI	**6376**	11314.2	18490.52
	SVDI	10835	**10156.6**	**17268.33**

mance than PCAI on the MNIST and IMDB datasets, which can be attributed to SVDI's omission of the standardization step, resulting in quicker processing time.

5 Conclusion and Future Works

This study explores the application of the SVDI framework for image classification and text classification tasks that involve sparse data. The experimental setup consists of the dimensionality reduction of fully observed features and subsequent imputation of missing data using the reduced feature set. This approach enables us to effectively handle the high-dimensional and sparse data common in image and text classification while also addressing the issue of missing data that frequently arises in real-world datasets. The average RMSE, average accuracy, and running time are used as evaluation metrics to compare the performance of SVDI with other dimension reduction imputation methods, such as PCAI. After conducting experiments, it illustrates that the SVDI method improves the speed and accuracy of the imputation process when compared to the PCAI method.

For future work, the proposed method can be experimented on visible and thermal infrared image datasets, such as KTFE [27] and USTC-NVIE [31]. The objective is to comprehensively investigate these frameworks to gain deeper insights into the underlying factors contributing to this discrepancy and develop viable solutions to address these challenges.

Acknowledgments. This research is funded by Vietnam National University Ho Chi Minh City in Vietnam under the funding/grant number DS2023-18-01.

References

1. Alter, O., Brown, P.: Processing and modeling genome-wide expression data using singular value decomposition. In: Proceedings of SPIE - The International Society for Optical Engineering, vol. 4266 (2001)

2. Awan, S.E., Bennamoun, M., Sohel, F., Sanfilippo, F., Dwivedi, G.: Imputation of missing data with class imbalance using conditional generative adversarial networks. Neurocomputing **453**, 164–171 (2021)
3. Berry, M., Dumais, S., Gavin, W.: O'brien, using linear algebra for intelligent information retrieval. SIAM Rev. **37**, 573–595 (1995)
4. van Buuren, S., Groothuis-Oudshoorn, K.: mice: multivariate imputation by chained equations in R. J. Stat. Softw. **45**(3), 1–67 (2011). https://doi.org/10.18637/jss.v045.i03. https://www.jstatsoft.org/index.php/jss/article/view/v045i03
5. García-Peña, M., Arciniegas-Alarcón, S., Krzanowski, W.J., Duarte, D.: Missing-value imputation using the robust singular-value decomposition: proposals and numerical evaluation. Crop Sci. **61**(5), 3288–3300 (2021)
6. Gelman, A., Hill, J.: Data analysis using regression and multilevel/hierarchical models (2007)
7. Hassan, G.S., Ali, N.J., Abdulsahib, A.K., Mohammed, F.J., Gheni, H.M.: A missing data imputation method based on salp swarm algorithm for diabetes disease. Bull. Electric. Eng. Inf. **12**(3), 1700–1710 (2023)
8. Huang, J., Shen, H., Buja, A.: The analysis of two-way functional data using two-way regularized singular value decompositions. J. Am. Stat. Assoc. **104**, 1609–1620 (2009)
9. Jafrasteh, B., Hernández-Lobato, D., Lubián-López, S.P., Benavente-Fernández, I.: Gaussian processes for missing value imputation (2022)
10. Jerez, J.M., et al.: Missing data imputation using statistical and machine learning methods in a real breast cancer problem. Artif. Intell. Med. **50**(2), 105–115 (2010)
11. Khan, S.I., Hoque, A.S.M.L.: SICE: an improved missing data imputation technique. J. Big Data **7**(1), 1–21 (2020)
12. Lakshminarayan, K., Harp, S.A., Goldman, R.P., Samad, T., et al.: Imputation of missing data using machine learning techniques. In: KDD, vol. 96 (1996)
13. Little, R., Rubin, D.: Regression with missing XS - a review. J. Am. Stat. Assoc. **87**, 1227–1237 (1992)
14. Little, R., Rubin, D.: Modeling the drop-out mechanism in repeated-measures studies. J. Am. Stat. Assoc. **90**, 1112–1121 (1995)
15. Little, R., Rubin, D.: Statistical analysis with missing data (2014)
16. Liu, M., et al.: Handling missing values in healthcare data: a systematic review of deep learning-based imputation techniques. Artif. Intell. Med., 102587 (2023)
17. Gondara, L., Wang, K.: MIDA: multiple imputation using denoising autoencoders. In: Phung, D., Tseng, V.S., Webb, G.I., Ho, B., Ganji, M., Rashidi, L. (eds.) PAKDD 2018. LNCS (LNAI), vol. 10939, pp. 260–272. Springer, Cham (2018). https://doi.org/10.1007/978-3-319-93040-4_21
18. Lu, C., Zhu, C., Xu, C., Yan, S., Lin, Z.: Generalized singular value thresholding. In: Proceedings of the AAAI Conference on Artificial Intelligence, vol. 29 (2015)
19. Lüdtke, O., Robitzsch, A., Grund, S.: Multiple imputation of missing data in multilevel designs: a comparison of different strategies. Psychol. Methods **22**(1), 141 (2017)
20. Malarvizhi, R., Thanamani, A.S.: K-nearest neighbor in missing data imputation. Int. J. Eng. Res. Dev. **5**(1), 5–7 (2012)
21. Mazumder, R., Hastie, T., Tibshirani, R.: Spectral regularization algorithms for learning large incomplete matrices. J. Mach. Learn. Res. **11**(80), 2287–2322 (2010). http://jmlr.org/papers/v11/mazumder10a.html
22. Musil, C.M., Warner, C.B., Yobas, P.K., Jones, S.L.: A comparison of imputation techniques for handling missing data. West. J. Nurs. Res. **24**(7), 815–829 (2002)

23. Nguyen, H.D., Sakama, C., Sato, T., Inoue, K.: Computing logic programming semantics in linear algebra. In: Kaenampornpan, M., Malaka, R., Nguyen, D.D., Schwind, N. (eds.) MIWAI 2018. LNCS (LNAI), vol. 11248, pp. 32–48. Springer, Cham (2018). https://doi.org/10.1007/978-3-030-03014-8_3
24. Nguyen, H.D., Sakama, C., Sato, T., Inoue, K.: An efficient reasoning method on logic programming using partial evaluation in vector spaces. J. Log. Comput. **31**(5), 1298–1316 (2021)
25. Nguyen, T., Nguyen, D.H., Nguyen, H., Nguyen, B.T., Wade, B.A.: EPEM: efficient parameter estimation for multiple class monotone missing data. Inf. Sci. **567**, 1–22 (2021)
26. Nguyen, T., Nguyen-Duy, K.M., Nguyen, D.H.M., Nguyen, B.T., Wade, B.A.: DPER: direct parameter estimation for randomly missing data. Knowl.-Based Syst. **240**, 108082 (2022)
27. Nguyen, V., Tran, N., Nguyen, H., et al.: KTFEv2: multimodal facial emotion database and its analysis. IEEE Access **11**, 17811–17822 (2023)
28. Rubin, D.: Inference and missing data. Biometrika **63**, 5781–590 (1976)
29. Prasantha, H.S., Shashidhara, H.L., Murthy, K.B.: Image compression using SVD. In: International Conference on Computational Intelligence and Multimedia Applications, pp. 143–145 (2008)
30. Suthar, B., Patel, H., Goswami, A.: A survey: classification of imputation methods in data mining. Int. J. Emerg. Technol. Adv. Eng. **2**(1), 309–12 (2012)
31. Wang, S., Liu, Z., Lv, S., et al.: A natural visible and infrared facial expression database for expression recognition and emotion inference. IEEE Trans. Multimedia **12**(7), 682–691 (2010)
32. Woźnica, K., Biecek, P.: Does imputation matter? benchmark for predictive models. In: 37th International Conference on Machine Learning (2020)
33. Yang, D., Ma, Z., Buja, A.: A sparse SVD method for high-dimensional data. J. Comput. Graph. Stat. **23**, 923–942 (2014)
34. Yoon, J., Jordon, J., van der Schaar, M.: Gain: missing data imputation using generative adversarial nets (2018)
35. Zhai, R., Gutman, R.: A Bayesian singular value decomposition procedure for missing data imputation. J. Comput. Graph. Stat., 1–13 (2022)

Combination of Deep Learning and Ambiguity Rejection for Improving Image-Based Disease Diagnosis

Thanh-An Pham[1] and Van-Dung Hoang[2(✉)]

[1] Ho Chi Minh University of Banking, Ho Chi Minh City, Vietnam
anpt@buh.edu.vn

[2] Ho Chi Minh City University of Technology and Education, Ho Chi Minh City, Vietnam
dunghv@hcmute.edu.vn

Abstract. Artificial intelligent (AI) based medical image recognition plays important task to assist in many disease diagnosis systems. In medical diagnosis, the incorrect decision is very serious. The healthcare diagnosis guides the treatment plan, and it is significant impact on the patient's health outcomes. An incorrect diagnosis can lead to delays in treatment or even the wrong treatment being administered, which results in serious harm to the patient. In this article, we propose an approach to reject ambiguity samples in the classification results, which improve the accuracy of the medical image- based diseases diagnose. In this study, we also experimented using some well-known deep learning models such as MobileNet (lightweight architecture) and DenseNet (more complex and dense connected architecture). Additionally, we combine with some solutions to address the problem of the data imbalance such as focal loss and data augmentation techniques. In the classification stage, there are still significant misclassification results. Therefore, we present the solution for ambiguity rejection of uncertain samples. Experimental results show that the accuracy increases significantly after removing uncertain samples. The high removal rate of uncertain samples also affects to the diagnosing quality. This approach eliminates uncertain samples, which utilizes for improving the diagnosing quality from results of deep learning classification around 10% recall and 70% coverage rate, respectively.

Keywords: Ambiguity rejection · Classification · Feature extraction · Medical image processing

1 Introduction

Skin cancer is a prevalent and dangerous disease that requires high accurate diagnosis for effective treatment. Melanoma, a type of skin cancer, has become increasingly common in recent decades and affects people of all ages. Although melanoma accounts for only 1% of skin cancers, it causes the majority of skin cancer deaths. Early prediction of skin cancer are crucial for effective treatment and cauterization. Advanced technology, particularly in the field of artificial intelligence, has led to the development of practical

N. T. Nguyen et al. (Eds.): ACIIDS 2023, LNAI 13995, pp. 147–160, 2023.
https://doi.org/10.1007/978-981-99-5834-4_12

applications for medical and healthcare. Deep learning (DL) has been widely applied in various fields, which includes medical diagnosis and healthcare, robotics and automation, and intelligent assistance systems and so on. The high performance with handling variety tasks become a popular choice for solving specific problems. DLis particularly useful for image processing tasks, such as medical image analysis and diagnosis, due to its ability to learn and extract features in high performance. DL techniques have been shown to produce better results compared to traditional shallow learning approaches. It is abilited to handle large datasets with many trainable parameters. However, a major challenge in training DL models is small dataset, data imbalance. This problem is leaded to biased classification models, with high performance on majority categories and low performance on minority categories. For example, in the ISIC 2018 dataset, the NV category is large samples, while other categories are a little samples. This problem leads to the NV categories dominating the model during training, and low performance on other categories. To address this issue, some techniques such as data augmentation and focal loss approach are used to improve performance. Augmented data techniques is a common technique to balance the dataset by artificially increasing the number of samples in under-fitting categories. However, this technique leads to overfitting or making noisy samples into the dataset. Therefore, in this study, we only focus on reject uncertain samples, which may lead incorrect diagnosis, for improving accuracy of decision with high rate of sample coverage and reject accuracy.

2 Related Works

These are some of the popular and well-known DL models in image classification and pattern recognition. Each of them has strength points and characteristics that make suitable for different datasets and application fields. The GoogleNet approach [1] is known as a deep architecture with multiple layers, MobileNet [2] is designed to be lightweight and efficient for mobile devices, ResNet [3] and DenseNet [4] are ability to train very deep neural networks and overcome the vanishing gradient problem, while EfficientNet [5] has shown to be highly accurate and efficient for various image recognition tasks. These are selected models, which have greatly improved the flexibility and accuracy of image recognition systems [6, 7]. Generally, it selects the appropriate model for specific dataset with expected that the system achieves higher accuracy without the need for manual tuning or hand craft selection. This is particularly useful in applications where the dataset is changing or evolving, the classified system should adapt to new data. Overall, DL models are more accessible and effective for a wider range of applications in image recognition and beyond. In industrial aspects, DLs-based methods have been widely used in many applications such as video surveillance system [8]. These approaches aim to find the optimal configuration of hyperparameters for the DL model, such as learning rate, batch size, number of filters, etc. The search method randomly selects a combination of hyperparameters and evaluates the performance of the model. The grid search method searches for the best combination of hyperparameters within a predefined range. The Bayesian optimization algorithm uses prior knowledge to guide the search for the best hyperparameters [9–11]. These methods have been shown to be effective in finding optimal hyperparameters for DL models [12, 13]. There are various approaches to improving the performance of DL models for image recognition tasks. These include using state-of-the-art models, selecting models automatically based on the data, optimizing the structure and hyperparameters of the models, and data augmentation to address the problem of imbalanced data [14, 15]. The selection approach depends on the specific problem and available resources, and combination of different approaches is necessary to reach higher accuracy.

In the field of medical images-based cancer disease diagnosis, dermoscopy is a skin surface imaging microscopic technique technology. Numerous studies have demonstrated that DL models produce high diagnostic performance when compared to standard imaging, dermatologists [16]. The paper [17] analysis methods and experimental results on the ISIC Challenge 2018. They presented a two-stage method to segment lesion regions from medical images based optimized training method and applied some parts for post-processing. The lesion images were acquired with a variety of dermatoscope types, from all anatomic sites, or historical sample of patients presented for skin cancer screening, from several different institutions. Each lesion image contains exactly one main lesion. Inspired by synthetic minority oversampling technique [18]. This method focuses the minority category samples before performing up sampling, which supports for better consideration of the uneven distribution of the samples. In another approach, MC-SMOTE method [19] combines of over-sampling the minority categories and under-sampling the majority categories, which achieves higher classifier performance than just using under-sampling the majority categories. This method uniformly increases minority categories samples by utilizing k-mean method, e.g., wind turbine fault detection for applied to practical application.

Other recent developments in the field of pattern recognition and classification based on the use of attention mechanisms in DL models [20]. In this approach, it allows the classification models to focus on the most informative parts of input images rather than processing on the entire image as equally importance. Nowadays, attention mechanisms have been shown that its outperformers accuracy than the DL models based on convolutional network in various tasks such as pattern classification, object recognition, image captioning, and so on.

In other approach, some research works report methods for eliminating uncertain samples [21–23]. These solutions are the inspiration for proposed solutions in the problem of diagnosing diseases, which improve the accuracy of medical image classification. This approach is integrated reject option that enables the network to reject input samples that are difficult to classify with high confidence. The authors argue that this can lead to better performance in real-world applications where the cost of misclassification is high. The reject option is implemented using a binary decision tree that operates on the output of the network. The decision tree takes as input the predicted class probabilities and other features such as the maximum and minimum probabilities and decides whether to reject the input sample or classify it with one of the predefined classes. The methods achieved state-of-the-art performance on several benchmark datasets and performs particularly well on imbalanced datasets.

3 Proposed Methodology

3.1 Overview Approach

This method aims to improve the performance of a DL by optimizing its architecture and ambiguity rejection. The general processing architecture, illustrated in Fig. 1, includes three major stages that should be investigated and customized: feature extraction, fully connected network for the classifier, and ambiguity rejection.

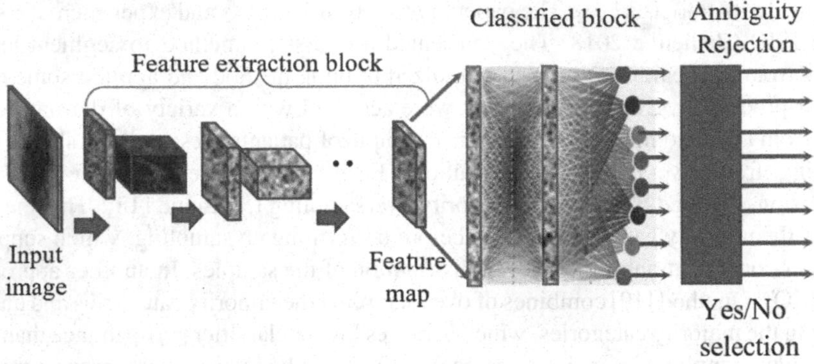

Fig. 1. General training flowchart of a DCNN based classification architecture.

3.2 Feature Extraction and Classification

In the first stage of feature extraction, the DL model is adjusting the training parameters, and refining the loss formulation. The approach has been evaluated empirically using various convolutional neural network (CNN) backbones for feature extraction tasks on different criteria. Our research does not focus on designing new deep learning architectures. Instead, we use the popular CNN model and customize fully connected layers for multiple category classification. There are many approaches to solve the feature extraction stage, such as using state-of-the-art backbone architectures with their pretrained parameters or initially constructing CNN architectures for selected searching the best model. The output feature maps are used as input for the classification stage. Experimental results prove the stability and efficiency on some predefined DCNN backbones, such as DenseNet and MobileNet family.

In this paper, two popular outstanding CNN architectures of DenseNets [4], MobileNets [2] were investigated. Among that, the family MobileNet architectures are known as lightweight model, which is efficiently model for limited resources. Two versions of MobileNet and MobileNetV3Large models were explored the performance ratio. The transfer learning was applied from a pretrained ImageNet model to ISIC2018 dataset for finetuning network hyperparameters. In contrast, DenseNets are more accurate and efficient, which are two versions of DenseNet121 and DenseNet201. The DenseNet is transferred learning from the pretrained model using ImageNet, without including the

Table 1. The list of backbones and their parameters

Backbone name	Number of layers	Number of parameters
MobileNet	90	3.757.255
MobileNetV3Large	280	4.885.895
DenseNet121	431	7.565.895
DenseNet201	711	19.309.127

last top layer, and the feature map is taken from its last layer named "ReLU". These architectures with trainable parameters are illustrated in Table 1.

In the classification stage, there are various approaches, such as using fully connected neural network (FCNN), support vector machines (SVM), or other machine learning approaches, which are appropriately applied. In this study, the FCNN for multiple classification, which takes the input feature maps from the feature extraction stage to classify. To avoid overfitting problems, we add some special layers to this neural network architecture, such as dropout layers. The optimal architecture was estimated using the trial-and-error method. Finally, the architecture consists of two dense connected layers with 1,024 nodes and 512 nodes following by activated layer. The activation function results to dropout layer with the ratio of 50% probabilities. The final output layer with c nodes following softmax activation function.

3.3 Imbalanced Data Processing

As mentioned above, to address imbalanced data issue, we investigated several solutions, such as data augmentation (AU) method and focusing on hard samples using focal loss (FL) [24] approach. The AU technique is also explored in this study. Augmentation processing involves applying image processing techniques such as geometric and artificial color transformations to augment data samples of minority categories and to concentrate on misclassification samples. This technique helps to address the problem of data imbalance. The method is suitable for multi-skin disease classification and effectively addresses issues of underfitting and overfitting, which is happened due to the imbalance of samples between the major categories and minor categories. Some image processing techniques are applied such as color normalization and geometrical transformations, which applied to the training dataset. We used color processing and affined transformations such as rotation, flip, skews, zoom, and crop. The augmented data was generated with random parameters within a predefined period, and each new sample was created and fixed for all methods. That means our approach is different to the image data generator, such as Tensorflow and PyTorch libraries, which generate new data from the original dataset for each epoch. In the data generator processing, training data is different each time a trained model, different methods. The image data generator is used to avoid overfitting, but it is difficult to show compared results of different methods because generated training dataset is different each time. The data augmentation method was used to balance the dataset between all categories with the expectation of improving the correct rates. The main problem with this approach is that it produces a huge training dataset from the original one, which requires high hardware requirements and significantly increases computational time. The details of the parameters used to generate the dataset are presented in Table 2.

Table 2. The details of parameters for data augmented processing.

Transformation	Random value
Rotation	[−10, 10]
Flip	Left-right, up-down
Contrast	[0.7, 1]
Tx	[−10, 10]
Ty	[−10, 10]
Zx	[0.8, 1]
Zy	[0.8, 1]
Shear	[−5, 5]

In this paper, we also investigate the weighting mechanism by FL [24] that affects the efficiency of the model for different categories of data. This approach deals the problem of data imbalance without data augmentation processing. Different to data augmentation, the loss functions (LFs) applied for multiple classification, but it may less effective because the performance metrics for this problem are composed of indicators such as one versus all accuracy, sensitivity/recall, and specificity. The training task aims to optimize the model's parameters to achieve the lowest loss cost across all datasets, thereby increasing classification performance. However, this approach leads to a seesaw problem where majority categories are more influential than minority categories, resulting in lower weighting towards performance scores.

3.4 Ambiguity Rejection

Normally, a multi-class classification model can be defined as a set of probabilities $P = \{p_1, p_2, .., p_m\}$ where each p_i denotes predicted probability of classifier of the m categories, p_i is the predicted probability of the ith category and the output of the classifier is defined as a function $f(x) = argmax(p_i)$, with $i \in \{1,2,..,m\}$.When we use a per-class confidence thresholds ambiguity rejection module to reject confusion region, the function $f(x)$ is adjusted as the Formula below.

$$f(x) = \begin{cases} reject, \; if \; p_i \leq \delta_i, \forall i \in \{1, 2, .., m\} \\ argmax(p_i), \; i \in \{1, 2, .., m\} \; otherwise \end{cases} \tag{1}$$

where $\delta = \{\delta_1, \delta_2, .., \delta_m\}$ denotes confidence thresholds, of which δ_i is the threshold of ith category (c_i). δ set is usually obtained from a training sample so that the correctly classified accuracy on test dataset is greater than or equal to the pre-set select accuracy e.g., 95%.

In this study, we use the validation dataset to determine the threshold δ_i of the class c_i. More specifically, from the validation dataset, by using classifier, we calculate a set of probabilities P of each sample in this dataset. For a given class c_i, we determine the potential thresholds ($\delta_{possible}$), which are the unique values of the list probability p_i. The most importance question is how to choose the best threshold for the class c_i. For a given threshold $\delta_i \in \delta_{possible}$ of the class c_i, we determine rejected samples by the Eq. 1. For example, we have n rejected samples and there are k samples that are failures (corrected classified by our model). The probability of having more than k failures is ProbFailure(k,n). A given δ_i is acceptable when ProbFailure(k,n) is greater than 1-β, β is a given significance level. For each acceptable δ, we calculate select accuracry and coverage respectively, and threshold of the class c_i is the one with the highest select accuracry and coverage. In this research, ProbFailure(k,n) is estimate by using Binomial Cumlative Distribution function in Eq. 2 as the following formula.

$$binom.cdf\,(k, n, p) = \sum\nolimits_{i=0}^{k} \binom{n}{i} p^i (1-p)^{n-i} \tag{2}$$

where n denotes the number of rejected samples, k denotes the number of failures in n rejected samples, and p denotes the probability that a given rejected sample is failure. A given rejected sample is failure as random, so $p = 0.5$.

4 Experimental Results and Analysis

4.1 Materials and Preprocessing

In this study, the ISIC2018 [25, 26] skin cancer dataset is used to experiment and evaluate the solution. Due to this dataset is still used for a competition then the ground truth labels of testing images are not available. Therefore, the experiment and comparison are based on the training and validation datasets. The original dataset for training contains 10,015 samples and 193 samples for evaluation. The dataset consists of 7 categories, which include Melanoma (MEL), Melanocytic nevus (NV), Basal cell carcinoma (BCC), Actinic keratosis (AKIEC), Benign keratosis (BKL), Dermatofibroma (DF), and Vascular lesion (VASC). The image samples are uniformed 450×600 resolution. For evaluation, the original validation dataset is used as the validation1 dataset. The original training dataset is split into 80% for training and 20% for evaluation as the validation2. Details about the dataset used in this experiment is presented in Table 3.

Table 3. Details of the experimental dataset

	MEL	NV	BCC	AKIEC	BKL	DF	VASC	Total
Training	890	5364	411	262	879	92	114	8012
Validation1	21	123	15	8	22	1	3	193
Validation2	223	1341	103	65	220	23	28	2003
Augmentation	5340	5364	5343	5240	5274	5336	5358	37255

4.2 Evaluation Metrics

To evaluate the performance of the studied methods on the task of feature extraction and classification, we assessed using popular effectiveness measures such as Recall (REC), Accuracy (ACC), Precision (PRE), Specificity (SPE), and F1. Notice that the accuracy metric of multiple classification is different to that of binary classification problem. The accuracy is estimated based on the one versus all retained classes. For each category, the samples are treated as positive samples and other retained classes are treated as negative samples in the binary classification problem. So, the accuracy score criterion differs between binary and multiple classification. However, some other metrics are the same as in binary classification. The effectiveness measured metrics are computed as follows:

$$ACC_i = (TP_i + TN_i)Ns \tag{3}$$

$$ACC = \frac{1}{Ns}\sum_{i=1}^{c} n_i * ACC_i \tag{4}$$

$$Recall = TP/(TP + FN) \tag{5}$$

$$PRE = TP/(TP + FP) \tag{6}$$

$$SPE = TN/(TN + FP) \tag{7}$$

$$F_1 = TP/[TP + \frac{1}{2}(FP + FN)] \tag{8}$$

where Ns is the total number of samples in dataset, $Ns = TP_i + FP_i + FN_i + TN_i$ where TP_i and FP_i are the number of true positive and false positive samples belonging to the category i^{th}, respectively; FN_{i} and $\mathrm{TN_i}$ are the number of false negative and true negative samples belonging to the category i^{th}, respectively. The number of samples of the class i^{th} is n_i. In that approach the accuracy of each class c^{th} is calculated by $\mathrm{TP_c}$/total instances of the class c^{th}. However, this performance measurement is same with Recall ratio. Therefore, we used the above formulation for estimating the accuracy rate.

4.3 Evaluation Results and Analysis

In this study, we experimentalize and analyze feature extraction and classification task using the category cross entropy and FL, AU method and then ambiguity sample rejection for improving high confident disease diagnosis. In amount of solutions for data imbalance treatment, the AU requires higher computational cost for model training due to that it generates more significant new samples for balancing training dataset. We also customized two kinds of feature extraction backbones, such as MobileNet, DenseNet family. These kinds of backbones are representative for different approaches. The MobileNet backbone represents for a small and compact architecture. It is suitable for applying to limited resource computing systems. The DenseNet backbone represents for the dense connected network with a heaving trainable parameter. In general, MobileNets are lightweight architectures, which consist of several million of trainable parameters. However, they achieve high accuracy with different applications. The MobileNet models are efficient mechanisms based on the depth-wise separable convolutions. The DenseNet architecture with dense connection layers through dense blocks. The network layers relate to matching feature-map sizes directly with each other. Each layer obtains additional inputs from all preceding layers and passes on its feature maps to all subsequent layers. The experimental results on the evaluated dataset show that DenseNet121 + FL method reach outperformer on validation dataset1 at 88.08% recall and 94.18 accuracy rate, as depicted in Table 5 of appendix section. Meanwhile, DensseNet201 and FL method reach the best result on validation dataset2 with all criteria. So, the DensNet network family is more stable results comparing to other methods (Fig. 2). Meanwhile, CC method get the lowest with 76.68% Recall at 88.77% accuracy. In overall, the DenseNet family and FL response the best results on ISIC2018 dataset, as illustrated in Fig. 3.

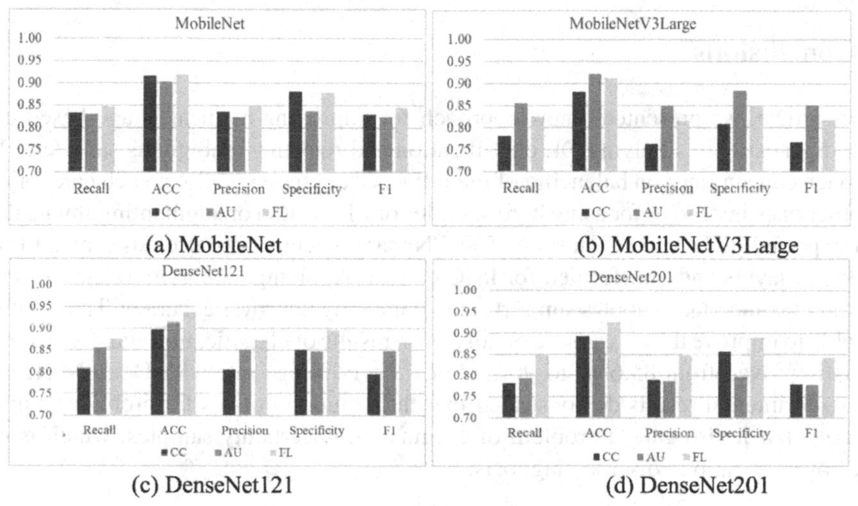

Fig. 2. Experimental results on both evaluation datasets

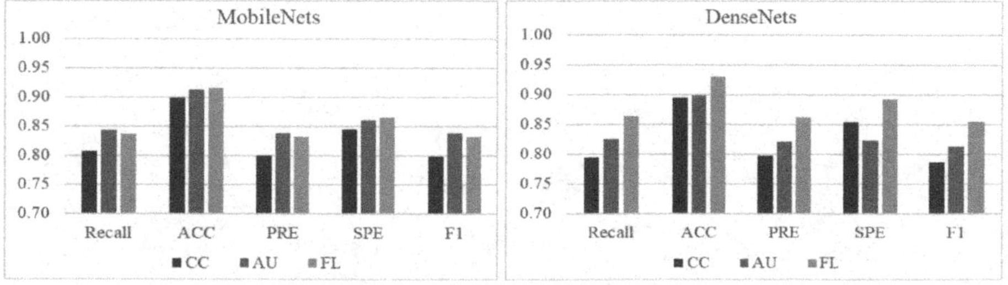

Fig. 3. Average evaluated result of MobileNets and DenseNets family on both validation sets.

In ambiguity reject stage, we adjust δ set so that select accuracy is high, around 95.0% corresponds to error rate at 5%, to ensure acceptable error rate in real-world applications and to compare performances of methods. The validation dataset1 and validation dataset2 are used determine the threshold $\delta_{possible}$ with expected to reach accepted select_recall rate with highest coverage ratio of the class of each category. Some experimental results are shown in Table 4. Ambiguity rejection with $\delta = 0.1$, selected recall ratio was reached about 96.25% at 75% correct coverage ratio with DenseNet121 + AU. Meanwhile, MobileNetV3Large + CC archives 93.19% select recall at 66.24 correct coverage ratio only, as depicted in Table 4 (a). Experimental results also illustrated that the determined coefficient of delta = 0.3, the DenseNet201 + FL achieved the highest precision with 94.91% select_recall at 81.06% correct coverage ratio, while the MobileNet + CC achieved the lowest accuracy with 91.69% select_recall at 80.93% correct coverage ratio, as illustrated in Table 4 (b). According to experimental result shows that CC loss function archives the lowest recall ratio in both situation classification and ambiguity rejection.

5 Conclusions

In this article, we presented a new approach for improving medical image-based diseases diagnosing by applying DL classification and rejecting ambiguous samples. Our approach concentrates on balancing of the influence coefficient ratio of each category to the other ones instead of focusing hard samples of LF method or augmenting image data with expected higher precision ratio. The CNN architecture was also customized fully connected layers and transformed for ISIC dataset. Applying ambiguity rejection stage to removing uncertain samples support for significantly improves accuracy. The solution was able to improve the diagnosing quality from results of classification stage, e.g. recall rate is improved from 85.63% to 96.25% at 75% coverage rate with DenseNet121 + AU, Experimental results demonstrated that this solution utilizes for archiving higher accuracy, but it also gaps a problem of eliminating uncertainty samples, which is not fully coverage ratio in disease diagnosis.

Table 4. Experimental results of ambiguity rejection on both evaluation datasets

(a) Thresh_func is b_cdf and delta=0.1

MobileNet- based ambiguity rejection (delta=0.1)

	base_recall	select_recall	coverage	reject_acc
CC	0.8331	0.9405	0.7481	0.4790
AU	0.8311	0.9544	0.7435	0.4930
FL	0.8478	0.9513	0.7498	0.4583

MobileNetV3Large-based ambiguity rejection (delta=0.1)

	base_recall	select_recall	coverage	reject_acc
CC	0.7806	0.9319	0.6624	0.5088
AU	0.8546	0.9551	0.7772	0.4864
FL	0.8255	0.9375	0.7523	0.5166

DenseNet121- based ambiguity rejection (delta=0.1)

	base_recall	select_recall	coverage	reject_acc
CC	0.8059	0.9535	0.6572	0.4719
AU	0.8563	0.9625	0.7500	0.4586
FL	0.8765	0.9544	0.7950	0.4286

DenseNet201- based ambiguity rejection (delta=0.1)

	base_recall	select_recall	coverage	reject_acc
CC	0.7822	0.9524	0.6546	0.5426
AU	0.7932	0.9496	0.6588	0.4952
FL	0.8498	0.9567	0.7841	0.5165

(b) Thresh_func is b_cdf and delta=0.3

MobileNet- based ambiguity rejection (delta=0.3)

	base_recall	select_recall	coverage	reject_acc
CC	0.8331	0.9169	0.8093	0.5186
AU	0.8311	0.9446	0.7719	0.5139
FL	0.8478	0.9353	0.8208	0.5457

MobileNetV3Large-based ambiguity rejection (delta=0.3)

	base_recall	select_recall	coverage	reject_acc
CC	0.7806	0.9273	0.6875	0.5340
AU	0.8546	0.9354	0.8236	0.5156
FL	0.8255	0.9188	0.7988	0.5528

DenseNet121- based ambiguity rejection (delta=0.3)

	base_recall	select_recall	coverage	reject_acc
CC	0.8059	0.9389	0.7006	0.5015
AU	0.8563	0.9334	0.8193	0.4928
FL	0.8765	0.9300	0.8942	0.6092

DenseNet201- based ambiguity rejection (delta=0.3)

	base_recall	select_recall	coverage	reject_acc
CC	0.7822	0.9315	0.7036	0.5725
AU	0.7932	0.9343	0.6899	0.5120
FL	0.8498	0.9491	0.8106	0.5479

Appendix

Table 5. Detail of classified results on validation dataset1, validation dataset2. The average result is formed (result1 on dataset1 + result1 on dataset1)/2.

MobileNet: Validation on dataset 1

	Recall	ACC	Precision	Specificity	F1
CC	0.8290	0.9241	0.8368	0.9051	0.8260
AU	0.7876	0.8829	0.7790	0.8072	0.7769
FL	0.8394	0.9169	0.8455	0.8802	0.8377

MobileNet: Validation on dataset 2

	Recall	ACC	Precision	Specificity	F1
CC	0.8372	0.9068	0.8339	0.8557	0.8332
AU	0.8747	0.9234	0.8700	0.8660	0.8712
FL	0.8562	0.9196	0.8523	0.8775	0.8521

MobileNet: Average of results

	Recall	ACC	Precision	Specificity	F1
CC	0.8331	0.9154	0.8354	0.8804	0.8296
AU	0.8311	0.9032	0.8245	0.8366	0.8241
FL	0.8478	0.9183	0.8489	0.8788	0.8449

MobileNetV3Large: Validation on dataset 1

	Recall	ACC	Precision	Specificity	F1
CC	0.7668	0.8777	0.7418	0.8206	0.7505
AU	0.8446	0.9237	0.8396	0.8985	0.8402
FL	0.8238	0.9199	0.8161	0.8871	0.8175

MobileNetV3Large: Validation on dataset 2

	Recall	ACC	Precision	Specificity	F1
CC	0.7943	0.8840	0.7840	0.7956	0.7816
AU	0.8647	0.9203	0.8614	0.8689	0.8619
FL	0.8273	0.9045	0.8148	0.8144	0.8174

MobileNetV3Large: Average of results

	Recall	ACC	Precision	Specificity	F1
CC	0.7806	0.8809	0.7629	0.8081	0.7661
AU	0.8546	0.9220	0.8505	0.8837	0.8511
FL	0.8255	0.9122	0.8154	0.8507	0.8175

DenseNet121: Validation on dataset 1

	Recall	ACC	Precision	Specificity	F1
CC	0.8031	0.9006	0.8060	0.8669	0.7892
AU	0.8549	0.9160	0.8531	0.8644	0.8472
FL	0.8808	0.9418	0.8761	0.9013	0.8683

DenseNet121: Validation on dataset 2

	Recall	ACC	Precision	Specificity	F1
CC	0.8088	0.8937	0.8051	0.8337	0.8007
AU	0.8577	0.9133	0.8518	0.8302	0.8497
FL	0.8722	0.9307	0.8709	0.8914	0.8690

DenseNet121: Average of results

	Recall	ACC	Precision	Specificity	F1
CC	0.8059	0.8972	0.8056	0.8503	0.7950
AU	0.8563	0.9147	0.8525	0.8473	0.8484
FL	0.8765	0.9363	0.8735	0.8964	0.8686

DenseNet201: Validation on dataset 1

	Recall	ACC	Precision	Specificity	F1
CC	0.7876	0.9070	0.8020	0.8914	0.7849
AU	0.7876	0.8788	0.7865	0.7980	0.7691
FL	0.8238	0.9173	0.8214	0.8793	0.8094

DenseNet201: Validation on dataset 2

	Recall	ACC	Precision	Specificity	F1
CC	0.7768	0.8757	0.7762	0.8200	0.7719
AU	0.7988	0.8835	0.7885	0.7967	0.7858
FL	0.8757	0.9328	0.8771	0.8983	0.8727

DenseNet201: Average of results

	Recall	ACC	Precision	Specificity	F1
CC	0.7822	0.8913	0.7891	0.8557	0.7784
AU	0.7932	0.8811	0.7875	0.7973	0.7775
FL	0.8498	0.9250	0.8492	0.8888	0.8411

Table 6. Detail of ambiguity rejection results on validation dataset1 and validation dataset2

MobileNet- based ambiguity rejection on dataset1 (delta=0.3)

	base_recall	select_recall	coverage	reject_acc
CC	0.8290	0.9281	0.7927	0.5500
AU	0.7876	0.9833	0.6218	0.5342
FL	0.8394	0.9481	0.7979	0.5897

MobileNet- based ambiguity rejection on dataset2 (delta=0.3)

	base_recall	select_recall	coverage	reject_acc
CC	0.8372	0.9057	0.8258	0.4871
AU	0.8747	0.9058	0.9221	0.4936
FL	0.8562	0.9225	0.8437	0.5016

MobileNetV3Large-based ambiguity rejection on dataset1 (delta=0.3)

	base_recall	select_recall	coverage	reject_acc
CC	0.7668	0.9520	0.6477	0.5735
AU	0.8446	0.9474	0.7876	0.5366
FL	0.8238	0.9182	0.8238	0.6176

MobileNetV3Large- based ambiguity rejection on dataset2 (delta=0.3)

	base_recall	select_recall	coverage	reject_acc
CC	0.7943	0.9025	0.7274	0.4945
AU	0.8647	0.9233	0.8597	0.4947
FL	0.8273	0.9194	0.7738	0.4879

DenseNet121- based ambiguity rejection on dataset1 (delta=0.3)

	base_recall	select_recall	coverage	reject_acc
CC	0.8031	0.9609	0.6632	0.5077
AU	0.8549	0.9416	0.7979	0.4872
FL	0.8808	0.9326	0.9223	0.7333

DenseNet121- based ambiguity rejection on dataset2 (delta=0.3)

	base_recall	select_recall	coverage	reject_acc
CC	0.8088	0.9168	0.7379	0.4952
AU	0.8577	0.9252	0.8407	0.4984
FL	0.8722	0.9274	0.8662	0.4851

DenseNet201- based ambiguity rejection on dataset1 (delta=0.3)

	base_recall	select_recall	coverage	reject_acc
CC	0.7876	0.9632	0.7047	0.6316
AU	0.7876	0.9746	0.6114	0.5067
FL	0.8238	0.9720	0.7409	0.6000

DenseNet201- based ambiguity rejection on dataset2 (delta=0.3)

	base_recall	select_recall	coverage	reject_acc
CC	0.7768	0.8998	0.7024	0.5134
AU	0.7988	0.8941	0.7683	0.5172
FL	0.8757	0.9263	0.8802	0.4958

References

1. Szegedy, C., et al.: Going deeper with convolutions. In: Proceedings of the IEEE Conference on Computer Vision and Pattern Recognition, pp. 1–9 (2015)
2. Howard, A.G., et al.: Mobilenets: Efficient Convolutional Neural Networks for Mobile Vision Applications. arXiv preprint arXiv:1704.04861 (2017)
3. He, K., Zhang, X., Ren, S., Sun, J.: Deep residual learning for image recognition. In: Proceedings of the IEEE conference on computer vision and pattern recognition, pp. 770–778 (2016)
4. Huang, G., Liu, Z., Van Der Maaten, L., Weinberger, K.Q.: Densely connected convolutional networks. In: Proceedings of the IEEE Conference on Computer Vision and Pattern Recognition, pp. 4700–4708 (2017)
5. Tan, M., Le, Q.: Efficientnet: Rethinking model scaling for convolutional neural networks. In: International Conference on Machine Learning, pp. 6105–6114 (2019)
6. Li, L., Talwalkar, A.: Random Search and Reproducibility for Neural Architecture Search. arXiv preprint arXiv:1902.07638 (2019)
7. Bertrand, H., Ardon, R., Perrot, M., Bloch, I.: Hyperparameter optimization of deep neural networks: combining hyperband with Bayesian model selection. Conférence sur l'Apprentissage Automatique (2017)
8. Hoang, V.-T., Huang, D.-S., Jo, K.-H.: 3-D facial landmarks detection for intelligent video systems. IEEE Trans. Industr. Inf. **17**, 578–586 (2020)

9. Kotthoff, L., Thornton, C., Hoos, H.H., Hutter, F., Leyton-Brown, K.: Auto-WEKA 2.0: automatic model selection and hyperparameter optimization in WEKA. J. Machine Learning Res. **18**, 826–830 (2017)
10. Tran, D.-P., Nguyen, G.-N., Hoang, V.-D.: Hyperparameter optimization for improving recognition efficiency of an adaptive learning system. IEEE Access **8**, 160569–160580 (2020)
11. Dikov, G., van der Smagt, P., Bayer, J.: Bayesian Learning of Neural Network Architectures. arXiv preprint arXiv:1901.04436 (2019)
12. Huang, C., Lucey, S., Ramanan, D.: Learning policies for adaptive tracking with deep feature cascades. Proceedings of the IEEE International Conference on Computer Vision, pp. 105–114 (2017)
13. Long, M., Cao, Y., Cao, Z., Wang, J., Jordan, M.I.: Transferable representation learning with deep adaptation networks. IEEE Trans. Pattern Anal. Mach. Intell. **41**, 3071–3085 (2018)
14. Le, N.Q.K., Huynh, T.-T., Yapp, E.K.Y., Yeh, H.-Y.: Identification of clathrin proteins by incorporating hyperparameter optimization in deep learning and PSSM profiles. Comput. Methods Programs Biomed. **177**, 81–88 (2019)
15. Snoek, J., et al.: Scalable bayesian optimization using deep neural networks. International Conference on Machine Learning, pp. 2171–2180 (2015)
16. Carli, P., et al.: Pattern analysis, not simplified algorithms, is the most reliable method for teaching dermoscopy for melanoma diagnosis to residents in dermatology. Br. J. Dermatol. **148**, 981–984 (2003)
17. Qian, C., et al.: A Two-Stage Method for Skin Lesion Analysis. arXiv preprint arXiv:1809.03917 (2018)
18. Yi, H., Jiang, Q., Yan, X., Wang, B.: Imbalanced classification based on minority clustering synthetic minority oversampling technique with wind turbine fault detection application. IEEE Trans. Industr. Inf. **17**, 5867–5875 (2020)
19. Chawla, N.V., Bowyer, K.W., Hall, L.O., Kegelmeyer, W.P.: SMOTE: synthetic minority over-sampling technique. J. Artificial Intell. Res. **16**, 321–357 (2002)
20. Niu, Z., Zhong, G., Yu, H.: A review on the attention mechanism of deep learning. Neurocomputing **452**, 48–62 (2021)
21. Geifman, Y., El-Yaniv, R.: Selectivenet: A deep neural network with an integrated reject option. International Conference on Machine Learning, pp. 2151–2159 (2019)
22. Franc, V., Prusa, D., Voracek, V.: Optimal strategies for reject option classifiers. J. Mach. Learn. Res. **24**, 1–49 (2023)
23. Kashani Motlagh, N., Davis, J., Anderson, T., Gwinnup, J.: Learning when to say "i don't know. Advances in Visual Computing: 17th International Symposium, ISVC 2022, San Diego, CA, USA, October 3–5, 2022, Proceedings, Part I, pp. 196–210 (2022)
24. Lin, T.-Y., Goyal, P., Girshick, R., He, K., Dollár, P.: Focal loss for dense object detection. Proceedings of the IEEE Conference on Computer Vision and Pattern Recognition, pp. 2980–2988 (2017)
25. Codella, N., et al.: Skin Lesion Analysis Toward Melanoma Detection 2018: A Challenge Hosted by the International Skin Imaging Collaboration (isic). arXiv preprint arXiv:1902.03368 (2019)
26. Tschandl, P., Rosendahl, C., Kittler, H.: The HAM10000 dataset, a large collection of multi-source dermatoscopic images of common pigmented skin lesions. Scientific data **5**, 1–9 (2018)

Data Mining and Machine Learning

Towards Developing an Automated Chatbot for Predicting Legal Case Outcomes: A Deep Learning Approach

Shafiq Alam[1(✉)], Rohit Pande[2], Muhammad Sohaib Ayub[3], and Muhammad Asad Khan[4]

[1] School of Management, Massey University, Auckland, New Zealand
salam1@massey.ac.nz

[2] Department of Computer Science, Lahore University of Management Sciences, Lahore, Pakistan
15030039@lums.edu.pk

[3] Duco Consultancy Limited, Gurgaon, India

[4] Department of Telecommunication, Hazara University, Mansehra, Pakistan
asadkhan@hu.edu.pk

Abstract. The accurate prediction of legal case outcomes is crucial for effective legal advocacy, which relies on a deep understanding of past cases. Our research aims to develop an automated chatbot for predicting the outcomes of employment-related legal cases using deep learning techniques. We compare and significantly improve on mining the New Zealand Employment Relations Authority (NZERA) dataset, using various deep learning models such as Latent Dirichlet Allocation (LDA) with different activation functions of Recurrent Neural Network (RNN) to determine their predictive performance. Our study's findings show that SoftSign-based RNN-LDA models have the highest accuracy and consistency in predicting outcomes.

Keywords: Legal advocacy · Predictive models · Semantic analysis · Deep learning

1 Introduction

Legal advocacy relies heavily on predicting the outcomes of new cases, which requires a deep understanding of the details contained within past cases [9]. In fact, one of the main skills in legal advocacy is the ability to study past cases and make informed decisions for predicting the potential outcome of new legal cases based on that knowledge. Therefore, it is crucial to develop knowledge based on the specifics of past cases in order to forecast future cases accurately. With this in mind, an automated system that can learn from past cases and make predictions for future cases could be incredibly valuable for both the general public and legal practitioners [9]. This kind of system has the potential to offer initial assessments of new cases, taking into account the provided circumstances.

N. T. Nguyen et al. (Eds.): ACIIDS 2023, LNAI 13995, pp. 163–174, 2023.
https://doi.org/10.1007/978-981-99-5834-4_13

The aims of this research included; retrieving and processing past employment case documents from the NZERA; conducting a comprehensive semantic analysis of these documents to identify patterns and relationships within the text; using various deep-learning models to predict the outcomes of these cases and comparing and assess various matrices of these models.

To accomplish these objectives, the study extracted employment case documents from an online dataset, conducted feature selection, applied semantic analysis through Latent Dirichlet Allocation (LDA), and developed deep neural network models to predict employment case outcomes. The accuracy of the models was then evaluated and compared.

This paper extends and significantly improves upon the previous work [13] by a significant increase in data, additional matrices, and improved results. This study's contribution lies in its unique approach to analyzing employment case documents, which has not been widely explored in prior research. Moreover, it is the first known effort to combine feature selection, semantic analysis, and deep learning models to predict employment case outcomes, potentially transforming legal research and fostering future developments in the field. The paper comprises a review of prior research, details of the experimentation, results, and discussion, key findings and their implications, and potential future research.

2 Related Work

The prior research has primarily utilized automated analysis of legal text to extract meaning and predict outcomes for generic cases, as shown in Table 1.

The first application of automated analysis of legal text involves identifying the semantics within case documents using unsupervised machine learning, including LDA on legal documents in China [4], and [5] found that it underperformed against Latent Semantic Analysis (LSA) when analyzing Singaporean Supreme Court judgments. [15] used different models for sequence labeling of Lahore High Court judgments. In contrast, [2] used summarization algorithms such as to analyze Indian Supreme Court case judgments and obtained mixed results.

The second application is to predict legal case outcomes through supervised shallow and deep learning methods, which include support vector machines, random forests, decision trees, and gradient-boosted machines [12,16,18]. Efficient data analysis is essential for informed decision-making and extracting meaningful insights from human-sourced data [1]. Recurrent neural networks [20] have also been used to achieve a low F-measure of 0.36, precision of 0.34, and recall of 0.42 for analyzing Chinese civil court cases. Another notable work [17] compared the performance of Recurrent Neural Network (RNN) and Convolutional Neural Network (CNN) on US Supreme Court decisions and found that CNN outperformed RNN with an accuracy of 72.4% to 68.6%, respectively. [7] used various machine learning and deep learning models to predict Brazilian court decisions. Similarly, [11] employed Support Vector Machine (SVM) to analyze European Court of Human Rights cases achieving an accuracy of 65% with their model.

Table 1. Techniques and their purpose in legal NLP.

Techniques	Purpose
LDA [4]	Identifying semantics in Chinese legal documents
LDA and LSA [5]	Identifying semantics in Singaporean Supreme Court judgments
Hidden Markov Models, Maximum Entropy Markov Models, and Conditional Random Fields [15]	Sequence labeling of Lahore High Court judgments to extract topics
CaseSummarizer, LetSum, and GraphicalModel [2]	Analyzing Indian Supreme Court case judgments using summarization algorithms
Support Vector Machines, Random Forests, Decision Trees, and Gradient-Boosted Machines [12,16,18]	Predicting legal case outcomes using shallow learning methods
Recurrent Neural Networks [20]	Predicting legal case outcomes using deep learning methods
RNN and CNN [17]	Analyzing US Supreme Court decisions
SVM [11]	Analyzing European Court of Human Rights cases
RNN [6,9,14]	Predicting legal case outcomes for Chinese Judgement Online (CJO) cases
SVM, CNN, and RNN [22]	Predicting law articles, charges, and terms of penalty in criminal cases from the Supreme Court of China
RNN-based MANN [8]	Predicting law articles, charges, and prison terms in criminal cases from the Supreme Court of China
CNN [21]	Predicting law articles, charges, and terms of penalty in Chinese criminal cases

In the local domain of China [6] employed RNN on China Judgement Online (CJO) cases and achieved an accuracy of 76.3% while [14] utilized RNN to achieve an accuracy rate of 90.01% and [9] developed an RNN-based model named AutoJudge, achieving an accuracy of 82.2% for the same dataset. Furthermore, [22] studied criminal cases from the Supreme Court of China and used SVM, CNN, and RNN to predict different outcomes, stating that CNN had a higher accuracy than SVM, while their RNN models failed to converge. Their best CNN models achieved an accuracy of 84.7%, 83.6%, and 40.0% for predicting law articles, charges, and terms of penalty, respectively. On the other hand, [8] conducted the same experiment using an RNN-based Multichannel Attentive Neural Network (MANN) and achieved improved accuracy of 91.3%,

95.5%, and 69.3% for predicting law articles, charges, and prison terms, respectively. Additionally, [21] achieved high accuracy rates of 97.6%, 97.6%, and 78.2% for predicting law articles, charges, and terms of penalty using CNN.

3 Experimentation

The section describes the methodology used to analyze New Zealand Employment Relations Authority cases. It covers data retrieval, processing, semantic analysis, model training, and evaluation, including techniques such as LDA for semantic analysis and tokenization for secondary data processing.

3.1 Data Retrieval

To begin, we retrieved NZERA case documents from the online dataset of Employment Law [3]. Our dataset contained 12,389 case documents spanning from January 1st, 2005 to May 22nd, 2022. However, certain cases were excluded due to inconsistencies in their URL naming.

3.2 Data Preprocessing

Our data preprocessing included the following steps:

Paragraph Extraction: Each paragraph in the case documents was found to contain a unique semantic feature, which required identification for feature extraction. However, the PDF format of the raw data did not preserve the structure of the paragraphs during text extraction. To solve this, regular expressions were used to identify and extract each paragraph as an individual feature. Additionally, all numerical data was retained as it contained meaningful information within the cases. The dataset for the study did not have any annotations or metadata, so individual features of each case had to be manually derived from the raw text. The first step was to identify paragraphs or sections representing the document preamble (P) and the case determinations (D) expressed by the presiding authority. However, as most cases were interim court hearings without final determinations, only 30.66% of the documents (3,230 cases) could be identified as having the D feature. Feature selection was performed through keyword searches. Two types of data were derived: (1) full documents (FD) including both P and D features and (2) full documents with determinations redacted ($FD - D$) to enable the independent assessment of case circumstances alone

Manual Document Labelling: The metadata of the collected case documents didn't contain the case outcomes, so binary labels for the cases had to be manually added by reading the documents. Cases dismissed by the authority were counted as losses for the applicant. Out of 3,230 documents, 260 cases (130 victories and 130 losses) were chosen for classification with an attempt to balance the number of cases between the two labels to achieve fairness in classification [10].

3.3 Semantic Analysis

We analyzed the semantics of 12, 311 case documents using LDA, an unsupervised topic-detection method. The analysis was conducted by testing LDA with different numbers of topics, a maximum iteration of 5, and a learning offset of 50. A text feature extraction method based on term frequency was used to identify distinct features and a variety of top words were selected from each topic to create 10 topic-clusters, each containing several words.

3.4 Secondary Data Processing

The data was processed further after the identification of topic-clusters through semantic analysis. The processing involved measuring document similarity using cosine similarity, reducing document size by only keeping words related to the assigned topic-cluster, tokenizing the data, and converting it into 128-dimensional word embeddings to prepare it for model training.

3.5 Model Training

After conducting semantic analysis to identify topic-clusters, the data was further processed and utilized for supervised learning using three deep neural network models. Each model was tested on both FD and $FD-D$ features. The model employed the Gated Recurrent Unit (GRU) variant of the Recurrent Neural Network (RNN) on a TensorFlow platform and was trained for 26 epochs. The model was tested using two different activation functions: Sigmoid and SoftSign. Additionally, RNN served as the base model to assess the performance of various Latent Dirichlet Allocation (LDA) parameters, such as K, n, and d, as discussed in Sect. 3.3.

3.6 Model Evaluation

We evaluated LDA-RNN models in the single run and cross-validated the models with 10-folds to assess their accuracy, precision, recall, and F1-score.

4 Results and Discussion

The section describes an experiment with two stages: semantic analysis and predictive analysis. In the semantic analysis stage, a recurrent neural network (RNN) was used to classify topics based on Latent Dirichlet Allocation (LDA) models with varying complexity. The best-performing model had five topics, 5, 000 features, and the top 300 words. In the predictive analysis stage, the LDA model was used to train two variations of RNN with different configurations of the number of MultiRNN cells and the inclusion of determinations. The best-performing RNN model had SoftSign activation function, FD-D features, and three cells for single-run accuracy, while the best 10-fold cross-validation accuracy was achieved using RNN with SoftSign activation function, FD-D features, and one cell, which had the highest precision and recall.

4.1 Semantic Analysis

Semantics within the text are analyzed in the first stage using RNN with topic-clusters created by LDA comprising various features, top words, and topics.

Various Number of Features ***(d):*** In the initial experiments, the accuracy of the model was assessed using RNN with topic-clusters created by LDA comprising d features. These LDA models had 5 clusters, each limited to the top 300 words. The results presented in Table 2 revealed that when the model identified 5,000 features within the NZERA data, the cross-validation accuracy was significantly higher at 0.7299 compared to when it identified 2,000 features. This indicates that the model performed better with a higher number of features.

Table 2. Accuracy of LDA-RNN with various numbers of features (d), top words (n) and LDA topics (k).

Parameter	Single Run	Cross Validation
d = 2,000	0.5814	0.6652
d = 5,000	**0.8095**	**0.7299**
n = 50	0.7442	0.7203
n = 100	0.6512	0.6656
n = 200	0.7442	**0.7900**
n = 300	**0.8095**	0.7299
n = 400	0.7209	0.6470
k = 4	0.6977	0.7110
k = 5	**0.8095**	**0.7299**
k = 6	0.6047	0.6154

Various Top Words ***(n):*** The next step of the experiments involved testing the performance of LDA-created topic-clusters that were restricted to different values of n, representing the number of top words in each cluster. These models consisted of 5 clusters created from a corpus of 5,000 features. According to the results presented in Table 2, the top-performing models had n values of either 200 or 300. While the former yielded the highest cross-validation accuracy of 0.79, the latter produced the best single-run accuracy. These findings suggest that LDA models limited to the top 200 or 300 words in each topic-cluster were more accurate than those with higher or lower n values.

Various LDA Topics ***(K):*** We also evaluated LDA models with different values of K, which represents the number of topics, using 5,000 features limited to the top 300 words in each topic-cluster. The results presented in Table 2 indicate that the LDA model with $K = 5$ provided the highest single-run accuracy (0.8095) and cross-validation accuracy (0.7299). This implies that using five topics in LDA models produced better results. Based on the results, which showed that LDA-created topic-clusters with 5 topics, 5,000 features, and limited to the top 300 words provided favorable outcomes with NZERA data, the subsequent experiments were carried out using LDA models with this configuration.

4.2 Predictive Analysis

In the subsequent phase, the LDA model with the aforementioned parameters was employed to train classification models using two variations of RNN. The results of these experiments with various configurations of the number of MultiRNN cells and inclusion of determinations are provided in Table 3.

Table 3. RNN Model Performance Using SoftSign and Signmoid Activation Function.

Text Features	Activation Function	RNN Cells	Single Run				10-fold Cross-Validation			
			Accuracy	Precision	Recall	F1-Score	Accuracy	Precision	Recall	F1-Score
FD	SoftSign	1	0.6923	0.6897	**0.7407**	0.7143	0.6462	0.6618	0.6109	0.6295
		2	0.7115	0.7727	0.6296	0.6939	**0.6923**	0.6928	0.6822	**0.6834**
		3	0.6923	0.7619	0.5926	0.6667	0.6308	0.6346	0.6514	0.6310
		4	0.6538	0.7647	0.4815	0.5909	0.6115	0.6212	0.6109	0.5847
	Sigmoid	1	0.7115	0.7500	0.6667	0.7059	0.6608	0.6456	0.6612	0.6450
		2	0.7500	0.8182	0.6667	0.7347	0.6500	0.6812	0.5706	0.6151
		3	0.5577	0.7000	0.2593	0.3784	0.6200	0.6401	0.6192	0.6031
		4	0.6346	0.7222	0.4815	0.5778	0.6269	0.6332	0.6817	0.6392
$FD-D$	SoftSign	1	0.6538	0.6957	0.5926	0.6400	0.6808	**0.7010**	0.6579	0.6722
		2	0.7115	0.7727	0.6296	0.6939	0.6599	0.6670	**0.6824**	0.6672
		3	**0.7692**	0.8000	**0.7407**	**0.7692**	0.6376	0.6578	0.6607	0.6426
		4	0.6154	0.6522	0.5556	0.6000	0.6038	0.6225	0.5457	0.5752
	Sigmoid	1	0.5385	0.5652	0.4815	0.5200	0.6442	0.6478	0.6409	0.6366
		2	0.6346	0.6333	0.7037	0.6667	0.6084	0.6143	0.5931	0.5948
		3	0.6923	0.7391	0.6296	0.6800	0.6215	0.6553	0.6042	0.6104
		4	0.6923	**0.8235**	0.5185	0.6364	0.6423	0.6401	0.6356	0.6281

Analysis of *FD*. The study compared the performance of recurrent neural network models with various configurations in predicting the outcomes of NZERA cases based on the evidence within the case circumstances. Sigmoid activation function with 2 MultiRNN cells demonstrated the best performance with 75% accuracy and 82% precision, whereas RNN-Softsign shows the best performance with 2 cells in 10-fold cross-validation. The 10-fold cross-validation results show that Softsign is most effective in realistically predicting the outcome of NZERA cases.

Analysis of *FD-D*. The performance results show that the RNN-SoftSign based configurations without the evidence within the case circumstances ($FD-D$) were most accurate and consistent in predicting the outcomes of cases. These results demonstrate the potential for using LDA in combination with deep neural networks to predict case outcomes, even before they are officially determined, based solely on the case circumstances.

Analysis of Single Run Vs 10-Fold Cross Validation. Figure 1 shows the performance matrices for the single run and 10-fold cross-validation using various RNN configurations.

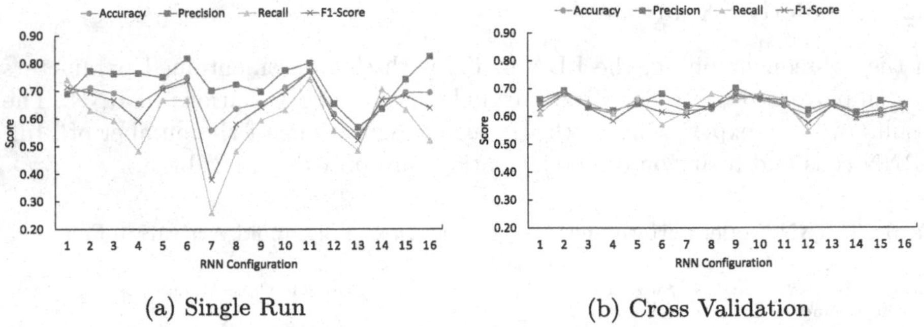

(a) Single Run (b) Cross Validation

Fig. 1. Single run and cross-validation performance matrices using various RNN configurations.

Figure 2 shows the comparison of single run vs cross-validation performance, showing that the single run has slightly higher values for all the metrics compared to cross-validation. This is expected, as the single run uses all the data for training and testing, while cross-validation uses only a subset of the data for testing and the rest for training. Therefore, the single run is more likely to overfit the data, resulting in higher metrics values. Cross-validation, on the other hand, provides a more realistic estimate of the performance.

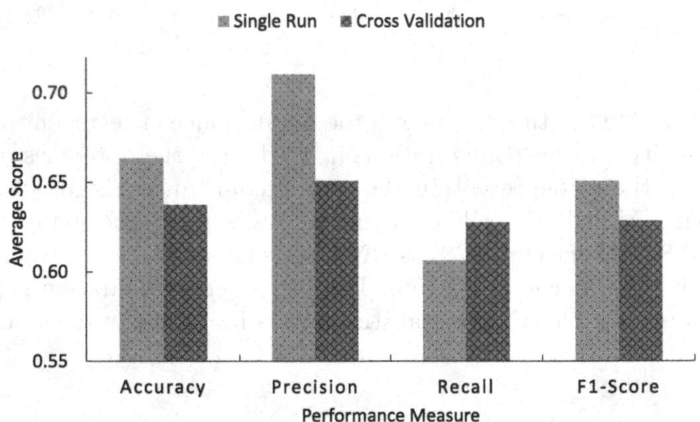

Fig. 2. Comparison of single run and 10-fold cross-validation using an average of the performance matrices.

4.3 CNN with LDA

In this stage of the study, we tested the performance of CNN with both FD and *FD - D*. The CNN model used a batch size of 64 and a dropout rate of 0.5. These experiments used 5 LDA clusters which were created using 5, 000 features and 300 top words. A plot showing the accuracy of this model is provided in Fig. 3. The results are provided in Table 4.

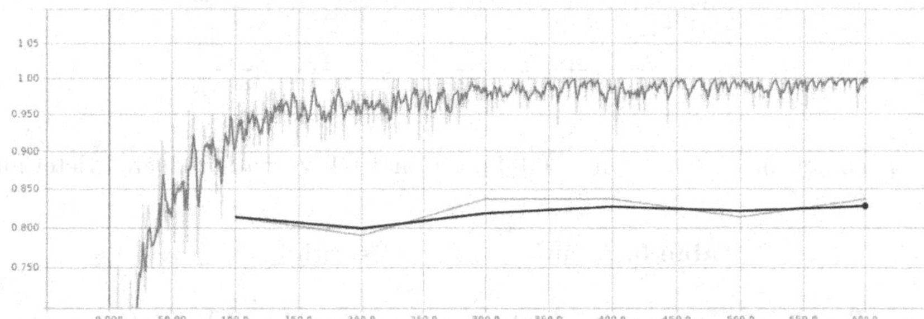

Fig. 3. Train-Test Accuracies with CNN. It shows the accuracy of training data in orange, and test data, in blue. (Color figure online)

Table 4. Accuracy of CNN with LDA.

Features	Single Run	Cross-validation
Full Documents (*FD*)	**0.8372**	**0.7526**
Full Documents with Redacted Determinations (*FD - D*)	0.7674	0.7206

The results show that the model analyses *FD* with consistently higher accuracy than it does for *FD - D*. Although the cross-validation accuracy of *FD-D* at 0.7206% can be considered reasonably high, this model may not perform very well when predicting outcomes based on case circumstances alone.

4.4 CAPSULES with LDA

In this section, we have discussed the performance of Capsules with *FD* and *FD-D*. The LDA implementation had 5 clusters created with 5, 000 features and 300 top words. These experiments used multidimensional GloVe embeddings [19] instead of word2vec. The performance of 50-dimensional, 128-dimensional, and 300-dimensional embeddings was tested. A sample plot of the accuracy is provided in Fig. 4 and the results in Table 5.

The results show that when analyzing *FD*, 300-dimensional embeddings provided the highest cross-validation accuracy of 0.7758. Meanwhile, 50-dimensional embeddings provided the highest accuracy of 0.8140 with the analysis of *FD*. Both 50-dimensional and 128-dimensional embeddings performed equally well with Capsules when predicting outcomes from case circumstances alone.

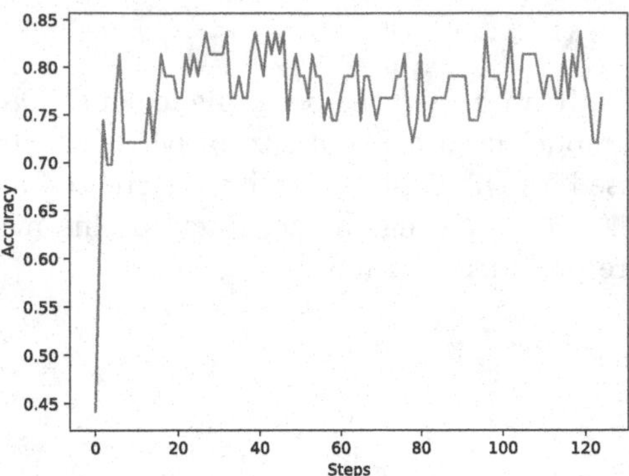

Fig. 4. Accuracy in capsules with 300-dimensional GloVe with redacted determinations.

Table 5. Accuracy of Capsules with LDA.

Embedding Dimensions	Features	Single Run	Cross Validation
50	*FD*	**0.8140**	0.7571
	FD - *D*	0.7674	0.7615
128	*FD*	0.6512	0.7617
	FD - *D*	0.7674	0.7519
300	*FD*	0.7907	**0.7758**
	FD - *D*	0.7476	0.7429

5 Conclusions and Recommendations

The main goal of our research is to develop an automated chatbot that predicts the outcomes of legal cases related to employment relationships, analyze the semantics of legal case documents of the Employment Relations Authority of New Zealand, and compares the performance of various deep learning models in predicting these outcomes.

We retrieve the original data and preprocess it by labeling the extracted text features. We then conduct a comprehensive semantic analysis of the data using LDA to identify patterns and relationships within the text. After that, we implement multiple deep-learning models to forecast the results of these cases. Our research concludes that LDA models with 5 topics and 5,000 features, restricted to the top 300 words, exhibit exceptional performance.

This research contributes a novel approach to the analysis of employment case documents by combining feature selection, semantic analysis, and deep learning models. The results show that LDA models based on RNN-SoftSign demonstrated superior accuracy and consistency and were proficient in making accurate predictions solely based on case circumstances. These findings hold promise for the use of automated chatbots for legal advice and preliminary assessments.

However, to enhance the performance of the model, additional research is necessary, such as testing alternative algorithms and adjusting LDA hyperparameters.

Acknowledgment. The authors acknowledge Citizen AI and Massey University, New Zealand, for providing support and funding the project.

References

1. Ali, S., Ahmad, M., Hassan, U.U., Khan, M.A., Alam, S., Khan, I.: Efficient data analytics on augmented similarity triplets. In: International Conference on Big Data, pp. 5871–5880. IEEE (2022)
2. Bhattacharya, P., Hiware, K., Rajgaria, S., Pochhi, N., Ghosh, K., Ghosh, S.: A comparative study of summarization algorithms applied to legal case judgments. In: Azzopardi, L., Stein, B., Fuhr, N., Mayr, P., Hauff, C., Hiemstra, D. (eds.) ECIR 2019. LNCS, vol. 11437, pp. 413–428. Springer, Cham (2019). https://doi.org/10.1007/978-3-030-15712-8_27
3. Employment New Zealand: Employment Law Database (2018). https://www.employment.govt.nz/elaw-search
4. Hao, Z., Wei, X., Hu, H.: A comparative method of legal documents based on LDA. In: Abawajy, J., Choo, K.-K.R., Islam, R., Xu, Z., Atiquzzaman, M. (eds.) ATCI 2018. AISC, vol. 842, pp. 271–280. Springer, Cham (2019). https://doi.org/10.1007/978-3-319-98776-7_29
5. Howe, J.S.T., Khang, L.H., Chai, I.E.: Legal area classification: a comparative study of text classifiers on singapore supreme court judgments. CoRR abs/1904.06470 (2019)
6. Jiang, X., Ye, H., Luo, Z., Chao, W., Ma, W.: Interpretable rationale augmented charge prediction system. In: International Conference on Computational Linguistics: System Demonstrations, pp. 146–151 (2018)
7. Lage-Freitas, A., Allende-Cid, H., Santana, O., Oliveira-Lage, L.: Predicting Brazilian court decisions. PeerJ Comput. Sci. **8**, e904 (2022)
8. Li, S., Zhang, H., Ye, L., Guo, X., Fang, B.: MANN: a multichannel attentive neural network for legal judgment prediction. IEEE Access **7**, 151144–151155 (2019)
9. Long, S., Tu, C., Liu, Z., Sun, M.: Automatic judgment prediction via legal reading comprehension. In: Sun, M., Huang, X., Ji, H., Liu, Z., Liu, Y. (eds.) CCL 2019. LNCS (LNAI), vol. 11856, pp. 558–572. Springer, Cham (2019). https://doi.org/10.1007/978-3-030-32381-3_45
10. Mansoor, H., Ali, S., Alam, S., Khan, M.A., Hassan, U.U., Khan, I.: Impact of missing data imputation on the fairness and accuracy of graph node classifiers. In: International Conference on Big Data, pp. 5988–5997. IEEE (2022)
11. Masha, M., Michel, V., Martijn, W.: Using machine learning to predict decisions of the European Court of Human Rights. Artif. Intell. Law **28**(2), 237–266 (2020)
12. Nay, J.J.: Predicting and understanding law-making with word vectors and an ensemble model. PLoS ONE **12**(5), e0176999 (2017)
13. Pande, R., Alam, S.: Predicting the outcome of judicial cases using semantic analysis. In: Symposium Series on Computational Intelligence (SSCI), pp. 1757–1761. IEEE (2020)
14. Shang, L., et al.: Prison term prediction on criminal case description with deep learning. Comput. Mater. Continua **62**(3), 1217–1231 (2020)

15. Sharafat, S., Nasar, Z., Jaffry, S.W.: Legal data mining from civil judgments. In: Bajwa, I.S., Kamareddine, F., Costa, A. (eds.) INTAP 2018. CCIS, vol. 932, pp. 426–436. Springer, Singapore (2019). https://doi.org/10.1007/978-981-13-6052-7_37
16. Sulea, O.M., Zampieri, M., Vela, M., van Genabith, J.: Predicting the law area and decisions of french supreme court cases. CoRR abs/1708.01681 (2017)
17. Undavia, S., Meyers, A., Ortega, J.E.: A comparative study of classifying legal documents with neural networks. In: Federated Conference on Computer Science and Information Systems, pp. 511–518. IEEE (2018)
18. Virtucio, M.B.L., Aborot, J.A., Abonita, J.K.C., et al.: Predicting decisions of the philippine supreme court using natural language processing and machine learning. In: Computer Software and Applications Conference (COMPSAC), vol. 2, pp. 130–135. IEEE (2018)
19. Wang, Y., Sun, A., Han, J., Liu, Y., Zhu, X.: Sentiment analysis by capsules. In: World Wide Web Conference, pp. 1165–1174. International World Wide Web Conferences Steering Committee (2018)
20. Xi, R., Zhenxing, K.: Hierarchical RNN for information extraction from lawsuit documents. In: Proceedings of the International MultiConference of Engineers and Computer Scientists, vol. 1 (2018)
21. Xiao, C., et al.: CAIL 2018: a large-scale legal dataset for judgment prediction. arXiv preprint arXiv:1807.02478 (2018)
22. Zhong, H., Guo, Z., Tu, C., Xiao, C., Liu, Z., Sun, M.: Legal judgment prediction via topological learning. In: Conference on Empirical Methods in Natural Language Processing, pp. 3540–3549 (2018)

Fuzzy-Based Factor Evaluation System for Momentum Overweight Trading Strategy

Chi-Fang Chao[1(✉)], Mu-En Wu[1], and Ming-Hua Hsieh[2]

[1] Department of Information and Finance Management, National Taipei University of Technology, Taipei, Taiwan
s9860320@gmail.com

[2] Department of Risk Management and Insurance, National Chengchi University, Taipei, Taiwan

Abstract. Futures are not only used for offset hedging but also widely used in speculative trading. Due to the high leverage characteristic, managing risks with trading strategies is ver important. Currently, there is no systematic approach to evaluate the suitability of trading strategies in futures. In this article, we propose a novel trading strategy factor evaluation system based on fuzzy set theory as a solution to address this issue. The system is composed of random trading algorithm, profitability indicator, and fuzzy quantification module. First, the objective of utilizing a random trading algorithm is to eliminate the impact of market information and extract the characteristics of underlying futures. Then the profitability indicator used to evaluate the performance of probability distribution of profitability of strategies. Fuzzy quantitation module maps profitability indicator to fuzzy degrees between 0 to 1, which make the performance more comparable. The fuzzy degree of the system provide investor the suitability and profitability of strategies to make the trading decision. Experimental results demonstrate the consistency of correlation coefficients in both training set and testing set, indicating the effectiveness and robustness of the evaluation system. The upcoming research concentrates on improving the adaptability of system to other futures and dealing with the impact of uncertainty in membership function.

Keywords: Fuzzy Theory · Profitability Indicator · Trading Strategy · Random Trading

1 Introduction

1.1 Background and Motivation

Derivative financial instruments are mainly used in risk hedging and speculative trading, and futures is one of the most representative instruments. Futures

N. T. Nguyen et al. (Eds.): ACIIDS 2023, LNAI 13995, pp. 175–185, 2023.
https://doi.org/10.1007/978-981-99-5834-4_14

originated from the agricultural product markets that is used to solve the counterparty risk in forwards through the establishment of cleaning house. In retrospect, the evolution of futures are mainly according to the demands of both parties. Since 1972, Chicago Mercantile Exchange (CME) developed multiple financial futures, such as foreign exchange futures on 1972, interest rate futures on 1975, and index futures on 1982. In 1992, CME proposed CME Globex, the first futures e-trading system in the world, and led global futures industries into new era of electronic trading.

With the rapid development of information technology and the popularization of electronic equipment, in last decades, lots of speculative trader are able to use quantitative analysis to search for profit in markets. However, since futures trading on margin, traders only required to pay parts of the contract value as margin to take position, there exists leverage effect in futures trading. The high leverage allows trader to get higher return relative to principal, but also exposes trader to the risk of losing more than principal. Therefore, the most important thing to futures speculative traders is controlling the losses by establishing systematic quantitative trading strategies. Only when traders stay away from unbearable losses can they continue to survive in the market and provide the liquidity to reflect the price without slippage while pursuing profit.

The underlying assets of futures have a great variety, in general, there does not exist a speculative trading strategy could be applied to all futures contract. At present, most of speculative traders uses specific quantitative indicators, such as accumulated payoff, maximum drawdown, and sharpe ratio to measure the suitability of trading strategy for the contract, there is no systematic method to support traders in evaluating the profitability of strategies to futures contracts.

1.2 Purpose

The suitability of a trading strategy for a futures contract is not a binary classification problem, but a continuous regression issue. This study aims to propose a novel fuzzy system with a customized membership function based on type-2 fuzzy set to evaluate the suitability to provide support for traders to make a rational decision.

First, we employed random trading algorithm for simulation to generate the probability density function of profitability in order to eliminate the impact of market informative data. After that, we defined a profitability indicator to measure the suitability of trading strategy for futures contract. Finally, we designed a fuzzy quantification module to analyze the suitability and profitability.

2 Literature Review

In this section, we first review related work on development of trading strategy and profitability index used to measure the performance. Then, we survey the literature and mechanism of fuzzy set theory used in this paper.

2.1 Development of Trading Strategy and Profitability Indicator

Trading strategies can be broadly categorized into two types based on their trading behavior: momentum-type and contrarian-type strategies. Momentum-based trading strategies, such as the notable Turtle trading strategy introduced by Richard Dennis, emphasize taking positions during price breakouts and overweighting positions based on sustained trends [2], have received significant attention and adoption in both academia and industry. Ryan Jones proposed the fixed ratio overweight methodology to ensure consistent capital growth for each position in momentum trading [3]. In this study, we present a momentum-type trading strategy inspired by the Turtle trading approach and incorporate stop-loss and overweight mechanisms using the fixed ratio method.

Trading indicators are commonly used to evaluate the performance of trading strategies, focusing on measures of profitability and risk [4,5]. Profit Factor and Sharpe Ratio are widely employed indicators for evaluating both profitability and risk concurrently [6]. Profit Factor represents the profits relative to the losses, indicating the profitability per unit of loss. On the other hand, Sharpe Ratio considers the ratio of returns to the standard deviation of returns, providing a measure of profitability per unit of risk [12]. There are other indicators that provide valuable insights into trading performance. The Sortino ratio, similar to Sharpe ratio, focuses specifically on downside risk [13]. The Maximum Drawdown is an indicator that measures the largest decline in portfolio value from a peak to a subsequent trough [14]. The Calmar Ratio provides a measure of risk-adjusted return for a trading strategy [15]. While there are various trading indicators available, not all of them are applicable to the proposed system. Indicators like the Sortino ratio, maximum drawdown, and Calmar ratio primarily focus on only downside risks and may not provide a comprehensive evaluation of the overall risk. Additionally, the profit factor is not suitable as it assumes random trading rather than real-world trading scenarios. Therefore, we have chosen to utilize the Sharpe Ratio, which objectively evaluates both upward and downward risks, and have made adjustments to align it more effectively with our specific needs.

$$SR = \frac{R_p - R_f}{\sigma_p} \tag{1}$$

Equation 1 represents the formula for Sharpe Ratio, which calculates the ratio as expected return of portfolio (R_p) minus the risk-free rate (R_f) divided by the standard deviation of the portfolio return (σ_p).

2.2 Fuzzy Set Theory

Fuzzy set theory [7] provides a mathematical framework to handle uncertainty and imprecision. In this theory, sets are known as fuzzy sets A, and their membership values are determined by a membership function μ. The membership function assigns a degree of membership to each element in the set X, allowing for a more nuanced representation of uncertainty. Unlike binary values of 0 or 1, the membership values can range between 0 and 1, indicating the degree of membership or likelihood that an element belongs to a set. This concept could be demonstrated as $\mu_A : X \rightarrow [0,1]$.

$$\mu_A : X \rightarrow [0,1] \tag{2}$$

The theory has extensive applications across various fields, including financial trading [11], artificial intelligence [8], computer science [9], decision theory [10]. In financial trading, for instance, where uncertainties in market conditions and investor sentiment exist, fuzzy logic proves to be an effective tool for capturing and modeling these uncertainties.

Membership functions are essential for quantitatively representing imprecise or uncertain data within fuzzy sets. Depending on the characteristics of the fuzzy set, different shapes of membership functions, such as triangles [16] or trapezoids [17], can be employed. These shapes feature increasing and decreasing slopes, as well as a peak where the membership value is equal to 1, allowing for the accurate representation of gradual transitions and uncertainties in the data.

3 Proposed Trading Strategy Factor Valuation System

We developed a random trading algorithm to evaluate the profitability of a specific futures trading strategy independently of market information. It was applied to a momentum overweight strategy (MOS) based on turtle trading and a fixed overweight method. To assess trading performance, we introduced a profitability indicator based on the probability density function of accumulated payoff. Additionally, a fuzzy quantification was established using fuzzy set theory to interpret futures characteristics. The quantification utilized the profitability indicator to generate degree of fuzzy indicating the suitability of the momentum overweight strategy.

This section provides an overview of the RTA for MOS with long-side trading. The algorithm utilizes a random process to determine whether to take a position at a given time point. If a position is taken, the algorithm records the price and keeps track of the price movement. Overweighting and stop-loss prices are assigned based on predefined level δ and criteria, as shown in Eq. 3. The strategy allows for overweighting positions if asset price exceeds a specified threshold and the maximum overweight times L has not been reached. While the positions are covered if the asset price exceeds the stop-loss threshold.

$$\begin{aligned} \text{Overweighting Threshold} : p^{ow}_{n+1} &= P_n \times (1+\delta) \\ \text{Stop-Loss Threshold} : p^{sl}_{n+1} &= P_n \times (1-\delta) \end{aligned} \tag{3}$$

$$\text{Payoff}_n = \sum_{t=1}^{T} \text{payoff}_t \tag{4a}$$

$$\text{payoff}_t = \begin{cases} \sum_{P_i \in P} (P^c - P_i), & \text{if } R_t = 1, \\ 0, & \text{if } R_t = 0. \end{cases} \tag{4b}$$

The payoff at each time point t is determined by the value of R_t, which represents the trading result (1 for a triggered signal, 0 for no signal triggered). If R_t is equal to 1, the payoff is computed by subtracting the purchase price P_i from the current closing price P^c and summing it over all positions P_i in P. If R_t is 0, indicating no trading at that time point, the payoff is 0. The accumulated payoff in each random simulation is denoted as Payoff_n, which is calculated by summing the payoffs at each time point t. The calculation of accumulated payoff and payoff at each time point is shown in Eq. 4.

Profitability Indicator is a performance measurement for analyzing the payoff distribution and inspired from the Sharpe ratio, which could be defined as the ratio of the mean of payoff distribution to its standard deviation, as shown in Eq. 5. It provides a comprehensive assessment of performance that is independent of the principal and not affected by variations in its amount.

$$\text{PI} = \frac{\frac{1}{N}\sum_{n=1}^{N}\text{Payoff}_n}{\sqrt{\frac{1}{N}\sum_{n=1}^{N}(\text{Payoff}_n - \bar{\text{Payoff}})^2}}, \quad n \in \mathbb{R} \text{ and } 1 \leq n \leq N \tag{5}$$

A fuzzy quantification module based on fuzzy set theory is developed to interpret the features of futures contracts. It employs an L-function membership function for both the long-side and short-side momentum overweight (MOS) strategies. This membership function, as shown in Eq. 6, is defined by the parameter α, which controls the gradient.

$$\mu(x) = \begin{cases} 0 & \text{if } x \leq -\alpha \\ \frac{x+\alpha}{2\alpha} & \text{if } -\alpha < x < \alpha \\ 1 & \text{if } x \geq \alpha \end{cases} \tag{6}$$

In this membership function, the degree of fuzzy is determined by the input X, representing the profitability indicator of futures. When x is less than or equal to $-\alpha$, the fuzzy degree is 0, indicating that the MOS strategy is not suitable. For x between $-\alpha$ and α, the fuzzy degree increases linearly, reaching 1 when x is greater than or equal to α. This gradual transition captures the nuanced evaluation of strategy suitability.

The fuzzy quantification module utilizes these fuzzy degrees to assess the suitability of the MOS strategy. By assigning appropriate fuzzy degrees based on the profitability indicator, the module provides a quantified evaluation of the strategy's fitness for different futures contracts.

4 Experiment Results

We first state datasets used in this paper. Then, demonstrate the simulation results of random trading and calculate the profitability indicator. Finally, evaluate the effectiveness of fuzzy quantification module.

4.1 Data Usage

In this study, we apply the MOS to TAIFEX Futures (TX) data and evaluate the effectiveness of the strategy using a fuzzy quantification module. The data frequency includes intervals of 1 min, 5 min, 15 min, and 60 min. The dataset covers the time period from January 2020 to December 2022, which is divided into a training set (January 2020 to December 2021) and a testing set (January 2022 to December 2022).

4.2 Simulation Results of Random Trading and Profitability Indicator

We initially perform intraday trading to conduct a light-computation simulation and establish three simple trading strategies as benchmarks for evaluating the efficiency of MOS.

The first strategy involves adopting a holding approach (Strategy A), where a long position is taken at the beginning of the trading day and covered at the end of the day. The second and third strategies incorporate stop-loss and take-profit mechanisms. In Strategy B, both the stop-loss and take-profit levels are set at 20 points. In contrast, Strategy C has a take-profit level of 30 points, which is higher than the stop-loss level of 20 points. MOS proposed in the research incorporates overweight and stop-loss mechanism and employs the same degree for both the overweight and stop-loss threshold. Descriptive statistics of the accumulated payoff from 10,000 random trading simulations are presented in Table 1. The results demonstrate that Strategies B and C are significantly more profitable than Strategy A, after incorporating the stop-loss and take-profit mechanisms. Moreover, the momentum overweight strategy outperforms all the benchmark strategies. The payoff distribution is depicted in Fig. 1.

Table 1. Descriptive Statistics of Payoff for strategies

	Strategy A	Strategy B	Strategy C	MOS
Mean	−12.456	3599.539	4316.587	14355.172
Standard Error	28.387	7.165	7.953	24.412
Standard Deviation	2838.730	716.512	795.312	2441.156
PI	−0.004	5.024	5.428	5.880

The MOS strategy can be divided into two variants: long side and short side. The long-side MOS involves taking positions with the expectation of price increases, while the short-side MOS involves taking positions with the expectation of price decreases.

In terms of data frequency, the MOS strategy can be applied to different time intervals, including 1-minute, 15-minute, 30-minute, and 60-minute intervals.

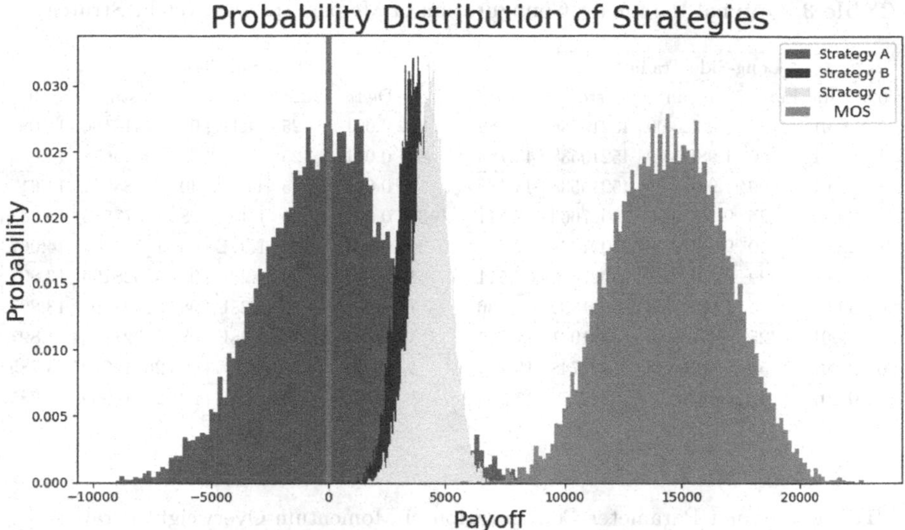

Fig. 1. Probability Distribution of Payoff for Strategies

The choice of data frequency depends on the trading objectives and the desired level of granularity in capturing market dynamics.

The MOS offers flexibility in adjusting two crucial parameters: the momentum level (δ), which determines the threshold for initiating overweight and stop-loss actions, and the maximum overweight times (L). These parameters are essential in fine-tuning the strategy to optimize trading performance.

To assess the effectiveness of the MOS strategy, we applied it to TX futures data over the specified period (as described in Sect. 4.1) and conducted 10,000 random simulations. By systematically varying the parameter settings, we thoroughly evaluated the performance and profitability of the MOS strategy under diverse market conditions.

The profitability indicator was obtained by applying the random trading algorithm to the momentum overweight strategy with the parameters specified in Table 2. The performance of the long-side MOS and short-side MOS strategies is presented in Table 3 for the optimal parameter combinations, and in Table 4 for the worst parameter combinations, respectively.

Table 2. Parameter Information for MOS Strategy

Parameter	Value Range
Data Frequency ($Freq$)	1-min, 15-min, 30-min, and 60-min
Momentum Threshold (δ)	0.1%, 0.5%, and 1.0%
Maximum Overweight Times (L)	0, 25, 50, and 99

Table 3. Optimal Parameter Configuration of Momentum Overweight Strategy

Long-Side Trading						Short-Side Trading					
Freq	Delta	Limit	mean	std	PI	Freq	Delta	Limit	mean	std	PI
1	0.01	25	1.98E+09	45210538	43.788	30	0.001	25	1.1E+08	6416186	17.083
1	0.01	50	1.98E+09	45210538	43.788	60	0.001	25	90532072	5625658	16.093
1	0.01	99	1.98E+09	45210538	43.788	15	0.001	25	96302440	6588202	14.617
1	0.005	25	9.39E+08	25027663	37.511	30	0.001	50	1.49E+08	10377505	14.383
1	0.005	50	9.39E+08	25027663	37.511	30	0.001	99	1.51E+08	10769304	14.067
1	0.005	99	9.39E+08	25027663	37.511	1	0.01	25	5.28E+08	38028193	13.897
60	0.001	25	1.03E+08	5501892	18.636	1	0.01	50	5.28E+08	38028193	13.897
30	0.001	25	84269440	4751081	17.737	1	0.01	99	5.28E+08	38028193	13.897
60	0.001	50	1.18E+08	7497748	15.715	1	0.005	25	3.6E+08	26114969	13.785
30	0.001	50	94630278	6244376	15.154	1	0.005	50	3.6E+08	26114969	13.785

Table 4. Worst Parameter Configuration of Momentum Overweight Strategy

Long-Side Trading						Short-Side Trading					
Freq	Delta	Limit	mean	std	PI	Freq	Delta	Limit	mean	std	PI
1	0.001	0	-4.1E+08	1483809	-279.086	1	0.001	0	-4.2E+08	1493147	-282.016
15	0.001	0	-3.1E+07	391987.1	-78.629	15	0.001	0	-3.3E+07	390732.2	-85.378
30	0.001	0	-1.7E+07	281869	-60.32	1	0.005	0	-5.4E+08	6937042	-77.435
60	0.001	0	-9764018	201370.2	-48.488	30	0.001	0	-1.9E+07	287530.8	-67.138
1	0.005	0	-3E+08	6826440	-43.79	60	0.001	0	-1.1E+07	207675.5	-54.627
1	0.01	0	-4.9E+08	13490246	-36.054	1	0.01	0	-3.5E+08	13340219	-25.91
1	0.001	25	-3E+08	10501565	-28.947	1	0.001	25	-3E+08	13471641	-22.496
1	0.001	50	-3E+08	10748294	-27.827	15	0.005	0	-3.7E+07	1796878	-20.422
1	0.001	99	-3E+08	10748294	-27.827	30	0.005	0	-1.9E+07	1289881	-14.683
15	0.005	0	-2E+07	1793817	-11.211	60	0.005	0	-1.1E+07	916597.7	-11.701

The analysis of the results reveals several key findings. Firstly, it is observed that the absence of an overweight mechanism (i.e., $L = 0$) significantly diminishes the performance of both long-term and short-term trading strategies, despite the presence of an effective stop-loss mechanism. This highlights the importance of incorporating an overweight mechanism to maximize the strategy's potential. Secondly, when examining strategies with unequal overweight limits (L), it is found that they consistently demonstrate a similar profitability indicator. This suggests that reaching the overweight limit can be challenging, regardless of the specific limit value chosen. In our study, a total of 48 parameter combinations were tested for both the long and short sides. Among these combinations, 33 yielded a positive profitability indicator, indicating that they generated positive average returns and are deemed suitable for the momentum overweight strategy.

4.3 Effectiveness of Fuzzy Quantification Module

This section provides an overview of the parameter optimization process for the membership function in the fuzzy quantification module and demonstrates its performance in measuring the profitability of strategies. The correlation coefficient (C.C.) is selected as the indicator to optimize the parameters of the membership function. To optimize the parameters, a search space ranging from 0 to 150 with increments of 1 is considered.

The objective is to find the parameter values that yield the highest correlation coefficient between the fuzzy degrees and the trading performance of the momentum overweight strategy. Following the optimization process, the optimal parameters for the fuzzy quantification module are determined as $\alpha = 100$ for the long-side momentum overweight strategy and $\beta = 100$ for the short-side momentum overweight strategy. These parameter values are found to maximize the correlation between the fuzzy degrees and the trading performance. Table 5 presents the correlation coefficients between the fuzzy degrees and the trading performance of the momentum overweight strategy. It is observed that the average correlation coefficient during the testing period reaches 0.57, which is only slightly lower than the average correlation coefficient of 0.61 during the training period. These results demonstrate the effectiveness of the evaluation system proposed in this study in providing traders with support to determine the suitability and profitability of trading strategies on futures, even without relying on market informative data.

Table 5. Correlation Coefficient between Fuzzy Degree and Trading Performance

	Long-side MOS	Short-side MOS
Training period	0.6239	0.6072
Testing period	0.5755	0.5741

5 Conclusions

In recent years, futures trading has evolved beyond its traditional role in hedging to become a prominent platform for speculative trading. However, the high leverage associated with futures trading exposes traders to significant risks, often exceeding their principal. Consequently, it becomes essential for traders to adopt quantitative strategies that effectively control and manage these risks.

Currently, there is a lack of a systematic approach to assist traders in evaluating the suitability of trading strategies for futures contracts. This article proposes a comprehensive solution to address this issue. It begins by utilizing a Random Trading Algorithm to extract the inherent characteristics of futures contracts in the absence of market information. Subsequently, a profitability indicator is developed based on an extension of the Sharpe ratio, which quantifies both the payoff and risk associated with the trading strategies. To determine the suitability of a particular strategy, a fuzzy quantification module is employed, utilizing fuzzy degrees generated by the system.

The experimental results showcase the effectiveness of this system in supporting investors in making informed decisions regarding their trading strategies for TAIFEX Futures. Future research will focus on enhancing the flexibility of the system to accommodate other futures contracts and analyzing the impact of the uncertainty associated with the membership function.

References

1. Castillo, O., Amador-Angulo, L., Castro, J.R., Garcia-Valdez, M.: A comparative study of type-1 fuzzy logic systems, interval type-2 fuzzy logic systems and generalized type-2 fuzzy logic systems in control problems. Inf. Sci. **354**, 257–274 (2016)
2. Wu, X., Chen, H., Wang, J., Troiano, L., Loia, V., Fujita, H.: Adaptive stock trading strategies with deep reinforcement learning methods. Inf. Sci. **358**, 142–158 (2020)
3. Jones, R.: The trading game: playing by the numbers to make millions. John Wiley and Sons (1999)
4. Jeng, L.A., Metrick, A., Zeckhauser, R.: Estimating the returns to insider trading: a performance-evaluation perspective. Rev. Econ. Stat. **85**(2), 453–471 (2003)
5. Harvey, C.R., Liu, Y.: Evaluating trading strategies. J. Portfolio Manag. **40**(5), 108–118 (2014)
6. Yong, B.X., Abdul Rahim, M.R., Abdullah, A.S.: A Stock Market Trading System Using Deep Neural Network. In: Mohamed Ali, M.S., Wahid, H., Mohd Subha, N.A., Sahlan, S., Md. Yunus, M.A., Wahap, A.R. (eds.) AsiaSim 2017. CCIS, vol. 751, pp. 356–364. Springer, Singapore (2017). https://doi.org/10.1007/978-981-10-6463-0_31
7. Zadeh, L.A.: Fuzzy logic. Computer **21**(4), 83–93 (1988)
8. Afzal, F., Yunfei, S., Nazir, M., Bhatti, S.M.: A review of artificial intelligence based risk assessment methods for capturing complexity-risk interdependencies: cost overrun in construction projects. Int. J. Manag. Projects Bus. **14**(2), 300–328 (2021)
9. Mansouri, N., Zade, B.M.H., Javidi, M.M.: Hybrid task scheduling strategy for cloud computing by modified particle swarm optimization and fuzzy theory. Comput. Ind. Eng. **130**, 597–633 (2019)
10. Kaya, I., Çolak, M., Terzi, F.: A comprehensive review of fuzzy multi criteria decision making methodologies for energy policy making. Energy Strategy Rev. **24**, 207–228 (2019)
11. Wu, M.E., Syu, J.H., Lin, J.C.W., Ho, J.M.: Effective fuzzy system for qualifying the characteristics of stocks by random trading. IEEE Trans. Fuzzy Syst. **30**(8), 3152–3165 (2021)
12. Sharpe, W.F.: The sharpe ratio. Streetwise Best J. Portfolio Manag. **3**, 169–185 (1998)
13. Sortino, F.A., Price, L.N.: Performance measurement in a downside risk framework. J. Investing **3**(3), 59–64 (1994)
14. Magdon-Ismail, M., Atiya, A.F.: Maximum drawdown. Risk Mag. **17**(10), 99–102 (2004)

15. Ali, O.A.M., Ali, A.Y., Sumait, B.S.: Comparison between the effects of different types of membership functions on fuzzy logic controller performance. Int. J. **76**, 76–83 (2015)
16. Pedrycz, W.: Why triangular membership functions? Fuzzy Sets Syst. **64**(1), 21–30 (1994)
17. Barua, A., Mudunuri, L.S., Kosheleva, Olga.: Why trapezoidal and triangular membership functions work so well: towards a theoretical explanation. J. Uncertain Syst. **8**, 164–168 (2014)

Enhancing Abnormal-Behavior-Based Stock Trend Prediction Algorithm with Cost-Sensitive Learning Using Genetic Algorithms

Chun-Hao Chen[1(✉)], Szu-Chi Wang[2], Mu-En Wu[2], and Kawuu W. Lin[1]

[1] Department of Computer Science and Information Engineering, National Kaohsiung University of Science and Technology, Kaohsiung, Taiwan
{chench,linwc}@nkust.edu.tw

[2] Department of Information and Finance Management, National Taipei University of Technology, Taipei, Taiwan
t109AB8020@ntut.org.tw, mnwu@ntut.edu.tw

Abstract. Stock market trend forecasting is an important and popular topic in academia or industry, and many studies have been proposed to handle this problem. Because stock market transactions are dynamic and nonlinear systems, it is difficult to predict through a single feature. Therefore, in the previous work, an algorithm was proposed to build prediction model by using abnormal behaviors extracted from the given data as features and transfer learning. However, when using transfer learning to generate more training instances, it may cause negative transfer. Besides, hyperparameter for the previous model should be tune to find a more effective prediction model. To solve the two problems, we propose an enhanced stock trend prediction model in this paper. For the first one, we modify the original fitness function by combining the concept of the cost-sensitive learning. As a result, the proposed approach can have the ability to avoid instances that could lead to negative transfer. For the second problem, we utilize the genetic algorithms for searching the suitable hyperparameter to construct the prediction model. At last, experiments were made to show the effectiveness of the proposed approach.

Keywords: Classification · Cost-sensitive learning · Genetic algorithm · Negative transfer · Stock trend prediction

1 Introduction

Predicting stock market trends has always been a fascinating topic for both industry and academia. Accurately predicting stock price trends helps investors to make good trading plans for buying and selling stocks for reducing the risk of surprises. However, financial time series data is not only noisy and subjects to economic, political, and unexpected events [1], but also makes the financial series

N. T. Nguyen et al. (Eds.): ACIIDS 2023, LNAI 13995, pp. 186–196, 2023.
https://doi.org/10.1007/978-981-99-5834-4_15

highly nonlinear due to highly nonlinear. As a consequence, predicting stock trends has always been a challenging task [10,19]. For instance, Nti *et al.* [16] proposed a stock market forecasting framework with multi-source information fusion, which is framed by a deep hybrid neural network architecture that are the convolutional neural network (CNN) and stacked Long short-term memory (LSTM), and their experiments also indicated that the stock market is really affected by multiple factors and does not rely on a single feature data source.

Since the financial crisis in 2007, many researchers, risk and financial practitioners have tried to detect abnormal situations in the stock market, and abnormal detection has been widely used in various fields [14]. For example, Lin *et al.* used candlestick charts and ensemble model frameworks to effectively address data noise to improve prediction accuracy through anomalous datasets [11]. In addition, Chen et al. proposed an anomaly-patterns based model for prediction stock trends [3]. In their work, they improved the accuracy of stock forecasting with anomaly patterns by specifying multi-faceted special events. Because anomaly patterns are relatively small, transfer learning which could be used to increase training instances from other similar datasets [17] was employed to handle data insufficient problem. Then, the original training and the generated instances were used to build the prediction model.

However, it still has problems to be solved in previous work [3]. First, when using transfer learning to generate more instances for training, it may cause negative transfer, which means that the accuracy of the built model can not get benefits from the transferred instances. That is not the case we expected. Besides, hyperparameter for the anomaly-patterns based model should also be tune to find more effective prediction model. To solve those problems, in this paper, we propose an enhanced algorithm for handling those two problems. For the first problem, we modify the original fitness function according to cost-sensitive learning. As a result, the proposed approach can have the ability to avoid instances that leading to negative transfer. For the second problem, we employ the genetic algorithms (GA) for searching the suitable hyperparameter used for the prediction model.

In the proposed approach, it first encodes hyperparameters, including transfer learning strategy, number of companies used for selecting extra training instances, and the defined anomaly patterns, including a chromosome. Then, according to the original fitness function, the cost-sensitive concept is utilized to enhance it. As a result, when inappropriate instances are selected to build the model, the fitness value of a chromosome is become smaller than the value calculated by the original one. Genetic operators are then applied to form offspring. After evolution, the final best chromosome is the provided to build the final stock trend prediction model. Finally, experimental results on real datasets were also made to show the effectiveness of the proposed approach.

2 Related Work

2.1 Transfer Learning and Negative Transfer

Transfer learning (TL) is a machine learning method that mainly transfers the trained model, knowledge or parameters (e.g., source domain) to another model (e.g., target domain), so that the target domain can obtain better learning result. Due to the expensive and time-consuming nature of data acquisition and labeling, it is difficult to construct large and well-labeled datasets, so use transfer learning to solve the problem of insufficient training data. Tan *et al.* [21] divided deep transfer learning into four categories: instances-based , mapping-based , network-based and adversarial-based deep transfer learning.

Based on Tan *et al.* [21] research on deep transfer learning, it has been proved that in most practical applications, combined with deep transfer learning methods can obtain better results and solve the problem of lack of training samples. Chen *et al.* combined transfer learning and proposed the use of the classical domain adaptive object detection algorithm (DA Faster R-CNN) to solve the problem of small data samples of aircraft object detection missions with remote sensing images [4]. And, with the wide application of transfer learning and deep learning, related research in the field of time series and stock forecasting is also increasing [8,13,15]. For example, Jeong *et al.* first added a deep neural network (DNN) regressor to predict the number of shares, and also used various movement strategies for Q values to analyze which action strategies were conducive to profiting in a chaotic market. Finally, transfer learning methods are proposed to prevent overfitting due to insufficient financial data [10].

While transfer learning can reduce training time, improve deep learning performance, and training with fewer data sets, there are still some limitations to be overcome. The effectiveness of transfer learning needs to meet its basic assumptions: (1) the source and target domains are highly correlated; (2) The data distribution of the source and target domains is relatively close; (3) There are models that can be applied to both. If the above assumptions are not met, the use of transfer learning methods will lead to a decrease in performance or accuracy, which is called negative transfer. However, negative transfer has been a long-standing and challenging problem for transfer learning [23,24]. Zhang *et al.* conducted a comprehensive investigation of negative transfers and summarized nearly fifty different ways to overcome negative transfers from four aspects: Secure Transfer, Domain Similarity Estimation, Distant Transfer, and Negative Transfer Mitigation [25].

2.2 Cost Sensitive Learning

Cost sensitive learning is the cost of considering misclassification [6] in order to minimize the total cost, not the number of errors. While cost-sensitive learning typically handles different misclassifications in different ways, fewer classes often differ from the misclassification costs of more classes. For example, in medical diagnosis, the cost of classifying a person with heart disease as a non-disease patient is more serious than classifying a person without a disease as a person with a heart disease, because the first situation is that the patient will lack the opportunity to be treated effectively in real time, and the second is a waste of medical resources. Therefore cost-sensitive learning plays an important role in addressing category imbalances at the data level and algorithmic level [7]. The solution to this problem is to set different costs through the cost matrix when training the classifier, and then let the classifier learn and adjust the results of the classification on its own [5,12,22].

Jiang *et al.* proposed the cost-sensitive parallel learning framework (CPLF) to enhance insurance operations with deep learning methods that do not require preprocessing [9]. The approach consists of end-to-end cost-sensitive parallel neural networks with heterogeneous real data. The designing cost-sensitive matrices automatically generates robust models for learning a small number of classifications. The parameters of the cost-sensitive matrix and hybrid neural network alternate during training but are optimized together. Alsubaie *et al.* proposed the cost-sensitive fine-tuned naïve Bayes (CSFTNB) to help to select the minimum number of stock market technical indicators to predict the direction of stock trends [2,20]. The proposed CSFTNB classifier is to enhance the classical Bayer classifier through the fine-tuning phase, where the value probability is modified or fine-tuned to reduce the classification cost.

3 Proposed Enhanced GA-Based Algorithm

In this section, the framework of the proposed approach is given firstly. Then, details of components of the proposed approach are described.

3.1 Framework of the Proposed Approach

This subsection will explore the framework that combines stock anomaly patterns, genetic algorithm transfer learning and cost-sensitive learning for for constructing cost-sensitive-based stock trend prediction model optimization technique. The proposed framework is shown in Fig. 1.

From Fig. 1, the framework is divided into two phases, namely offline learning and online simulated trading. In the offline learning, it further divides into two processes that are data preprocessing and optimization process. In data preprocessing, the stock price related data of the semiconductor industry in Taiwan's securities market are collected from the Taiwan Economic News Database (TEJ), including trading volume, stock price, three major legal person chips and margin data. Then, according to the anomaly patterns, three trends, observation and trend days that are defined in [3], anomaly pattern dataset is constructed for building the prediction model. Every instance in the dataset contains anomoly patterns appeared in observation period and its trend label that could be the uptrend, downtrend, or consolidation trend. The second process is using GA to find suitable hyperparameter and to select appropriate anomaly patterns for constructing trend prediction model. It first encodes model type, transfer learning

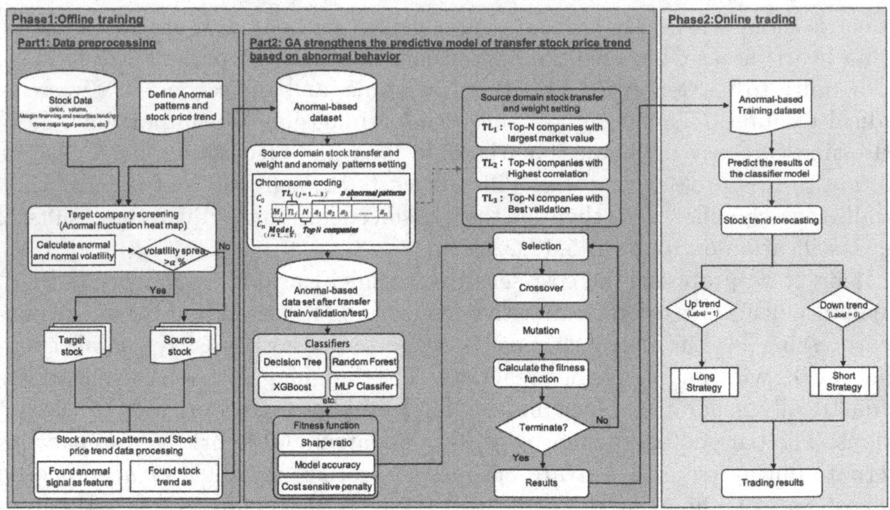

Fig. 1. The framework of the proposed algorithm.

strategy type, number of source companies, and n anomaly patterns into a chromosome. Then, the used type of transfer learning strategy and number of source companies represented in a chromosome, more training instances are extracted. The extracted instances and the original instances are merged to build prediction model in accordance with the used model type. Fitness value of a chromosome is calculated by the mode accuracy, Sharpe ratio, and cost-sensitive penalty. The genetic operations are then performed to generate new chromosomes. The evolution process is repeated until termination conditions are reached. At last, the chromosome with the best fitness value is provided to users making trading plans. In online simulated trading, by using the built trend prediction model, every anomaly-patterns based instance will get a buying or selling signal from the model. When a buying signal is generated, it will long the stock. On the contrary, it will short the stock. In the following, the components of the proposed GA-based approach are described.

3.2 Chromosome Representation

Since this paper aims to select the target domain for optimal transfer learning and to avoid negative transfer, the encoding scheme is shown in Fig. 2.

In Fig. 2, there are four parts in the chromosome representation, including model type, transfer learning strategy, number of the top-N companies, and anomaly patterns. The model type has five different machine learning models can be used and represented as M1, M2, M3, M4, and M5, where M1 stands for decision three (DT), M2 stands for random forest (RF), M3 stands for the support vector machine (SVM), M4 stands for neural network (NN), and M5 stands for XGBoost. In transfer learning strategy, it has three different transfer learn-

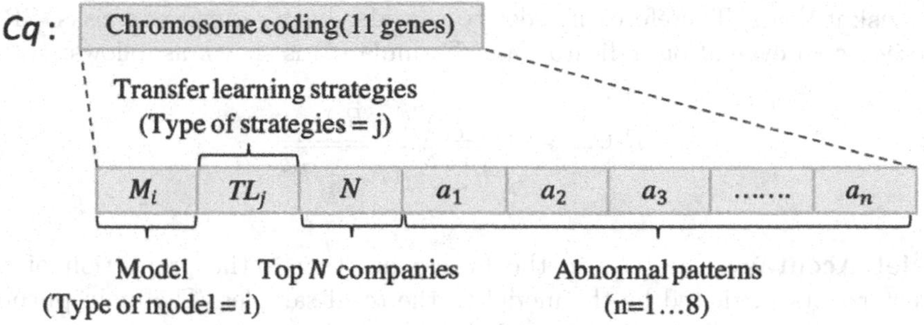

Fig. 2. Encoding scheme for a chromosome C_q.

ing strategies as represented as TL1, TL2, and TL3. TL1 means selecting the representative companies in the industry as the source domain. TL2 means using correlation coefficients to select companies as the source domain. TL3 represents selecting companies those have high validation values when building their own prediction models as the source domain. Number of the top-N companies means top N best companies based on the transfer learning strategies can be used as source domains to generate more training instances. In the anomaly patterns, eights used patterns are encoded as a bit string, where 0 means not to choose and 1 means choose it as prediction attribute.

3.3 Fitness Evaluation

In GA, every candidate solution is evaluated by calculating its fitness value using the designed fitness function to evaluate the merits of each chromosome. Then, the best solution can be carried out after evolution process. Since this study intends to construct a stock price trend prediction model, three factors are needed to be considered: return, risk and model accuracy. However, only considering these three factors is not be enough to avoid the phenomenon of negative transfer. To solve that, the concept of cost-sensitive learning is utilized to enhance the fitness function to deduce the negative transfer phenomenon. In sum, the fitness function $f(C_q)$ of the proposed algorithm is shown in the following formula (1):

$$f(C_q) = SR(C_q) \times Acc(C_q) \times CSP(C_q), \tag{1}$$

where $SR(C_q)$, $Acc(C_q)$, and $CSP(C_q)$ are Sharpe ratio(SR), model accuracy (Acc), and cost-sensitive penalty (CSP) of a chromosome. They are described as follows.

Sharpe Ratio. $SR(C_q)$ in the fitness function is SR, also known as sharp index, which is a commonly used indicator of reward and risk evaluation. The purpose of SR is to calculate how much return a portfolio will generate for each unit of total risk it bears. Therefore, in order to consider both returns and risks, SR is selected as an evaluation indicator, and Formula (2) is shown as follows:

$$SR(C_q) = \frac{\frac{1}{N}(\sum_{i=1}^{N} R_i)}{\sqrt{\frac{1}{N}\sum_{i=1}^{N}(R_i - \bar{R})^2}} \tag{2}$$

Model Accuracy. $Acc(C_q)$ in the fitness function is the proportion of the correct results predicted by the model to the total samples. The main purpose is to evaluate the effect of the model. In this study, the accuracy of the stock trend prediction model is defined as Formula (3) as follows:

$$Acc(C_q) = \frac{(TP+TN)}{(TP+FP+FN+TN)} \tag{3}$$

Cost-Sensitive Penalty. $CSP(C_q)$ in the fitness function means the cost-sensitive penalty value. In practical applications, there are many cases where the cost of misclassification is unequal, so the cost-sensitive learning can set different costs through the cost matrix when training the classifier, so that the classifier can learn and identify what the good sources are during the evolution process. The cost-sensitive penalty of a chromosome is designed and shown in Formula (4) as shown:

$$CSP(C_q) = \frac{(TP+TN)}{(TP + w_1 \times FP + w_2 \times FN + TN)} \tag{4}$$

It can be seen from the above Formula (4) that when the number of false positives and false negatives in the established model increases, it means that negative transfer may occur. Therefore, the effect will be amplified by the two weight values, w_1 and w_2, so that the chromosomes containing too many pseudo-positives and pseudo-negatives will gradually disappear during the evolutionary process.

3.4 Genetic Operation

Genetic operations include selection, crossover and mutation. A collection of chromosomes from a generation is called a population. It is like eliminating individuals with low survival values through competition with each other and the test of the environment. The higher the storage value, the more offspring can be copied, and then two mating (Crossover) can be selected from it. The mechanism of mutation is added in order to produce a next generation with a higher degree of diversity. In this way, the process of stacking generations allows the group to optimize from generation to generation and gradually approach the optimal solution. In this paper, the roulette wheel selection, one-point crossover, and one-point mutation are used in the proposed algorithm.

4 Experimental Results and Analysis

4.1 Dataset Description and Environment Setting

This study focuses on the Taiwan stock market, stock history data taken from the Taiwan Economic Journal (TEJ) database. In addition, the research targets of this study are mainly companies in the semiconductor industry, and the classification of industries and sub-industries is based on the industry classification of TEJ. The datasets are the original daily data from 2008/01/01 to 2020/12/31. In the following experiments, proportions of the training and validation, and testing datasets are eighty and twenty percents, and the time periods of them are from 2008 to 2019 and 2019 to 2021, respectively. The backtesting of the simulated stock trading strategy of this study is mainly in the form of a fixed amount. Each fixed entry amount is set to NT$ 100 thousands. Transaction cost is considered in the proposed approach. The backtesting method is mainly calculated based on the results predicted by the constructed model by the proposed approach. If the prediction is an upward trend, the long strategy is executed. After an anomaly occurs in a certain observation time range, it is bought at the opening price of the next day. And after the predefined days of trend range, it is sold at the closing price. On the contrary, if the forecast is a downward trend, the short strategy is executed.

4.2 Comparison Results of the Proposed and the Existing Approaches

In order to reduce the phenomenon of negative transfer, this paper introduces cost-sensitive learning idea to enhance evaluation function in addition to considering the reward, risk and model accuracy in the proposed approach. Experiments below were made to verify the effectiveness of the proposed approach in terms of cumulative returns. Take the two target domain companies of the memory-related sub-industries, NTC(2408) and UNIFOSA(8277), as examples. The cumulative returns along with increasing of tradings in accordance with the optimized model using the proposed approach and other approaches are shown in Figs. 3 and 4.

From Figs. 3 and 4, they show two important information. The first one is that when comparing the proposed approach with the existing approaches, the cumulative returns of the proposed approach are higher than the existing approaches. The second one is that cumulative returns of the proposed approach with is better than that without cost-sensitive learning. These two results show that the proposed approach can not only reduce the negative transfer in the training phase but also find suitable hyperparameter and anomaly patterns as well for building a better prediction model.

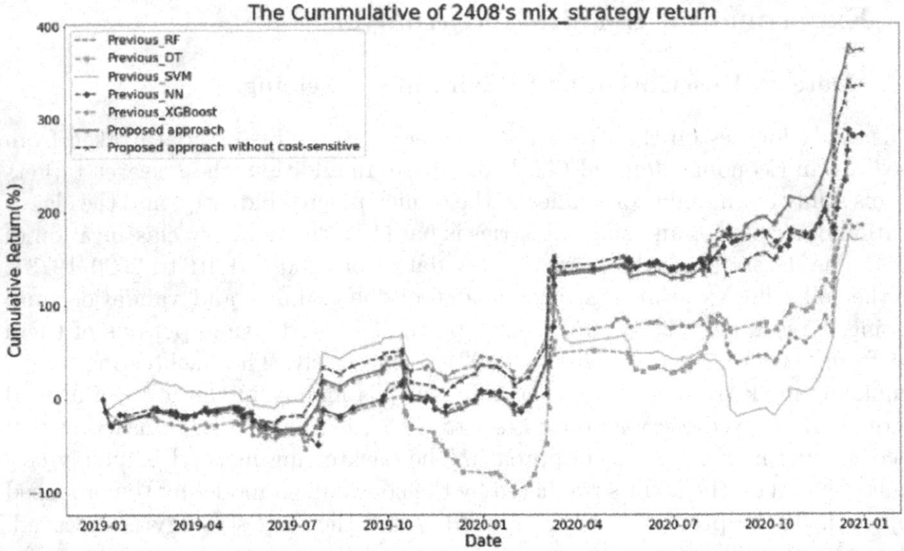

Fig. 3. Cumulative returns of NTC(2408) as target domain when comparing to other existing approaches.

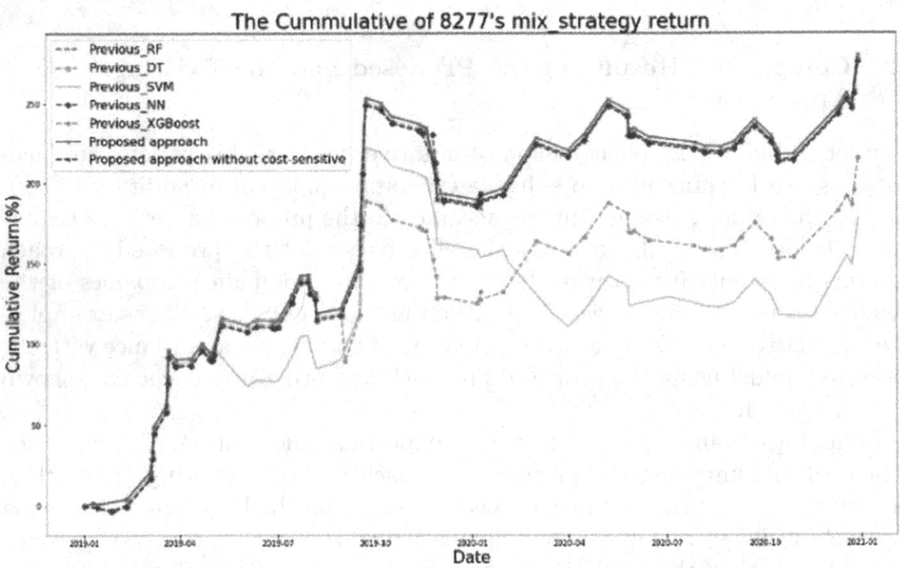

Fig. 4. Cumulative returns of UNIFOSA(8277) as target domain when comparing to other existing approaches.

5 Conclusion

This paper proposes to combine stock market abnormal behavior, transfer learning, and genetic algorithms to build a stock price trend prediction model optimization technology that can not only reduce negative instances are selected from the source domains as training instances but also find suitable hyperparameter and anomaly patterns to construct the trend prediction mode for trading. Experimental results on two target companies showed that the proposed approach can reach higher cumulative returns that the existing approaches. In addition, the returns of proposed approach using the enhanced fitness function are also higher than that with original fitness function, which means that considering the cost-sensitive learning is useful to select instances effectively to avoid negative transfer problem. In the future, more experiments will be conducted to show more positive evidences of the proposed approach, or try other classification models, such as long short term memory network (LSTM), artificial neural network (ANN) and other models to predict stock price trends.

Acknowledgements. This research was supported by the National Science and Technology Council of the Republic of China under grant NSTC 111-2221-E-992-088-MY2.

References

1. Abu-Mostafa, Y.S., Atiya, A.F.: Introduction to financial forecasting. Appl. Intell. **6**(3), 205–213 (1996) https://doi.org/10.1007/BF00126626
2. Alsubaie, Y., El Hindi, K., Alsalman, H.: Cost-sensitive prediction of stock price direction: selection of technical indicators. IEEE Access **7**, 146876–146892 (2019)
3. Chen, C.-H., Lin, Y.-T., Hung, S.-T., Wu, M.-E.: Forecasting Stock Trend Based on the Constructed Anomaly-Patterns Based Decision Tree. In: Nguyen, N.T., Chittayasothorn, S., Niyato, D., Trawiński, B. (eds.) ACIIDS 2021. LNCS (LNAI), vol. 12672, pp. 606–615. Springer, Cham (2021). https://doi.org/10.1007/978-3-030-73280-6_48
4. Chen, J., Sun, J., Li, Y., Hou, C.: Object detection in remote sensing images based on deep transfer learning. Multimedia Tools Appl. **81**(9), 12093–12109 (2021). https://doi.org/10.1007/s11042-021-10833-z
5. Elkan, C.: The foundations of cost-sensitive learning, In: International Joint Conference on Artificial Intelligence, vol. 17, no. 1, pp. 973–978 (2001)
6. Fernández, A., García, S., Galar, M., Prati, R.C., Krawczyk, B., Herrera, F.: Cost-sensitive learning. In: Learning from Imbalanced Data Sets, pp. 63–78. Springer, Cham (2018). https://doi.org/10.1007/978-3-319-98074-4_4
7. Galar, M., Fernandez, A., Barrenechea, E., Bustince, H., Herrera, F.: A review on ensembles for the class imbalance problem: bagging-, boosting-, and hybrid-based approaches. IEEE Trans. Syst. **42**(4), 463–484 (2011)
8. He, Q.-Q., Pang, P.C.-I., Si, Y.-W.: Transfer Learning for Financial Time Series Forecasting. In: Nayak, A.C., Sharma, A. (eds.) PRICAI 2019. LNCS (LNAI), vol. 11671, pp. 24–36. Springer, Cham (2019). https://doi.org/10.1007/978-3-030-29911-8_3

9. Jiang, X., Pan, S., Long, G., Xiong, F., Jiang, J., Zhang, C.: Cost-sensitive parallel learning framework for insurance intelligence operation. IEEE Trans. Ind. Electron. **66**(12), 9713–9723 (2018)
10. Jeong, G., Kim, H.Y.: Improving financial trading decisions using deep Q-learning: predicting the number of shares, action strategies, and transfer learning. Expert Syst. Appl. **117**, 125–138 (2019)
11. Lin, Y., Liu, S., Yang, H., Wu, H.: Stock Trend Prediction Using Candlestick Charting and Ensemble Machine Learning Techniques With a Novelty Feature Engineering Scheme. IEEE Access **9**, 101433–101446 (2021)
12. Lomax, S., Vadera, S.: A cost-sensitive decision tree learning algorithm based on a multi-armed bandit framework. Comput. J. **60**(7), 941–956 (2017)
13. Li, X., Xie, H., Lau, R.Y., Wong, T.L., Wang, F.L.: Stock prediction via sentimental transfer learning, IEEE Access **6**,73110–73118 (2018)
14. Malkiel, B. G.: The efficient market hypothesis and its critics. J. Econ. Perspect. **17**(1), 59–82 (2003)
15. Nguyen, T.T., Yoon, S.: OA novel approach to short-term stock price movement prediction using transfer learning, Appl. Sci. **9**(22), 4745 (2019)
16. Nti, I.K., Adekoya, A.F., Weyori, B.A.: A novel multi-source information-fusion predictive framework based on deep neural networks for accuracy enhancement in stock market prediction. J. Big Data **8**(1), 1–28 (2021). https://doi.org/10.1186/s40537-020-00400-y
17. Pan, S.J., Yang, Q.: SA survey on transfer learning. IEEE Trans. Knowl. Data Eng. **22**(10), 1345–1359 (2009)
18. Prachyachuwong, K., Vateekul, P.: Stock Trend Prediction Using Deep Learning Approach on Technical Indicator and Industrial Specific Information. Industrial Specific Information. Information **12**(6), 250 (2021)
19. Ren, R., Wu, D.D., Liu, T.: Forecasting stock market movement direction using sentiment analysis and support vector machine. IEEE Syst. J. textbf13(1), 760–770 (2018)
20. Teixeira, L.A., De Oliveira, A.L.I.: A method for automatic stock trading combining technical analysis and nearest neighbor classification. Expert Syst. with Appl. **37**(10), 6885–6890 (2010)
21. Tan, C., Sun, F., Kong, T., Zhang, W., Yang, C., Liu, C.: A survey on deep transfer learning, In: International Conference on Artificial Neural Networks, pp. 270–279 (2018)
22. Wan, J., Wang, Y.: OCost-sensitive label propagation for semi-supervised face recognition, IEEE Trans Inf. Forensics Secur. **14**(7), 1729–1743 (2018)
23. Weiss, K., Khoshgoftaar, T.M., Wang, D.: A survey of transfer learning. J. Big Data **3**(1), 1–40 (2016)
24. Wang, Z., Dai, Z., Póczos, B., Carbonell, J.: Characterizing and avoiding negative transfer, In: Proceedings of the IEEE/CVF Conference on Computer Vision and Pattern Recognition, pp. 11293–11302 (2019)
25. Zhang, W., Deng, L., Zhang, L., Wu, D.: Overcoming negative transfer: a survey, arXiv Preprint, https://arxiv.org/abs/2009.00909 (2020)

Leveraging Natural Language Processing in Persuasive Marketing

Evripides Christodoulou and Andreas Gregoriades(✉)

Cyprus University of Technology, Limassol, Cyprus
ep.xristodoulou@edu.cut.ac.cy, andreas.gregoriades@cut.ac.cy

Abstract. The language used in marketing communication influences consumers' attitudes towards products or services and likelihood of purchasing or recommending these to others. Knowing the personality of the consumer is important in persuasion marketing and can be inferred from the abundance of consumer information available online. This paper utilizes text classification to extract consumers' personality from electronic word-of-mouth (e-WOM) and topic modelling to identify consumers' opinions. The aim is to optimize marketing communication through personalized messages that abide to targeted consumers' personalities. The method is based on the theory of self-congruence, stipulating that consumers are inclined to purchase a brand that reflects their own personalities. Consumer reviews are obtained from TripAdvisor and their textual part is expressed as a proportion of different discussion themes identified through topic modelling. The personality of each reviewer is recognised using the textual part of their eWOM and a deep learning model trained on labelled text using the personality model of Myers-Briggs Type Indicator (MBTI). Four XGBoost (eXtreme Gradient Boosting) classifiers are trained, one for each of the four MBTI personality traits, using as predictors the topic embeddings and output the personality type of consumers. An explainable AI technique, namely, Shapley Additive Explanations (SHAP), is used to explain how the topics discussed by consumers in eWOM are related to their personality. Patterns from each XGBoost model are collated into a table showing how topics can be exploited by marketers during advertisement message design to appeal to specific consumer personalities. Preliminary results are compared against persuasion marketing and consumer behavior literature.

Keywords: Persuasion Marketing · Personality extraction · Electronic word of mouth · XGBoost

1 Introduction

Utilizing consumers' personality during the design of advertising messages attracted significant attention in persuasion marketing. Marketers leverage human psychology to convey a message to consumers and influence them in a positive way to purchase a product or service. Information regarding consumers' personality, preferences and needs can be identified from consumers' online behavior such as clicks, likes, reviews etc. When this is done ethically it can bring mutual benefits to both consumers and businesses.

N. T. Nguyen et al. (Eds.): ACIIDS 2023, LNAI 13995, pp. 197–209, 2023.
https://doi.org/10.1007/978-981-99-5834-4_16

Evidence from the application of personality in Facebook ads suggest that the effectiveness of marketing messages can be increased by aligning these with consumer's psychological profile. Similar work showed that people's personality can predict whether someone accepts online ads or not [12] and can influence their video ad preferences. Explanation of this behaviour is based on self-congruence theory stipulating that consumers are inclined to purchase a brand that reflects their own personalities [18] or have similar preferences with people with similar personality to them [4]. The theory suggests that consumer preferences are determined by the cognitive match between particular aspects of consumer's opinion about themselves and the brand's personality which is usually expressed through ads and company messaging [26]. Tailoring a message to adapt to the recipients' psychological characteristics can increase the acceptance of the message [6]. This behaviour has been observed in different domains, [7] and represents the underlying theory used in this paper.

Electronic word-of-mouth (e-WOM) and specifically the textual part of eWOM provide valuable information that can be harnessed by machine leaning techniques to infer consumers personalities, opinions regarding a product or service [9, 12]. This work focuses on extracting personality and user opinions of consumers from text and identifying patterns between personality traits and consumer opinions using XGBoost to support marketers during the design of advertising messages. The contribution of this work is a method that exploits eWOM and integrates explainable AI with text classification to support marketers in designing persuasive advertising messages. The method exploits patterns (Shapley values) that emerge after training personality classifiers with eWOM embeddings (topics) to identify relationships between personality and topics discussed in eWOM. The proposed approach differs from other trait-based persuasion methods in literature that use experiments (and thus are time consuming), through an automated analysis and exploitation of eWOM data.

The paper is organized as follows. The next section present the literature on personality based persuasion marketing and the various automated methods for recognizing personality from text. Subsequent sections elaborate on methodology, topic modelling, text classification and explainable AI. The patterns that are identified from the XGBoost models are discussed. The paper concludes with limitations and future directions.

2 Literature Review on Persuasion Marketing and Personality

Consumers go through different stages before making a final purchase decision. Ads aim to create awareness of the product and purchase intention. An important concept in advertising messages is how persuasive the message is. The aim is that these messages will be remembered by consumers and influence their purchasing behaviour. Such marketing activities add significant cost to the product which could reach up to 34% of a product price.

Persuasion marketing focuses on using psychological concepts to influence consumer behavior and in particular to convince consumers to buy goods and services that will give them satisfaction [21]. The underlying theory of personality-based persuasive marketing posits that people with different personalities react differently to different marketing messages [5]. Thus, messages are adapted to the consumer's personality using framing that is likely to be valued by persons with given characteristics. In the field of online persuasion marketing, there is a lack of research methods that use e-WOM data as a primary source of information. E-WOM content can be utilise to extract valuable insights for persuasion marketing [27] and can assist in improving consumer decision-making, which ultimately benefits both businesses and customers. In our previous work we utilised eWOM to extract topics that resonate more with consumers from different cultural background to assist digital content marketing [14].

One of the underlying theories of persuasion is that of self-concept, which has been known to influence consumer behaviour. This theory stipulates that individuals have a unique and relatively stable set of beliefs, interests and attitudes about themselves that are linked to their personality. Self-congruity theory proposes that when a brand verifies or enhances a person's self-concept, the person creates positive opinion and greater intention in purchasing that brand. By understanding an individual's self-concept, marketers can create messages that align with their beliefs and attitudes, making the message more likely to be accepted and acted upon. Self-congruity is also defined as the matching of a brand's personality to the consumers' self-concept, with brand personality defined as the set of human-like characteristics associated with a brand [1]. Information about individuals' self concept can be extracted from words and phrases used by people using text analytics as shown in [31]. An example of how the matching of consumer and brands personality affect purchasing behaviour is reported in [20] showing that when people spend money on products that fit their personality, they are more satisfied. Similarly, Mulyanegara et al. [22] found that highly extraverted people prefer sociable brands (brands which are friendly, creative, and outgoing), while Chen et al. [12] showed that personality also can indicate which advertising channel (social media, TV, radio) is more appropriate for each trait. Research also shows that customers tend to buy from brands with a clear messaging strategy and strong personality congruence to their own while also they tend to respond more positively to a marketing message when the message is tailored to their personality [33].

Human personality is therefore a key in persuasion marketing and is defined as a complex set of characteristics and behaviours that influence an individual's motivations, preferences, and behaviours. Personality has been found to remain relatively stable throughout adulthood [30] and thus constitutes a valuable property for marketers. There are several personality models that are widely used in psychology and marketing. The Myers-Briggs Type Indicator (MBTI) which is based on Carl Jung's theories has four dimensions: Extroversion or Introversion, which examines if the person is more energized by their external world (extroverts enjoy social interactions) or their inner world (introversion). Sensing or Intuition, which examines whether a person perceives information using his/her senses or uses intuition. Sensors are task oriented and like tasks that are specific and have clear sequence. Intuitive people are creative and like unusual and new activities and dislike routine. Thinking or Feeling dimension examines if a person tends to make decisions using logic (thinking) or tends to be more sensitive and concerned with values and harmony in relationships(feeling). Judgment or Perception examines if a person makes sense of the world by organizing it or if he/she stays open to new information. Individuals who are judging-oriented are structured and make formal decisions, while perceiving-oriented individuals tend to plan less and adapt better to change. Another personality model that is widely used is the Big Five or the Five Factor Model, that uses five dimensions: openness, conscientiousness, extraversion, agreeableness, and neuroticism.In this work we utilise the MBTI model since it yielded a petter personality performance in our text classification task as shown in our previous work[10].

2.1 Automated Techniques for Recognizing Personality

There are two main methods for recognising the personality of consumers: questionnaires and automated methods. Questionnaires are more accurate in assessing personality, but the process is tedious and expensive. Automated methods, on the other hand, utilize existing data such as consumer generated text, images, videos, clicks and likes (behavioural data) to predict personality [11]. Text based personality recognition gains considerable attention due to recent good results. The underlying theory is that words can reveal psychological states and the personality of the author. There are two main categories of techniques for extracting personality traits from text: feature-based machine learning and deep learning methods. Feature-based extracts tokens from text in the form of unigrams or n-grams (open vocabulary approach) and finds connections between these and personality traits (supervised approach). Alternatively a closed vocabulary approach use prespecified words that have been found to explain personality such as the well-known, Linguistic Inquiry and Word Count [3]. Deep learning techniques utilize language models trained on large corpora of text in an unsupervised manner, fine tuned on a text classification task to classify personality in text. Text embeddings from these models vectorise the text and through an attention mechanism [29] assign weights to words based on the context. Such methods enable the extraction of semantic content from text [16] and provide improved performance to vocabulary based techniques. The use of a pre-trained model as a starting point for a new task is useful in situations where there is a lack of sufficient labelled data to train a deep learning model from scratch. Fine-tuning involves adjusting the pre-trained model to the specific task by adding a feedforward layer on top of the model and using a labelled dataset for training the classifier. A popular architecture in this category is the Bidirectional Encoder Representations from Transformers (BERT) model, which utilizes transformer neural networks and an attention-mechanism to extract semantics and make predictions [15]. BERT has been successfully used in various natural language processing tasks, including language translation and text classification (used in this study). Previous research has shown that this pre-training and fine-tuning approach outperforms traditional text classification methods [28] and has been applied with success in personality recognition. In this work labelled textual data is used to train different BERT classifiers per personality trait as explained in our previous work [13]. The personality model that is chosen is the MBTI due to its superior performance in classifying personality in comparison to BIG-five. In this work the classifiers were initially trained using labelled data and subsequently used to recognise the personality of consumers based on their reviews.

3 Methodology

The personality based persuasion marketing method proposed in this work combines: 1) topic modeling to identify the main consumer opinions expressed in eWOM 2) automated consumers' personality recognition from eWOM's text 3) Explainable AI for extracting patterns from learned models. Both topics and personality traits of consumers are extracted from a restaurant reviews dataset obtained from Trip Advisor using a dedicated scraper. The information extracted from these is used to train four binary classification models (XGBoots) with consumer topics as input and personality traits as output. The XGBoots patterns that emerge are used to discover relationships between personality and consumer opinions and which can be utilized by marketers during designing of advertising messages. The method's process flow is depicted in Fig. 1 and is composed of 5 steps.

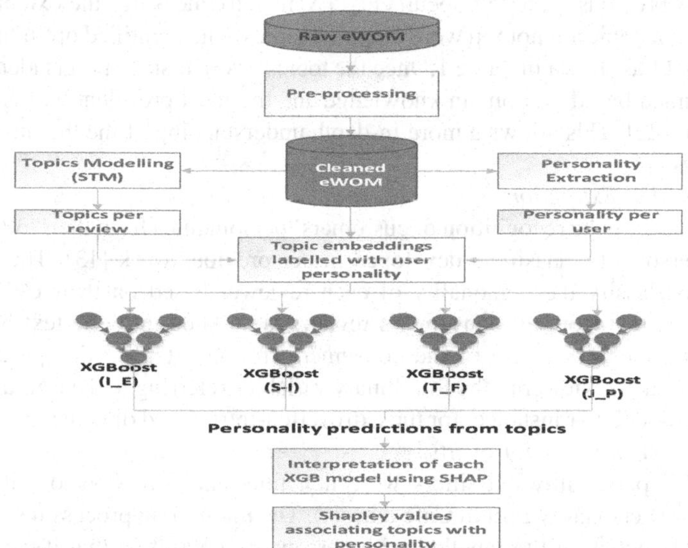

Fig. 1. Method's process flow

Step1: Topic Modelling

The initial dataset is filtered to include only highly positive reviews with rating score > 4 out of 5, thus keeping only eWOM from satisfied customers in the corpus. The data is then preprocessed and prepared for subsequent tasks. The preprocessing includes elimination of punctuations and URLs, lowering of text, removal of stop-words (for topic modeling), tokenization and expansion of text abbreviations (i.e., don't to do not). To identify the discussion topics that associate more with each personality only highly positive and negative reviews were utilised, since negative reviews also contain information that marketers need to avoid while designing their advertising messages.

Topic modeling is a technique used to identify themes in our review-corpus, by finding sets of words that frequently occur together in reviews. In this work topic modelling is used to vectorize the corpus. There are several techniques for topic analysis, including Latent Dirichlet Allocation and Structural Topic Modeling (STM). In this study, the STM approach was used to identify key themes discussed by consumers since it yielded more meaningful topics in contrast to LDA when the reviews' rating was used as meta data.

The process of learning the topic model begins with data preprocessing, which involves removing common stop-words, as well as other irrelevant information such as punctuation. The data is then tokenized, or broken down into individual word tokens, and stemmed, or converted to its root form. The optimal number of topics (K) for the dataset is determined based on the semantic coherence, held out likelihood, and exclusivity of different models. Coherence refers to the degree of semantic similarity between high-scoring words in the topic, while held out likelihood is testing the trained model using a set of previously unseen documents. Exclusivity measures the extent to which top words in one topic are not top words in other topics. The identified optimum number of topics was 11 as shown in Table 1.Once the topics' word distribution is identified, the topics are named based on domain knowledge and the most prevalent words that characterize them [22]. This allows a more in-depth understanding of the themes discussed in the reviews.

Step2: Personality Extraction

This step focuses on the recognition of customers' personalities from their eWOM using the MBTI personality classifiers developed in our previous work [13]. The classifiers were used to classify the personality of each reviewer based on their eWOM's text only. Thus, the text from all consumer's reviews is collated into one text block prior to recognizing the personality of the consumer. The output from this process is the association of each reviewer to the four binary variables referring to the four dimensions of the MBTI model. For instance, for the extroverted-introverted dimension, "one" refer to extroverts and "zero" to introverts.

The BERT personality classifiers were first fine-tuned on a personality labelled dataset of text (Personality Forum Café dataset). The fine-tuning process involves training the BERT model on the labelled with personality dataset so that it can learn the specific patterns of language that are associated with different personality traits. The main advantage of using BERT is the transfer of knowledge (language representations) learned from large corpora to the specific task of personality classification. Additionally, the bidirectional nature of the model allows it to take into account the context of the words in the text.

The trained BERT classifiers were used to label the personality of each reviewer in our dataset based on each consumer's eWOM. Thus, reviews generated by individual consumers were collated into one long text and analysed collectively. This was necessary, since some reviewers produced more than one reviews, and thus these were accumulated to produce sufficient text based on which to extract the personality. The identified personality per reviewer was used as extra information associated with all his/her reviews.

Step3: XGBoost training
The associations between customers' opinions and personality traits are identified through four binary XGBoost personality trait classification models. Each model was trained to predict one of the four dimensions of the MBTI model. XGBoost algorithm is used due to its excellent performance on a variety of similar problems [8]. XGBoost combines multiple decision trees in a linear fashion, with each tree being built based on the output of the previous tree. They are popular since they can prevent overfitting through the use of regularization and are computationally efficient [8].

In this work, four XGBoost models are trained using as input the topics' thetas per review and as class variable the consumers' personality that is extracted from BERT. Four models are trained one for each MBTI personality trait. The XGBoost models are optimized using hyperparameter tuning of the following parameters: max_depth, reg_alpha,learning_rate, scale_pos_weight, min_child_weight, and validated using a 70/30 train/test data split. A popular performance optimization metric used for binary classifiers is the area under a receiver operating characteristic (AUC) curve, which measures the overall performance of binary classifiers, and F1 score which combines the precision and recall of a classifier into a single metric by taking their harmonic mean. AUC was used as an optimization metric during hyperparameter tuning using GridSearch. The predictive performance of each model is an essential validation step to reassure the correctness of the patterns that emerge from these models. Therefore, the minimum acceptable AUC performance was 80%. All models satisfied this criterion.
Step4: Extraction of XGBoost patterns using an explainable AI technique.

To extract the patterns from each of the four XGBoost models the SHAP (SHapley Additive exPlanations) explainable AI technique [19] is used. SHAP is a popular technique for explaining complex machine learning models such as XGBoost that is based on the principles of game theory with players being the features of the model engaging each other in a cooperative game. In this case study the features are topics associated with each review. SHAP explain the output of a model by assigning importance scores to models' features. These represent the contribution of each feature to the output. Contributions are used to infer how each feature influence the model's output. SHAP provide various visualizations that help in understanding the model's logic on a global or local scale.
Step 5: Identify associations between opinions and personality traits.
Explanations of the XGBoost models' patterns are expressed in terms of shapley values for each feature(topic) of the model. The SHAP summary plots are utilised since they provide an overview of feature importance scores in a graph (the order of the features on the y axis determining their importance to the prediction). The shapley values, used to generate the plots, are utilised to identify the association between topics and personality traits, and collated in Table 2. Values in the generated table shows the strength of each association. Topics with shapley values less than a prespecified threshold (0.04) were ignored since their contribution was insignificant. The generated table can be used by marketing departments to optimize their ads' performance by tailoring messages to resonate with the targeted consumer personality trait.

4 Results

The data utilised in this study include 85,000 reviews in English from UK consumers who visited restaurants in Cyprus between 2015 and 2020 and shared their experiences online. The data was collected using a dedicated web scraper. The case study focus on UK customers (largest percentage of tourists in Cyprus) to eliminate confounding influence of consumers from different cultural backgrounds. The data include 510 unique users. To ensure that the personality of each consumer is evaluated correctly from their eWOM text, their reviews are initially collated into a new feature prior to personality extraction. Reviews by users that generated only one review and the length of that was less than 50 words were eliminated since it is difficult to accurately extract personality from such small text. This resulted in a dataset with 310 unique users. During topic modelling only positive reviews were considered which reduced the dataset even further. During topic modelling, each review was considered as a document, in contrast to the personality extraction process where the aggregated text was used. This was necessary since each review refers to different topics' proportions.

Table 1. Specified names for the topics that emerged from topic modelling.

Topic Names	Subset of words associated with each topic
Lunch/Breakfast	breakfast, lunch, menu
Service quality	best, service, quality
Tasty Dishes	delicious, dishes, quality
Couple food	wife, visit, leisure
Value for money	euros, advert, value
Location and Atmosphere	atmosphere, location, amazing
Friendly Staff	staff, friendly, helpful
Beach place	sea, drink, beach
Return over years	visited, times, before
Local cuisine	meze, traditional, cypriot
Relaxation	wine, taste, relax

The STM topic model was developed by first determining the optimal number of topics in the corpus. This was done by evaluating the semantic coherence of the topic model by varying the numbers of topics [23, 25]. This process identified that the optimum number of topics for this corpus is 11. Topics' names were determined by manually reviewing the prevalent words in each topic's distribution in relation to the literature. The resulting topic names are listed in Table 1. The percentage of each topic mentioned in each review, known as the "theta values," were used as features during model training.

Consumers' personality extraction is performed using a fine-tuned text classification model (BERT) trained and validated on text with MBTI personality labels [13]. The optimum number of tokens for the training of the BERT model was 512, which denotes the use of the first 512 words(or sub-word tokens) of the text written by consumers. Four personality classifiers were trained using secondary labelled data (personality caffe), each achieving an AUC and F1 performance above 80%. These performance metrics provided us with the required confidence for labelling the unseen by the classifier eWOM data. Each classifier labeled consumers' text in terms of: Introversion-Extroversion, Intuitive-Sensing, Thinker-Feeler and Judger-Perceiver. Text aggregation maximized the text used by the classifiers to recognize the personality of consumers. Figure 2 shows descriptive results associating a personality type (extraversion-introversion) with the average number of words in their reviews. This shows that extraverted consumers are more expressive and thus use more words in their reviews. This abides with the literature and thus provided a preliminary validation of the personality model.

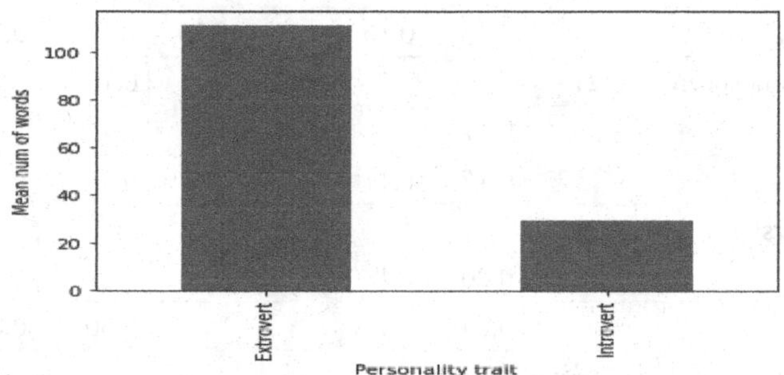

Fig. 2. Personality trait "extraverted-introverteed" against average num of words in messages

Four XGBoost models are trained(70%) and validated(30%) each refering to different personality trait. Topics embeddings per review are used as model features and personality trait as target variable. The SHAP (TreeSHAP) technique [19] is used to interpret the logic of each of the four XGBoost models. The SHAP summary plot shows the association between topics and personality. Dots on the SHAP summary plot represent single observations (reviews). The horizontal axis shows the SHAP values which refer to the average marginal contribution of each feature to the output. This is calculated by considering all possible coalitions of features (in game theoretic terms) and how these contribute to the output. The color of the dots defines if that observation has a higher or a lower value (in this case is the topic theta values), when compared to other observations. SHAP values of less than 0 indicates a negative contribution to the output, equal to 0 indicates no contribution, and greater than 0 indicates a positive contribution. For example, Fig. 3 shows the global explanations of Intuitive/Sensing XGBoost model. The probability of a customer being "intuitive" is high if the majority of the red dots of a feature are placed on the left hand side of the vertical line (negative SHAP values) and "sensor" if they are on the right-hand side.

A SHAP summary plot is generated for each personality trait (XGBoost). The shapley values of topics' contribution to personality are extracted from the SHAP dataframe and filtered based on a threshold value (0.04). From this process the Table 2 is generated showing all personality traits and their topics' associations. The numbers in the table indicate the strength of each personality-topic association and refers to the absolute shapley value. The higher this value, the stronger the link between the two.

Table 2. Shapley values greater than a threshold, showing associations of topics with personality traits.

Topic Names	INTR	EXT	INT	SEN	THI	FEEL	JUD	PERC
Lunch/Breakfast		0.11			0.12			0.12
Service quality	0.1			0.08	0.37		0.10	
Tasty Dishes			0.09				0.04	
Couple food		0.14	0.12					
Value for money			0.16		0.20		0.13	
Location and Atmosphere	0.21			0.13		0.10		0.14
Friendly Staff		0.17		0.11				0.08
Beach bar		0.15	0.11			0.11		
Return over years								
Local cuisine		0.20	0.18					0.30
Relaxation		0.04	0.12			0.06	0.05	

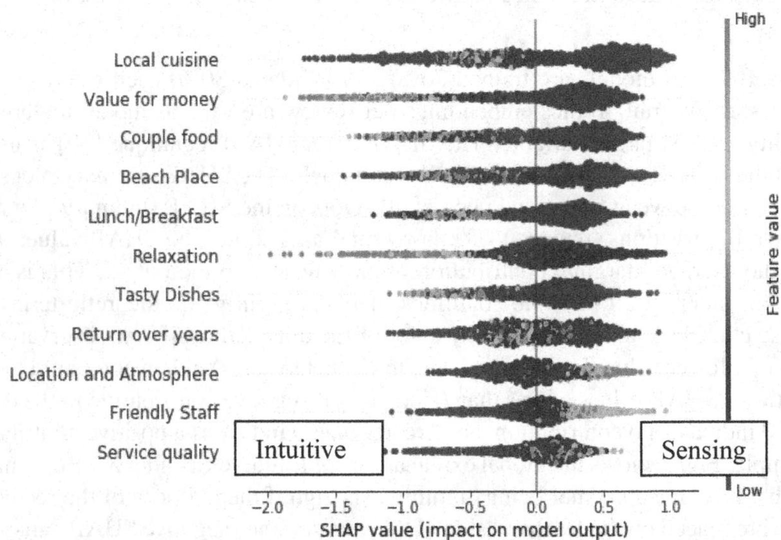

Fig. 3. Topics' associations with Intutive-Sensing personality trait. Red dots indicate high topic discussion.

Based on Table 2, the designer of an advertisement can utilize words associated with the personality trait of the consumer segment that is targeting. For instance, if the target consumers are extroverts, then the ads' content should highlight the motivational concerns of this personality trait in a way that resonates with their characteristics. Thus, by utilizing the recommendations from Table 2 the designer can enhance an ad message by incorporating information from topics associated with extroverted consumers and thus addressing the topics "location" and "busy place" that are relevant to this trait. Based on this recommendation the message could have a spin on how busy and cosmopolitan the area around the restaurant is. Therefore, the ad could include connotations such as "At bar-restaurant X you'll always be where the excitement is". Alternatively, if the target consumers are thinkers, then the message could highlight the price dimension since it emerged as an important topic for this trait.

The results show that there are thematic association between different traits with introverts showing an inclination to properties such as safe options and quality of experience [17, 24]. Intuitive people as past research suggests [2], are open to new experiences, with topic "Local cuisine" indicating the need for exploration and trying new traditional food, while the opposite occurs for "sensing" consumers, where they are more skeptical and prefer "service quality" and "good location". Similarly, "thinkers" tend to evaluate their options and don't like to take risks, thus the topics "value for money", "service quality" would resonate well with them. Feelers do act impulsively wanting to live the experience and value their feelings [24], thus, the topics "beech place", "romantic atmosphere" and "relaxation" would resonate well with them since they are highly associated with feelings. Judgers prefer to judge (tasty dishes, value for money) before committing to a decision while the perceivers as their name suggest, are more open to the experiences (local dishes) and don't like to plan ahead [17, 24].

5 Conclusions

This study illustrate an application of machine learning and natural language processing in personality based persuasion marketing. It expand upon previous research and provide information that can be used in persuasion marketing strategies to attract targeted personality groups, while aiming for mutual benefits for consumers and businesses by increasing consumer satisfaction and business revenue. Preliminary results from this work abide with the literature of persuasion marketing and reinforce our confidence of their validity. However, it is important to note that personality is not the only factor that influences how people respond to marketing messages, and there are other factors such as: values, beliefs, culture, and context that are also important. Thus, as part of our future work, our aim is to expand our method while also addressing these.

References

1. Aaker, J.L.: Dimensions of brand personality. J. Mark Res. **34**, 347–356 (1997)
2. Bardi, A., Schwartz, S.H.: Values and behavior: strength and structure of relations. Personal Soc. Psychol. Bull. **29**, 1207–1220 (2003)
3. Boyd, R.L., Pennebaker, J.W.: A way with words: using language for psychological science in the modern era. Consum Psychol. a Soc. Media World, pp. 222–236 (2015)
4. Byrne, D., et al.: The ubiquitous relationship: attitude similarity and attraction: a cross-cultural study. Hum. Relations **24**, 201–207 (1971)

5. Campbell, M.C., Kirmani, A.: Consumers' use of Persuasion knowledge: the effects of accessibility and cognitive capacity on perceptions of an influence agent. J. Consum. Res. **27**, 69–83 (2000)
6. Cesario, J., Grant, H., Higgins, E.T.: Regulatory fit and persuasion: transfer from "feeling right." J. Pers. Soc. Psychol. **86**, 388–404 (2004)
7. Cesario, J., Higgins, E.T.: Making message recipients "feel right": how nonverbal cues can increase persuasion. Psychol. Sci. **19**, 415–420 (2008)
8. Chen, T., Guestrin, C.: XGBoost: a scalable tree boosting system. In: Proceedings of the ACM SIGKDD, pp. 785–794 (2016)
9. Chevalier, J.A., Mayzlin, D.: The effect of word of mouth on sales: online book reviews. J. Mark Res. **43**, 345–354 (2006)
10. Christodoulou, E., Gregoriades, A., Herodotou, H., Pampaka, M.: Combination of user and venue personality with topic modelling in restaurant recommender systems. Rectour, RecSys. **3219**, 21–36 (2022)
11. Dhelim, S., Aung, N., Bouras, M.A., Ning, H., Cambria, E.: A survey on personality-aware recommendation systems. Artif. Intell. Rev. **55**, 2409–2454 (2022)
12. Godes, D., Mayzlin, D.: Using online conversations to study word-of-mouth communication. Mark Sci. **23**, 545–560 (2004)
13. Gregoriades, A., Pampaka, M., Christodoulou, E., Gregoriades, A., Herodotou, H., Pampaka, M.: Extracting user preferences and personality from text for restaurant recommendation. RecSys. 2022 (2022)
14. Gregoriades, A., Pampaka, M., Herodotou, H., Christodoulou, E.: Supporting digital content marketing and messaging through topic modelling and decision trees. Expert Syst. Appl. 115546 (2021)
15. Jun, H., Peng, L., Changhui, J., Pengzheng, L., Shenke, W., Kejia, Z.: Personality classification based on bert model. Proceedings 2021 ICESIT, pp. 150–152 (2021)
16. Kardakis, S., Perikos, I., Grivokostopoulou, F., Hatzilygeroudis, I.: Examining attention mechanisms in deep learning models for sentiment analysis. Appl. Sci. **11** (2021)
17. Kassarjian, H.H.: Personality and consumer behavior: a review. J. Mark Res. **8**, 409–418 (1971)
18. Lee, J.W.: Relationship between consumer personality and brand personality as self-concept: From the case of Korean automobile brands. Acad. Mark Stud. J. **13**, 25–44 (2009)
19. Lundberg, S., Lee, S.-I.: A unified approach to interpreting model predictions. Adv. Neural Inf. Process Syst. **30**, 4768–4777 (2017)
20. Matz, S.C., Gladstone, J.J., Stillwell, D.: Money buys happiness when spending fits our personality. Psychol. Sci. **27**, 715–725 (2016)
21. Miles, C.: Persuasion, marketing communication, and the metaphor of magic. Eur. J. Mark **47**, 2002–2019 (2013)
22. Mulyanegara, R.C., Tsarenko, Y., Anderson, A.: The big five and brand personality: investigating the impact of consumer personality on preferences towards particular brand personality. J. Brand Manag. **16**, 234–247 (2009)
23. Nikolenko, S.I., Koltcov, S., Koltsova, O.: Topic modelling for qualitative studies. J. Inf. Sci. **43**, 88–102 (2017)
24. Pelau, C., Serban, D., Chinie, A.C.: The influence of personality types on the impulsive buying behavior of a consumer. Proc. Int. Conf. Bus Excell. **12**, 751–759 (2018)
25. Roberts, M.E., Stewart, B.M., Tingley, D., Lucas, C., Leder-Luis, J., et al.: Structural topic models for open-ended survey responses. Am. J. Pol. Sci. **58**, 1064–1082 (2014)
26. Sirgy, M.J.: Using self-congruity and ideal congruity to predict purchase motivation. J. Bus. Res. **13**, 195–206 (1985)
27. Sotiriadis, M.D., van Zyl, C.: Electronic word-of-mouth and online reviews in tourism services: the use of twitter by tourists. Electron. Commer. Res. **13**, 103–124 (2013)

28. Sun, C., Qiu, X., Xu, Y., Huang, X.: How to Fine-Tune BERT for Text Classification? (2019)
29. Vaswani, A., et al.: Attention is all you need. In: Advances in Neural Information Processing Systems. Curran Associates, Inc. (2017)
30. Wang, H., Zuo, Y., Li, H., Wu, J.: Cross-domain recommendation with user personality. Knowledge-Based Syst. **213**, 106664 (2021)
31. Wilkie, D.C.H., Rao Hill, S.: Beyond brand personality. a multidimensional perspective of self-congruence. J. Mark Manag. **38**, 1529–1560 (2022)

Direction of the Difference Between Bayesian Model Averaging and the Best-Fit Model on Scarce-Data Low-Correlation Churn Prediction

Paul J. Darwen(✉)

James Cook University, 349 Queen Street, Brisbane, QLD, Australia
paul.darwen@jcu.edu.au

Abstract. On a scarce-data customer churn prediction problem, using the tiny differences between the predictions of (1) the single-best model and (2) the ensemble from Bayesian model averaging, gives greater accuracy than state-of-the-art approaches such as XGBoost. The proposed approach reflects the cost-benefit aspect of many such problems: for customer churn, incentives to stay are expensive, so what's needed is a short list of customers with a high probability of churning. It works even though in every test case, the predicted outcome is always the same from both the best-fit model and Bayesian model averaging. The approach suits many scarce-data prediction problems in commerce and medicine.

Keywords: ensembles · customer churn · small data · scarce data · Bayesian model averaging · committee of committees · cold start problem

1 Introduction and Motivation

Many prediction problems suffer from scarce data: few rows of data, and the few input columns have poor discriminant value, giving low correlation. Such small-data low-dimensional problems include medical triage prediction [10], the diagnosis of rare cancers [8], and the hospital readmission problem [9,12,14].

As a typical example of such scarce-data problems, here the task is to predict which e-commerce customers will churn (change to a different provider). Churn affects mobile phone providers [1], online games [15], e-commerce web sites [18, 22], and any service where the customer has a low cost of switching providers.

Predicting customer behaviour is still not a solved problem, even for big-data problems like mobile phone customer churn [19]. There are many more businesses with low-dimensional small-data prediction problems than businesses with the luxury of big data sets. This includes fields with rapid changes (such as fashion and phones) in which data more than a few months old is irrelevant.

N. T. Nguyen et al. (Eds.): ACIIDS 2023, LNAI 13995, pp. 210–223, 2023.
https://doi.org/10.1007/978-981-99-5834-4_17

1.1 Previous Work and Current Contribution

The class of problems studied here are small-data, low-correlation prediction tasks. Ensemble approaches have been used for other problems or in other ways [6,24], such as high-correlation (instead of low-correlation) big data [4], or only to do feature selection [17], or with small ensemble sizes like eight models [16] instead of thousands, or to track changes over time [2,21]. And of course customer churn prediction has been attempted in many ways [15,19,22], usually presuming that huge data sets are available [1].

Predicting from small-data low-correlation data has much in common with the cold-start problem in recommender systems. The approach presented here relies heavily on selecting suitable inputs/attributes, as do some studies of the cold-start problem [3,25].

The approach presented here has been used on predicting river floods [7] and cancer diagnosis [8], but this study is the first to apply the approach to an e-commerce prediction problem. The results will show that for a customer churn prediction problem typical for millions of retailers, the proposed approach gives better accuracy than any other algorithm in the SciKit library for Python.

This paper's approach uses the tiny differences between the predictions of (1) the single-best model, and (2) the ensemble from Bayesian model averaging. The proposed approach is not merely comparing an ensemble method with a single model because both predict the same result, in almost all cases. Section 4 will show that the proposed method does better than XGBoost, which was the best in a recent study about mobile phone churn prediction [1].

2 Background

2.1 A Scarce-Data Low-Correlation Customer Churn Problem

The churn prediction problem comes from an e-commerce site that sells a single product. The data set[1] only covers the browsing history of each customer. There is no demographic or other personal information.

The business case is to identify a short list of customers who are most likely to churn, in order to persuade them to stay, using discounts or similar inducements. The list must be short, because there is a tight budget for these expensive offers. Naturally the seller is reluctant to offer such inducements to a customer who wasn't planning to churn after all. The profit margin is thin, so it only makes sense to offer the inducement if the prediction is correct around 90% of the time.

What measure of accuracy to use? If the classifier predicts the customer will not churn, then no inducement will be offered, so true negative (TN) and false negative (FN) can be ignored. If the classifier predicts a customer will churn, and we offer the inducement, then it would be an expensive waste if the customer had no intention of churning (a false positive, FP). So the chosen measure of accuracy for this problem will be precision = TP/(TP + FP).

[1] See https://www.kaggle.com/huzaiftila/customer-churn-prediction-analysis.

2.2 Description of the Data, and Data Cleaning

Big spenders or "whales" are the most desirable customers. Whales who regularly view the web site tend to keep visiting, and rarely churn. The cases of interest are the big spenders who rarely visit the web site. From the original 91,698 visitors over 5 weeks, the client is only interested in big-spending but rarely-visiting customers: those who spent 2,076 rupees or more in their first 5 weeks, but only had 7 or fewer visits. This gives a small data set of 510 rarely-visiting, big-spending customers. It's close to balanced, with 51.6% of customers churning and 48.4% staying.

Some redundant columns are deleted. It has the sales amounts for each week, so binary "week 1" through "week 4" (to say if they bought anything) are redundant. Similarly, the last three columns (binary for rare, frequent, or regular visitors) are redundant because the column "Visitors Type" gives the same information (and we convert it to 0, 1, and 2 to make it numeric). Similarly deleted are the three columns for high, regular, and low value which merely repeat the column "Customer Value".

The values of each input column are rank normalized from 0 to +1: an input's smallest value becomes 0, the median becomes 0.5, and the biggest becomes 1.

In this paper, accuracy is calculated from leave-one-out cross-validation [13, page 242]. From the data set of 510 rows, we create an equal number of training data sets, each one having 509 rows, with the left-out row becoming the single point of test data for the model built with that training set. This gives an average accuracy without random sampling.

Interaction variables are new inputs created by multiplying the values in existing input columns[2]. For each of the 510 training sets created by leave-one-out cross validation, each pair of inputs is multiplied to create every possible 2-way interaction variable.

Feature selection uses brute-force sampling, generating many combinations of inputs, and looking for the best feature set from the many sets thus generated. In practice, the feature selection only picks less than 10 inputs, from the many on offer: most interaction variables are no better than random.

2.3 Bayesian Model Averaging: Same Prediction as Best Model

Bayesian model averaging is an ensemble approach. It enumerates all possible models, and each model's prediction is weighted by the model's posterior probability of being correct, given the data [11, page 8] [8, section II.C]. The weighted vote from all the models gives the prediction.

Unfortunately, for the problem studied here and for many other scarce-data problems [8], Bayesian model averaging always predicts the same outcome as the single best-fit model. In contrast, the proposed approach makes use of the tiny differences in the predicted likelihoods: the following Sects. 2.4 and 2.5 describe Bayesian model averaging as used here.

[2] Some software packages do this automatically, such as SPSS and Unistat.

2.4 A Likelihood Function for Bayesian Model Averaging

To calculate the probability of a model's correctness, each model needs to a likelihood function. The one used here follows earlier work on predicting floods [7] and lung cancers [8]. Each model includes a linear regression model, with a slope coefficient for each input, plus a coefficient for an intercept. To put a pipe on top of that line (or rather, hyperplane), one more model coefficient is the standard deviation of a bell curve cross-section, to make the model into a likelihood function, as needed for Bayesian model averaging.

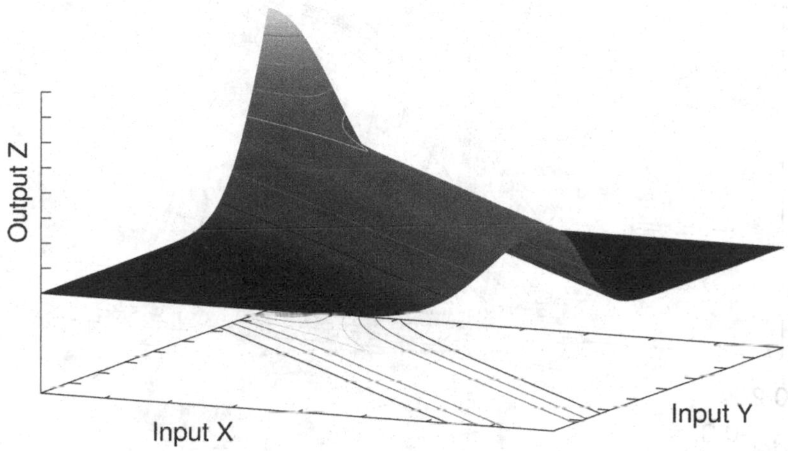

Fig. 1. In Bayesian model averaging, each model calculates likelihoods. This stylized view shows one half of the "buried trombone" likelihood function used here. A model is a line (or hyperplane) from linear regression, covered by a bell curve that inflates at both ends. Each input is ranked, so values go from zero to one, and the bell curve is truncated below zero or above one, requiring the untruncated remainder to get bigger, to guarantee that the integral of each cross-section will always sum to one.

The likelihood predicted by a model is the vertical axis. Back in Sect. 2.2, preprocessing ranked each input's values from 0 to +1 (0 being the smallest, 1 being the biggest, and 0.5 being the median), so a model's pipe-shaped likelihood function must be truncated at 0 and 1. As it approaches a truncation, the each cross-section of the bell curve must integrate to one, so the trombone-shaped blow-up in Fig. 1 is needed.

2.5 Two Approximations to Bayesian Model Averaging

Bayesian model averaging wants to *integrate* over *all* possible models, but this integral can't be done symbolically with a software model. This section describes two approximations to get around those two requirements.

Firstly, for each set of n input features, the continuous integration is approximated with an n-dimensional grid over the space of model parameters. The integral thus becomes a sum over thousands of models.

Secondly, even with a discretized grid, it's too many to enumerate *all* models with non-zero posterior probability of being correct. Ignoring models with a near-zero posterior probability should only give a small error. So in the proposed method in Sect. 3, the enumeration of models in that grid is terminated after it has found a large number of models: 40,000 for 4 inputs, 50,000 for 5, 60,000 for 6, and 75,000 for 7 inputs. Note that number of inputs chosen by feature selection can be different on different data sets, because feature selection is run independently on each data set from the leave-one-out cross-validation.

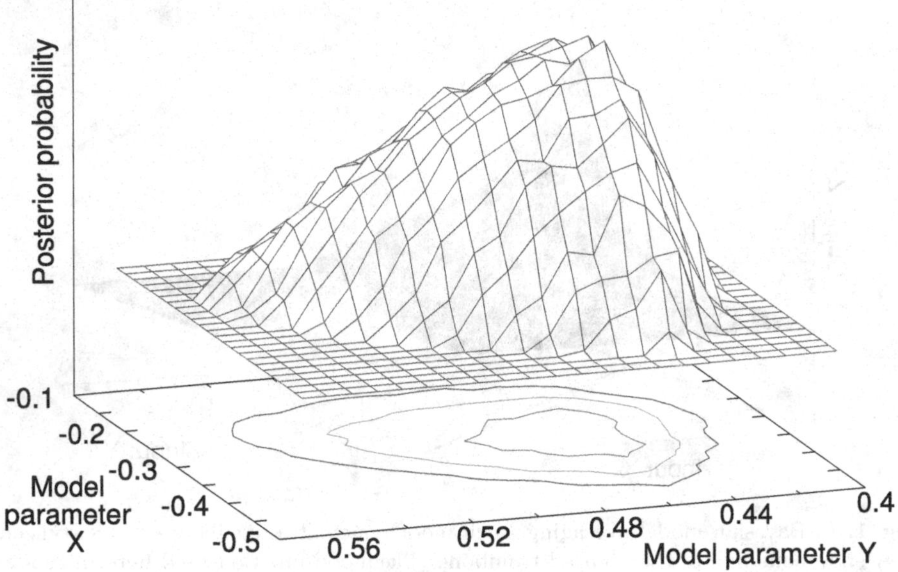

Fig. 2. For the leave-one-out data set created by leaving out customer 144, this shows a posterior landscape: for models whose first two parameters are x and y, the z-axis is the posterior probability that the model is correct, given the data. The distribution is skewed, so the mode will be different from the mean, i.e., the model with the highest probability of being correct (the best-fit model is the peak) will predict differently from the weighted average vote of all models. This happens with other prediction problems, such as for lung cancer [8, Figure 2] and floods [7, Figure 4].

For example, Fig. 2 shows a posterior landscape from a leave-one-out data set with 5 inputs, and the search ended at 50,219 models. This approximation only considers the mountain, but ignores the lower area around it.

For one of the training sets created by leave-one-out-cross validation, Fig. 2 shows a 2-dimensional cross-section of the posterior distribution of models. This data set leaves out customer 144, who becomes the single point of test data. The x and y axes are two model parameters, and take their values from a discretized grid, to give a range of models of varying fit to the data. For each model in x-y space, z is that model's posterior probability of being correct, given the training data. The best-fit model has the highest probability of being correct, so it's the model at the peak. Sloping down from that peak are thousands of models with a worse fit to the data, and thus a lower posterior probability of being correct.

The posterior distribution in Fig. 2 is skewed: this causes the prediction of the best-fit model to be different from the prediction of the weighted average vote of all the models in the distribution. Those two predictions are indeed different in Fig. 3, but only by a tiny amount: the solid line shows the prediction of the most-probable, best-fit model (it's higher at $x = 1$ predicting that churn is more likely) and the dotted line shows the weighted average vote of all models in the landscape of Fig. 2 (which also says that churn is more likely). The punchline is that the difference points to the right answer: the vote's prediction is both lower at $x = 0$ and higher at $x = 1$ than the best-fit model.

The y-axis in Fig. 3 is the probability density of different outcomes, and the x-axis shows the two outcomes: 0 means no churn, 1 is churn, and x-values in between can be interpreted as probabilities [5]. Probability density can exceed 1, but it can never be negative, and must integrate to exactly 1.

The skewness in Fig. 2 is mild enough that both methods predicted that churn is the more likely outcome, and in this example they were both right, this particular customer did indeed churn. In fact, for this customer churn problem, Bayesian model averaging always predicts the same outcome as the best fit model, just like in Fig. 3. So it's no more accurate, and has the disadvantage of using more computer time.

Instead of condemning Bayesian model averaging, Sect. 3 will use it merely to find the direction of the tiny differences in predictions, as seen in Fig. 3. Increasing how large the difference needs to be, can give better accuracy.

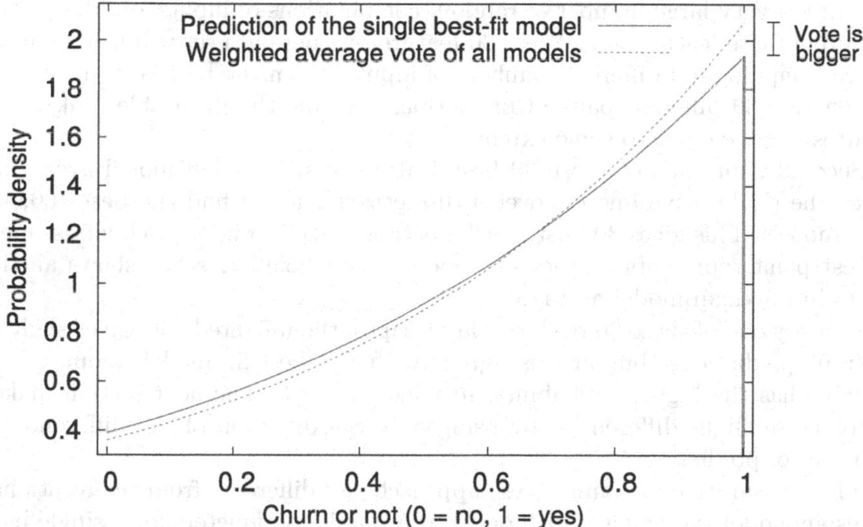

Fig. 3. The skewed posterior landscape in Fig. 2 means the prediction of the best-fit model will be different from the prediction of the weighted vote of all the models. But that difference is small. Both predict this customer is more likely to churn ($x = 1$) than not churn ($x = 0$). This particular customer did indeed churn. Intriguingly, the direction of the difference is leaning the right way: the weighted average vote is both higher at $x = 1$ and lower at $x = 0$. The proposed approach uses those tiny differences.

3 The Proposed Prediction Method

3.1 Feature Selection Difficulties in Scarce-Data Problems

Low-dimensional low-correlation small-data problems have few inputs (i.e., few columns or features), and the data set is short (i.e., few rows or individuals), often just a multiple of 10 or 20 of the small number of inputs.

Any two points in two-dimensional space can have a line that fits them perfectly, and any three points in three-dimensional space can have a plane that fits them perfectly, and so on. So a data set with few rows will have are many combinations of input columns that will fit the data moderately well, purely due to bad luck. That is, feature selection becomes problematic because fewer rows of data increase the probability that an uninformative or random input appears to have a statistical relationship with the output.

To average out the inevitable junk inputs, the proposed approach has a second layer of variation: instead of choosing just a single feature set of inputs and generating a single posterior landscape, it does so for the best 30 feature sets[3]. That's not 30 features, but 30 feature *sets* for each prediction. This will be done for every one of the 510 training sets created by leave-one-out cross validation on the data set of 510 customers, in the following two steps.

Firstly, leave-one-out cross-validation generates 510 data sets (one for each of the 510 rows of data), and then creates all possible 2-way interaction variables, to give thousands of inputs to choose from. For each data set, brute-force sampling generates a very large number of random combinations of inputs, and keeps the best 30 of these feature sets. These 30 feature sets per data set will usually have different inputs, and different numbers of inputs. From the best 30 feature sets, we generate 30 different pairs of predictions, causing the inevitable junk input features to cancel out to some extent.

Secondly, for each of those 30 best feature sets, Bayesian model averaging varies the model's parameters over a discretized grid, to find the best 100,000 or so models. This gives 30 posterior landscapes, like in Fig. 2. Each landscape's highest point represents the best-fit model. The whole landscape shows all the voters in Bayesian model averaging.

In every one of these 30 posterior landscapes, the weighted vote gives slightly different predictions, but always agrees with the best-fit model about which outcome has the higher probability, just like in Fig. 3. The next section makes use of those slight differences, by using only the direction of the difference in each pair of predictions.

This committee-of-committees approach is different from conventional Bayesian model averaging, which only varies model parameters for a single best set of input features. Using 30 different committees, each with thousands of models, doesn't aim to reduce CPU time, but to average out random errors from inputs that are not informative, just like in real elections [20].

[3] Why 30 feature sets? Kreft's 30-30 rule says that 30 groups of at least 30 each is reasonable. Here each group has thousands of models, so 30 of those groups should be enough. More than 30 would be desirable, but the CPU cost is already large.

3.2 Predict from the Direction of the Differences

In Fig. 3, the likelihood of churn is highest at $x = 1$. This indicates that churn is predicted by both the single best model, and by the weighted vote from Bayesian model averaging. The difference in predicted likelihood is tiny.

Consider the following two possibilities. One possibility: if all 30 pairs of predictions give essentially the *same* likelihood for churn, the standard deviation of the 30 differences will be small. This indicates the data in the 30 feature sets are in agreement about which outcome is more likely.

The other possibility: if enough of the 30 pairs disagree, and give sufficiently *different* likelihoods for churn, then the standard deviation of the 30 differences will be bigger. This suggests the 30 feature sets disagree, with some predicting one outcome, and some predicting the other outcome: that's perfectly reasonable, because all will have a few junk inputs.

So consider this proposal for a prediction algorithm. Leave-one-out cross-validation gives 510 different data sets, and each of the 510 data sets has its 30 best feature sets, and each feature set gives both (1) a single best-fit model, which is the highest point in (2) a posterior landscape of thousands of models.

- If the standard deviation of the 30 differences in the likelihood of churn is too small, no prediction is made.
- If the standard deviation of the 30 differences is big enough, then the prediction goes by the average (not the standard deviation) of the differences:
  - If the Bayesian approach (on average) predicts a higher likelihood for output 1 (i.e., the average difference is positive), then predict 1 (churn).
  - If the average difference is negative or zero, then no prediction.

The general idea of the proposed approach is that if the mass of voters have a small but consistent disagreement with the well-informed expert, then take the direction of those differences to get the prediction. It's like a country of 30 different states, where each state's voters cast their choice, and then the vote tally is compared with the percentage given by the best-informed voter for that state. When the votes disagree with the best-informed voter by more than some pre-set percentage (even if they both favor the same candidate), then whichever way the voter tally was different, that difference goes towards that candidate. Imagine if in all 30 states, both the vote tally and the best-informed voter favor the same candidate, but the standard deviation of the differences was big enough; then the direction of the tiny differences is what chooses the winning candidate.

4 Results: Raising a Threshold for Fewer Predictions

The comparison algorithms are all from the Python library SciKit. Each algorithm has a *threshold,* i.e., a parameter which makes it more likely to choose yes or no. In Fig. 4, logistic regression with a threshold halfway between yes = 1 and no = 0 would not favor either outcome. A threshold closer to "yes" would result fewer "yes" predictions, but which are more likely to be "yes", because that y-axis is often interpreted to be a probability.

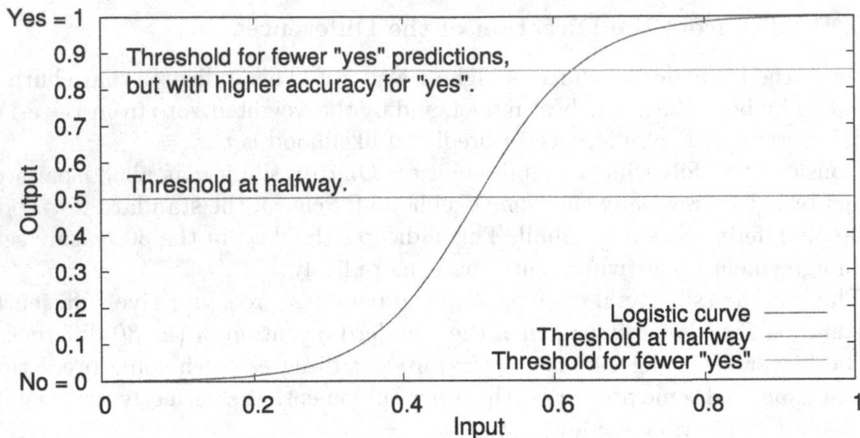

Fig. 4. In this stylized view of logistic regression, the default threshold is in the middle, with no bias towards "yes" or "no". Setting the threshold higher will cause it to make fewer "yes" predictions, for those predictions should be more likely to be correct for "yes". The results below will similarly vary each algorithm's threshold, to go from many "yes" predictions, to fewer predictions but with higher accuracy for "yes".

In the results below, each algorithm's threshold is varied from the neutral middle value (where it is equally likely to predict churn or not) upwards, so that it makes fewer "yes" predictions but are more likely to be correct, until it gets so high that it makes few or no "yes" predictions. For the proposed method, that threshold parameter is the required standard deviation of the difference in the likelihoods of (1) the single best model and (2) the whole ensemble.

4.1 The Bigger the Difference, the Higher the Precision

Figure 5 shows that increasing the required standard deviation of the difference gives better accuracy. The error bars are the 95% confidence intervals for binomial proportions from the Clopper-Pearson method[4]. In Fig. 5, the x-axis says how big must be the standard deviation of the difference between the likelihoods for the two methods, to dare make a prediction. The further to the right, the greater the precision, but for fewer predictions, so the confidence intervals get wider. It gives a short list of customers most likely to churn.

From Fig. 5, by increasing the required direction of the difference, that precision for churn can be raised above 90%, but for a short list of customers. Taking Fig. 5 and using the nonlinear least-squares Marquardt-Levenberg (and using as uncertainties the Clopper-Pearson confidence intervals of the binomial proportions), gives a p-value below 0.01, indicating a relationship exists.

[4] Clopper-Pearson does not approximate a binomial distribution with a normal distribution, instead it gives exact results for small samples.

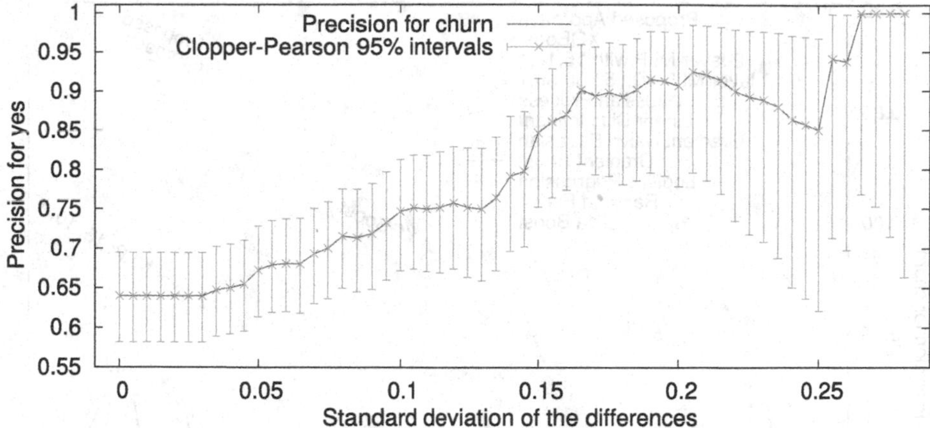

Fig. 5. The x-axis shows how big the standard deviation of the difference between 30 pairs of likelihoods needs to be, to venture a prediction. Moving right by increasing how different the two likelihoods need to be, raises precision but for fewer customers.

4.2 Comparison with XGBoost, Deep Learning, and Others

The proposed approach compares favourably with popular algorithms such as XGBoost and deep learning. Varying each algorithm's threshold (to give fewer but more confident predictions) gives Fig. 6, where x is the size of the short list produced, so that as the threshold becomes more demanding, the short list gets shorter. The y-axis shows how many customers on that short list do churn, representing the precision $= TP \div (TP + FP)$. All algorithms used leave-one-out cross-validation to select the training and test data, so there is no sampling error.

The business case is that around half of the 510 customers will churn, and the profit margin is such that it's only worth offering an incentive to a customer if the chance of them churning is close to 90%. The marketing budget can give an incentive to about 60 customers.

Given these requirements, Table 1 picks the threshold of each algorithm, such that the short list[5] is the smallest size that is at least 60. In Table 1, the proposed method has the highest precision: the threshold that gives the smallest short list of size at least 60 covers 63 customers, of whom 88.89% go on to churn.

XGBoost is slightly behind with precision of 84.13%, and the third-best result is 76.67% from a shallow neural network with two hidden layers, of 15 and 10 neurons[6] Trying many different numbers of layers (and neurons in each layer) suggests that adding more layers or neurons tends to reduce accuracy, perhaps

[5] For some algorithms, a tiny increase in the threshold gives more than one extra customer, so the short list is not always exactly 60.

[6] Other MLP learning parameters were very ordinary: the optimizer was `RMSpro`, the activation function was `relu` for the hidden layers and `sigmoid` for the output, the initializer was `GlorotUniform`.

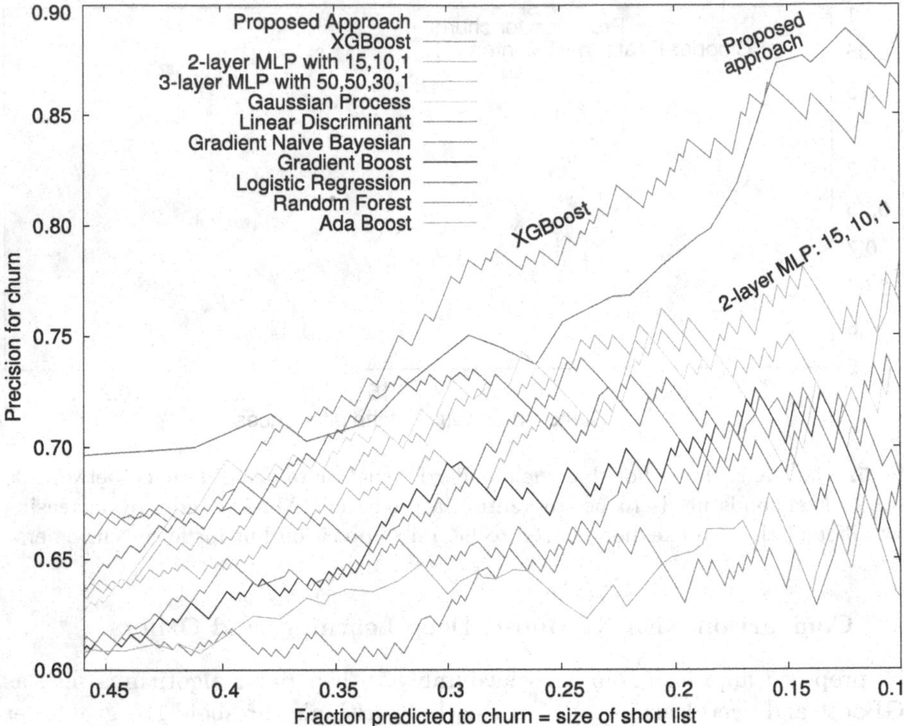

Fig. 6. The client wants a short list of at least 60 of the 510 customers, such that about 90% of them do churn. The y-axis shows how many are in the short list, which gets shorter by adjusting the algorithm's threshold. The x-axis shows how many customers on that short list did in fact churn, being precision = TP/(TP + FP). The proposed method gives a short list from which 88.89% did churn. Second-best is XGBoost, and third-best is a neural network with 15 and 10 neurons in the two hidden layers.

because the data set is so small that there just isn't enough signal to tune a large number of parameters: the next-best of all these attempts was three hidden layers with 50, 50, and 30 neurons, which did slightly worse than SciKit's default neural network algorithm at 73.33%. This is comparable to another study that used deep learning for e-commerce customer churn prediction [18] which achieved precision of 78%.

The one-tailed 2-sample Z-test for proportions on the difference between the proposed method and XGBoost in Table 1 does not show a significant difference, with a p-value of 0.284, but the client still likes 88.9% more than 84.1%.

Comparing the proposed method with the next-best algorithm (a shallow neural network), the Z-test gives a p-value less than 0.05 which suggests a significant improvement. It's possible that more tinkering might help, but adding more layers to the neural network tends to give lower accuracy in Table 1. Random

Table 1. The proposed method narrowly out-performs XGBoost, and is significantly better ($p < 0.05$) than deep learning with either small or large neural networks.

Algorithm	Precision	TP	FP	Short list
Proposed Approach	0.8889	56	7	63
XGBoost	0.8413	53	10	62
2-layer neural network: 15,10,1 nodes	0.7667	46	14	60
3-layer neural network: 50,50,30,1 nodes	0.7167	43	17	60
Gaussian Process	0.7167	43	17	60
Linear Discriminant	0.7167	43	17	60
Gaussian NB	0.7000	42	18	60
Gradient Boost	0.6833	41	19	60
Logistic Regression	0.6833	41	19	60
Random Forest	0.6667	40	20	60
4-layer neural network: 15,12,10,9,8,1 nodes	0.6667	40	20	60
Ada Boost	0.6452	40	22	62

Forest and Gradient Boost work impressively on many other problems, but in this scarce-data task they are at 66.67% and 68.33% respectively.

As mentioned in Sect. 2.2, the data set used here is close to balanced, with 51.6% of customers churning and 48.4% staying, so no claims are made for how well it might cope with unbalanced data sets.

5 Discussion and Conclusion

The proposed approach achieves more accuracy than state-of-the-art algorithms including XGBoost and deep learning, on a real small-data low-dimensional prediction problem whose inputs are weak discriminants, the kind of problem faced by millions of organizations of all sizes.

The result here is broadly in agreement with a previous study about predicting who has cancer [8]. There are many other scarce-data prediction problems, with inputs that are poor discriminants: diagnosing unusual diseases, medical triage prediction [10], and hospital readmission [9,12,14], to name a few.

Like the customer churn problem covered here, many medical prediction problems also have a cost-benefit aspect. For example, a hospital needs a short list of patients who are most in need of care (to match the limited number of available beds), so the accuracy of that short list matters, not accuracy overall.

For most hospitals, these are scarce-data problems: while common diseases like influenza infect millions of people each year, those diseases behave differently from place to place and year to year, so 5-year-old data on influenza in Norway's sub-arctic [23] could be misleading for influenza in Australia's dry outback, due to the evolution of the virus, and differences in weather, housing, and immunities.

Acknowledgments. The author thanks Lachlan Butler, Harrison Burrows, and Juan Moredo for technical support. The author also thanks the anonymous peer reviewers for critically reading the manuscript and suggesting substantial improvements.

References

1. Ahmad, A.K., Jafar, A., Aljoumaa, K.: Customer churn prediction in telecom using machine learning in big data platform. J. Big Data **6**(1), 1–24 (2019). https://doi.org/10.1186/s40537-019-0191-6
2. Albuquerque, R.A.S., Costa, A.F.J., dos Santos, E.M.: A decision-based dynamic ensemble selection method for concept drift. In: 31st IEEE International Conference on Tools with Artificial Intelligence, ICTAI 2019, pp. 1132–1139. IEEE Computer Society, November 2019. https://doi.org/10.1109/ICTAI.2019.00158
3. Berger, P., Kompan, M.: User modeling for churn prediction in E-commerce. IEEE Intell. Syst. **34**(2), 44–52 (2019). https://doi.org/10.1109/MIS.2019.2895788
4. Braytee, A., Anaissi, A., Kennedy, P.J.: Sparse feature learning using ensemble model for highly-correlated high-dimensional data. In: Cheng, L., Leung, A.C.S., Ozawa, S. (eds.) ICONIP 2018. LNCS, vol. 11303, pp. 423–434. Springer, Cham (2018). https://doi.org/10.1007/978-3-030-04182-3_37
5. Caves, C.M., Fuchs, C.A., Schack, R.: Quantum probabilities as Bayesian probabilities. Phys. Rev. A **65**(2), 22305 (2002). https://doi.org/10.1103/PhysRevA.65.022305
6. Cerqueira, V., Torgo, L., Pinto, F., Soares, C.: Arbitrage of forecasting experts. Mach. Learn. **108**(6), 913–944 (2019). https://doi.org/10.1007/s10994-018-05774-y
7. Darwen, P.J.: The varying success of Bayesian model averaging: an empirical study of flood prediction. In: Sundaram, S. (ed.) 2018 IEEE Symposium Series on Computational Intelligence (SSCI), pp. 1764–1771. IEEE, November 2018. https://doi.org/10.1109/SSCI.2018.8628939
8. Darwen, P.J.: Cost-effective prediction in medicine and marketing: only the difference between Bayesian model averaging and the single best-fit model. In: 31st IEEE International Conference on Tools with Artificial Intelligence, ICTAI 2019, pp. 1274–1279. IEEE Computer Society, November 2019. https://doi.org/10.1109/ICTAI.2019.00178
9. Garmendia-Mujika, A., Graña, M., Lopez-Guede, J.M., Rios, S.: Neural and statistical predictors for time to readmission in emergency departments: a case study. Neurocomputing **354**, 3–9 (2019). https://doi.org/10.1016/j.neucom.2018.05.135
10. Garmendia-Mujika, A., Rios, S.A., Lopez-Guede, J.M., Graña, M.: Triage prediction in pediatric patients with respiratory problems. Neurocomputing **326**, 161–167 (2019). https://doi.org/10.1016/j.neucom.2017.01.122
11. Gelman, A., Carlin, J.B., Stern, H.S., Rubin, D.B.: Bayesian Data Analysis. Texts in Statistical Science Series, 2 edn. Chapman-Hall, Boca Raton (2004). https://doi.org/10.1201/9780429258411
12. Hammoudeh, A., Al-Naymat, G., Ghannam, I., Obied, N.: Predicting hospital readmission among diabetics using deep learning. Procedia Comput. Sci. **141**, 484–489 (2018). https://doi.org/10.1016/j.procs.2018.10.138
13. Hastie, T., Tibshirani, R., Friedman, J., Franklin, J.: The Elements of Statistical Learning: Data Mining, Inference, and Prediction, 2nd edn. Springer, New York (2009). https://doi.org/10.1007/978-0-387-84858-7

14. Hooijenga, D., Phan, R., Augusto, V., Xie, X., Redjaline, A.: Discriminant analysis and feature selection for emergency department readmission prediction. In: 2018 IEEE Symposium Series on Computational Intelligence (SSCI), pp. 836–842. IEEE, November 2018. https://doi.org/10.1109/SSCI.2018.8628938
15. Liu, X., Xie, M., Wen, X., Chen, R., Ge, Y., Duffield, N., Wang, N.: A semi-supervised and inductive embedding model for churn prediction of large-scale mobile games. In: 2018 IEEE International Conference on Data Mining (ICDM), pp. 277–286. IEEE Computer Society (2018). https://doi.org/10.1109/ICDM.2018.00043
16. Najafi, M., Moradkhani, H., Jung, I.: Assessing the uncertainties of hydrologic model selection in climate change impact studies. Hydrol. Process. **25**(18), 2814–2826 (2011). https://doi.org/10.1002/hyp.8043
17. de O. Nunes, R., Dantas, C.A., Canuto, A.P., Xavier, J.C.: Dynamic feature selection for classifier ensembles. In: 2018 7th Brazilian Conference on Intelligent Systems (BRACIS), pp. 468–473. IEEE Computer Society, October 2018. https://doi.org/10.1109/BRACIS.2018.00087
18. Pondel, M., Wuczyński, M., Gryncewicz, W., Łysik, Ł., Hernes, M., Rot, A., Kozina, A.: Deep learning for customer churn prediction in e-commerce decision support. In: Proceedings of the 24th International Conference on Business Information Systems, pp. 3–12. TIB Open Publishing (2021). https://doi.org/10.52825/bis.v1i.42
19. Praseeda, C., Shivakumar, B.: Fuzzy particle swarm optimization (FPSO) based feature selection and hybrid kernel distance based possibilistic fuzzy local information C-means (HKD-PFLICM) clustering for churn prediction in telecom industry. SN Appl. Sci. **3**(6), 1–18 (2021). https://doi.org/10.1007/s42452-021-04576-7
20. Rapeli, L.: Does sophistication affect electoral outcomes? Gov. Oppos. **53**(2), 181–204 (2018). https://doi.org/10.1017/gov.2016.23
21. Saadallah, A., Priebe, F., Morik, K.: A drift-based dynamic ensemble members selection using clustering for time series forecasting. In: Brefeld, U., Fromont, E., Hotho, A., Knobbe, A., Maathuis, M., Robardet, C. (eds.) ECML PKDD 2019. LNCS (LNAI), vol. 11906, pp. 678–694. Springer, Cham (2020). https://doi.org/10.1007/978-3-030-46150-8_40
22. Subramanya, K.B., Somani, A.: Enhanced feature mining and classifier models to predict customer churn for an E-retailer. In: Confluence 2017: 7th International Conference on Cloud Computing, Data Science and Engineering, pp. 531–536. IEEE (2017). https://doi.org/10.1109/CONFLUENCE.2017.7943208
23. Tazmini, K., Nymo, S.H., Louch, W.E., Ranhoff, A.H., Øie, E.: Electrolyte imbalances in an unselected population in an emergency department: a retrospective cohort study. PLoS ONE **14**(4), e0215673 (2019). https://doi.org/10.1371/journal.pone.0215673
24. Wurl, A., Falkner, A.A., Haselböck, A., Mazak, A., Sperl, S.: Combining prediction methods for hardware asset management. In: Proceedings of the 7th International Conference on Data Science, Technology and Applications - DATA 2018, pp. 13–23. SciTePress (2018). https://doi.org/10.5220/0006859100130023
25. Zhu, Y., et al.: Addressing the item cold-start problem by attribute-driven active learning. IEEE Trans. Knowl. Data Eng. **32**(4), 631–644 (2020). https://doi.org/10.1109/TKDE.2019.2891530

Tree-Based Unified Temporal Erasable-Itemset Mining

Tzung-Pei Hong[1,2](✉), Jia-Xiang Li[2], Yu-Chuan Tsai[3], and Wei-Ming Huang[4]

[1] Department of Computer Science and Information Engineering, National University of Kaohsiung, Kaohsiung, Taiwan
tphong@nuk.edu.tw

[2] Department of Computer Science and Engineering, National Sun Yat-sen University, Kaohsiung, Taiwan

[3] Library and Information Center, National University of Kaohsiung, Kaohsiung, Taiwan
yjtsai@nuk.edu.tw

[4] Department of Electrical and Control, China Steel Inc., Kaohsiung, Taiwan

Abstract. Erasable itemset mining is an important research area for manufacturers, as it aids in identifying less profitable materials in product datasets to facilitate better decision-making for managers. It allows for improved trade-offs between manufacturing and procuring activities. Traditional erasable itemset mining does not account for the time factor, which is critical for time-sensitive industries, with product time periods significantly impacting a company's profitability. Therefore, we previously proposed an Apriori-based unified temporal erasable itemset mining approach, which could consider different user scenarios to solve this issue. In this work, we design a tree-based algorithm to raise the mining efficiency. It applies a lower-bound strategy to reduce the candidate erasable itemsets and the number of database scanning. According to the experimental results, our proposed algorithm has better performance on execution time and memory usage than the previous work.

Keywords: Data Mining · Erasable Itemset Mining · Temporal Erasable Itemset Mining · Tree Structure · Lower-bound Strategy

1 Introduction

With the recent rapid development of big data, data mining has become very popular in recent years [1]. As data becomes more valuable, extracting useful information from massive amounts of data is an important issue for every enterprise. Deng et al. proposed erasable itemset mining for manufacturing requirements [3]. Given the need to cut expenses, this helps managers decide which less profitable materials can be removed. Some researchers were devoted to improving the performance of erasable itemset mining with different data structures [2, 11–13], such as hash tables, tree structures, and so on. For example, Le et al. [11] used hash tables to store product profits. In [12], the gain of each node was stored in a tree structure in a hash table. Using the tree structure can reduce the number of database scanning [2, 13].

N. T. Nguyen et al. (Eds.): ACIIDS 2023, LNAI 13995, pp. 224–233, 2023.
https://doi.org/10.1007/978-981-99-5834-4_18

In time-sensitive industries, the time factor is essential to the effectiveness of mined results. Hong et al. proposed the concept of temporal erasable itemset mining [6], which added temporality to traditional erasable itemset mining. After adding time features to erasable itemset mining, the method no longer exhibits downward closure and thus generates an excessive number of candidates of erasable itemsets, leading to resource consumption. Hong et al. [9] proposed an Apriori-based UALB algorithm to reduce the number of candidates. In this work, we propose a tree-based algorithm, which applies the lower-bound strategy to reducing the number of candidates and database scanning. We evaluate the performance of execution time and memory usage between our proposed algorithm and previous work. According to the experimental results, our proposed algorithm has better performance on the evaluation criteria.

The rest of the paper is organized as follows. Section 2 presents a review of related works. Section 3 describes the research problem. Section 4 presents the proposed algorithm. Section 5 shows the performance evaluation results. Section 6 concludes this paper.

2 Related Works

Many recent works have been devoted to erasable itemset mining studies and improving the mining performance [4–9, 11, 12, 14–17]. Chan et al. [5] proposed a tree-based algorithm, MERIT, which leverages the WPPC-tree and includes NC-set, a data structure that enables convenient access to itemset information. Vo and Coenen proposed MERIT and dMERIT+ [12], which address missing erasable itemsets in the MERIT algorithm and make it more efficient. dMERIT+ utilizes the dNC-Set data structure and a hash table to increase the mining efficiency. Besides tree-based methods, researchers have proposed list-based algorithms, for instance, VME [4] and MEI [11], to enhance performance. Although VME algorithm leverages the PID_list data structure to minimize database scans, it consumes significant memory for large databases to store the PID_list. Hong et al. [8] used a bit vector to store the relationship between items and products, which has better performance than VME in memory usage. The MEI algorithm [11], proposed by Le et al., used the Pidset and dPidset, where Pidset collects item information, i.e., which products use the item as a component, and dPidset is the difference set between two Pidsets. MEI utilizes a hash table to store product profits, increasing the speed of Pidset accessing the product profits. With the development of erasable itemset mining, extended methods have emerged, such as mining top-rank-k erasable itemsets [17], erasable itemset mining on dynamic incremental databases [14], incremental weighted erasable itemset mining [15], and metamorphic testing for erasable itemset mining [7].

In the context of real factory manufacturing plants, time is a crucial consideration when extracting erasable itemsets. To address this issue, Hong et al. [6] proposed temporal erasable itemset mining and the concept of the lifespan of each item. In [9], Hong et al. proposed the UALB algorithm, which used the lower-bound strategy to satisfy the downward closure property for reducing the number of candidates.

3 Problem Description

The goal of temporal erasable itemset mining methods is to find temporal erasable itemsets with high performance. Let *TPD* be a temporal product dataset with material items of the product P_i, $items_{P_i}$, and the profit of product P_i, $profit_{P_i}$. So that each data row T_i consists of a product P_i with a set of items, $items_{P_i}$, and the profit of a product P_i, $items_{P_i}$. In [6], the seven kinds of the lifespan of an itemset X (LSP_X), itemset's start time slot (*IST*), itemset's end time slot (*IET*), temporal gain, lifespan gain, temporal gain ratio, and temporal erasable itemset were defined.

3.1 The Temporal Erasable Itemset Mining

Definition 1 (*Temporal gain*). The temporal gain of an itemset X, denoted $tgain_X$, is defined as:

$$\mathrm{tgain_X} = \sum_{\{P_i \mid P_i\ \varepsilon\ PSET_X\}} \mathrm{Profit}_{P_i} \tag{1}$$

where $PSET_X$. Represents the products that use at least one item in X during the LSP_X.

Definition 2 (*Period gain*). The total product profits in time slot T_i, denoted $pdgain_i$, is defined as

$$\mathrm{pdgain_j} = \sum_{\{P_{i,j} \mid (j\varepsilon T_j)\}} \mathrm{Profit}_{P_i}, \tag{2}$$

where $P_{i,j}$ is product i in time slot j, T_j is the time slot in *TPD*, and j is the corresponding time slot ID, $0 \leq j \leq n$. And we extend to the lifespan gain of an itemset.

Definition 3 (*Lifespan gain*). The total product profits in the lifespan of itemset X, denoted $lspgain_X$, is defined as

$$\mathrm{lspgain_X} = \sum_{\{T_i \mid (T_i \varepsilon\ LSP_X)\}} \mathrm{pdgain_i}, \tag{3}$$

where LSP_X is the lifespan of itemset X. After finding the temporal gain and lifespan gain of the itemset, we determine the temporal gain ratio to measure the importance of the itemset in its lifespan.

Definition 4 (*Temporal gain ratio*). The temporal gain ratio of an itemset X is formally defined as

$$\mathrm{tgr_X} = \frac{\mathrm{tgain_X}}{\mathrm{lspgain_X}}, \tag{4}$$

which represents the proportion of profits brought by itemset X in its lifespan.

Definition 5 (*Temporal erasable itemset*). According to a user-specified maximum temporal gain-ratio threshold λ, when the temporal gain ratio of a candidate temporal erasable itemset is less than or equal to the threshold, the itemset is a temporal erasable itemset (*TEI*).

Definition 6 (*Temporal erasable itemset mining*). In temporal erasable itemset mining, the user provides a *TPD*, a maximum temporal gain-ratio threshold λ, and lifespan option *LO*. To obtain temporal erasable itemsets, the temporal gain of each itemset is initially computed from the *TPD*. Then, based on the *LO*, the lifespan gain of each itemset is derived. The temporal gain ratio of each itemset is obtained by dividing the temporal gain by the lifespan gain. Temporal erasable itemsets are identified by comparing their temporal gain ratio to λ. Finally, all identified temporal erasable itemsets within the *TPD* are presented as the output.

Example 1. Table 1 shows a small production database with a time slot. Given both the gain ratio threshold and temporal gain ratio threshold are 40%, is item E an erasable itemset or not? In the traditional erasable itemset mining, the gain ratio of item E is (1350 + 1350)/8450 = 32%. However, item E only appears in two time slots, T2 and T3. When we consider the lifespan of item E, the temporal gain ratio of item E is (1350 + 1350)/4500 = 60%. According to the given lifespan of itemsets, item E is not an erasable itemset, but it is a temporal erasable itemset.

According to the definition of lifespan option *(LO)* [6], we use the itemset {E, F} as an example to describe the different lifespan options. The first one is the whole database. The second one is from the first start time slot of itemset {E, F} to the end, so that is from time slot T2 to T4. The third one is from the beginning of the database to the last end time slot of the itemset {E, F}, so that is from T1 to T4. The fourth option is from the earliest itemset start time slot to the earliest itemset end time slot. Therefore, it is from T2 to T3 for itemset {E, F}. The fifth option is from the latest itemset start time slot to the end of the database so that it is from T3 to T4 for itemset {E, F}. The sixth option of the itemset {E, F} is from the beginning of the database to the latest itemset end time slot, so that it is from T1 to T4. The last lifespan option of itemset {E, F} is from the latest itemset start time slot to the latest itemset end time slot, so that it is from T3 to T4. In this work, we focus on how to solve the temporal erasable itemset mining for all possible lifespan options.

3.2 Lower-Bound Strategy

According to the temporal erasable itemset mining algorithm, UTE [6], it does not satisfy the downward closure property. It generates too many candidate erasable itemsets. In [12], Hong et al. proposed a lower-bound strategy to reduce the search space of candidates. The definitions of the lower-bound gain, lower-bound gain ratio, and two theorems are listed as follows [9].

Definition 7 (*Start time slot of the last item*). The time slot of the last item in the material set *I* to appear in the *TPD*, which we denote as IST_{last}. No new items appear in *TPD*

Table 1. A small production database

Period	PID	Items	Profit
T1	P1	A C	500
T1	P2	B C D	250
T2	P1	A C	500
T2	P2	B C D	250
T2	P3	D E	1350
T3	P2	B C D	250
T3	P3	D E	1350
T3	P4	C D F	800
T4	P1	A C	500
T4	P4	C D F	800
T4	P5	A G	1900

after IST_{last}. This is defined as.

$$\mathrm{IST}_{\mathrm{last}} = \underset{z\,\varepsilon\,\mathrm{I}}{\mathrm{Max}}\mathrm{IST}_{\mathrm{z}}, \tag{5}$$

where the material set *I* is the set of all items used in *TPD,* and IST_{last} is the last item start time slot of all the item start time slots in *I*.

Definition 8 (*End time slot of the first item*). The end time slot of the first expired item in the material set *I* in the *TPD*, which we denote as IET_{first}. No items expire in *TPD* before IET_{first}. It is defined as.

$$\mathrm{IET}_{\mathrm{first}} = \underset{z\,\varepsilon\,\mathrm{I}}{\mathrm{Min}}\mathrm{IET}_{\mathrm{z}}, \tag{6}$$

where IET_{first} is the earliest item end time slot of all the item end time slots in *I*.

Definition 9 (*Lower-bound gain*). The lower-bound gain of itemset *X*, denoted $lbgain_X$, is defined as.

$$\mathrm{lbgain}_{\mathrm{X}} = \sum_{\{\mathrm{P_i} \mid \mathrm{P_i}\,\varepsilon \mathrm{LBPSET_X}\}} \mathrm{Profit}_{\mathrm{P_i}}, \tag{7}$$

where $LBPSET_X$ is the set of products that use at least one item in *X* as a material during the IST_{last} to IET_{first}. Compared with temporal gain, the lower-bound gain represents the profit that the itemset brings during the period of all items' lifespan overlapping.

Definition 10 (*Lower-bound gain ratio*). The lower-bound gain ratio of itemset *X*, denoted $lbgr_X$, is defined as

$$\mathrm{lbgr}_{\mathrm{X}} = \frac{\mathrm{lbgain}_{\mathrm{X}}}{\sum_{T_i\,\varepsilon\,\mathrm{TP}} \mathrm{pdgain_i}}, \tag{8}$$

where *TP* is the set collecting all time slots in *TPD.* If $lbgr_X$ is less than or equal to the maximum temporal gain-ratio threshold λ, itemset *X* is called a lower-bound temporal erasable itemset (*LBTEI*).

Theorem 1. The proposed lower-bound gain ratio is downward closed. Assume that $A \subseteq B$, if B is an *LBTEI*, then A is also an *LBTEI*. □

Theorem 2. If itemset A is a temporal erasable itemset (*TEI*), then A must also be a lower-bound temporal erasable itemset (*LBTEI*). □

4 Proposal Algorithm

According to the above theorem, finding lower-bound temporal erasable itemsets can avoid generating many candidates. In this section, we propose the two-phase unified tree-based lower-bound temporal erasable-itemset mining algorithm (UTLB) that uses the nodes to store the information of itemsets to avoid scanning datasets and is workable under seven lifespan options. The tree data structure used by UTLB, we called UTLB-tree. Each node in the tree contains itemset information, including itemset name, time slot IDs where the itemset appears, pidset, and gain value. A pidset is a set of products that use at least one item in the itemset as a component. An example node of the UTLB tree is shown in Fig. 1 below.

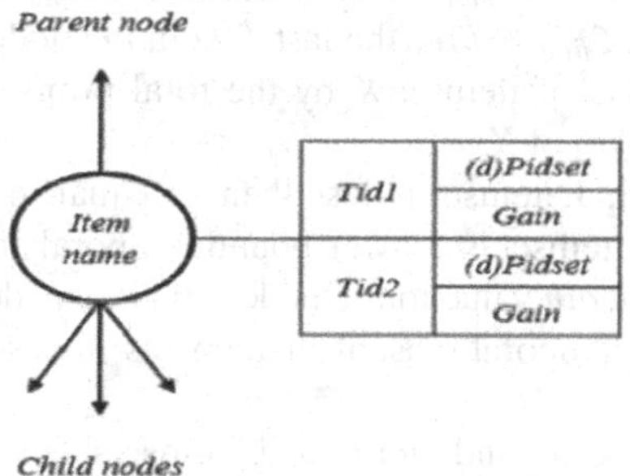

Fig. 1. Node of UTLB-tree

It can be observed from Fig. 1 that the pidset and gain value in the node are mapped to the corresponding *TID*. In this way, the UTLB algorithm can determine whether a lower-bound temporal erasable itemset is an actual temporal erasable itemset without scanning the dataset. In addition to the tree structure, UTLB also uses four tables to complete the entire mining task, *PG* (*Period_Gain*) table, *PP* (*Product_Profit*) table, *LSP* (*Lifespan*) table, and *TR* (*Temporal_Result*) table. The *PG*, *PP*, and *LSP* tables store the intermediate results in the first phase. The *PG* table, a hash table of period gains, stores the corresponding *pdgain* of each time slot using its TID as an index. The second table, called the *PP* table, which is a hash table of product profits, uses the product ID as an index to store the corresponding product profit. The third table, the *LSP* table, stores the lifespan of each item. A row in the table consists of an item, its *IST*, and its *IET*. And the *TR* table stores temporal erasable itemsets and outputs to the user in the second phase. The detailed steps of the proposed algorithm are described below.

Algorithm

Unified Tree-based Lower-Bound Temporal Erasable Itemset Mining Algorithm - UTLB

Input

(1) *TPD*: A temporal product dataset,
(2) λ: A user-specified maximum temporal gain-ratio threshold, and
(3) *LO*: A user-specified lifespan option.

Output

All the temporal erasable itemsets (*TEIs*) satisfy the given threshold (λ) under the given lifespan option (*LO*).

Step 1. For each time slot T_i in the given *TPD*, sum all the profits of the products appearing in T_i as $pdgain_i$ and sequentially putting it with their indices in the *PG* table.
Step 2. Build the *PP* table by sequentially putting the profit of product P_i with their indices in the table.
Step 3. For each item I_j in the given *TPD*, find its pidset and gain value of the time slots in which it occurs, and sequentially insert the I_j as the node into the tree.
Step 4. Obtain the IST_{last} and IET_{first} from the nodes as follows: $IST_{last} = Max$(the first *TID* in the node I_j), and $IET_{first} = Min$(the last *TID* in the node I_j).
Step 5. Divide *lbgain* value of itemset *X* by the total profit of the dataset to get the lower-bound gain ratio (*lbgr*) of *X*.
Step 6. If the *lbgr* of each 1-itemset is less than or equal to the maximum temporal gain-ratio threshold λ, the itemset is a lower-bound temporal erasable 1-itemset: keep it in the tree and insert the *lbgain* value to the node; otherwise, delete it from the tree.
Step 7. Sort lower-bound temporal erasable 1-itemsets in descending order according to the number of *TIDs*.
Step 8. Use the sorted lower-bound temporal 1-itemsets as input and use the *Expand* procedure to generate all lower-bound temporal erasable *k*-itemsets.
For each lower-bound temporal erasable itemset *X* in the tree, determine whether it is a temporal erasable itemset by the following steps:
Step 9. According to *LO*, find the desired lifespan of *X* by the *TIDs* stored in the nodes of lower-bound temporal erasable 1-itemset.
Step 10. From the *PG* table, sum up all $pdgain_i$ values over the desired lifespan of *X* as its lifespan gain ($lspgain_X$).
Step 11. Find the actual temporal gain ($tgain_X$) of *X* by summing up the gain value in the lifespan.
Step 12. Calculate the temporal gain ratio *tgr* of *X* by dividing $tgain_X$ by $lspgain_X$.
Step 13. If the tgr_X is less than or equal to the λ, the itemset is a *TEI* and add it to the *TR* table; otherwise, skip it.

After **Step 13**, all the itemsets in the *TR* table are the final results. In Step 8, the *Expand* procedure uses the depth-first search algorithm to generate nodes from lower-bound temporal erasable *k*-itemsets as input to generate lower-bound temporal $(k + 1)$-itemsets, where $k \geqq 1$, and derives the candidate itemset's information from the parent nodes.

5 Experimental Evaluation

We evaluated the performance of the proposed UTLB algorithm and compared it with two previous works, UTE [6] and UALB [9] algorithms, on different criteria. All the algorithms used in the experiments were implemented in Python 3.6.5 and ran on a computer with an Intel i5-8250U CPU, 12 GB RAM, and the operating system Windows 11. We conducted the experiments with a set of synthetic datasets, which were generated from the IBM generator [10], I0.05KDiKT10, including 40K to 120K ($40 \leq i \leq 120$) products. We fixed the number of items and the average item length per product to analyze the algorithm's performance as the number of products in the dataset increases. We used the log-normal distribution to simulate the product profits and the normal distribution to generate time slot IDs.

We used the execution time and memory usage as the performance criteria on different sizes of synthetic datasets, IjKD100KT10. In the numerical experiments, we assumed that all experiments were based on the seven lifespan options. The user-defined maximum temporal gain ratio threshold was set at 10%.

Figure 2 shows that as the number of products increases, UTE has the longest execution time, followed by UALB and, finally, UTLB. The reason for the execution time of UTE and UALB to be significantly raised is that they both require multiple scans of the dataset. Especially when the dataset size is 120K, UTE and UALB execution times are more than 100 times slower than UTLB. Since UALB and UTLB use the same lower-bound strategy, the number of candidate itemsets generated is the same, so we only use a blue line, UTLB, to represent both methods' results.

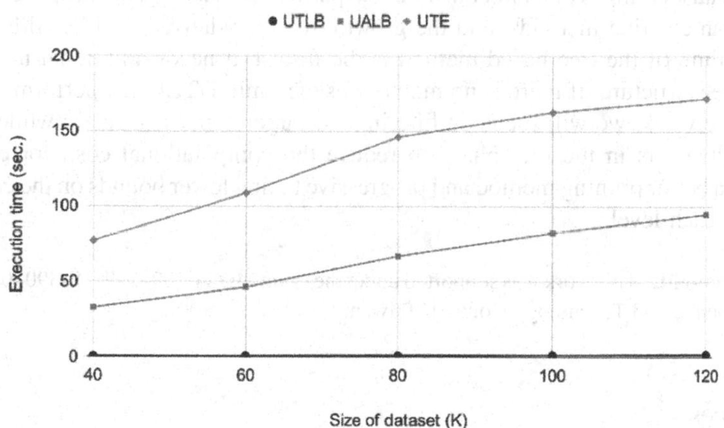

Fig. 2. Execution time for I0.05KDiKT10

Figure 3 shows that the memory usage of the three methods increases with the dataset size, among which UTLB has the largest memory usage, then UTE and UALB. The main reason is that UTLB stores the items' information in the nodes, which requires more memory than other Apriori-based methods.

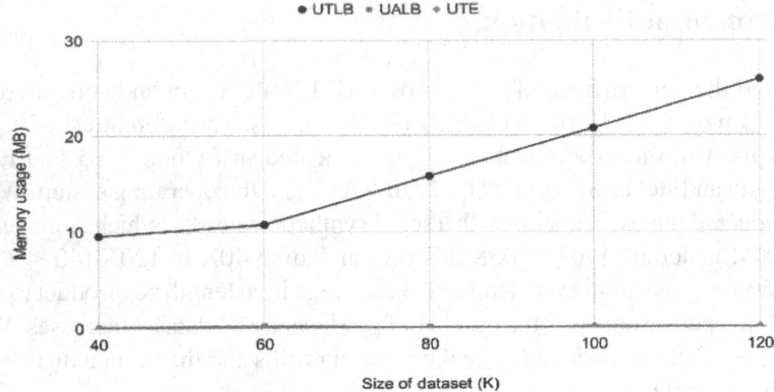

Fig. 3. Memory usage for I0.05KDiKT10

6 Conclusion and Future Work

In this paper, we have proposed a tree-based lower-bound strategy temporal erasable itemset mining algorithm. Because temporal erasable itemset mining does not exhibit downward closure, traditional erasable itemset mining algorithms cannot be used. The existing methods need multiple dataset scans, increasing computation times. For a shorter execution time, we propose the UTLB algorithm based on a tree structure. In addition to using the lower-bound strategy, UTLB uses nodes to store the information, which is used to determine temporal erasable itemsets. Comparing UTE, UALB, and UTLB on synthetic datasets shows that under different experimental settings, the proposed method is faster than existing methods, and the growth trend is relatively stable. Although the execution time of the tree-based method is the fastest, it needs additional memory to store the tree structure. If there is no memory usage limit, UTLB can perform well.

In future work, we will focus on filtering out target itemsets directly while finding candidate itemsets in the first phase to reduce the computational cost, for example, proposing a better pruning method and progressive taking lower bounds on the candidate itemsets at each level.

Acknowledgment. This work was supported under the grant MOST 109–2221-E-390–015-MY3, National Science and Technology Council, Taiwan.

References

1. Agrawal, R., Imieliński, T., Swami, A.: Mining association rules between sets of items in large databases. In: Proceedings of the 1993 ACM SIGMOD International Conference on Management of Data, pp. 207–216 (1993)
2. Baek, Y., et al.: Erasable pattern mining based on tree structures with damped window over data streams. Eng. Appl. Artif. Intell. **94**, 103735 (2020)
3. Deng, Z.H., Fang, G.D., Wang, Z.H., Xu, X.R.: Mining erasable itemsets. In: Proceedings of the 2009 International Conference on Machine Learning and Cybernetics, vol. 1, pp. 67–73 (2009)

4. Deng, Z.H., Xu, X.R.: An efficient algorithm for mining erasable itemsets. In: Proceedings of the International Conference on Advanced Data Mining and Applications, pp. 214–225 (2010)
5. Deng, Z.H., Xu, X.R.: Fast mining erasable itemsets using NC_sets. In: Expert Systems with Applications, vol. 39, pp. 4453–4463 (2012)
6. Hong, T.P., Chang, H., Li, S.M., Tsai, Y.C.: A unified temporal erasable itemset mining approach. In: Proceedings of the 2021 International Conference on Technologies and Applications of Artificial Intelligence (TAAI), pp. 194–198 (2021)
7. Hong, T.P., Chui, C.C., Su, J.H., Chen, C.H.: Applicable metamorphic testing for erasable itemset mining. IEEE Access **10**, 38545–38554 (2022)
8. Hong, T.P., Huang, W.M., Lan, G.C., Chiang, M.C., Lin, C.W.: A bitmap approach for mining erasable itemsets. IEEE Access **9**, 106029–106038 (2021)
9. Hong, T.P., Li, J.X., Tsai, Y.C.: Unified temporal erasable itemset mining with a lower-bound strategy. In: Proceedings of the 2022 IEEE International Conference on Big Data, pp. 6207–6211 (2022)
10. IBM Quest Data Mining Projection, Quest Synthetic Data Generation Code (1996). http://www.almaden.ibm.com/cs/quest/syndata.html
11. Le, T., Vo, B.: MEI: an efficient algorithm for mining erasable itemsets. Eng. Appl. Artif. Intell. **27**, 155–166 (2014)
12. Le, T., Vo, B., Coenen, F.: An efficient algorithm for mining erasable itemsets using the difference of NC-Sets. In: Proceedings of the 2013 IEEE International Conference on Systems, Man, and Cybernetics, pp. 2270–2274 (2013)
13. Le, T., Vo, B., Fournier-Viger, P., Lee, M.Y., Baik, S.W.: SPPC: a new tree structure for mining erasable patterns in data streams. Appl. Intell. **49**, 478–495 (2019)
14. Lee, G., Yun, U.: Single-pass based efficient erasable pattern mining using list data structure on dynamic incremental databases. Futur. Gener. Comput. Syst. **80**, 12–28 (2018)
15. Lee, G., Yun, U., Ryang, H., Kim, D.: Erasable itemset mining over incremental databases with weight conditions. Eng. Appl. Artif. Intell. **52**, 213–234 (2016)
16. Lee, C., et al.: An efficient approach for mining maximized erasable utility patterns. Inf. Sci. **609**, 1288–1308 (2022)
17. Nguyen, G., Le, T., Vo, B., Le, B.: A new approach for mining top-rank-k erasable itemsets. In: Nguyen, N.T., Attachoo, B., Trawiński, B., Somboonviwat, K. (ed.) Intelligent Information and Database Systems. ACIIDS 2014. Lecture Notes in Computer Science, vol. 8397. Springer, Cham (2014). https://doi.org/10.1007/978-3-319-05476-6_8

Design Recovery of Data Model Hidden in JSON File

Bogumiła Hnatkowska(✉)

Wroclaw University of Science and Technology, Wroclaw, Poland
bogumila.hnatkowska@pwr.edu.pl

Abstract. JSON is a commonly used standard for data interchange and data storage. The hierarchical file structure makes it difficult to understand the real dependencies between the stored data. Data analyzability, reusability, and modifiability are crucial during the maintenance stage. Existing tools for data visualization reflect only the file structure and, therefore, can be classified as working at the redocumentation level. The paper proposes a method that is able to retrieve a design of the data model of JSON data. The design is presented in the form of a UML class diagram. The method has been implemented in a prototype tool. The tool performance has been compared with alternative solutions showing its advantage regarding selected metrics.

Keywords: JSON · Design Recovery · Class Diagram

1 Introduction

JSON (JavaScript Object Notation) is an open standard file format used for many purposes, e.g., data interchange [1] and data storage (like MongoDB [2]). The language is human-readable and easy to process by computer. However, the structure of the JSON document is hierarchical, which can conceal the real dependencies between data, making its maintenance difficult. Maintenance requires the element to be analyzable, reusable, and modifiable [3]; those features are essential for necessary changes.

The reverse engineering process can reveal the dependencies between the structural JSON elements. Such a process can produce a representation of the artifact in another form (redocumentation) or at a higher level of abstraction (design recovery) [4]. Most of the tools processing JSON documents work at the redocumentation level (e.g., [5–9]), where the resulting form of representation can be treated as an alternate view of the input document. The first three are source code generators, while the last two enable diagram generation. On the other hand, design recovery is the process in which domain knowledge and deduction are added to the observations of the considered subject "to identify meaningful higher-level abstractions beyond those obtained directly by examining the element itself" [4]. It aims to reproduce the information that allows one to understand the examined subject fully. Only a few tools try to perform the design recovery for JSON files and present it with a graphical notation like ERD or a class diagram, e.g., [10–14]; however, the results provided by these tools are far from ideal. None of them can process

N. T. Nguyen et al. (Eds.): ACIIDS 2023, LNAI 13995, pp. 234–247, 2023.
https://doi.org/10.1007/978-981-99-5834-4_19

a complex JSON file containing a table of tables of objects of different types or correctly infer multiplicities of class properties. Most cannot find and reuse the same structure repeated in many places of a JSON file.

The paper proposes a method of design recovery of the data model hidden in a JSON file. The result is presented as a class diagram defined at the conceptual level [15]. The method is insensitive to the file structure, providing the same diagram for the same data independently of the hierarchy used. It solves the problems observed in the existing tools. The method has been implemented in a prototype tool and compared with existing ones with a carefully selected case study and test cases. The runnable version of the tool, together with a set of usage examples, is available at [16].

The rest of the paper is structured as follows. Section 2 presents related works. Section 3 presents a motivating example. The design recovery method is described in Sect. 4. Section 5 describes how the method has been evaluated. The last section, 6, concludes the paper.

2 Related Works

Visualization of a JSON file is offered by a few tools, e.g., [8–14]. The first two visualize data at the instance level. The next three can draw diagrams at the meta-level using an ERD diagram [10] or a UML class diagram [11–14]. Unfortunately, none of them can process complex JSON structures like a table of tables, infer inheritance relationships between classes, or infer proper class names. They present the structure of the JSON file in a readable manner but cannot be thought of as complete design recovery tools. The tool [17] serves for visualization of the JSON schema. The schema describes the constraints imposed on its instances; however, it is defined at the documentation level even if provided. As a standard, JSON Schema does not directly support modeling inheritance or many-to-many relationships between objects. Therefore, using tools for schema deduction, e.g., JSON-schema-inferrer[1] or schema-guru[2], is not a solution. JSON to source code converters, e.g., [5–7], belong to the same category, providing structure re-documentations.

According to the best author's knowledge, no tool presents the structure of a JSON file at a higher abstraction level, inferring important information directly from JSON data.

3 Motivating Example

A JSON file can be treated as a hierarchical database with a root element. A person translating a many-to-many relationship must select a parenting element and agree that some data will be repeated in different places of the JSON structure.

Let us assume one wants to create a JSON file data consistent with a class diagram from Fig. 1. The diagram introduces two classes (*Course*, *Student*) and one many-to-many association. Classes have several value properties with different types (e.g., Integer,

---

1 https://github.com/saasquatch/json-schema-inferrer.

2 https://github.com/snowplow/schema-guru.

String) and multiplicities. The root of the JSON file can be either an array of *Course* instances (Listing 1) or an array of *Student* instances (Listing 2). The diagrams generated for both files should look the same, except for class names that are not directly delivered.

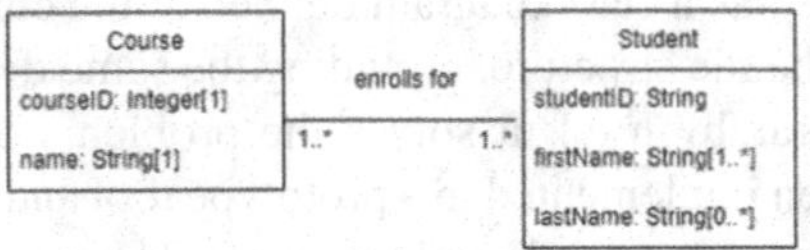

Fig. 1. Example class diagram.

```
[{ "courseID": 1, "name": "Math",
    "students": [
    {"studentID":"1S", "firstName":["Ann","Eve"],"lastName":["Smith"]},
    {"studentID":"2S", "firstName":["Shahir"], "lastName":[]}]},
  { "courseID": 2, "name": "Music",
    "students": [
    {"studentID":"1S", "firstName":["Ann","Eve"],"lastName":["Smith"]},
    {"studentID":"2S", "firstName":["Shahir"],"lastName":[]}]}
]
```

Listing. 1. JSON for a motivating example – version 1.

```
[{"studentID": "1S","firstName": ["Ann", "Eve"], "lastName": ["Smith"],
   "courses": [{"courseID": 1, "name": "Math"},
               {"courseID": 2, "name": "Music"}]},
  {"studentID": "2S", "firstName": ["Shahir"], "lastName": [],
   "courses": [{"courseID": 1, "name": "Math"},
               {"courseID": 2, "name": "Music"}]}
]
```

Listing. 2. JSON for a motivating example – version 2.

For the JSON files given in Listings 1 and 2, the Software Ideas Modeler [14] generated the diagrams shown in Fig. 2 (note: the original diagrams were rewritten for readability purposes). The tool correctly recognized classes (except for using a plural name, e.g., *Students*) and their value properties (including types and partially their multiplicities). However, it was utterly lost in relationships. For each JSON file, the tool discovered three one-to-many relationships (two represented by compositions and one defined by a reference attribute, e.g., students: Students [*], instead of one many-to-many association.

The JSON discoverer tool [13] generated the class diagrams shown in Fig. 3. The multiplicities of value properties are entirely neglected. Instead of one many-to-many association, the tool produced one composition. The names of classes were discovered correctly.

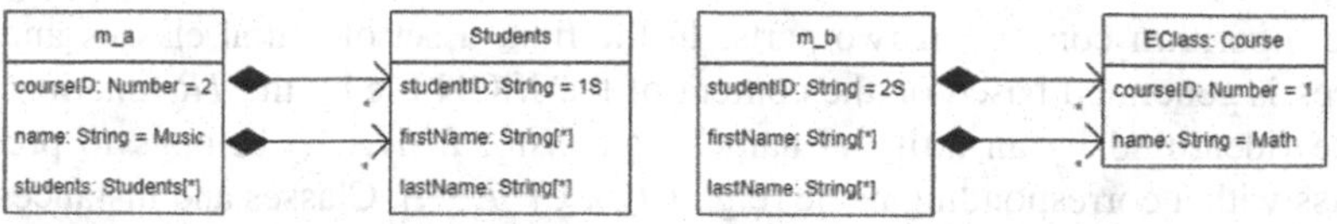

Fig. 2. Class diagrams generated for the motivating example by the Software Ideas Modeler.

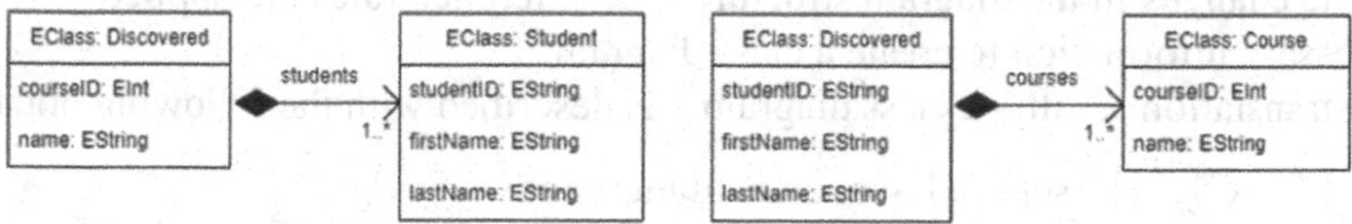

Fig. 3. Class diagrams generated for the motivating example by the JSON discoverer.

As one can expect, code generators could not recognize the many-to-many association. For this trivial example, some could correctly identify class names and their value properties (e.g. [7]) except for the multiplicity constraints.

4 Design Recovery Method

4.1 Method Assumptions

The design recovery method requires a complex JSON expression, at least one object or array, as input. It produces nothing for singular values, e.g., 3. Null values or empty arrays are expected for the elements (e.g., object fields) which should appear in the class diagram. There are no other constraints on the input. Field names are assumed to be written in English, but this is unnecessary.

The recovered data design is presented as a UML class diagram (written in the plant UML notation – https://plantuml.com/). The diagram can contain the following:

- concrete classes with appropriately – if possible – inferred names in English
- class attributes with inferred types and multiplicities
- binary associations with inferred names and multiplicities at both ends
- self-associations
- generalizations between concrete classes

Only concrete classes (with instances) are considered to avoid a possible class explosion. Therefore, a generalization between two classes will be created only if all the parent class attributes appear in the child class. A similar remark refers to self-associations. As automatic recognition of self-associations represented by a foreign key in the general case is impossible, one will be created only if the whole structure of the class is repeated, e.g., {"id": 1, "next": {"id": 2, "next": null}}. Association navigability will be used to reflect the data relationships in the JSON file.

4.2 Translation Algorithm

The algorithm aims to transform a JSON expression into a class diagram that should reflect a given application domain.

The algorithm consists of two parts. In the first, a set of initial classes and their instances is generated based on the content of the JSON file by the *TR* function. Each instance (identified by an artificial name, e.g., *Inst0*, *Inst1*, …) belongs to precisely one class with a corresponding name (e.g., *Cl_0*, *Cl_1*, …). Classes and instances have properties defined (classes at the specification level, instances at the value level). In the second part, the number of classes is reduced, and the instances are refactored to reflect the changes in the diagram structure. The inference rules are applied to recover the necessary information to create a class diagram.

The translation result – a class diagram – is described with the following notation:

- $C = \{C_1, C_2, \ldots\}$ – set of classes and enumerations
- C_i = *{name, A, I, role}*:
  - *name* – the name of class C_i
  - $A = \{A_1, A_2, \ldots\}$ – set of attributes defined in C_i
  - $I = \{I_1, I_2, \ldots\}$ – set of instances that belong to C_i
  - *role* = *{root, array, object}*

    - A_i = (*name, type, role, mult*) – attribute A_i schema definition containing the attribute name, type (e.g., Boolean, String), role (*role* = {*value, literal, reference, reverse*}), and multiplicity (*mult* = {*zero_one, one, many, one_many*}); a *reference*/*reverse* attributes are used to represent binary associations on the diagram
    - I_i = {*name*, V} – instance I_i description containing the instance name and the set V of attributes valuations
    - $V = \{V(A_1), V(A_2), \ldots\}$ – set of attributes valuations kept by an instance; $V(A_i) = (A_i, \{v_1, \ldots, v_k\})$, where v_i is a literal, e.g., 2, "John"
- $G = \{G_1, G_2, \ldots\}$ – set of generalization relationships, where $G_i = (C_i, C_j)$, C_i – the parent class, and C_j – the child class

A "dot" notation is used to retrieve specific values from its context, e.g., C_i.*role*, C_i.*name*.

The *TR* is a recursive function that translates a JSON expression to a specific class and links it to a context provided as a transformation parameter. An initial context is an empty class marked as root, i.e., C_0 = {*name* = "Cl0", $A = \{\}$, $I = \{I_0\}$, *role* = *root*} with one instance I_0 = {*name* = "Inst0", $V = \{\}$}.

The *TR* function is presented in parts for readability purposes. Each part translates a specific JSON construct. All new class/instance names and most attribute names are artificial. The rule of generating new names ensures that the new name differs from all already generated. The generated name has a *gen* prefix in the function definition. Initially, the attribute multiplicity is set to one. It will be changed by the inference rules.

An auxiliary function *type*(v) returns *Void* for v = null, *Boolean* for true/false literals, *Integer* for an integer value (e.g., 1), *Real* for a real value (e.g., 2.5), String for other literals (e.g., "Eve").

Array Translations
Translation of an array of values $[e_1, \ldots, e_k]$ in the *Ctx* context is defined as:

$TR([e_1, \ldots, e_k], Ctx) = C_x$
$C_x = \{name = gen, A = \{A_1, \ldots, A_k\}, I = \{I_1\}, role = array\}$
$A_i = \{name = gen, \text{type} = type(e_i), role = value, mult = one\}$ for each e_i being an elementary value
$A_j = \{name = gen, type = TR(e_j, C_x), role = reference, mult = one\}$ for each e_j not being an elementary value

$$I_1 = \left\{ name = gen, V = \bigcup_i (A_i, \{e_i\}) \cup \bigcup_j (A_j, \{instance \in TR(e_j, C_x)\}) \right\}$$

The context class *Ctx* and the created C_x class are linked with newly created attributes:

$A_{ctx} = \{name = C_x.name, type = C_x, role = reference, mult = one\}$ and $A_{ctx} \in Ctx.A$
$A_c = \{name = Ctx.name, type = Ctx, role = reverse, mult = one\}$ and $A_c \in C_x$.

Attribute instance valuations are added to the *Ctx* and C_x instances, respectively.

Object Translations
Translation of an object $\{n_1: e_1, \ldots, n_k: e_k\}$ in the *Ctx* context is defined as:

$TR(\{n_1: e_1, \ldots, n_k: e_k\}, Ctx) = C_x$
$C_x = \{name = gen, A = \{A_1, \ldots, A_k\}, I = \{I_1\}, role = object\}$
$A_i = \{name = n_i, type = type(e_i), role = value, mult = one\}$ for each e_i being an elementary value
$A_j = \{name = n_j, type = TR(e_j, C_x), role = reference, mult = one\}$ for each e_j not being an elementary value

$$I_1 = \left\{ name = gen, V = \bigcup_i (A_i, \{e_i\}) \cup \bigcup_j (A_j, \{instance \in TR(e_j, C_x)\}) \right\}$$

As previously, reference attributes are created to connect the C_x class with its context *Ctx*, and instance attribute valuations are updated respectively.

Class Reduction and Inference Rules
The number of classes the *TR* function generates can be very high as a new class is created for every array and object. Therefore, a class reduction is necessary. The reduction process (*REDUCE*) merges equivalent elements (classes or properties) and removes non-informative classes, e.g., the root class or so-called proxy classes. The reduced diagram

is an input for the inference process (*INFER*), which enriches the generated diagram with required data, e.g., generalizations, self-associations, and multiplicities of properties. It can also merge some classes (side effect of inference).

Both processes have many stages. Some stages refer to the definitions given below.

Two classes are considered *partially equivalent* if their value attributes have the same names and types (*Void* is regarded as the same as any type).

Two classes are considered *equivalent* if they are partially equivalent and their reference/reverse attributes have the same types.

Two reference attributes are considered *equivalent* if they have the same type.

Two reverse attributes are considered *equivalent* if they have the same type.

Two instances of the same class are considered *equivalent* if they have equal valuations for all value attributes.

A class is called empty if it lacks attributes other than the reverse.

A class is called a *proxy class* if it intermediates between two other classes, i.e., it satisfies the following conditions:

- an array class (role)
- without value attributes
- with only one reference attribute A_k
- with only one reverse attribute A_j

Definition of the REDUSE Process.

1. Remove empty classes – remove any empty class C and the attribute in its context class *Ctx* referencing C; update the instances accordingly.
2. Collapse value list:
   a. Remove any class C satisfying the conditions:
      (1) marked as an array (role),
      (2) all value attributes have names artificially generated (gen).
   b. Find the context class (*Ctx*) for the C class. In the *Ctx* class, change the type of the attribute referencing to C for T (the type of value attributes or its supertype *Object*, if they are different) and the attribute multiplicity to many.
   c. Update instances of *Ctx* to store values of the instances of the removed class C.
3. Merge equivalent classes:
   a. While a pair of equivalent classes (C_i, C_j) belongs to the set of classes, merge them:
      (1) Move reference and reverse attributes from C_j to C_i (after this step, the set of attributes in C_i may contain equivalent attributes).
      (2) Move instances of C_j to C_i and update their attribute valuations (they should refer to the attributes of the C_i class).
      (3) Change references in all class attributes leading to C_j for C_i.
      (4) Remove the C_j class.
      (5) While a pair of equivalent attributes (A_m, A_n) exists in the set of C_i attributes, merge them:
         i. For each instance $inst \in C_i$, change the attribute valuations of A_n for A_m.
         ii. If A_m has a *Void* type, assign to it the A_n type.
         iii. Remove A_n.

4. Remove proxy classes:
   a. For any proxy class C_i:
      (1) Find the context for the proxy class (set in *type*(A_j)) and the target class (set in *type*(A_k)).
      (2) Change the type of the context class attribute A_{proxy} referring to C_i with the target class and the target class attribute referring to C_i with the context class.
      (3) Change the multiplicity of the context class attribute A_{proxy} to many.
      (4) Update instances of context and target classes accordingly to changes in their attributes.
      (5) Remove C_i.
5. Remove the root class.

Definition of the INFER Process.

1. Infer types for class attributes with *Void* type – based on the attribute valuations of class instances, if all the valuations are of the same type T, type T is selected; otherwise, String.
2. Infer self-relations:
   While a pair of partially equivalent classes (C_i, C_j) belongs to the set of classes C, and C_i refers to C_j with an attribute, merge them (see 3a (1) – (5)).
3. Infer generalizations:
   a. Sort classes in descending order based on the number of value attributes and store results in S.
   b. For each class $C_i \in S$:

      For each class $C_j \in C$ such that $C_j \neq C_i$ and $C_j.A \subset C_i.A$:
      Create generalization $G = (C_i, C_j)$.

4. Infer class names – for each reference attribute A_i, find a lemma version of the attribute name (using the Sandford tagger), and set it as the name of the referenced class ($C = Ai.type$), if C is the object class and all reference attributes pointing to C have either a generated name or the same name as A_i.
5. Merge equivalent instances:
   For each class $C_i \in C$:
   While a pair of equivalent instances (I_i, I_j) belongs to $C_i.I$, merge them:
   (1) Change references in all class attribute valuations leading to I_j for I_i.
   (2) Remove I_i.
6. Infer attributes multiplicities:
   For each class $C_i \in C$:
   For each attribute $A_k \in C_i.A$ infer attribute multiplicity based on information included in C_i instances.

Examples of Final Transformations

Table 1 presents examples of transformation results for basic test cases. It serves for the method illustrating purposes as well as for results comparison with existing tools.

5 Method Evaluation

5.1 Short Description

The method has been evaluated with the use of a complex case study. A class diagram has been prepared to make the assessment objective, covering all interesting constructs, i.e., classes with different properties and a variety of multiplicities, an association class, enumeration, generalization relationship, composition, and many-to-many association – see Fig. 4.

Table 1. Examples of transformation results for basic test cases.

Test Id	JSON	Translation results
1	[1, 2]	 cl_0 cl_1:Integer[1..*]
2	{"grades" : [3, 4]}	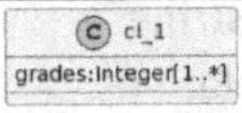 cl_1 grades:Integer[1..*]
3	{"name": "Eve", "age": 20}	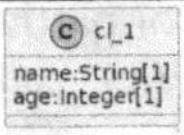 cl_1 name:String[1] age:Integer[1]
4	{"name": "Eve", "grades": [3,4, "five"]}	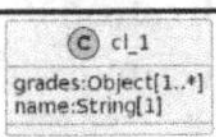 cl_1 grades:Object[1..*] name:String[1]
5	{"id": 1, "next": {"id": 2, "next" : null}}	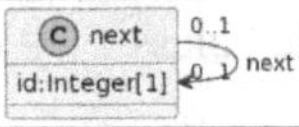 next id:Integer[1] 0..1 next 0..1
6	[[{"name":"Eve"}], [{"name":"Adam", "surname":"Smith"}]]	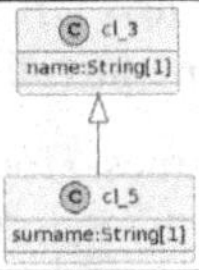 cl_3 name:String[1] cl_5 surname:String[1]
7	Motivatingexample	Fig. 1

The diagram next has been transformed into a JSON file and filled with example data by an independent expert. The expert split the diagram into two subgraphs, represented by three arrays in the JSON file.

The first subgraph covered classes marked with blue and red colors (*Mariage*, *Person*, *Address*, and *Gender*). The second subgraph covered classes written in black and red (*Address*, *Bank*, *Account*, *BankCustomer*, *Gender*). The *Gender* class was defined in a separate table. The *Address* was present in both subgraphs.

The first assumption was that the expert prepared more than one version of the file with the same data, but it appeared that none of the existing tools (except the one presented in this paper) could process a table of tables correctly. Therefore, the two subgraphs (without *Gender* class) were physically separated into two JSON files.

The JSON files were processed by available tools ([13, 14] – class diagram generators, and [7] – one of the code generators) as well as by the tool implementing the proposed method, and the results of the transformation were compared with selected metrics. The

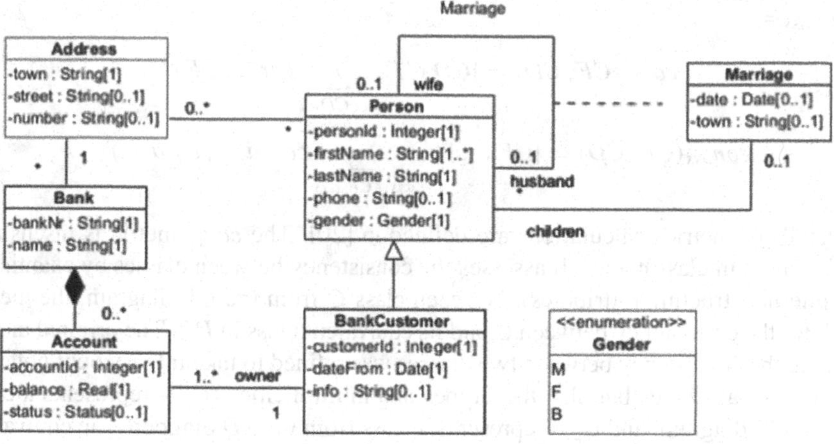

Fig. 4. Case study used for the method evaluation.

files and the class diagrams created by the proposed method are available under the link [16].

5.2 Consistency Metrics

Many approaches allow assessing the similarity between two class diagrams, e.g., [18, 19]. The paper [18] introduced three types of similarity metrics, referring to naming information (at the class level), internal information (at the properties level), and neighborhood information (structural dependencies). The first two are uninteresting as the name of classes is retrieved from the JSON file, and the properties are limited to attributes. A simpler similarity metric was proposed in [19]. To compare the two diagrams, the authors consider the similarity of classes and relationships. However, in both cases, the syntactic component (names resemblance) is important, which makes the metrics calculation complex. As the names of attributes are determined by the transformation method, the metric proposed in [19] is inadequate.

Eventually, the consistency metric *CON* [20] has been adapted as it was constructed for a similar task – checking how well a diagram generated from a set of data (*CF*) is consistent with another class diagram (*CD*) representing the domain description the data should come from. The metric takes values from the 0–1 range. A higher value means better consistency.

$$CON(CF, CD) = \alpha con_{cl}(CF, CD) + \beta con_{ass}(CF, CD), \alpha = \beta = 0.5. \quad (1)$$

Informally, diagram D1 is consistent with diagram D2, if D_1 has all the information (attributes, associations) from D_2, and the multiplicities of the properties are consistent with those defined by D_2.

The *CON* metric uses two auxiliary metrics, assessing the consistency of classes (con_{cl}), and consistency of associations (con_{ass}). If *CF* diagram doesn't have association, the $con_{ass} = 1$.

$$con_{cl}(CF, CD) = |CD.cl|^{-1} \sum_{c \in CD.cl} con(c, CF) \tag{2}$$

$$con_{ass}(CF, CD) = |CF.ass|^{-1} \sum_{ass_1 \in CF.ass} con(ass_1, eq(ass_1)) \tag{3}$$

Details of metrics calculations are defined in [20]. The con_{cl} metric is insensitive to differences in class names. It assesses the consistency between classes by examining their internal structure (attributes). For each class *C* from the *CF* diagram, the metric calculates the consistency between *C* and its equivalent class in *DF*. The original metric assessing the consistency between two classes was refined to take into account not only the names of attributes but also their types and multiplicities (C_1 – represents a class from the *CD* diagram, and C_2 – represents a class from the *FD* diagram – an equivalent to C_1 class):

$$con(C_1, C_2) = \frac{\sum_i^{|C_1.attr|} \sum_j^{|C_2.attr|} con(C_1.attr[i], C_2.attr[j])}{|C_1.attr|} \tag{4}$$

$$con(a_1, a_2) = \begin{cases} 1 \ if a_1.name = a_2.name, a_1.type = a_2.type, a_1.mult = a_2.mult \\ 0.5 \ if a_1.name = a_2.name, a_1.type = a_2.type, a_1.mult \subset a_2.mult \\ 0 \quad otherwise \end{cases} \tag{5}$$

The con_{ass} metric calculates consistency between each association ass_1 in *DF* and its equivalent association ass_2 in *CD* using the formula (6).

$$con(ass_1, ass_2) = \frac{con(ass_1.ends[0], ass_2.ends[0]) + con(ass_1.ends[1], ass_2.ends[1])}{2} \tag{6}$$

Association end e_1 is consistent with association end e_2, if they link equivalent classes, and their multiplicities are the same ($con(e_1, e_2) = 1.0$); the metric returns zero otherwise.

5.3 Method Performance

The method performance was assessed with the case study and a set of test-cases (see Table 1). The original JSON file with the case study data (test-case 10) was a source of two JSON files covering the subgraphs described in Sect. 5.1 (test-case 8 – the first subgraph, test-case 9 – the second subgraph).

The results of *CON* metrics and their components were calculated manually (see Table 2). For the code generator the way of consistency metric calculation was less restrictive, as a declaration of IList < string > was considered as consistent with the multiplicity *, and 1..*.

Example generated diagrams for test-case 8 are shown in Fig. 5. As one can observe, none of the tools was able to generate enumeration or an association class (there are some

Table 2. *CON* metrics comparison for considered test cases.

Test ID	Design Recovery			Ideas Modeler			JSON Discoverer			Source code generator		
	con_{cl}	con_{ass}	CON	con_{cl}	con_{ass}	CON	con_{cl}	con_{ass}	CON	con_{cl}	con_{ass}	CON
1	1	1	1	Error			1	1	1	Error		
2	1	1	1	0	1	0.5	0.5	1	0.75	1	1	1
3	1	1	1	1	1	1	1	1	1	1	1	1
4	1	1	1	0.5	1	0.75	0.5	1	0.75	1	1	1
5	1	1	1	1	0	0.5	0	0.5	0.25	1	0	0.5
6	1	1	1	0.75	1	0.88	0	0	0	Error		
7	1	1	1	1	0.25	0.63	0.83	0.5	0.67	1	0.5	0.75
8	0.76	1	0.88	0.54	0.33	0.44	0.44	0.38	0.41	0.54	0.25	0.39
9	0.94	1	0.97	0.78	0.36	0.57	0.68	0.66	0.67	0.75	0.5	0.62
10	0.72	0.75	0.73	Only the first array processed (row 9)			Error			Error		

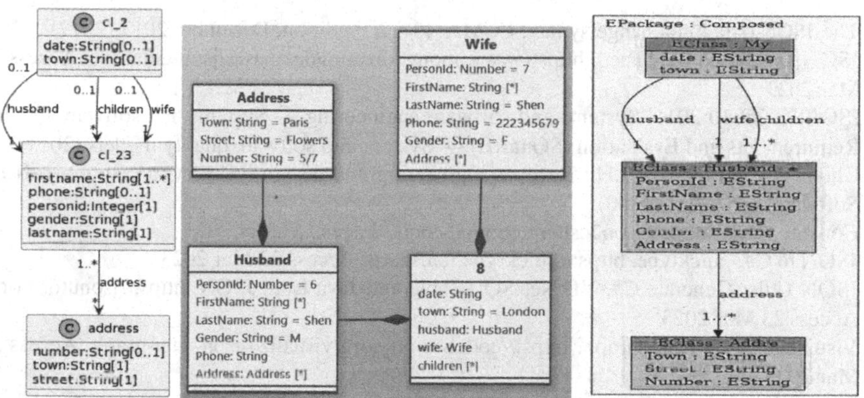

Fig. 5. Example class diagrams generated for test case 8 by Design Recovery (on the left), Ideas Modeler (in the middle) and JSON Discoverer (on the right) tools, respectively.

equivalents, i.e. *Cl_2* or *My* classes). The source code generator offered the worse solution, with a separate class for each role (husband, wife, child). All tools have problems with the data values which was described as String.

Nevertheless, the *Design Recovery* tool gave the best results in terms of consistency with the original class diagram. It was also able to process complex JSON files none of the competitors could handle. Only that tool was able to deal with class names properly.

Other tools create class names in a naïve way simply by cutting the word endings ('es' or 's', e.g. Addre [7, 13]).

To check the ability of the tools to process big data structures (1 GB), another test (11) was prepared. The JSON file contains data from three classes with the following dependences (root "1..*" – "1..*" reseller, reserell "1..*" – "1..*" customer). Only the *Design Recovery* tool discovered these dependencies correctly. The other tools missed many to many associations.

6 Summary and Further Works

The main contribution of the paper is a design recovery method for a data model extracted from a JSON file. The method is an improvement of existing ones, which are sensitive to the file structure. The method examines data included in the file to infer necessary information, for example, about inheritance or the properties multiplicities.

Conducted tests and case study confirmed the method advantage over existing tools in terms of consistency of the produced class diagram with a domain diagram for which the translated JSON file has been created.

The method can be improved to enable finding compositions, association classes, enumerations or complex data types (e.g. dates).

References

1. The JSON data interchange syntax, ECMA-404, 2nd edition, December 2017 (2017)
2. JSON Databases Explained. https://www.mongodb.com/databases/json-database. Access 23 Mar 2023
3. ISO/IEC 25010:2011: Systems and software engineering — Systems and software Quality Requirements and Evaluation (SQuaRE) — System and software quality models (2011)
4. Chikofsky, E.J., Cross, J.H.: Reverse engineering and design recovery: a taxonomy. IEEE Software 1990, **7**(1) (1990)
5. Convert JSON. https://json2csharp.com/all-tools. Access 23 Mar 2023
6. JSON to C#, guicktype. https://quicktype.io/csharp. Access 23 Mar 2023
7. JSON Utils: Generatc C#, VB.Net SQL Table and Java from JSON. https://jsonutils.com/. Access 23 Mar 2023
8. Visualize JSON Data Graph. https://codebeautify.org/visualize-json-data-graph. Access 23 Mar 2023
9. https://plantuml.com/json. Access 23 Mar 2023
10. DataFinz - No Code Data Integration Platform | Build ODS | ERD, datafinz.com. Access 23 Mar 2023
11. Cánovas Izquierdo, J.L., Cabot, J.: Discovering implicit schemas in JSON data. In: Daniel, F., Dolog, P., Li, Q. (eds) Web Engineering. ICWE 2013. Lecture Notes in Computer Science, vol 7977. Springer, Berlin, Heidelberg (2013)
12. Izquierdo, J.L.C., Cabot, J.: Composing JSON-based web APIs. In: Casteleyn, S., Rossi, G., Winckler, M. (eds) Web Engineering. ICWE 2014. Lecture Notes in Computer Science, vol 8541. Springer, Cham (2014)
13. JSON discoverer (uoc.edu). Access 23 Mar 2023
14. Diagram CASE Tool for Software Modeling & Analysis - UML, BPMN, ERD (softwareideas.net). https://www.softwareideas.net/. Access 23 Mar 2023
15. Hnatkowska, B., Walkowiak-Gall, A.: Towards definition of a unified domain meta-model. In: Engineering software systems: research and practice, Kosiuczenko, P., Zieliński, Z. (eds.) Advnces in Intelligent Systems and Computing, Vol. 830, pp. 86–100, Springer (2019)

16. Hnatkowska, B.: https://github.com/bhnatkowska/JSONDesignRecovery
17. Cánovas, J.: JSONSchema To UML: Tool to Generate UML diagrams from JSON Schema Definitions, Modeling Languages (2018). Accessed 23 Mar 2023
18. Mojeeb, A.-R.A.-K., Moataz, A.: UML class diagrams: similarity aspects and matching. Lecture Notes on Software Eng. **4**(1) (2016)
19. Fauzan, R., Siahaan, D., Rochimah, S., Triandini, E.: Class diagram similarity measurement: a different approach. ICITISEE 2018, Yogyakarta, Indonesia, pp. 215-219 (2018)
20. Hnatkowska, B., Huzar, Z., Tuzinkiewicz, L.: Consistency assessment of datasets in the context of a problem domain, In: Advances and trends in artificial intelligence: from theory to practice: IEA/AIE 2021, Fujita, H., others (eds.), Springer Nature, pp. 112–125 (2021)

Accurate Lightweight Calibration Methods for Mobile Low-Cost Particulate Matter Sensors

Per-Martin Jørstad[1], Marek Wojcikowski[2], Tuan-Vu Cao[3], Jean-Marie Lepioufle[3], Krystian Wojtkiewicz[4], and Phuong Hoai Ha[1(✉)]

[1] UiT The Arctic University of Norway, Tromsø, Norway
{per.m.jorstad,phuong.hoai.ha}@uit.no

[2] Gdansk University of Technology, Gdansk, Poland
marwojci1@pg.edu.pl

[3] Norwegian Institute for Air Research, Oslo, Norway
{tvc,jml}@nilu.no

[4] Wrocław University of Science and Technology, Wrocław, Poland
krystian.wojtkiewicz@pwr.edu.pl

Abstract. Monitoring air pollution is a critical step towards improving public health, particularly when it comes to identifying the primary air pollutants that can have an impact on human health. Among these pollutants, particulate matter (PM) with a diameter of up to 2.5 µm (or PM2.5) is of particular concern, making it important to continuously and accurately monitor pollution related to PM. The emergence of mobile low-cost PM sensors has made it possible to monitor PM levels continuously in a greater number of locations. However, the accuracy of mobile low-cost PM sensors is often questionable as it depends on geographical factors such as local atmospheric conditions.

This paper presents new calibration methods for mobile low-cost PM sensors that can correct inaccurate measurements from the sensors in real-time. Our new methods leverage Neural Architecture Search (NAS) to improve the accuracy and efficiency of calibration models for mobile low-cost PM sensors. The experimental evaluation shows that the new methods reduce accuracy error by more than 26% compared with the state-of-the-art methods. Moreover, the new methods are lightweight, taking less than 2.5 ms to correct each PM measurement on Intel Neural Compute Stick 2, an AI-accelerator for edge devices deployed in air pollution monitoring platforms.

This work was supported in part by the National Centre for Research and Development (grant NOR/POLNOR/HAPADS/0049/2019-00), Research Council of Norway (grant 270053) and European Commission (grant 101086541). The authors would like to thank the Agency of Regional Atmospheric Monitoring Gdansk-Gdynia-Sopot in Poland for providing data from the reference air pollution measurement stations.

N. T. Nguyen et al. (Eds.): ACIIDS 2023, LNAI 13995, pp. 248–260, 2023.
https://doi.org/10.1007/978-981-99-5834-4_20

1 Introduction

Air pollution monitoring is crucial to improving public health. According to the World Health Organization (WHO), air pollution places 99% of the global population at risk of developing various diseases and causes 6.7 million deaths each year [14]. The primary air pollutants responsible for affecting human health include Particulate Matter (PM) and nitrogen dioxide (NO2). Out of the contaminants listed, particles with a diameter of up to 2.5 μm (also known as $PM_{2.5}$) are particularly hazardous. They have the ability to breach the body's natural barriers and enter the bloodstream, leading to cardiovascular and/or respiratory problems [21]. As a result, continuous and accurate monitoring of pollution related to PM is of great significance [4,13]. Dependable measurements of PM are essential for creating early warning systems that can provide information on sudden spikes or prolonged high levels of pollutants.

Current PM monitoring relies on a limited number of costly air monitoring stations due to the inaccuracy of mobile low-cost PM sensors (MLCS). Although the fixed stations can provide high quality measurements, their complex measurement techniques result in very high purchase and maintenance cost, preventing them from being deployed widely and monitoring many locations. The mobile low cost PM sensors, on the other hand, enable covering more locations quickly and cheaply. The MLCS sensors enable localizing hot spots and mapping the spatial and temporal dynamics of air pollution, particularly in urban areas. However, the accuracy of the MLCS sensors is often questionable since it depends on geographical factors such as local atmospheric conditions and pollutant concentration levels [3]. Because of the small size and low cost requirements of the MLCS sensors, the MLCS sensors cannot be equipped with built-in calibrator, air filtering equipment, temperature controller (cooler and heater) nor relative humidity controllers that are used in the air monitoring stations.

Improving the accuracy of mobile low-cost PM sensors with in-situ software-based calibration models is an emerging research area [11,17,20,22]. The accuracy of the calibration methods can be evaluated using root mean square error RMSE ($\mu g/m^3$), which varies from 2.82 [20] to 8.48 [22]. The seminal work [20] has developed a kriging-based model to correct the measurement from MLCS sensors. The kriging-based correction model is established using optimized lower and upper bounds, which are tailored to the environmental conditions for which both PM measurements from the reference station and sensor are available. Therefore, the model's corrective capability is restricted to these specified ranges. Alternatively, machine learning-based (ML-based) methods have been proposed to improve the MLCS measurement accuracy [9,11]. The PM measurement accuracy of the ML-based methods is still limited with reported RMSE ($\mu g/m^3$) from 4.21 to 5.45 [11].

In this work, we propose new ML-based methods (called NAS-RS and NAS-RE) that significantly improve the PM measurement accuracy of mobile low cost sensors, achieving RMSE ($\mu g/m^3$) of 1.89, reducing accuracy error by more than 26% compared with the state-of-the-art methods. Moreover, our new methods are lightweight and their correction of each PM measurement takes less than

2.5 ms on Intel Neural Compute Stick 2 (NCS2) [6], an AI-accelerator for edge devices deployed in air pollution monitoring platforms [1]. Our new methods leverage Neural Architecture Search (NAS) [24] to improve the accuracy of calibration models for MLCS. NAS is a growing field within the domain of deep learning model design. NAS aims to find the best architecture for a new model by automating architecture engineering, avoiding the time-consuming and error-prone manual development of neural architectures. Even though NAS has several potential benefits, it is still a young field [10,15,16,24]. There are few active NAS frameworks for the mainstream audience, making NAS difficult to leverage by users on their practical problems. Furthermore, the few libraries that do exist, often have a tendency to simplify the NAS-process to hyper-parameter tuning [23]. Although this works, to a degree, it does not fully utilize NAS capabilities. In this work, we leverage NAS to find the most optimal architecture for a regression model to correct the PM measurement from MLCS that can run on edge devices (e.g., Intel Neural Compute Stick 2 (NCS2) [6]). Furthermore, to examine the different features of NAS, we compare two NAS methods: a random search method (NAS-RS) and a highly sophisticated evolution method called regularized evolution (NAS-RE).

The rest of the paper is organized as follows. Section 2 describes the design and implementation of the new calibration methods. Section 3 presents the model training and testing results. Section 4 discusses the results and Sect. 5 concludes the paper.

2 Design and Implementation

2.1 Data Preprocessing

Since the input data consists of several sensor measurements, with many of vastly different ranges, the model was given standardized data as input. That is, the model was given data where all the input values had a mean of 0 and a standard deviation of 1. Equation 1 show the standardization formula.

$$y = \frac{(x - \mu_x)}{\sqrt{\sigma_x^2}} \tag{1}$$

where x is some input feature, μ_x denotes the mean of the input features, and σ_x^2 denotes the input feature's variance.

2.2 Base Model Design for MLCS Calibration

To correct the inaccurate measurement from MLCS, we first build a small regression model as the base model. Figure 1 shows this base model. It consists of three fully connected layers, each with a width - set empirically - to 64 and initialized using Kaiming's initialization [5]. Following each fully connected layer is ReLU activation function to act as a non-linearity. Both the input and the output of the model is configurable.

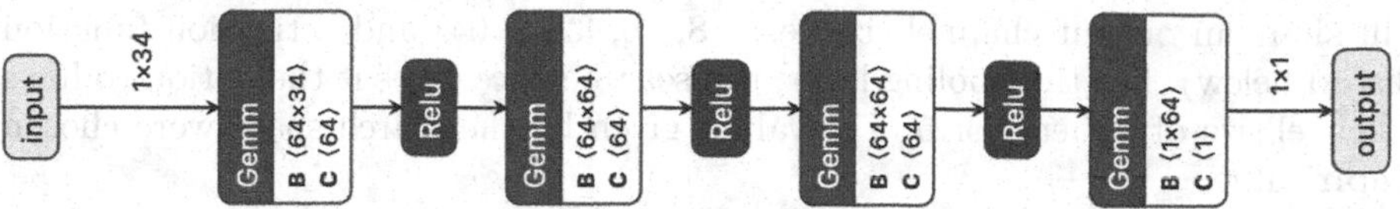

Fig. 1. Base Model.

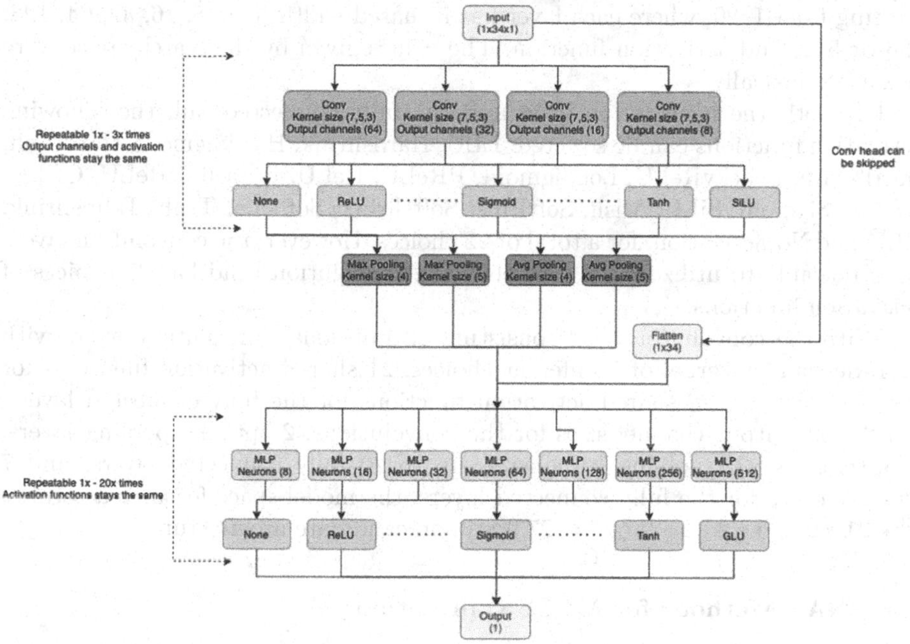

Fig. 2. Model Search Space.

2.3 NAS Model Search Space for MLCS Calibration

Figure 2 depicts our NAS model search space created for the base regression model. It is essentially a jumbo net expanding the depth of the network with an optional convolution head and a bigger fully connected tail.

The convolution head is the first section of the expanded network and consists of up to three convolution layers followed by a single max-pooling or average-pooling layer. The head functions as an optional and lightweight feature extractor motivated by the design of popular computer vision models like the VGG16 [19]. Its purpose is essentially to try and find features that the fully connected tail can use later for the actual regression. Since the head is optional, it can be skipped, allowing the data to flow directly to the fully connected tail.

In the head, the search space gives the convolution layers an independent choice of kernel size and a shared choice of activation function and output channel. This means that each layer will have its own kernel size (e.g., 3, 5, or 7), but will share an output channel size (e.g., 8, 16, 32 or 64) and activation function (listed below). For the pooling layer, the search space gives it the option to have a kernel size of either 4 or 5. The values given by the search space were chosen empirically.

Following the convolution head is the fully connected tail. As in the base model, the fully connected layers are to do the regression and yield the final output. As shown in Fig. 2, the search space allows the tail to have a depth ranging from 1–20, where each layer has a shared width (e.g., 8, 16, 32, 64, 128, 256 or 512) and activation function. The values given by the search space were chosen empirically.

For both the convolution head and the fully connected tail, the following activation functions can be selected: ELU, Hardshrink, HardSigmoid, Hardtanh, HardSwish, LeackyReLU, LogSigmoid, PReLU, ReLU, ReLU6, RReLU, CELU, GELU, Sigmoid, SiLU, Mish, SoftPlus, Softshrink, Softsign, Tanh, Tahnshrink, GLU and No-activation, for a total of 22 choices. However, the convolution layers were not able to utilize GLU, and thus the convolution head has 21 choices of activation functions.

With the convolution head consisting of 3 optional convolution layers with an independent kernel of 3 different choices, 21 shared activation functions for the convolutions, 22 shared activation functions for the fully connected layers, 4 different output channel sizes for the convolutions, 2 optional pooling layers, 2 options for the pooling kernel-size, up to 20 fully connected layers, and 7 width choices for the fully connected layers, the model space features a total of $3^3 * 21 * 22 * 4 * 2 * 2 * 20 * 7 = 27.9$ million candidate architectures!

2.4 NAS Methods for MLCS Calibration

We leverage two NAS methods for MLCS calibration, namely Brute Force NAS and Evolution-based NAS, to get insights into the affects of NAS methods on MLCS calibration. NAS can be described as a gradient-based method that exploits reinforcement learning (RL) to iteratively find better networks. The method was first introduced in [24], and is based on the idea that most DNNs have the structure and connectivity that can be expressed by a variable-length string. This allows most DNN architectures to be defined by generative mechanisms that can define such a string. If the mechanism is trainable, it can be taught to generate better and better architectures over time.

The NAS process consists of two main parts: a static search space, containing all possible model configurations; and an iterative training loop that uses the search space to train the generative mechanism (usually called the controller). More specifically, a round of NAS can be expressed as follows:

1. The controller samples a network (referred as child network) from the search space.
2. The child network is trained and tested using a predefined training and validation set.

3. The child network produces a validation score that the controller can use to update itself.
4. The controller receives reward signal and updates itself, hopefully getting slightly better. After the update, the controller once again samples a network from the search space to create a new child network. Then, the entire process repeats again.

Brute Force NAS. Brute force NAS considers algorithms that search their search space unintelligently without using a controller for guidance, e.g., Grid-Search [12], a method that systematically tries all options one-by-one. A variant of Brute Force NAS is Random-Search NAS (NAS-RS), which chooses a candidate from the search space at random. Although unintelligent, NAS-RS has been reported to yield worthwhile results [2].

Evolution-Based NAS. Evolution-based NAS is a method inspired by real life biological evolution. Hence, instead of using a controller, it aims to directly improve the model space by letting good models mutate and bad models die. By doing this repeatedly, only good models will be the ones standing at the end [16].

An example of this approach can be found in Regularized Evolution [16] (RE, but also known as Aging Evolution - AE). It follows the same principle as other evolution based methods. However, instead of killing the worst performing models in the population (or search space), it rather kills the oldest one. The change acts as some sort of regularization, preventing overfitting.

3 Evaluation

To train and test both the base model and the child networks of the NAS process, we utilized a dataset of measurements from the HAPADS air pollution monitoring platform (see Fig. 3) [1]. The HAPADS platform deploys the mobile low cost SPS30 PM sensor [18] and provides the dataset of PM measurements for evaluation in this paper. The dataset contains 765 air quality measurements by the HAPADS platform over one month period for an area in Gdansk, Poland. The dataset includes inaccurate measurements from the SPS30 PM sensor in the HAPADS platform and highly accurate measurements from a nearby modern weather station ARMAG. To form a regression problem, we used all measurement data from the sensor (e.g., PM, air pressure, humidity, temperature) in the HAPADS platform to predict the accurate PM2.5 measurements from the modern weather station ARMAG (i.e., the ground truth). During the development of the base model and NAS models, roughly 60% of the dataset was used as a training set while the remaining 40% was used for validation/testing. We used mean absolute error (MAE) and mean square error (MSE) for the loss and accuracy, respectively.

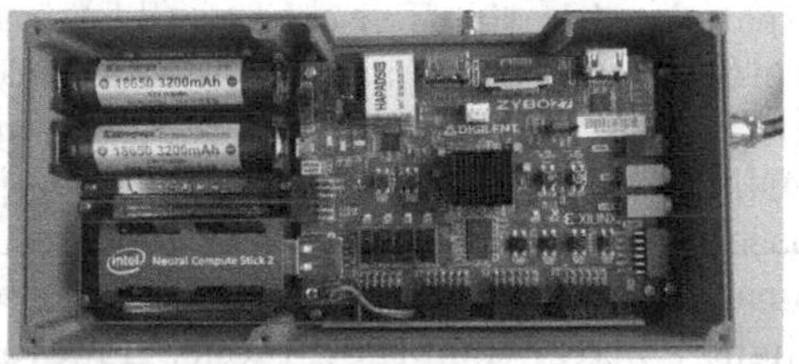

Fig. 3. HAPADS air pollution monitoring platform [1].

Table 1. Base model Mean Absolute Error (MAE) and Mean Square Error (MSE), where columns Avg and Std are the average and standard deviation.

	TensorFlow					PyTorch					Avg	Std
MAE	1.88	1.83	1.84	1.84	1.90	1.80	1.72	1.75	1.74	1.70	1.80	0.06
MSE	6.90	6.42	6.63	7.13	7.01	6.38	5.88	6.62	6.16	6.52	6.57	0.36

Table 2. Mean preprocessing time and mean model runtime on the test set, using an ONNX and OpenVINO IR version of the base model, where columns Avg and Std are the average and standard deviation.

	ONNX - PyTorch						OpenVINO IR - PyTorch						Avg	Std
	Warm Start			Cold Start			Warm Start			Cold Start				
Preprocessing (μs)	10.73	6.35	9.93	5.87	7.92	23.3	6.39	7.01	10.50	11.19	22.74	22.51	12.03	6.48
Model (ms)	1.78	1.75	1.75	1.70	1.85	2.00	1.76	1.78	1.80	1.74	1.86	1.83	1.80	0.07

3.1 Base Model

We trained the base model for 100 epochs on the train set using the Adam optimizer. We utilized a machine with Intel i7-7700 4-core CPU for training. For testing, we used the Intel Neural Compute Stick 2 (NCS2), an AI-accelerator for edge devices deployed in the HAPADS air pollution monitoring platform (see Fig. 3) [1].

Table 1 shows the test loss (MAE) and test accuracy (MSE) after ten runs of the base model. Among these runs, five runs were done using a model instance created using the TensorFlow (TF) framework while the remaining five runs were achieved using a model instance from the PyTorch (PT) framework. Model instances were trained from the bottom before testing and tested in both the ONNX and OpenVINO IR formats supported by NCS2. The results show that base model achieves relatively good results with a mean test MAE of 1.80 and mean test MSE of 6.57. The results also show that the PT implementation is slightly better than the TF one. However, the difference is quite small. Interestingly, there were no differences between running the models in the ONNX and OpenVINO IR formats.

Table 2 shows the mean latency of data pre-processing (in microseconds) and model (in millisecond) for each test sample. The test was done 12 times, 6 times using the OpenVINO IR model format and 6 times using the ONNX model format. In both cases, the model came from a PT instance. For both the OpenVINO IR and ONNX runs, 3 of the 6 runs were done from a cold state, namely from a nearly idle state (i.e., 3-min pause between each run). The other 3 runs came from a warm state, namely the underlying hardware and network had already processed 1000 samples unrelated to the test set before starting the test. There were also no pause between test runs while the model was in a warm state. The table shows that the average latency of data pre-processing is negligible to that of model (12.03 μs vs. 1.80 ms). Both the ONNX and OpenVINO IR implementations have similar average latency.

3.2 NAS Models

Setup. We evaluated NAS capabilities to find good architectures by exploring the model space (defined in Sect. 2.3) using Regularized Evolution (NAS-RE) and Random Search (NAS-RS) described in Sect. 2.4. During the evaluation, both search strategies found 200 architectures and they repeated 3 times. NAS-RE was run with a population size (model space size) of 100, a tournament size of 25 and mutation probability of 0.05 in a specific direction. NAS-RE could use the same configuration twice (i.e., deduplication set to false) while NAS-RS did not use a model configuration more than once (i.e., deduplication set to true) and ignored failed models, following the NNI framework recommendation [12]. The NAS processes were conducted using the NNI framework on a MacBook Pro with 2,2 GHz 4-core Intel Core i7 processor. The metrics used for the evaluation were the same as for the base model evaluation: MAE for loss and MSE for accuracy.

Results. Table 3 and Table 4 show the 10 best models after the 3 runs of NAS-RE and NAS-RS, respectively. The columns are divided into three main sections (Conv1D, Pool and MLP) and depict the choices of the 11 options discussed in Sect. 2.3 for each model, together with the model's final accuracy MSE on the test set.

The Conv1D section describes the convolution part of the convolution head. Columns "Num Layers", "Kernel Size", "Out Channels" and "Activation Function" refer to the numbers of convolution layers, the kernel size of each layer, the shared output channel size, and the shared activation function that follows each layer, respectively. For example, the model at the last row in Table 3 has 2 convolution layers. The first layer has a kernel size of 7 while the second has a kernel size of 3. Both layers have 8 output channels and use HardSigmoid as their activation function.

The Pool section describes the pooling part of the convolution head. In this section, columns "Kernel Size" and "Type" denote the filter size of the pooling window and the kind of pooling operation, respectively. For example, the model at the last row in Table 3 has an average pooling layer with the window size of 4.

Lastly, the MLP (Multilayer Perceptron) section describes the fully connected part of the models. Columns "Num Layers", "Num Neurons" and "Activation Function" refer to the number of MLP layers, the shared width of all layers and the shared activation function that follows each layer. For example, the model at the last row in Table 3 has a three MLP layers with a width of 32 and a GELU activation function.

The results show that the overall best model - in terms of accuracy - was a pure MLP network, found by NAS-RE (the top model in Table 3). The network has 18 layers of which each has 32 neurons and no activation function. Interestingly, the MSE of the network is only 0.17 better than that of the second best model, a combination of convolution and MLP networks also found by NAS-RE. Comparing the two search strategies, it is clear that NAS-RE achieves better

Table 3. Top 10 best models using NAS-RE.

Conv1D						Pool		MLP			Final MSE
Num Layers	Kernel Size			Out Channels	Activation Function	Kernel Size	Type	Num Neurons	Num Layers	Activation Function	
N/a	N/a	N/a	N/a	N/a	N/a	N/a	N/a	32	18	N/a	3.57
3	5	3	3	16	N/a	5	Max	8	3	Mish	3.74
N/a	N/a	N/a	N/a	N/a	N/a	N/a	N/a	512	2	Sigmoid	3.84
3	5	3	3	16	N/a	4	Max	8	3	Mish	3.92
2	7	3	N/a	8	Sigmoid	4	Avg	32	3	GELU	3.95
N/a	N/a	N/a	N/a	N/a	N/a	N/a	N/a	32	2	Sigmoid	3.97
2	7	7	N/a	8	HardTanh	5	Avg	512	10	LeakyReLU	3.99
2	7	3	N/a	8	Sigmoid	4	Avg	32	3	GELU	4.00
N/a	N/a	N/a	N/a	N/a	N/a	N/a	N/a	512	2	Sigmoid	4.01
2	7	3	N/a	8	HardSigmoid	4	Avg	32	3	GELU	4.02

Table 4. Top 10 best models using NAS-RS.

Conv1D						Pool		MLP			Final MSE
Num Layers	Kernel Size			Out Channels	Activation Function	Kernel Size	Type	Num Neurons	Num Layers	Activation Function	
N/a	N/a	N/a	N/a	N/a	N/a	N/a	N/a	8	1	HardSigmoid	3.94
3	7	7	7	8	HardTanh	5	Max	128	6	N/a	3.95
3	3	5	7	16	CELU	4	Max	512	4	SiLU	4.01
2	3	3	N/a	64	RReLU	4	Max	256	1	HardSigmoid	4.05
3	3	3	7	16	SiLU	5	Avg	256	8	N/a	4.10
2	7	7	N/a	16	HardSigmoid	4	Max	256	1	SiLU	3.97
2	7	7	N/a	8	HardTanh	5	Avg	512	10	LeakyReLU	4.2
N/a	N/a	N/a	N/a	N/a	N/a	N/a	N/a	8	8	N/a	4.23
N/a	N/a	N/a	N/a	N/a	N/a	N/a	N/a	8	14	N/a	4.26
3	7	5	5	16	N/a	4	Avg	8	3	ReLU	4.27

accuracy than NAS-RS. While NAS-RE achieved a top-5 MSE of 3.95, NAS-RS only achieved a top-1 MSE of 3.94 and a top-5 MSE of 4.10.

Figure 4 and Fig. 5 shows the PM measurements of the mobile low cost SPS30 PM sensor [18] (HAPADS) against the ground truth from the reference station (ARMAG) before and after corrected by our NAS-RE calibration method using the best model (i.e., top model in Table 3), respectively. The results show that our calibration methods can significantly improve the accuracy of the mobile low cost PM sensor. Table 5 shows the accuracy comparison of our new calibration methods (NAS-RE and NAS-RS) with state-of-the-art methods for mobile low-cost PM sensors.

3.3 Evaluation of NAS-RE and NAS-RS for Particulate Matter Monitoring Platform

To evaluate the viability of the new calibration models NAS-RE and NAS-RS in practice, we run the top NAS-RE and NAS-RS models on Intel Neural Compute Stick 2 (NCS2), an AI-accelerator for edge devices deployed in the HAPADS air pollution monitoring platform (see Fig. 3). Since HardSigmoid, the activation function used in the best model from NAS-RS, is not supported by the NCS2, we emulated HardSigmoid using HardTanh. Namely, we replaced HardSigmoid with HardTanh using a lower limit of -3 and upper limit of 3, then scaling the result by a factor of $1/6$ and adding 0.5 to it.

Table 6 and Table 7 show the test loss (MAE) and test accuracy (MSE) after 10 runs using the best NAS-RE model and best NAS-RS model, respectively. All model instances were trained from the bottom before testing and tested in

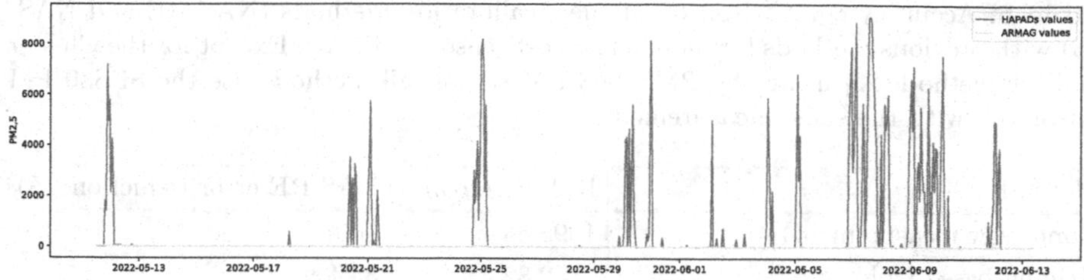

Fig. 4. The PM measurement of a low cost PM sensor (HAPADS) without correction and the ground truth from a reference station (ARMAG). Figure 5 zooms in on the ground truth.

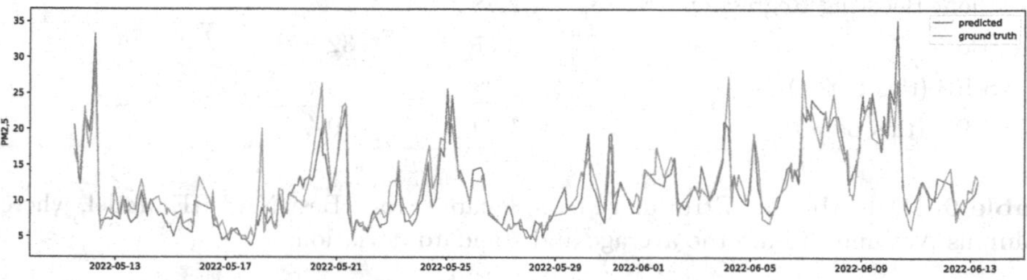

Fig. 5. The PM measurement of a low cost PM sensor (HAPADS) corrected by our NAS-RE method and the ground truth (ARMAG).

both the ONNX and OpenVINO IR formats. The tables show that the top-1 NAS-RE model actually performed worse than anticipated, having a mean MSE of 5.69. Meanwhile the top-1 NAS-RS model performed better than anticipated and reported a mean MSE of 3.90.

Tables 8 and 9 show the mean latency of data pre-processing (in microseconds) and model (in millisecond) for each test sample. The test was done 12 times, 6 times using the OpenVINO IR model format and 6 times using the ONNX model format. In both cases, the model came from a PyTorch instance. For both the OpenVINO IR and ONNX runs, 3 out of the 6 runs were done from a cold state, namely from a nearly idle state (3 min pause between each run). The other 3 runs came from a warm state, namely the underlying hardware and network had already processed 1000 samples unrelated to the test set before starting the test. There is also no pause between test runs while the model was in a warm state. The tables show that the average latency of data pre-processing is negligible to that of model (e.g., 10.62 µs vs. 1.74 ms in Table 9). Both the ONNX and OpenVINO IR implementations have similar model latency. The results show that the best NAS-RS model (with average model runtime of 1.74 ms) is faster than the best NAS-RE model (with average model runtime of 2.44 ms). Both models are fast enough to correcting PM measurements from the HAPADS platform where measurements occur every 10 min.

Table 5. Accuracy comparison of our new calibration methods (NAS-RE and NAS-RS) with previous methods for mobile low-cost sensors (MLCS). Except for the Cluster analysis method [22] using the PMS7003 PM sensor, all methods use the SPS30 PM sensor [18] with the same measurement.

Methods	RMSE ($\mu g/m^3$)	NAS-RE error reduction (%)
None (raw measurement)	1695.63	100%
Kriging-based [20]	2.82	33%
Cluster analysis [22]	8.48	78%
Linear regression	3.97	52%
Random forest [9]	2.55	26%
Gradient Boosting Regression Tree [8]	2.58	27%
AdaBoost [7]	2.77	32%
NAS-RS (this paper)	1.98	5%
NAS-RE (this paper)	1.89	0%

Table 6. Mean Absolute Error and Mean Square Error - Best NAS-RE Model, where columns Avg and Std are the average and standard deviation.

	OpenVINO/ONNX - PyTorch										Avg	Std
MAE	1.90	1.72	1.79	1.65	1.75	1.66	1.63	1.57	1.58	1.74	1.70	0.09
MSE	7.94	6.28	5.79	5.00	5.59	5.42	4.95	4.86	4.95	6.07	5.69	0.88

4 Discussion

Even though the top-1 NAS-RE model failed to bring satisfactory results on the Intel NSC2 (i.e., its high MSE of 5.69), it is clear that the best NAS-RS model succeeded and outperformed all tested models (i.e., its lowest MSE of 3.90). Why it is so effective is likely because of its simplicity. Having only 8 neurons, for instance, reduces the risk of overfitting, which is a problem with many deeper neural networks. Because of its shallowness, its HardSigmoid non-linearity could work optimally with less risk of running into the vanishing gradient problem.

In terms of latency, we see the same results as discussed above: the top-1 NAS-RE model is worse than the top-1 NAS-RS model that outperforms all models. Considering the total number of neurons in each model, this is not a surprising result. The baseline model has $3 * 64 = 192$ neurons in its hidden layers and the top-1 NAS-RE model has $32 * 18 = 576$ neurons in its hidden layers while the top-1 NAS-RS model only has 8 neurons in its hidden layer. Since fewer neurons imply fewer parameters and faster data throughput, it is clear that the model with the fewest neurons (i.e., NAS-RS) will be faster than all other models, at least when comparing MLPs.

Table 7. Mean Absolute Error and Mean Square Error - Best NAS-RS Model, where columns Avg and Std are the average and standard deviation.

	OpenVINO/ONNX - PyTorch										Avg	Std
MAE	1.35	1.49	1.42	1.48	1.39	1.35	1.49	1.43	1.36	1.41	1.42	0.05
MSE	3.69	4.09	3.85	4.07	3.96	3.66	4.19	3.91	3.82	3.85	3.90	0.16

Table 8. Mean preprocessing time and mean model runtime of the best NAS-RE model, where columns Avg and Std are the average and standard deviation.

	ONNX - PyTorch						OpenVINO IR - PyTorch						Avg	Std
	Warm Start			Cold Start			Warm Start			Cold Start				
Preprocessing (μs)	10.73	9.94	9.16	25.21	20.33	10.80	9.69	10.41	9.66	9.87	9.34	20.89	13.00	5.41
Model (ms)	2.27	2.30	2.33	2.52	2.76	2.41	2.28	2.29	2.69	2.54	2.60	2.36	2.44	0.16

Table 9. Mean preprocessing time and mean model runtime of the best NAS-RS model, where columns Avg and Std are the average and standard deviation.

	ONNX - PyTorch						OpenVINO IR - PyTorch						Avg	Std
	Warm Start			Cold Start			Warm Start			Cold Start				
Preprocessing (μs)	9.43	16.26	7.39	15.73	13.13	8.74	6.73	8.45	9.98	9.59	12.46	9.60	10.62	2.96
Model (ms)	1.72	1.76	1.71	1.73	1.74	1.91	1.74	1.81	1.63	1.78	1.75	1.68	1.74	0.06

5 Conclusion

We have introduced new accurate lightweight calibration methods for improving accuracy of measurements from mobile low-cost particulate matter sensors. The new methods utilize Neural Architecture Search (NAS) to improve the accuracy and efficiency of calibration models compared with the state-of-the-art methods. The experimental results show that the new calibration methods achieves the best root mean square error (RMSE) of 1.89, reducing accuracy error by more than 26% compared with the state-of-the-art methods. The new methods take less than 2.5 ms to correct an inaccurate measurement from a mobile low-cost PM sensors on the HAPADS air pollution monitoring platform, making the methods suitable for real-time calibration. The new accurate lightweight methods would contribute to resolving the accuracy challenge of mobile low-cost PM sensors, enabling them to be deployed in monitoring PM continuously and accurately in many locations.

References

1. HAPADS: Highly accurate and autonomous programmable platform for providing air pollution data services to drivers and the public. Accessed 21 Mar 2023
2. Bergstra, J., Bardenet, R., Bengio, Y., Kégl, B.: Algorithms for hyper-parameter optimization. In: Advances in Neural Information Processing Systems, vol. 24 (2011)

3. European Commission: Review of sensors for air quality monitoring, JRC technical report (2019)
4. Gressent, A., Malherbe, L., Colette, A., Rollin, H., Scimia, R.: Data fusion for air quality mapping using low-cost sensor observations: feasibility and added-value. Environ. Int. **143**, 105965 (2020)
5. He, K., Zhang, X., Ren, S., Sun, J.: Delving deep into rectifiers: surpassing human-level performance on imagenet classification (2015)
6. Intel: Intel neural compute stick 2 (intel NCS2). Accessed 26 Feb 2023
7. Scikit learn developers: An adaboost regressor. Accessed 26 Feb 2023
8. Scikit learn developers: Histogram-based gradient boosting regression tree. Accessed 26 Feb 2023
9. Lepioufle, J.-M., Marsteen, L., Johnsrud, M.: Error prediction of air quality at monitoring stations using random forest in a total error framework. Sensors **21**(6), 2160 (2021)
10. Liu, H., Simonyan, K., Yang, Y.: DARTS: Differentiable architecture search (2018)
11. Liu, H.-Y., Schneider, P., Haugen, R., Vogt, M.: Performance assessment of a low-cost PM2.5 sensor for a near four-month period in Oslo, Norway. Atmosphere **10**(2), 41 (2019)
12. Microsoft: Neural Network Intelligence (2021)
13. Méndez, M., Merayo, M.G., Núñez, M.: Machine learning algorithms to forecast air quality: a survey. Artif. Intell. Rev. **56**, 10031–10066 (2023)
14. World Health Organization: Air pollution data portal. Accessed 25 Feb 2023
15. Pham, H., Guan, M.Y., Zoph, B., Le, Q.V., Dean, J.: Efficient neural architecture search via parameter sharing (2018)
16. Real, E., Aggarwal, A., Huang, Y., Le, Q.V.: Regularized evolution for image classifier architecture search (2018)
17. Xie, S., et al.: Feasibility and acceptability of monitoring personal air pollution exposure with sensors for asthma self-management. Asthma Res. Pract. **7**(1), 13 (2021)
18. Sensirion: SPS30 particulate matter (PM) sensor. Accessed 26 Feb 2023
19. Simonyan, K., Zisserman, A.: Very deep convolutional networks for large-scale image recognition (2014)
20. Wojcikowski, M., et al.: A surrogate-assisted measurement correction method for accurate and low-cost monitoring of particulate matter pollutants. Measurement **200**, 111601 (2022)
21. Yang, L., Li, C., Tang, X.: The impact of PM2.5 on the host defense of respiratory system. Front. Cell Dev. Biol. **8**, 91 (2020)
22. Yun, J., Woo, J.: IoT-enabled particulate matter monitoring and forecasting method based on cluster analysis. IEEE Internet Things **8**(9), 7380–7393 (2021)
23. Zhang, Q., et al.: Retiarii: a deep learning exploratory-training framework. In: OSDI (2020)
24. Zoph, B., Le, Q.V.: Neural architecture search with reinforcement learning (2016)

Generating Music for Video Games with Real-Time Adaptation to Gameplay Pace

Marek Kopel(✉), Dawid Antczak, and Maciej Walczyński

Faculty of Information and Communication Technology, Wroclaw University of Science and Technology, wybrzeże Wyspiańskiego 27, 50-370 Wroclaw, Poland
marek.kopel@pwr.edu.pl

Abstract. The paper aimed to develop an automatic music generation method for video games that could create various types of music to enhance player immersion and experience, while also being a more cost-effective alternative to human-composed music. The issue of automatic music generation is an interdisciplinary research area that combines topics such as artificial intelligence, music theory, art history, sound engineering, signal processing, and psychology. As a result, the literature to review is vast. Musical compositions can be analyzed from various perspectives, and their applications are extensive, including films, games, and advertisements. Specific methods of music generation may perform better only in a narrow field and range, although we rarely encounter fully computer-generated music nowadays. The proposed algorithm, which utilizes RNN and 4 parameters to control the generation process, was implemented using PyTorch and real-time communication with the game was established using the WebSocket protocol. The algorithm was tested with 14 players who played four levels, each with different background music that was either composed or live-generated. The results showed that the generated music was enjoyed more by the players than the composed music. After implementing improvements from the first round of play tests, all of the generated music received better evaluations than the composed looped music in the second run.

Keywords: live music · machine learning · AI generated content

1 Introduction

The topic of automatic music generation is an interdisciplinary area of research that combines issues from fields such as artificial intelligence, music theory, art history, sound engineering, signal processing, and psychology.

Musical compositions can be analyzed from many perspectives, and their applications are very broad, including, for example, films, games, and advertisements. Different methods of music generation may work better in only a narrow field and range, although fully computer-generated music is still rare today.

N. T. Nguyen et al. (Eds.): ACIIDS 2023, LNAI 13995, pp. 261–272, 2023.
https://doi.org/10.1007/978-981-99-5834-4_21

Representation refers to the way in which music is described and determines the form of presentation of sounds and musical structures so that they are understandable for algorithms processing and generating music. The choice of representation and its encoding is closely related to the configuration of the input and output of the architecture used, i.e., the number of input and output variables and their corresponding types. Music has a hierarchy and structure, so it can be analyzed at different levels of abstraction, each of which has its own methods of representing music.

2 Related Works

Automatic music generation is not a new idea, however, the field is a bit of a niche [5, 7]. The description of solutions dedicated merely for video games follows.

MetaCompose. [14] is an extensible music generator used for creating real-time music for board games. It consists of three main components, including a composition generator, an affective composer, and an archive of previous compositions.

The composition generator includes three steps: creating chord sequences, melodies, and accompaniments. The chord sequence is created by randomly traversing a graph of typical chord changes. The authors use known features and elements of music that should be considered and avoided in melodies. These are based on guidelines for classical music composition and musical practice. Appropriate matching and penalty functions are constructed for genetic algorithms based on this. The solution combines FI-2POP and multi-objective optimization.

The affective composer transforms the composition in real-time according to a specific mood or affective state. The system is designed to respond to events in the game. It can change the affective state simply, change the current composition, or force a completely new composition. Before a player makes a move, the system determines how good their situation is on the board, how many move options they have, and how significant those moves are. Based on this, pleasure (valence) and arousal values are calculated.

Research with participants clearly indicated that all elements effectively combine to create harmonious, pleasant, and interesting compositions. Differences in player experiences with static and dynamic affective music were investigated, and based on the collected research data, players preferred the latter.

Mezzo. [3] is a computer program that procedurally generates music in real-time, inspired by the Romantic era, for video games. An interesting solution introduced is the main themes assigned to characters and other important game elements. When the appropriate characters, objects, or events appear on the screen, the system smoothly combines their themes and modifies them, taking into account the current state of the game, such as the player's health. Game elements must have assigned main themes of any length, but the suggested length is 2 to 4 bars. Another necessary input data is a sample chord progression. Based on these, new fragments will be created, stylistically similar to them, using a genetic algorithm. This allows for achieving a specific composer's style. The next necessary data is a small set (about 10) of chord progressions, from which new progressions are constructed in real-time using genetic algorithms.

Bardo Composer. [6] is a system that generates music for tabletop role-playing games. It recognizes words spoken by players, processes them, and generates music that reflects the current situation in the game. The system generates music in real-time based on the emotions conveyed by the players' words, as well as the genre and mood of the game. It can also recognize specific phrases or names to trigger a particular melody or musical phrase.

The system uses a combination of machine learning and rule-based methods to generate music. It recognizes the emotional content of words spoken by players using a pre-trained sentiment analysis model. Based on this analysis, it selects appropriate musical elements and combines them to create a composition. The system can generate music in a variety of styles, including classical, jazz, and electronic.

3 Requirements Analysis

In the case of gaming solutions, the music produced must not only be of high quality but also fit the specific game and the current situation within it. Each of the presented gaming solutions has options for controlling the generation process. The most commonly used approach is based on the pleasure and arousal values known from Russell's circumplex model [13] and the dependence on these characteristics of the music. The proposed solution must, to some extent, provide the possibility of controlling the next generated fragment by providing an interface to the game.

The use of deep learning techniques for music generation is rapidly developing, but embedding control and interactivity in these systems still remains a critical challenge [2]. Neural networks were not designed to enable easy control, but Meade et al. [11] and Oore et al. [12] have presented examples of music generation processes using recurrent neural networks with the use of conditioning signals.

3.1 The Model

Recurrent neural networks (RNNs) are widely used for modeling sequential data, including music. They are an extension of feed-forward networks, that can handle variable-length input sequences. The concept of RNN is simple: it accepts an input vector x and produces an output vector y. However, the output vector is influenced not only by the input data but also by the previous inputs stored as a recurrent hidden state of the network. Given a sequence $x = (x_1, ..., x_T)$, the hidden state of the network h_t is defined as 1:

$$h_t = \begin{cases} 0, & t = 0 \\ \phi(h_{h-1}, x_t), & t \neq 0 \end{cases} \quad (1)$$

where ϕ is the activation function.

The probability of a sequence $p(x_1, ..., x_T)$ can be expressed as 2:

$$p(x_1, ..., x_T) = p(x_1)p(x_2|x_1)p(x_3|x_1, x_2) \cdots p(x_T|x_1, ..., x_{T-1}) \quad (2)$$

However, the classical RNN faces a serious limitation known as the vanishing or exploding gradient problem [1]. The solution to this problem is variants of RNN: long short-term memory (LSTM) [8] and gated recurrent unit (GRU) [4], which is a simplified version of LSTM, where the gate mechanism allows controlling the transfer of information from older time steps to newer ones.

The solution described below is based on EmotionBox [15] using the GRU model. EmotionBox is used to generate emotionally charged music without requiring any emotionally labeled datasets. Low-level attributes of music, such as note density and pitch-class histogram, are used to determine arousal and valence of the music, respectively. These values are mapped onto emotions from the Russell's circumplex model. The authors were able to achieve satisfactory results compared to the method using labeled data, although they acknowledged that representing pleasure solely with key is not accurate enough and other elements of music should also be considered.

3.2 Event Encoding

The model takes polyphonic MIDI files as input, which are encoded as proposed by Oore et al. [12]. However, velocity-related events (VELOCITY) were omitted due to their small variance in the dataset. The final event dictionary consists of:

- 88 NOTE_ON(i) events encoding frequency values from the MIDI range of 21–108; they indicate the start of a sound with a certain frequency
- 88 NOTE_OFF(i) events encoding frequency values from the MIDI range of 21–108; they indicate the end of a sound with a certain frequency
- 32 TIME_SHIFT(i) events encoding time shifts

Successive TIME_SHIFT(i) events cause a time shift by a value in seconds given by the formula 3:

$$\frac{1.15^i}{65}, \; where \; i \in [0, 31] \quad (3)$$

which gives quantized values from about 15 ms to about 1.35 s. The data transformation process tries to approximate the original tempo as accurately as possible.

3.3 Conditioning Parameters

As previously mentioned, a vector c is added to each input vector x during the training stage. This vector provides metadata about the event and can be used by the user during music generation to control its features. The vector c has a length of 15, and each position encodes a manually extracted feature as follows:

- [1, 3] key (3 values)
- [4, 9] beat density (6 values)
- [10, 12] average number of simultaneously played notes (3 values)
- [13, 15] frequency entropy of notes (3 values)

Each of these parameters is computed during data processing for each event. The data for the event and the events following it in a time window are used for this purpose. Then, each value is encoded using one-hot encoding, and the resulting vectors are concatenated.

The key (as defined in 3.4) takes values from the range [0, 2]. However, the remaining parameters must first be grouped (discretized) using binning. For each parameter, appropriate intervals must be created so that the number of events that match them in the training set is similar. This requires prior analysis of processed data.

During generation, conditioning parameters should not be mandatory because the user may not require control over a specific feature. In this case, a way to encode the lack of a parameter must be defined. One approach is to extend the vector by adding 1 extra position, which takes the value of 1 in this case, as implemented in *EmotionBox* [15]. However, Meade et al. [11] note that using a uniform distribution vector $v = (\frac{1}{n}, \frac{1}{n}, ..., \frac{1}{n})$ also yields good results, with additional advantages. Therefore, this approach was chosen.

3.4 The Key

The key parameter was proposed to address the problem of controlling the generated music on the pleasure axis. Its purpose is to categorize songs into 3 groups and takes the values: 0 for major key, 1 for minor key, and 2 for unknown. It is similar to a pitch histogram [15], but provides less control during generation. The assumption is to allow the user not to enter the target pitch distribution but only to indicate whether they are interested in major, minor, or undefined key. For training purposes, music fragments are classified into one of these groups.

For each song fragment, an attempt can be made to determine its key. This is not a trivial task, but algorithms have been developed for this purpose. The most popular is the Krumhansl-Kessler algorithm [10], which is based on key profiles—a 12-value vector that shows the ideal distribution of notes in a given key. To create an analogous profile for C♯ key, it is necessary to shift the chart to the right. Similarly, to change the pitch by a specific number of semitones, it is necessary to shift the chart by the corresponding number of positions.

4 The Method Description

The proposed method for automatic music generation uses RNN with customizable input parameters. The train data are crucial for the quality of the RNN output. Two sources are used to train the network: **VGMusic**[1] from 31763 MIDI files are used and **ADL Piano MIDI**[2] from which 954 MIDI files tagged as soundtrack are added to the training set. The latter source contained only keyboard MIDI tracks, but the first one needed to be filtered only to those files where, after dropping all non-keyboard tracks, there were actually some tracks left. Then to create the final version of the training set all track were divided to max 30 s fragment, dropping those shorter than 15 s and those containing pauses longer than 5 s. Additionally copies of those fragments sped up or slowed down by 5% were also added to the set.

In the next phase the training set fragments were transposed to the same key (C major or A minor), first finding their original key with Krumhansl-Kessler algorithm implementation *music21*[3]. And finally the MIDI fragments were encoded as sequences of events *NOTE_ON* and *NOTE_OFF* with pitch values filtered to the range [21,108] (piano keyboard range) and *TIME_SHIFT* along with the conditioning parameters as described in Sect. 3.3.

Having the training set prepared, the training can begin. The process used backpropagation of the ADAM [9] optimizer with a learning rate of 0.001. The loss function was the cross entropy between generated event and one found in the data. The process used randomly chosen sequences of 200 events from the training set. The model batch size was 64. The training process was stopped when the loss function could not get a lower value for several dozen minutes. The state with minimum loss function value has been saved.

Live music generation is implemented using *PyTorch* with the trained model and *WebSocket* protocol for synchronous communication with the game. The game can send JSON file with the following input parameters:

- ***mode***—optional vector of length 3 for key (major, minor, unknown), default: normalized vector of the uniform distribution
- ***attackDensity***— optional vector of length 6 for non-polyphonic note density, default: normalized vector of the uniform distribution
- ***avgPitchesPlayed***—optional vector of length 3 for average of simultaneous notes, default: normalized vector of the uniform distribution
- ***entropy***—optional vector of length 3 for entropy, default: normalized vector of the uniform distribution
- ***temperature***—optional value τ for exploration-exploitation compromise, $\tau < 1$ give more expected and $\tau > 1$ - more unexpected events, default: 1

[1] https://www.vgmusic.com/.
[2] https://github.com/lucasnfe/adl-piano-midi.
[3] http://extras.humdrum.org/man/keycor/.

- ***requestedTimeLength***—optional number of seconds of requested music, default: 5 s
- ***reset***—Boolean value for resetting the hidden layers before the generation process

In response a MIDI file with a piece of generated music is received.

4.1 Test Application Game

The method *Python* implementation is used by the *Unity* tower defence type game, similar to Plants vs Zombie. The game is quite simple enough even for first time player and do not include any drastic gameplay changes - everything happens on one screen with static camera. The only element that should affect different aspects of generated music are waves of enemies. With each new wave the gameplay becomes more dynamic and after defeating it the lazy pace returns. Then to this the music can adapt. *Unity* library *MusicInterface* is used for the communication with *Python* application with requests sent every 3 s. Each request contains current game play state. The state is described by four basic metrics calculated from number of current enemies, defences used, level timeout and frequency of mouse inputs: *danger*, *attack power*, *intensity* and *chaos*. When sent to the *PyTorch* model, the metrics are mapped to the input parameters and that is how the music adaptation happens.

5 Experiment Setup

In order to test the method 3 research problems where formulated:

- Is dynamically generated music as a background for gameplay is better evaluated by players than the pre-composed music?
- Is a simple solution as turning up the music tempo and volume affecting players' experience in a positive way?
- Do players positively evaluate live changes for adapting the music to the gameplay pace?

The playtests are blind tests, meaning the testers only know they will evaluate music, but no information about the music preparation process was revealed. For the tests seven levels have been prepared, as shown in Table 1. The first three levels are introductory. During the tutorial levels - the game mechanics, the goal and the win conditions are explained to the player. During the third level testers can try playing the game themselves and ask questions, while the test coordinator can give them tips on improving their play performance. During the first three levels no background music is playing.

The next four levels are the actual play test levels. Each player is supposed to beat each level. In case of losing, the current level is restarted. Beating each level takes about 3.5 min. After beating each level the tester is ask to evaluate last level background music on scale 1–10 before continuing to the next level. During those last four levels music is chosen from the following:

Table 1. Level types used in the playtest experiment.

Code	Type	Description
LVL_001	tutorial	Introduction to simple game mechanics
LVL_002	tutorial	Introduction to advanced game mechanics
LVL_003	warm-up	Players can try the mechanics themselves
LVL_101	regular	Low difficulty level
LVL_102	regular	Low difficulty level
LVL_103	regular	Medium difficulty level
LVL_104	regular	Medium difficulty level

- SM—static composition, music without tempo or volume changes
- DM—dynamic composition, music with tempo and volume changes overtime
- G—music generated in real time, no changes in input parameters
- GI—music generated in real time, with changes in input parameters adapting to the game state/pace
- RT—29 s of precomposed music *Ravio's Theme* from *The Legend of Zelda: A Link Between Worlds*
- ZP—33 s of precomposed music *Zombie Panic* from *Zombies Ate My Neighbors*

Again, these are blind tests, so the track selection is random and testers don't know which music is playing. The last 2 elements from the list: RT and ZP are the only precomposed pieces and had been chosen to fit the generated music in the aspect of being a single keyboard track and their cut fragments can loop properly. Having two of those allow to use one in the original form (SM) and the other with the modifications (DM) in the latter level without revealing to the player with one is modified by adapting to the gameplay.

The precomposed SM levels use looped RT or ZP with original tempo ($s = 1$) and 75% of original volume ($v = 0,75$). The corresponding DM levels have tempo 4 and volume 5 calculated for each moment t based on $level\_progress = \frac{t}{t_{max}}$:

$$s_t = 0,8 + \frac{2}{5} \cdot level\_progress \tag{4}$$

$$v_t = 0,5 + \frac{1}{2} \cdot level\_progress \tag{5}$$

So along with the level progress the music tempo will grow from 0.8 to 1.2 of the original value and the volume - from 0.5 to 1.

6 Results

The first play test run involved 6 players. Their results are shown in the first 6 rows of Table 2. The evaluations for generated music are lower than precomposed music, especially the customized GI music. But the first run tester remarks allowed to improve the algorithm setup before the second run.

The main disadvantage of the generated music in the first run was its chaotic manner. The game is seen by the players as calm and relaxing, which didn't get along with dynamic changes in music and broke the immersion. Another problem were the music changes in effect of the gameplay change. Despite the interpolation technique the music changes were too violent.

Table 2. Evaluations of background music by testers 1–6 in first test run and 7–14 in 2nd run with the corresponding arithmetic averages in the last 2 rows. For column meaning see previous section.

Tester no.	SM RT	DM ZP	SM ZP	DM RT	G	GI
1	3	4	–	–	6	5
2	5	6	–	–	7	5
3	8	9	–	–	8	7
4	–	–	9	5	1	2
5	–	–	7	6	6	5
6	–	–	5	6	3	4
7	–	–	9	7	7	8
8	–	–	6	5	7	7
9	7	6	–	–	3	6
10	5	6	–	–	5	7
11	–	–	6	7	8	4
12	–	–	4	3	5	6
13	6	4	–	–	5	7
14	5	8	–	–	5	6
μ for 1–6 (1nd run)	5,33	6,33	7,00	5,67	5,17	4,67
μ for 7–14 (2nd run)	5,75	6,00	6,25	5,50	5,63	6,50

To fix the problems the temperature τ was reduced from 1.5 in the first run to 0.7 in the second run. This made the generated music more predictable. Also encoding of the input parameters has been changed from discreet groups to continuous probability distribution.

The second run of the play tests 8 players took part (rows 7–14 in Table 2). The evaluations from both runs show that different tester operate in different spectrum of the scale. Some use the upper range for the evaluations, while other evaluated around the mean of the scale.

The average evaluations of the second run are for the generated music G are comparable with the evaluations of precomposed music RT and ZP. The average evaluation of generated music adapting to the gameplay pace GI was the highest in the second run, after implementing the corrections discovered in the first run. This gives the answer to the first research problem, stated at the beginning of Sect. 5. For the second problem the answer to whether the customized DM music is liked better for the players than original precomposed SM music is ambiguous. The third problem focuses on whether players appreciate the music being adapted to the pace of the game play which is shown the difference between G and GI evaluations. Again, after the correction before the second run, the answer is in favour of the continuously changing GI music.

7 Conclusions

The main goal for the paper was to create a method for automatic music generation for video games. The method is supposed to create different types of music that would be enjoyed by players. It is supposed to enforce immersion and make the overall experience better. It is also supposed to be a cheaper alternative to the music composed by human.

The goal has been achieved and, as the experiment results show, the proposed algorithm generates music that on average is more enjoyed by test players than the music composed by humans for other games. The algorithm uses RNN and 4 parameters to control the generation process. The network model is implemented with *PyTorch*, a *Python* library.

The real time communication between the game and the algorithm is implemented using *WebSocket* protocol. The client application is implemented with *.NET* technology and integrated with previously developed *Unity* game. This way the pace of gameplay affects the music being generated in real time.

The game has been used in two runs of play tests with total 14 test players. Each player had to be beat 4 levels, each with different background music, both composed and live generated. Afterwards each player evaluated the music at each level. After implementing improvements discovered in the first run of play tests, in the second run, all of the generated music had better evaluations than the composed looped music.

7.1 Aspects to Consider in Future Research

Before the actual experiment game levels had to be created. To properly set the difficulty of those levels some pre-experiment play tests had to be run. The play testers used for this purpose could not have been used in the experiment. Of course the play testers are of different skill, so the same difficulty level may be experienced differently by them. And for those who find the game too hard, it may affect their evaluation of music. In the end the game was made simpler with many mechanics dropped. This allowed to shorten the play test time in both the explanation and trial game and the actual game play. Eventually the tutorial part took 10 min and the actual test - 15 min. However a free conversation with a tester after the test often turned out to be much more informative than and formal evaluation. Because of that conversations the time for dealing with each tester could grow up to 60 min.

The playtests were performed in person with each tester. This limitation was enforced by the fact, that RNN model was run locally, and it was too complex for testers to setup the test environment by themselves. But probably a cloud version of the client app and a browser version of the game could solve the problem of testing "in person" and allow for tele-testing. On the other hand a specialized methods, e.g. gaze tracking, face emotion analysis, electrodermal activity - that are planned in further research - would still require to organize the playtest in a specialized laboratory.

Most of the testers didn't notice the music was changing to fit the pace of the gameplay. Only afterward, when the mechanism was explained to them, they appreciated the idea. Still the only difference in the music they could tell was its tempo. They usually didn't notice the major-minor scale switching.

Since the difficulty level was relatively low (simplified game) the music generated - with input parameters reflecting the pace of the game, i.e. GI - was slow and sparse in terms of notes per bar. This may be the main cause of GI outcome being evaluated better than G music (static input parameters): the slow GI music is just more pleasant and better suits this type of game as a background music. Maybe if G input parameters were set to generate slow and sparse music, then there would not be any difference in the evaluations of G and GI.

Though most of the testers preferred slow music and calm gameplay, some were really excited and played in a chaotic way. In response the algorithms would generate GI music as more diverse in tempo and note density. The chaotic play was only determined by speed of mouse movement and density of mouse clicks. In the future research it is planned to use, among others, facial expression analysis to tell testers excitement, frustration or any emotion that can cause player chaotic behaviour during the game.

The quality of GI music is mostly dependent on the input parameter changes. The first test run gave much worse evaluations, because the parameters had been changing too often, which in effect generated music too violently dynamic. So, to reiterate, the way of changing of input parameter is crucial for the GI music quality.

Eventually, the evaluations are heavily affected by testers musical tastes. In some cases, for the same piece of music testers evaluations could have been all across the scale. This shows that for any kind of custom music the first criteria that should be taken into account to make listeners enjoy the music is their musical preferences. And this probably should be another input parameter for the music generation algorithm.

References

1. Bengio, Y., Simard, P., Frasconi, P.: Learning long-term dependencies with gradient descent is difficult. IEEE Trans. Neural Netw. **5**(2), 157–166 (1994)
2. Briot, J.P., Hadjeres, G., Pachet, F.D.: Deep learning techniques for music generation-a survey. arXiv preprint arXiv:1709.01620 (2017)
3. Brown, D.: Mezzo: an adaptive, real-time composition program for game soundtracks. In: Proceedings of the AAAI Conference on Artificial Intelligence and Interactive Digital Entertainment, vol. 8, pp. 68–72 (2012)

4. Cho, K., Van Merriënboer, B., Bahdanau, D., Bengio, Y.: On the properties of neural machine translation: Encoder-decoder approaches. arXiv preprint arXiv:1409.1259 (2014)
5. Colombo, F., Seeholzer, A., Gerstner, W.: Deep artificial composer: a creative neural network model for automated melody generation. In: Correia, J., Ciesielski, V., Liapis, A. (eds.) EvoMUSART 2017. LNCS, vol. 10198, pp. 81–96. Springer, Cham (2017). https://doi.org/10.1007/978-3-319-55750-2_6
6. Ferreira, L., Lelis, L., Whitehead, J.: Computer-generated music for tabletop role-playing games. In: Proceedings of the AAAI Conference on Artificial Intelligence and Interactive Digital Entertainment, vol. 16, pp. 59–65 (2020)
7. Gunawan, A., Iman, A., Suhartono, D.: Automatic music generator using recurrent neural network. Int. J. Comput. Intell. Syst. **13**, 645 (2020). https://doi.org/10.2991/ijcis.d.200519.001
8. Hochreiter, S., Schmidhuber, J.: Long short-term memory. Neural Comput. **9**(8), 1735–1780 (1997)
9. Kingma, D.P., Ba, J.: Adam: A method for stochastic optimization. arXiv preprint arXiv:1412.6980 (2014)
10. Krumhansl, C.L.: Cognitive Foundations of Musical Pitch. Oxford University Press, Oxford (2001)
11. Meade, N., Barreyre, N., Lowe, S.C., Oore, S.: Exploring conditioning for generative music systems with human-interpretable controls. arXiv preprint arXiv:1907.04352 (2019)
12. Oore, S., Simon, I., Dieleman, S., Eck, D., Simonyan, K.: This time with feeling: learning expressive musical performance. Neural Comput. Appl. **32**(4), 955–967 (2020)
13. Russell, J.A.: A circumplex model of affect. J. Pers. Soc. Psychol. **39**(6), 1161 (1980)
14. Scirea, M., Eklund, P., Togelius, J., Risi, S.: Evolving in-game mood-expressive music with metacompose. In: Proceedings of the Audio Mostly 2018 on Sound in Immersion and Emotion, pp. 1–8 (2018)
15. Zheng, K., et al.: Emotionbox: a music-element-driven emotional music generation system using recurrent neural network. arXiv preprint arXiv:2112.08561 (2021)

Detecting Sensitive Data with GANs and Fully Convolutional Networks

Marcin Korytkowski[1,2], Jakub Nowak[1], and Rafał Scherer[1,2](✉)

[1] Czestochowa University of Technology, al. Armii Krajowej 36, Czestochowa, Poland
{marcin.korytkowski,jakub.nowak}@pcz.pl
[2] Intigo Ltd., Haryana, India
rafal.scherer@pcz.pl
http://intigo.ai/

Abstract. The article presents a method of document anonymization using generative adversarial neural networks. Unlike other anonymization methods, in the presented work, the anonymization concerns sensitive data in the form of images placed in text documents. Specifically, it is based on the CycleGAN idea and uses the U-Net model as a generator. To train the model we built a dataset with text documents with embedded real-life images, and medical images. The method is characterized by a very high efficiency, which enables the detection of 99.8% of areas where the sensitive image is located.

1 Introduction

Nowadays, many entities and private persons want to protect their data against leakage. It can be information about both health and company secrets, e.g. research works. The subject of the processing of sensitive data is also extremely important in the context of EU regulations, e.g. Directive 95/46/EC of the European Parliament and of the Council of 24 October 1995 on the protection of individuals with regard to the processing of personal data and on the free movement of such data and the criminal and financial liability of persons creating and processing such collections of information. It is also worth noting that the theft of sensitive data may be used to assess the health condition of politicians or other decision-makers. In this paper, we present a system for detecting documents containing sensitive data, also in the cases where they are intentionally hidden there.

For obvious reasons, the classification of data into one of two classes: with and without sensitive data must be automatic. In a situation where nowadays even small entities process gigabytes of information daily, a human is not able

The work was supported by The National Centre for Research and Development (NCBR), the project no POIR.01.01.01-00-1431/19.
The project financed under the program of the Polish Minister of Science and Higher Education under the name "Regional Initiative of Excellence" in the years 2019–2023 project number 020/RID/2018/19 the amount of financing PLN 12,000,000.

N. T. Nguyen et al. (Eds.): ACIIDS 2023, LNAI 13995, pp. 273–283, 2023.
https://doi.org/10.1007/978-981-99-5834-4_22

to manually verify the content of processed files. In this article, we propose a solution that fully automates this process based on machine learning techniques. The task facing the system is to detect and remove sensitive data from the documents being processed. It is based on the CycleGAN idea and uses the U-Net fully convolutional neural model as a generator. To train the model we built a dataset with text documents with embedded real-life images, and medical images.

The rest of the paper is organized as follows. In Sect. 2 we describe shortly models used. Our method for is described in Sect. 3. Section 4 presents the results of experiments on the dataset of sensitive documents we created and Sect. 6 concludes the paper.

2 Related Work

Currently, anonymization with the use of neural networks is used primarily to detect specific phrases in the text [11,13]. Unlike the text, we will not analyze the context of the text, but the actual information contained in a visual form of images. The detection of sensitive data such as the human face has been very well developed, among others, thanks to the huge amount of data available on social network channels [1,2]. In the studies cited, very good results were achieved with the use of convolutional neural nets [16]. Simirarly, well structured data are well processed by various neural networks, even anomalies in data can be easily detected [5,6]. Our work is aimed at detecting the places of occurrence of sensitive data, such as: results of magnetic resonance imaging, X-rays, etc. However, in the case of this type of data, we usually face the problem of small amounts of data available for training. One of the ways to improve the operation of classification algorithms is by generating synthetic data on the basis of the available set [12]. In our research, we tackle the presented problem differently. We want a generative adversarial neural network (GAN) [7] to be able to distinguish between sensitive data itself. To put it simply, the GAN is supposed to generate images without sensitive data. The general GAN diagram is presented in Fig. 1 and of the proposed model in Fig. 2.

The presented solution is based on the CycleGAN network model [17]. In its original application, the network was designed to convert graphic images. Among other things, they trained the model to convert horse images to zebra images, and city landscapes at night to city landscapes by day. The great advantage of CycleGAN is that this model can be trained without paired examples, i.e. it does not require sample photos before and after conversion to train the model. For example, it is not necessary to provide the same image of a horse turned into a zebra. The model architecture consists of two generator models: one generator (GeneratorA) for generating images for the first class (ClassA) and a second generator (GeneratorB) for generating images for the second class (ClassB):

GeneratorA → ClassA (documents requiring anonymization),

GeneratorB → ClassB (documents that do not require anonymization).

Generator models perform image translation, which means that the image conversion process depends on the input image, particularly an image from another

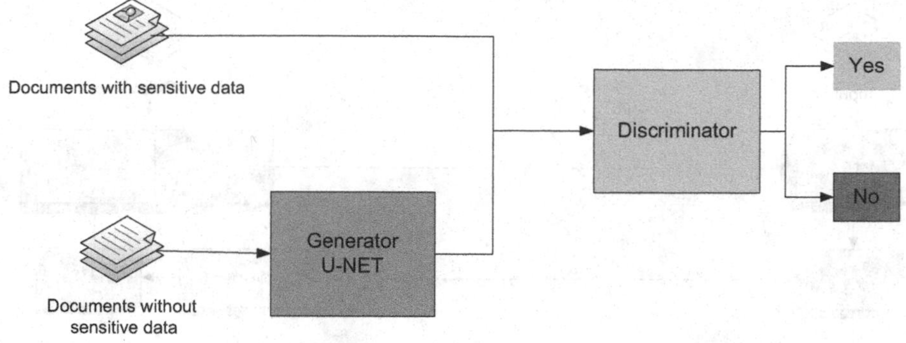

Fig. 1. General diagram of the GAN network.

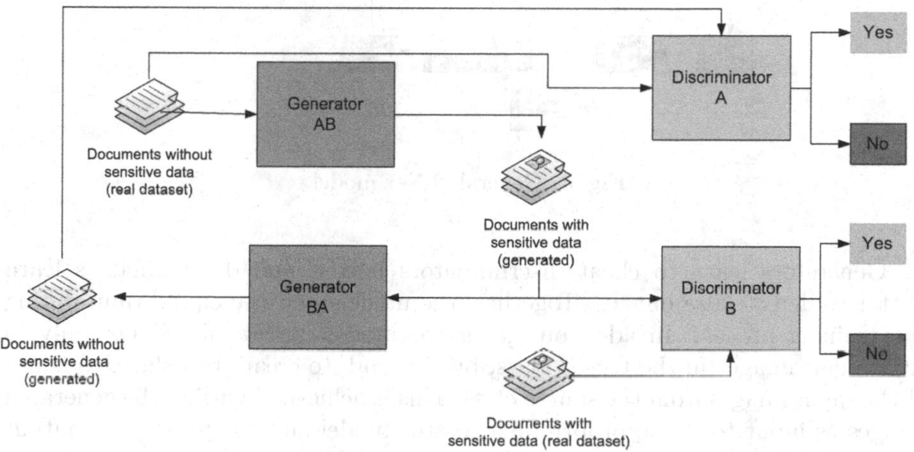

Fig. 2. General diagram of the proposed model: GAN network and the network used to build the training sets.

domain. Generator-A takes an image from ClassB as the input and ClassB takes an image from ClassA as the input:

ClassB → GeneratorA → ClassA,

ClassA → GeneratorB → ClassB.

Each generator has its own dedicated discriminator model. The first discriminator model (DiscriminatorA) takes the true images from ClassA and the generated images from GeneratorA and predicts whether they are true or false. The second discriminator model (DiscriminatorB) takes the true images from ClassB and the generated images from GeneratorB and predicts whether they are true or false. The discriminator and generator models are trained in an adversarial zero-sum process, much like normal GAN models. Our solution uses also the U-Net model [14] as a generator, which was initially used, inter alia, to detect neoplasms in medical images. The architecture stems from the so-called "fully convolutional network" [10] (Fig. 3).

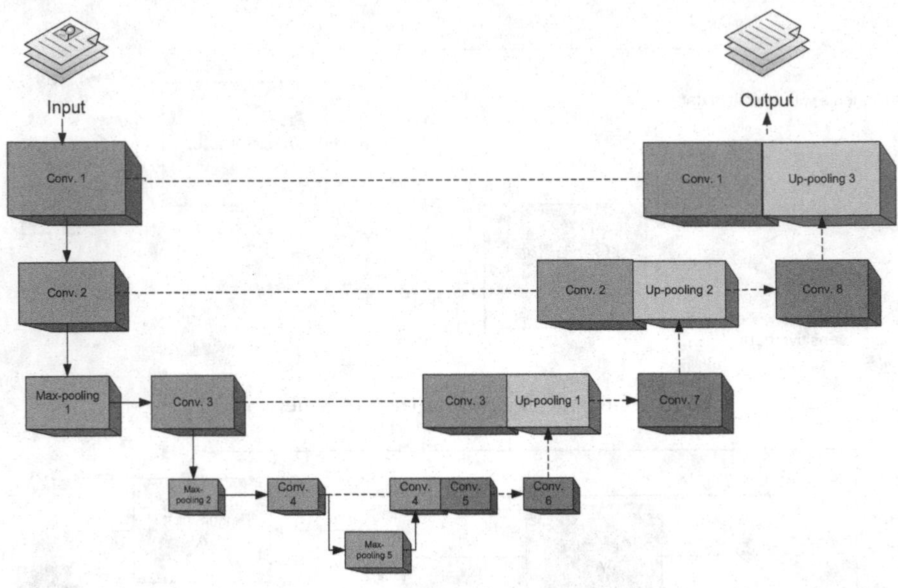

Fig. 3. General U-Net model.

Generators learn to cheat discriminators better, and discriminators learn better to detect false images. Together, the models find the equilibrium during the training process. In addition, generator models are regulated not only to create new images in the target class, but instead, to create translated versions of the input images from the source class. This is achieved by using the generated images as input to the appropriate generator model and comparing the output image with the original images. The transmission of the image by both generators is called a cycle. Together, each pair of generator models is trained to better recreate the original source image, which is referred to as cycle consistency.

3 Training the Anonymisation Model

In the presented model, the role of the generator is performed by a modified U-Net network inspired by the work [14]. In our structure, compared to the original concept, the activation functions have been changed from ReLU to SELU [9]. This change contributed to reducing the model's learning time from approximately 21.5 h to 19. However, in the conducted research, no significant impact on improving the learning outcome was observed. The U-Net model has also been downsized compared to the original by removing convolutional layers within the same dimension. In the original work, the creators used two convolutional layers after each max-pooling process, while in our solution, we limited it to one such operation. The reason for this approach was to create a network that processes input signals as quickly as possible, which directly translates to minimizing computational requirements.

The training sequence was built on the basis of text documents with the embedded medical data. Now we will describe the input data. The training dataset for the GAN requires two classes of objects — that require anonymisation and do not require anonymization. The diagram of the network is shown in Fig. 2. The non-anonymised part contained only a grayscale image obtained from a text document (e.g. WORD, PDF) along with random images from the ImageNET dataset [4] inserted in random places. The part requiring anonymisation was additionally provided with data in the form of X-ray images and computed tomography. The image was saved as a JPEG file with the size of 1024 × 1024 pixels. The chosen image size was dictated by the compromise between the model accuracy and processing speed.

Table 1. Experimental results for various input image sizes.

Input image size	Batch processing time For 100 images [seconds]	Image recognition result
256 × 256	8.3	79.8%
400 × 400	12.1	97.3%
1024 × 1024	21.6	99.8%
2048 × 2048	57.2	99.8%

Analyzing the results collected in Table 1, it can be concluded that satisfactory image recognition performance is already achieved at the image size of 400 × 400 px. Based on the obtained results, it is evident that to achieve the best outcomes, a network capable of analyzing images at a resolution of 1024 × 1024 px should be employed. Increasing the size twice to 2048 × 2048 px does not improve the already high classification accuracy significantly but increases the computation time of the system.

The method of generating training and testing images is presented in Fig. 4. In this case, the input signal is the text appropriately converted into bitmap maps derived from textual documents. During the generation of samples, consecutive textual documents are provided, which are converted into jpg images along with medical images from a medical database or the ImageNet dataset. When training the GAN network, three types of documents are distinguished: documents containing data in the form of images combined with medical photos (recognized as sensitive images), documents combined with images from the ImageNet database, or documents without any images, saved as files that do not contain sensitive data.

The portion of documents that did not required anonymisation did not contain images from the ImageNET database. For any document, the size of the image could not be greater than 80% of the width and height, and less than 20%. The inserted images were additionally scaled by a random factor.

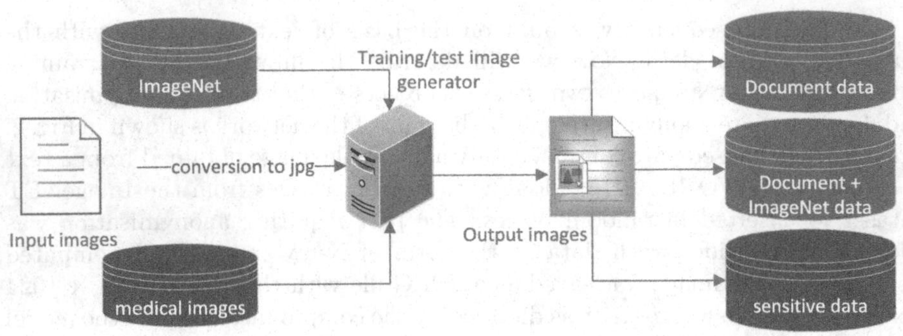

Fig. 4. Generating training and testing images.

The output of the GAN is also a grayscale image with the size of 1024 × 1024 pixels. In the GAN model, the discriminator as the evaluating part of the generator is also important.

In the conducted research, synthetic data requiring anonymisation were inserted in random places. The network in the learning process retrieved document scans without inserted sensitive images as ClassA. As ClassB, the network received documents with sensitive (e.g. medical) images.

On the basis of the conducted research, the network, apart from detecting the document with the newly generated sensitive image, was able to remove the sensitive image from text documents without losing such elements as a stamp or a barcode. Examples of such objects are presented in Fig. 5.

Fig. 5. Examples of objects placed in documents.

To prepare input data with sensitive medical objects we used our own X-ray images and images taken from the datasets described in [3,8] and [15]. Eventually, we created a dataset which composition is presented in Table 2.

Table 2. Dataset composition.

Type	Number of files	Size in MB (2048×2048)	Size in MB (1024×1024)	Size in MB (400×400)	Size in MB (256×256)
Number of files without images	20100	38592	4769	1923	588
Number of files with images from ImageNet	15000	29214	3632	1611	454
Number of files with sensitive medical images	15000	29112	3219	1022	410

4 Results

Through the operation of the GAN structure and the proprietary solution of inserting medical photos into the content of various text files, the first class of documents (containing sensitive data) was defined, which consisted of 30,000 graphic files. The second class of documents was created by artificially generating graphic files on the basis of the public ImageNET database (a total of 30,000 randomly selected files) of files.

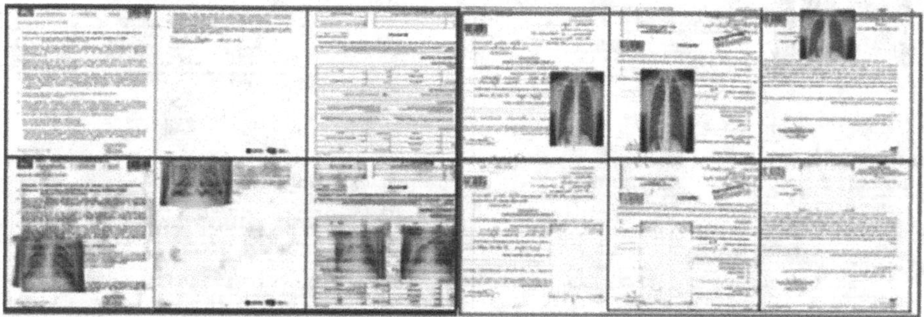

Fig. 6. Example images with sensitive objects inserted for training and removed by the model.

The first discriminator model (Fig. 2) (DiscriminatorA) takes the true images from ClassA and the generated images from GeneratorA and predicts whether they are true or false. The second discriminator model (DiscriminatorB) takes the true images from ClassB and the generated images from GeneratorB and predicts whether they are true or false. The discriminator and generator models are trained in an avdersarial zero-sum process, much like the standard GAN models. Figure 6 in the red box shows the effect of generating data by the network during the learning process. An extremely interesting feature of the trained GAN network, as described above, is that it can be used to remove sensitive medical data (a kind of graphical anonymisation). The effect of this structure used for this purpose is shown in Fig. 6 in a green frame. The input data is on the first line, and the net result (output) is on the bottom line. The percentage of correctly selected sensitive data was 99.8%. The percentage of incorrectly selected insensitive data was 29.3%. That was calculated based on the number of similar pixels.

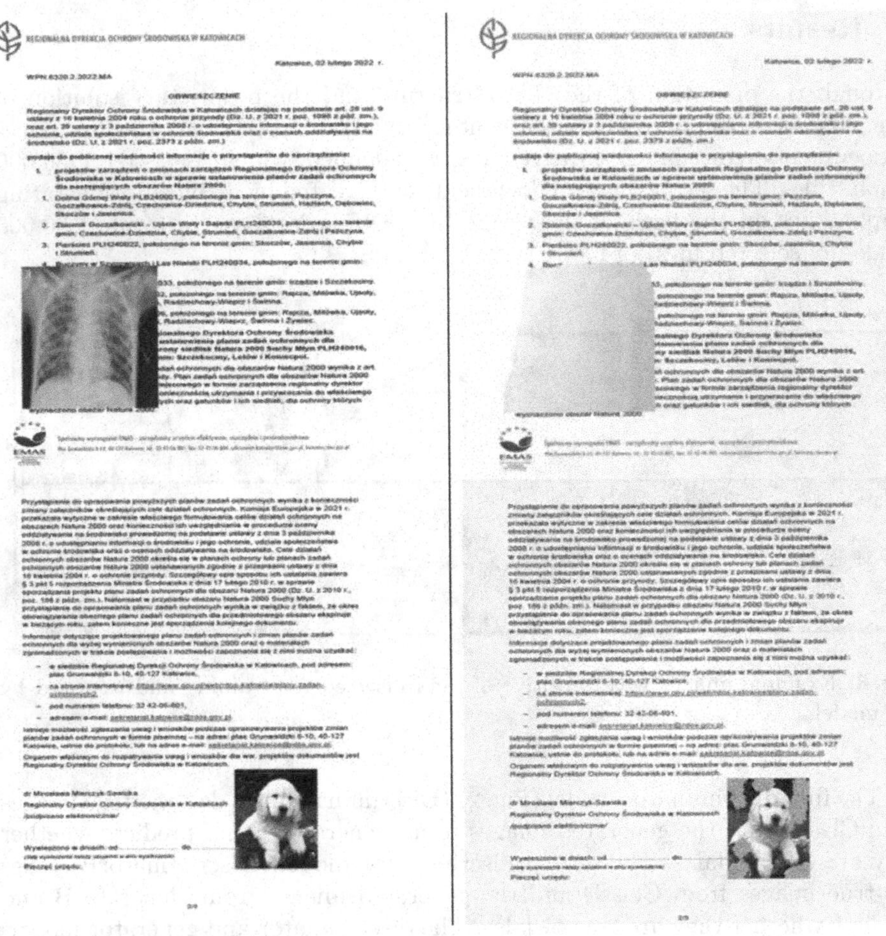

Fig. 7. Example input and output documents with sensitive and non-sensitive objects. We can observe the effect of anonimisation by the proposed system.

5 Automatic Sensitive Data Detection System

The previously described neural network model has practical applications in a document anonymization system. The primary task here is to protect client resources by verifying the presence of sensitive information within file resources in the context of GDPR or sensitive information for the company in terms of protecting confidential information.

It should be noted that the system only utilizes a portion of the trained CycleGAN model responsible for generating anonymized images. The remaining part of the network was implemented solely for the purpose of training the system. One practical challenge in the system's operation is the transmission of documents containing sensitive data. Therefore, several practical solutions have been devised to mitigate this risk. The concepts of such systems are presented in Fig. 8.

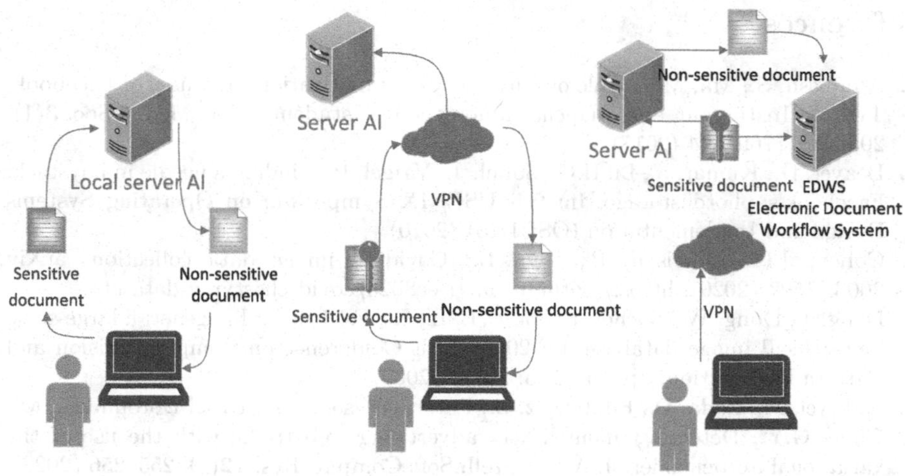

Fig. 8. Application of generative adversarial networks in anonymization systems.

The first solution pertains to a scenario where both the user's computer and the AI server with the GAN network operate within the same local network. In such cases, it is possible to directly transmit the documents to the AI server. In the second scenario, when the client cannot provide the necessary infrastructure for document processing, a remote AI server will act as the client's service. Communication with the user will take place through a dedicated VPN tunnel, and every sensitive document will be automatically encrypted. In this solution, there is no need for re-encrypting the anonymized documents. The last possible solution is when the client utilizes a document circulation system. In such cases, communication will occur between the system and the AI server. Depending on the infrastructure involved, an additional VPN connection between the servers providing the mentioned services or a local connection may be required.

6 Conclusions

We proposed a method to anonymise documents using generative adversarial neural networks and fully convolutional networks. The anonymisation concerns sensitive data in the form of images placed in text documents. It is based on the CycleGAN idea and uses the U-Net model as a generator. To train the model we built our own dataset with real-life MS Word and PDF text documents with embedded real-life images, and medical images. The method is characterized by a very high efficiency, which enables the detection of 99.8% of areas where the sensitive image is located. The method is able to remove the detected sensitive objects what is shown in Fig. 7.

References

1. Alhabash, S., Ma, M.: A tale of four platforms: motivations and uses of Facebook, Twitter, Instagram, and Snapchat among college students? Soc. Med.+ Soc. **3**(1), 2056305117691544 (2017)
2. Beaver, D., Kumar, S., Li, H.C., Sobel, J., Vajgel, P.: Finding a needle in haystack: Facebook's photo storage. In: 9th USENIX Symposium on Operating Systems Design and Implementation (OSDI 10) (2010)
3. Cohen, J.P., Morrison, P., Dao, L.: Covid-19 image data collection. arXiv 2003.11597 (2020). https://github.com/ieee8023/covid-chestxray-dataset
4. Deng, J., Dong, W., Socher, R., Li, L.J., Li, K., Fei-Fei, L.: Imagenet: a large-scale hierarchical image database. In: 2009 IEEE Conference on Computer Vision and Pattern Recognition, pp. 248–255. IEEE (2009)
5. Gabryel, M., Lada, D., Filutowicz, Z., Patora-Wysocka, Z., Kisiel-Dorohinicki, M., Chen, G.Y.: Detecting anomalies in advertising web traffic with the use of the variational autoencoder. J. Artif. Intell. Soft Comput. Res. **12**(4), 255–256 (2022). https://doi.org/10.2478/jaiscr-2022-0017
6. Gabryel, M., Scherer, M.M., Sulkowski, L., Damaševičius, R.: Decision making support system for managing advertisers by ad fraud detection. J. Artif. Intell. Soft Comput. Res. **11**(4), 331–339 (2021). https://doi.org/10.2478/jaiscr-2021-0020
7. Goodfellow, I., et al.: Generative adversarial nets. In: Ghahramani, Z., Welling, M., Cortes, C., Lawrence, N., Weinberger, K. (eds.) Advances in Neural Information Processing Systems, vol. 27. Curran Associates, Inc. (2014) https://proceedings.neurips.cc/paper/2014/file/5ca3e9b122f61f8f06494c97b1afccf3-Paper.pdf
8. Kermany, D.S., et al.: Identifying medical diagnoses and treatable diseases by image-based deep learning. Cell **172**(5), 1122–1131 (2018)
9. Klambauer, G., Unterthiner, T., Mayr, A., Hochreiter, S.: Self-normalizing neural networks. In: Advances in Neural Information Processing Systems, vol. 30 (2017)
10. Long, J., Shelhamer, E., Darrell, T.: Fully convolutional networks for semantic segmentation. In: Proceedings of the IEEE Conference on Computer Vision and Pattern Recognition, pp. 3431–3440 (2015)
11. Mosallanezhad, A., Beigi, G., Liu, H.: Deep reinforcement learning-based text anonymization against private-attribute inference. In: Proceedings of the 2019 Conference on Empirical Methods in Natural Language Processing and the 9th International Joint Conference on Natural Language Processing (EMNLP-IJCNLP), pp. 2360–2369. Association for Computational Linguistics, Hong Kong, China (2019). https://doi.org/10.18653/v1/D19-1240https://aclanthology.org/D19-1240
12. Röglin, J., Ziegeler, K., Kube, J., König, F., Hermann, K.G., Ortmann, S.: Improving classification results on a small medical dataset using a GAN; an outlook for dealing with rare disease datasets. Front. Comput. Sci., 102 (2022)
13. Romanov, A., Kurtukova, A., Shelupanov, A., Fedotova, A., Goncharov, V.: Authorship identification of a Russian-language text using support vector machine and deep neural networks. Future Internet **13**(1), 3 (2020)
14. Ronneberger, O., Fischer, P., Brox, T.: U-Net: convolutional networks for biomedical image segmentation. In: Navab, N., Hornegger, J., Wells, W.M., Frangi, A.F. (eds.) MICCAI 2015. LNCS, vol. 9351, pp. 234–241. Springer, Cham (2015). https://doi.org/10.1007/978-3-319-24574-4_28
15. Tschandl, P., Rosendahl, C., Kittler, H.: The ham10000 dataset, a large collection of multi-source dermatoscopic images of common pigmented skin lesions. Sci. Data **5**(1), 1–9 (2018)

16. Yamashita, R., Nishio, M., Do, R.K.G., Togashi, K.: Convolutional neural networks: an overview and application in radiology. Insights Imaging **9**(4), 611–629 (2018)
17. Zhu, J.Y., Park, T., Isola, P., Efros, A.A.: Unpaired image-to-image translation using cycle-consistent adversarial networks. In: Proceedings of the IEEE International Conference on Computer Vision, pp. 2223–2232 (2017)

An Unsupervised Deep Learning Framework for Anomaly Detection

Che-Wei Kuo and Josh Jia-Ching Ying(✉)

Department of Management Information Systems, National Chung Hsing University, Taichung, Taiwan, Republic of China

jashying@gmail.com

Abstract. In recent years, with the evolution of technology and hardware, people can per-form anomaly detection on machines by collecting immediate time series data, thereby realizing the vision of an unmanned chemical factory. However, the data is often collected from multiple sensors, and multivariate time series anomaly detection is a difficult and complex problem because of the different scales and the unclear interaction of each feature. In addition, there usually exist noises in the data, and those make it difficult to predict the trend of the data. Moreover, practically, it's hard to collect abnormal data, thus the imbalance is an important issue. Recently, with the rapid development of data science, unsupervised methods based on deep learning manner have gradually dominated the field of multivariate time series anomaly detection. In this paper, we propose a 3D-causal Temporal Convolutional Network based framework, namely *TCN3DPredictor*, to detect anomaly signals from sensors data. Our proposed *TCN3DPredictor* modifies multi-scale convolutional recurrent encoder-decoder by 3D-causal Temporal Convolutional Network which can learn the interaction and temporal correlation between features and even predict the next data. Based on the results of 3D-causal Temporal Convolutional Network, a new breed of statistical method is proposed in our proposed *TCN3DPredictor* to measure the anomaly score precisely. Through a series of experiments using dataset crawled from a computer numerical control (CNC) metal cutting machine tool in a precision machinery factory, we have validated the proposed *TCN3DPredictor* and shown that it has excellent effectiveness compared with state-of-the-art anomaly prediction methods under various conditions.

Keywords: Deep learning · Anomaly detection · Temporal convolution network

1 Introduction

Recently, big data analytic has swept the whole world and triggered the Industrial Revolution 4.0 [3]. More and more people use the data to train the models and get used to making decisions by it. For example, in recent years, more and

N. T. Nguyen et al. (Eds.): ACIIDS 2023, LNAI 13995, pp. 284–295, 2023.
https://doi.org/10.1007/978-981-99-5834-4_23

more marketing media have used AI to conduct precise marketing to customers to improve the adaptability of advertising information. However, these models aren't always reliable since we have no anomaly detection mechanism to check whether our data are normal or not and even use it to stop immediately when anomalies happen. Moreover, it might cause a huge crisis. For instance, the Taiwanese train crashed because of the speed control disabled, and it caused 18 people dead[1] Therefore, it's necessary that we need to develop an anomaly detection mechanism.

Before building the anomaly detection mechanism, we need to analyze what the anomaly detection is and what kinds of anomalies there are. Actually, anomaly detection can be seen as a binary clustering problem, which targets to identify the normal and abnormal data. In addition, anomalies can be grouped by time-based and non-time-based, which is corresponding to time series and non-time series data respectively. Moreover, there are specifically two types of time-based anomalies, point anomaly and sequence anomaly [2]. After that, we are going to focus on time sequence anomalies, and to bring out three of the most common challenges we should deal with. First of all, time plays a significant role for the past and future data, for example, Fig. 1 shows that there are two time series data, and the anomaly situation happened when the irregular waves appear. It causes the traditional clustering methods with-out considering the time relation to get ineffective, like the k-Nearest Neighbor (kNN) [6]. Second, practically, our data is usually collected from many sensors, hence it's a multivariate problem. Besides, there are possible interactive relationships between the features. Apart from the interaction, in practice, the data often contains lots of noises. The above factors all affect the performance for the most prominent time series techniques, like the Auto-Regressive Integrated Moving Average (ARIMA) [17], and the Long-Short Term Memory Networks (LSTM) [11]. Therefore, it's important that our proposed framework must consider the relationship between features and lower the influence of noises. Finally, the access of imbalanced data often happens because of the difficulty to obtain abnormal data in reality. Thus, it is not appropriate to use the supervised learning method.

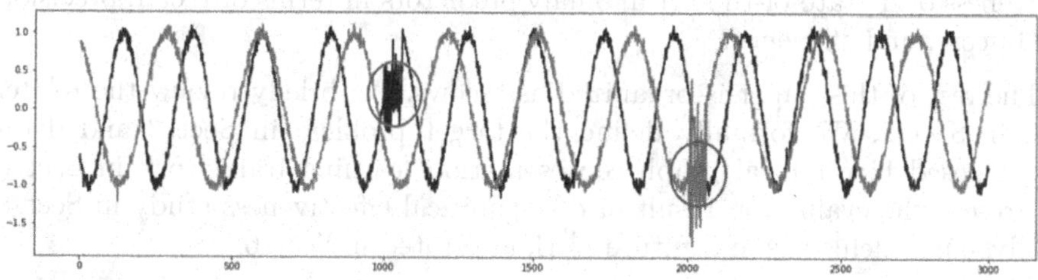

Fig. 1. Illustration of time series anomalies.

[1] BBC: Taiwan train crash driver disabled speed controls https://www.bbc.com/news/world-asia-45951475.

To jointly consider the aforementioned issues, Zheng et al. proposed a new idea, Multi-Scale Convolutional Recurrent Encoder-Decoder (MSCRED) [18] to diagnose the root causes of the anomalies. Specifically, it splits the problem into three parts. Initially, it calculates the signature matrix at each timestamp to save the correlations between features. Next, use the ConvLSTM model [14] to jointly learn the time and features relationship, furthermore to reconstruct the matrix at each timestamp. Finally, we develop an evaluation mechanism to calculate anomaly scores and set the threshold to determine whether the data is anomaly or not. However, there are three problems with the MSCRED method. First, using the reconstruction strategy can't detect the anomaly instantly. The second is the LSTM-based models are not sensitive to the drastic changes in data. And the final is that the evaluation mechanism of MSCRED is unfair to the learning ability of different features. Therefore, we propose a new breed of TCN-based anomaly prediction framework, *TCN3DPredictor*, with the prediction approach to deal with the problems from the original method. Meanwhile, we address the evaluation mechanism issue as well to deal with the unfairness for improving learning ability.

The contributions of our research are four-fold:

- We propose a new breed of TCN-based anomaly prediction framework, namely *TCN3DPredictor*, for predicting anomaly of a CNC machine. Meanwhile, our proposed *TCN3DPredictor* not only can deal with lack of anomaly signal issue but also utilize a suitable evaluation mechanism to produce a fair anomaly score.
- In order to detect the anomaly instantly and sense the drastic changes in data, we proposed a new variant of a 3D-causal Temporal Convolutional Network which can learn the characteristics of Multi-scale time series so that the anomaly can be detected instantly while the drastic changes occurs.
- To produce a precise anomaly score, we propose a statistical method to deal with the unfairness for improving learning ability.
- We use both synthetic data and real data crawled from a computer numerical control (CNC) metal cutting machine tool in the precision machinery factory[2] to evaluate the effectiveness of our proposal. The results show superior effectiveness over state-of-the-art anomaly predictors in terms of recall, precision, F1-score, and efficiency.

The rest of this paper is organized as follows. We briefly review the related work in Sect. 2. We formally define the target problem in Sect. 3 and detail the proposed hierarchical graph representation learning framework in Sect. 4. We present the evaluation result of our empirical effectiveness study in Sect. 5. Finally, our conclusions and future work are stated in Sect. 6.

2 Related Work

Typically, the deep-learning-based anomaly detection can be formalized as three sub-problems – features interaction, temporal relationship, and anomaly score

[2] https://www.goodwaycnc.com/exhtml_goodway/goodway_en/index.htm.

measuring, and dealt with them in order. First of all, the signature matrix was used to catch the interaction between features. Second, the ConvLSTM model [14] with the autoencoder method were utilized to jointly learn the temporal and the features relationship simultaneously and even reconstruct the signature matrix. And finally, the anomaly score was measured by calculating how many entries between the original and reconstructed signature matrix are out of tolerance. Zheng et al. proposed a new idea, Multi-Scale Convolutional Recurrent Encoder-Decoder (MSCRED) [18], to jointly consider the temporal and features relationship, and also proposed a new method to deal with the noise in the data. Liang et al. proposed the multi-time scale deep convolutional generative adversarial network (MTS-DCGAN) method [10], which attempted the deep convolutional generative adversarial network (DCGAN) method [13] to learn the relationship and reconstruct the signature matrix for the second sub-problem. And also, Zhao et al. proposed the multi-layer convolutional recurrent autoencoded anomaly detector (MCRAAD) method [19], using the statistical method to deal with the issue of how to determine the tolerance value in the third sub-problem. Nevertheless, there are two problems with all of the above methods. First, for anomaly score measuring, only using one value as the tolerance error for all the entries is unfair for the learning abilities of different features. Second, it's late to catch the anomaly at the first time for reconstruction strategy. In order to improve the difference of learning ability of features, Tayeh et al. proposed the unsupervised Attention-Based Convolutional Long Short-Term Memory (ConvLSTM) Autoencoder with Dynamic Thresholding (ACLAE-DT) architecture [16], which uses statistical methods to set the thresholds for each element of the residual matrix that comes from comparing the original and reconstructed signature matrix, and uses them to detect anomalies. However, the strategy for setting the thresholds for each element in the residual matrix separately must cause excessive memory usage as the number of features increases.

3 Problem Definition

Given the multivariate time series data X with m features and length t, i.e. $X = [X_1, X_2, ..., X_m]^T \in \mathbb{R}^{m \times t}$ and we assume that the anomaly data is not included in the given multivariate time series data. Our goal is to determine whether the data collected in the future is anomaly as soon as possible. Before determining whether the data collected in the future is anomaly, the the correlation coefficient [5,15] multivariate time series would be produced first. Specifically, we take the tth data and the last w data before the tth data as the multivariate time sequence at timestamp t. i.e. the multivariate time sequence with length $w+1$ is formulated as follows:

$$\begin{gathered} x_1^{t-w}, x_1^{t-w+1}, \ldots, x_1^{t-1}, x_1^t \\ x_2^{t-w}, x_2^{t-w+1}, \ldots, x_2^{t-1}, x_2^t \\ \vdots \\ x_m^{t-w}, x_m^{t-w+1}, \ldots, x_m^{t-1}, x_m^t \end{gathered} \tag{1}$$

And the signature matrix M^t at t is defined by

$$m_{i,j}^t = \frac{\sum_{k=t-\omega}^{t} x_i x_j}{\omega + 1}, \tag{2}$$

where $m_{i,j} \in M^t$ represents the relationship between feature i and j.

Actually, we can take three distinct lengths as the short-term, middle-term, and long-term time relationship respectively to represent the different time correlation. In this work, we take (30, 60, 90) as the three channels to represent the time relationship. Thus, for each timestamp, we have three signature matrices.

4 Our Proposed *TCN3DPredictor* Framework

In this section, we will introduce our proposed framework, *TCN3DPredictor*, which can be divided into two parts. The first part the 3D-causal Temporal Convolutional Network, which is to jointly learn the time and features relationship, and even to predict the next time stamp's matrix for determining whether the next stamp's data is anomaly or not immediately. Meanwhile, the time and features relationship would be formulated as Signature Matrix, which is to capture the correlation of features. The second part is the evaluation mechanism, which is to set a standard to determine anomaly. Specifically, we are going to calculate the anomaly score to measure how large the anomaly is and make a threshold as the standard to make the decision.

4.1 3D-Causal Temporal Convolutional Network

According to Bai et al.'s work [1], when forecasting time series, the result of taking time as a kind of feature to perform convolution operations outperform the result of the LSTM model. In addition, he also proposed a new time series forecasting method - the Temporal Convolutional Network (TCN), which was referenced to the WaveNet [12], taking causal convolution to avoid using data unknown at the current time point, and using the idea of dilated convolution to expand the view of the input data. However, the original TCN model can only be used for single-dimensional time series data, not the matrix time series mentioned above. Fortunately, Guo et al. [4] improved the original TCN model and proposed the 3D-Causal temporal convolution, which effectively solved the problem that was only applicable to single-dimensional time series. Based on the concept of Guo et al.'s 3D-causal convolution [4], we propose a new TCN3D model which can deal with multiple-dimensional time series.

Given a signature matrix time series $M = [M^0, M^1, \ldots M^{t-1}]$ of length t, $M \in \mathbb{R}^{c \times t \times n \times n}$, where n is the number of the features, our goal is to forecast the next signature matrix M^t. Specifically, we use a three-dimensional filter to convolute it. In detail, the first dimension of the filter would be convolute with time, and the other dimensions are convolute with the matrix, as shown in the Fig. 2.

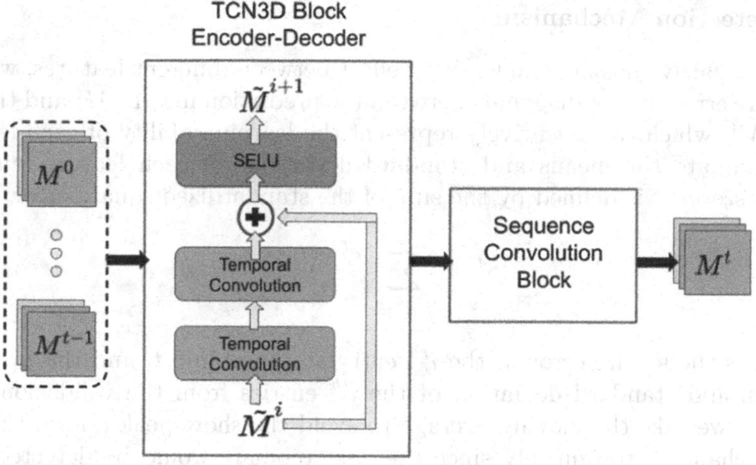

Fig. 2. Illustration of our proposed 3D-Causal temporal convolution (TCN3D).

In the process of temporal convolution, in order to expand the visual field utilization of the input data, our dilation and padding strategies for the ith layer are 2^i and $(filter\_size - 1) \times 2^i$ respectively, and in order to avoid overfitting in this process, we would drop out the output. Finally, in order to avoid too much information loss in the temporal convolution process, we use a three-dimensional filter with a size of 1 for residual block [7]. Formally, each TCN3D block can be formulated as follows:

$$\tilde{M}^0 = M, \tag{3}$$

$$f_{out}^i = \sigma(f^i(\tilde{M}^i)), \tag{4}$$

$$F_{out}^i = \sigma(F^i(f_{out}^i)), \tag{5}$$

$$\tilde{M}^{i+1} = \sigma(F_{out}^i + residual(\tilde{M}^i)), \tag{6}$$

where $\tilde{M}^i$ is the feature map after i^{th} TCN3D block, f^i, F^i are the first and second temporal convolution operation on i^{th} layer, and σ is the activation function, we take the scaled exponential linear units (SELU) [9] as our activation function here because of the properties [8] of self-normalization and convergent effectiveness when the conditions are almost satisfied. In this paper, we stacked 4 layers of TCN3D blocks as the encoder, and also stacked 4 layers of TCN3D blocks as the decoder. Finally, after the TCN3D model, we take the feature map at different times as different channels, and use a three-dimensional filter with a size of 1 to aggregate them as the final output, i.e. the prediction signature matrix $\hat{M}^t$.

In order to recognize the anomaly immediately, we compare the prediction matrix $\hat{M}^t$ with M^t. Thus, we take the mean square error (MSE) as the loss function to measure the learning effect. In this paper, for each timestamp, we take the signature matrices for 20 timestamps ahead as input data, and our goal is to predict the signature matrices for the next timestamp.

4.2 Detection Mechanism

In order to fairly consider the learning effect between different features, we take the square errors of the diagonals between the prediction matrix $\hat{M}^t$ and the real matrix M^t, which can Intuitively represent the learning ability of each feature. Then, compute the means and standard deviations of each feature, and our anomaly score S^t is defined by the sum of the standardized square error, i.e.

$$S^t = \sum_{i \in diag.} \frac{e_i^t - \mu_i}{\sigma_i} \tag{7}$$

where e_i^t is the square error of the ith entry at timestamp t, and the μ_i, σ_i are the mean and standard deviation of the ith entries from the validation data. Moreover, we take the moving average to avoid the short peak due to the fluctuations changing too quickly since the true anomaly would be detected for a while. Additionally, to catch the anomaly information sensitively, we take the short-term signature matrix, i.e. the matrix with sequence length 30, to calculate the anomaly score. And the threshold τ is set by the maximum of anomaly scores in the validation data multiplied by *beta*, where $\beta \in [1, 2]$, i.e. the threshold τ is little higher than the anomaly score under normal conditions.

$$\tau = \beta \times \underset{t \in validationset}{\text{argmax}} (S^t), \beta \in [1, 2] \tag{8}$$

5 Experiments

In this section, we conduct our experiment by comparing the predicted data with the real data. And our task is to detect anomalies within the first 15 timestamps when the anomalies appeal. Moreover, our experiments are all conducted in Python 3.8.10, using PyTorch 1.9.0 on the machine with GeForce GTX 1080 11 GB, Intel(R) Core(TM) i7-8700 CPU @ 3.20 GHz, and 64 GB RAM.

5.1 Dataset

And in this experiment, we used a synthetic dataset and a real dataset.

Synthetic Dataset. Synthetic dataset is generated by several sine functions with amplitude 1 and noise scale λ, concretely the formula is defined by

$$g_i(t) = sin(\frac{t - t_i}{\omega_i}) + \lambda\epsilon \tag{9}$$

where g_i denotes the generator function of the ith feature, t_i, ω_i denote the initial phase and frequency respectively, and $\epsilon \sim N(0, 1)$. In this work, we randomly generator 5 time series data by $t_i \in [-120, 120]$, $\omega_i \in [30, 50]$ and $\lambda = 0.05$, and each data includes 20000 time point. Then, we split it into three datasets, the first

10000 timestamps are training data, from 10001 to 14000 are the validation data, and the others are testing data. To generally test the ability to detect anomalies with different features, we make 10 versions of testing data by randomly injecting about 10 anomalies with duration belonging to three scales, i.e. [30, 60, 90], and anomaly scale 0.8.

Real Dataset. The real dataset was obtained from a computer numerical control (CNC) metal cutting machine tool in a precision machinery factory in Taichung Science and Technology Park. In this experiment, according to the advice from experts, we collected data such as load, current, temperature, and rotational speed of the machine in the process of metal cutting, and used it to detect whether there is any abnormality in the tool.

5.2 Baseline Methods

Our proposal is externally compared with the following methods, which can be grouped by two categories, the base deep learning type and the variant of MSCRED framework type. Besides, we make internal comparisons as well.

Basic Deep Learning Models: These methods learn how to predict the next data directly, and the strategy for anomaly score is to take the mean square error as the anomaly score. We use the Long Short Term Memory Networks based Encoder-Decoder (LSTM-ED) [11] model and the Temporal Convolution Network (TCN) [1] model.

Variants of MSCRED Framework: These methods are extensions of the MSCRED architecture, and in this classification, in addition to the original Multi-Scale Convolutional Recurrent Encoder-Decoder (MSCRED) method [18], we also use three other methods. The multi-time scale deep convolutional generative adversarial network (MTS-DCGAN) method [10], which uses the DCGAN as the prediction model. The multi-layer convolutional recurrent autoencoded anomaly detector (MCRAAD) method [19], which uses statistical methods to calculate its tolerance value problem. And the Autoencoder with Dynamic Thresholding (ACLAE-DT) method [16], which considers all elements together and sets threshold values for each element respectively.

5.3 Evaluation Results

To compare the performance under different lengths for moving averages (MA), we take three different lengths, including 1, 5, and 10 time point(s). We use three metrics, Recall, Precision, and F1-score to evaluate the performance of anomaly detection for each method. The results are reported in the Table 1, where the best scores are highlighted in bold-face.

Table 1. Results of anomaly detection for each method.

Method	Synthetic data			Real data		
	Recall	Precision	F1-score	Recall	Precision	F1-score
LSTM	0.0491	0.6585	0.0914	0.0745	**1.0000**	0.1387
TCN	**1.0000**	0.8818	0.9372	0.1500	**1.0000**	0.2609
MTS-DCGAN	0.0000	0.0000	-	0.0091	**1.0000**	0.0180
MSCRED	0.8520	0.8436	0.8478	0.4182	0.1631	0.2347
MCRAAD	0.9286	0.3957	0.5549	0.6455	0.1610	0.2577
ACLAE-DT	**1.0000**	0.2450	0.3936	**0.9364**	0.4055	0.5659
TCN3DPredictor (Our method) without MA1	**1.0000**	0.6777	0.8079	0.7327	0.5282	0.6139
TCN3DPredictor (Our method) with MA5	**1.0000**	0.8703	0.9307	0.8273	0.7374	0.7798
TCN3DPredictor (Our method) with MA10	**1.0000**	**0.9351**	**0.9665**	0.8655	0.7765	**0.8186**

In the result of synthetic data, we can observe that the TCN is the best of the base-line methods. And, Our proposal can get good results in f1-score metrics while taking into account both recall and precision, especially for our proposal with the length of moving average 10 gets the best score in all of the metrics. In the result of real data, We can observe that although the TCN achieves the second highest performance among all the compared methods, its performance is severely degraded compared to that in the synthetic data. Besides, compared with the performance in synthetic data, the ACLAE-DT method performs better on real data, and still maintains the highest recall score among all methods. Moreover, our proposal still dominantly leads the f1-score metrics. Apart from these, we find the MTS-DCGAN almost gets the worst performance in every metric because the GAN methods need a large amount of the data to learn the distribution for the time series, which is almost impossible in reality.

According to the Tables 1, although our proposal almost gets the highest scores in all of the variants of MSCRED framework, we still need to compare the other performances with the methods of this type to explore the differences between the methods in more detail. To specifically compare the other performances of the methods modified from the MSCRED method, we first compare the learning ability of models and the other detailed performance. And next, we are going to compare the performance of the detection mechanism.

Table 2 and Table 3 show the model performance for each model, and we can get three conclusions from it. First, from the convergence loss on the validation dataset, it can be seen that the TCN3D model has the best fitting effect among the three models, while the DCGAN model has a significantly poorer fitting effect compared with other models, which can also explain the reason why MTS-DCGAN can't achieve good results. Second, although the ConvLSTM model has the smallest number of parameters among the three, it is also the one with the largest memory, and the reason is that the LSTM model requires 4 channels to process long and short memory, so it must produce 4 times the expected amount of convolution features. In contrast, since the TCN3D model fully utilizes each convolutional feature, the required memory is the least of the three. Third, compared to the ConvLSTM model, the TCN3D model takes significantly less time to train or predict.

Table 2. The comparison of model performance on synthetic data.

Method	Backbone Model	Converged loss on validation set	Parameter	Total Memory (MB)	Training Time per epoch (sec)	Testing Time per data (sec)
MSCRED MCRAAD ACLAE-DT	ConvLSTM	0.00049	123747	6	80	0.04438
MTS-DCGAN	DCGAN	0.28589	693317	2.7	4	0.00026
TCN3DPredictor (Our method)	TCN3D	0.00003	326513	1	29	0.00179

Table 3. The comparison of model performance on real data.

Method	Backbone Model	Converged loss on validation set	Parameter	Total Memory (MB)	Training Time per epoch (sec)	Testing Time per data (sec)
MSCRED MCRAAD ACLAE-DT	ConvLSTM	0.17915	123747	6	49	0.04646
MTS-DCGAN	DCGAN	0.96747	693317	2.7	2	0.00048
TCN3DPredictor (Our method)	TCN3D	0.01179	326513	1	18	0.00194

Besides, we also compare the performance of the detection mechanism of each method, as the Table 4. It can be observed that the verification method of ACLAE-DT achieves the best results in the performance of recall, whether in synthetic dataset or real dataset, but these results cannot be balanced with the performance of precision. That is, in the performance of anomaly detection, although less false negatives are obtained because of the lower threshold, it also causes too many false positives. In contrast, our proposal is more able to achieve good results in a balanced manner while taking both recall and precision simultaneously into account.

Table 4. The performance for each detection mechanism.

Model	Detection method	Synthetic data			Real data		
		Recall	Precision	F1-score	Recall	Precision	F1-score
ConvLSTM	MSCRED	0.8520	0.8436	0.8478	0.4182	0.1631	0.2347
	ACLAE-DT	1.0000	0.2450	0.3936	0.9364	0.4055	0.5659
	TCN3DPredictor (Our method)	0.8265	0.9759	0.8950	0.8545	0.7581	0.8034
TCN3D	MSCRED	0.6735	0.9851	0.8000	0.5364	0.1578	0.2438
	ACLAE-DT	1.0000	0.2361	0.3821	0.9909	0.2472	0.3956
	TCN3DPredictor (Our method)	1.0000	0.9351	0.9665	0.8655	0.7765	0.8186

6 Conclusion

In this paper, we have proposed a novel framework named *TCN3DPredictor* in which we adopt the notion of Guo et al.'s 3D-causal convolution [4] and modify the model such that multiple-dimensional time series can be processed. As a result, our proposed *TCN3DPredictor* framework has a better fitting effect on rapidly changing two-dimensional time series data than state-of-the-art encoder-decoder architecture for modeling multiple-dimensional time series. Furthermore, the size of our proposed model is quite compact so that efficiency of our proposed *TCN3DPredictor* framework outperforms that of state-of-the-art anomaly prediction methods. Meanwhile, we proposed a statistical method To produce a precise anomaly score which lead our proposed *TCN3DPredictor* framework can deal with the unfairness for improving learning ability. Through a series of experiments using dataset crawled from a computer numerical control (CNC) metal cutting machine tool in a precision machinery factory, we have shown our proposed *TCN3DPredictor* framework has excellent effectiveness compared with state-of-the-art anomaly prediction methods under various condition.

References

1. Bai, S., Kolter, J.Z., Koltun, V.: An empirical evaluation of generic convolutional and recurrent networks for sequence modeling. arXiv preprint arXiv:1803.01271 (2018)
2. Blázquez-García, A., Conde, A., Mori, U., Lozano, J.A.: A review on outlier/anomaly detection in time series data. ACM Comput. Surv. (CSUR) **54**(3), 1–33 (2021)
3. Dopico, M., Gómez, A., De la Fuente, D., García, N., Rosillo, R., Puche, J.: A vision of industry 4.0 from an artificial intelligence point of view. In: Proceedings on the International Conference on Artificial Intelligence (ICAI), p. 407. The Steering Committee of The World Congress in Computer Science, Computer (2016)
4. Guo, H., Zhang, D., Jiang, L., Poon, K.W., Lu, K.: ASTCN: an attentive spatial-temporal convolutional network for flow prediction. IEEE Internet Things J. **9**(5), 3215–3225 (2021)
5. Hallac, D., Vare, S., Boyd, S., Leskovec, J.: Toeplitz inverse covariance-based clustering of multivariate time series data. In: Proceedings of the 23rd ACM SIGKDD International Conference on Knowledge Discovery and Data Mining, pp. 215–223 (2017)
6. Hautamaki, V., Karkkainen, I., Franti, P.: Outlier detection using k-nearest neighbour graph. In: Proceedings of the 17th International Conference on Pattern Recognition, 2004, ICPR 2004, vol. 3, pp. 430–433. IEEE (2004)
7. He, K., Zhang, X., Ren, S., Sun, J.: Deep residual learning for image recognition. In: Proceedings of the IEEE Conference on Computer Vision and Pattern Recognition, pp. 770–778 (2016)
8. Jin, S., Mordasini, C.: Compositional imprints in density-distance-time: a rocky composition for close-in low-mass exoplanets from the location of the valley of evaporation. Astrophys. J. **853**(2), 163 (2018)
9. Klambauer, G., Unterthiner, T., Mayr, A., Hochreiter, S.: Self-normalizing neural networks. In: Advances in Neural Information Processing Systems, vol. 30 (2017)

10. Liang, H., Song, L., Wang, J., Guo, L., Li, X., Liang, J.: Robust unsupervised anomaly detection via multi-time scale DCGANs with forgetting mechanism for industrial multivariate time series. Neurocomputing **423**, 444–462 (2021)
11. Malhotra, P., Ramakrishnan, A., Anand, G., Vig, L., Agarwal, P., Shroff, G.: LSTM-based encoder-decoder for multi-sensor anomaly detection. arXiv preprint arXiv:1607.00148 (2016)
12. van den Oord, A., et al.: WaveNet: a generative model for raw audio. arXiv preprint arXiv:1609.03499 (2016)
13. Radford, A., Metz, L., Chintala, S.: Unsupervised representation learning with deep convolutional generative adversarial networks. arXiv preprint arXiv:1511.06434 (2015)
14. Shi, X., Chen, Z., Wang, H., Yeung, D.Y., Wong, W.K., Woo, W.C.: Convolutional LSTM network: a machine learning approach for precipitation nowcasting. In: Advances in Neural Information Processing Systems, vol. 28 (2015)
15. Song, D., Xia, N., Cheng, W., Chen, H., Tao, D.: Deep r-th root of rank supervised joint binary embedding for multivariate time series retrieval. In: Proceedings of the 24th ACM SIGKDD International Conference on Knowledge Discovery & Data Mining, pp. 2229–2238 (2018)
16. Tayeh, T., Aburakhia, S., Myers, R., Shami, A.: An attention-based ConvLSTM autoencoder with dynamic thresholding for unsupervised anomaly detection in multivariate time series. Mach. Learn. Knowl. Extr. **4**(2), 350–370 (2022)
17. Yaacob, A.H., Tan, I.K., Chien, S.F., Tan, H.K.: Arima based network anomaly detection. In: 2010 Second International Conference on Communication Software and Networks, pp. 205–209. IEEE (2010)
18. Zhang, C., et al.: A deep neural network for unsupervised anomaly detection and diagnosis in multivariate time series data. In: Proceedings of the AAAI Conference on Artificial Intelligence, vol. 33, pp. 1409–1416 (2019)
19. Zhao, P., Chang, X., Wang, M.: A novel multivariate time-series anomaly detection approach using an unsupervised deep neural network. IEEE Access **9**, 109025–109041 (2021)

Extracting Top-k High Utility Patterns from Multi-level Transaction Databases

Tuan M. Le[1,2], Trinh D. D. Nguyen[3], Loan T. T. Nguyen[1,2(✉)], Adrianna Kozierkiewicz[4], and N. T. Tung[5]

[1] School of Computer Science and Engineering, International University, Ho Chi Minh City, Vietnam
tuanlm22@mp.hcmiu.edu.vn, nttloan@hcmiu.edu.vn
[2] Vietnam National University, Ho Chi Minh City, Vietnam
[3] Faculty of Information Technology, Industrial University of Ho Chi Minh City, Ho Chi Minh City, Vietnam
20126291.trinh@student.iuh.edu.vn
[4] Faculty of Computer Science and Management, Wroclaw University of Science and Technology, Wrocław, Poland
Adrianna.kozierkiewicz@pwr.edu.pl
[5] Faculty of Information Technology, HUTECH University, Ho Chi Minh City, Vietnam
nt.tung@hutech.edu.vn

Abstract. Several approaches have been introduced to solve the problem of high utility pattern mining (HUPM). However, the proposed algorithms require a minimum utility threshold before execution. This task is impractical for end users as they do not know utility distributions in the transaction datasets. The output will contain too many patterns if this value is too low. In contrast, if the threshold is set too high, the result would be empty or insufficient for analysis. Recently, HUPM was extended to work with hierarchical transaction datasets. With the search space of the mining task expanded, selecting a proper threshold is far more challenging. To address this issue, we propose a top-*k* high utility pattern mining method from multi-level transactions databases. The users only need to specify a *k* value, denotes the desired number of patterns of interest. To the best of our knowledge, the method proposed in our work is the first to address this mining topic. Experiments on both real and synthetic hierarchical datasets were extensively conducted to evaluate the performance of the proposed algorithm.

Keywords: multi-level patterns · top- high utility patterns · taxonomy-based · hierarchical datasets

1 Introduction

In the field of Data Mining, one of the key tasks is pattern mining and it is an active research field. Pattern mining reveals the useful information from heaps of data, represented as datasets [1]. Several algorithms were developed to carry out the task efficiently on many types of data, such as transaction datasets, sequences, graphs, etc.

N. T. Nguyen et al. (Eds.): ACIIDS 2023, LNAI 13995, pp. 296–306, 2023.
https://doi.org/10.1007/978-981-99-5834-4_24

Frequent Pattern Mining (FPM) is a task aiming to extract set of frequently occurred patterns in customer transaction datasets. A pattern is called a Frequent Patterns (FPs) if its occurrence frequency no less than a minimum support threshold (σ). FPM was first proposed by Agrawal et al. in 1994 [2]. Several approaches to effectively solve the FPM problem were later proposed [1]. However, the drawback of FPM is that it treats all the items equally. Items appear in transactions have binary occurrence only.

To address the drawbacks from FPM [2], the task of High Utility Pattern Mining (HUPM) was introduced in [3]. In HUPM, each item is now having a value denote its importance, weight, or utility. The goal of HUPM is to reveal set of patterns that have utility no less than a minimum utility threshold (μ) and is called High Utility Patterns (HUPs). HUPM considers more information than FPM, thus its search space is much larger. Besides, the utility measure is neither monotonic nor anti-monotonic. Thus, the effective pruning strategies in FPM cannot be used to reduce the search space. To solve the HUPM task, many efficient algorithms were introduced [4].

Inherited from FPM [4], HUPM was leveraged to work with datasets containing item categorizations, or hierarchical datasets [5-9]. Items now can be generalized into categories or sub-categories. The structure of item categorizations is called the dataset's taxonomy. With taxonomy, HUPM can reveal more useful information. The output patterns can be either multi-level patterns (MLHUPs), containing items at the same taxonomy level, or cross-level patterns (CLHUPs), containing items from different levels.

With the search space of the HUPM task expanded further when adopting taxonomy, picking a proper μ threshold to use can be a difficult task. This threshold impacts the performance of the mining algorithms. Low μ value would generate too many patterns, increasing memory usage and mining time. Analyzing all the patterns is also time consuming. Conversely, selecting μ too high would yield no outputs. To select a proper μ, the user has to rerun the algorithm over the dataset several times until he/she is satisfied with the results. This process is not intuitive. Previously, an algorithm named TKC [6] was introduced to solve the issue in cross-level HUP mining. The output of TKC may contain interesting patterns for the decision maker. However, the run time of TKC is long, even with low *k* due to the huge search space of the task. Furthermore, as in many previous generalized patterns mining tasks in FPM, the decision makers may only interest in the patterns that contain sub-categories or categories at the same abstraction level to improve their marketing performance [10]. Considering cross-level HUPs are computationally expensive and might not be practical to their needs.

To solve this issue, this work integrates the top-*k* problem to the multi-level HUPM task. To the best of our knowledge, our work is the first to address the task of mining top-*k* multi-level high utility patterns from hierarchical datasets. Our contributions can be summarized as follows.

- We propose the top-*k* multi-level high utility pattern mining task.
- We also propose an algorithm name mlTKO to efficiently perform the task.
- Experiments are also carried out to demonstrate the effectiveness of the proposed algorithm on both real-life and synthetic datasets.

The remaining content of the paper is organized as follows. The next section surveys the recently related works in the field of HUPM. The third section provides the basic definitions and states the problem. The fourth section proposes an algorithm to carry out the top-k multi-level HUPM task. Experiments are done and discussed in the fifth section. Finally, conclusions and future plans and discussed in the sixth section.

2 Related Works

In recent years, several HUPM methods have been introduced [4]. Some notable works HUI-Miner [11], FHM [12], iMEFIM [13], etc. HUI-Miner converts datasets into a vertical format by using a structure called utility-list. Utility of a pattern can be directly computed using this structure, without rescanning the datasets. FHM suggested the EUCS structure to store the co-occurrence information to prune more candidates. iMEFIM suggests the concept of the dynamic utility of items and reducing further the cost of sparse dataset scans by proposing the pattern-to-TID projection technique.

These approaches require a user-input μ threshold to operate. However, selecting a proper threshold is a time-consuming process. With top-*k* algorithms, users feed the algorithms the number of patterns they want to retrieve from the datasets, denoted as k. The algorithms rely on k to determine a proper μ to obtain enough patterns if possible. TKU [14] and REPT [15] are the first approaches for this task, which are based on the two-phase model. TKO [14] and THUI [16] are one-phase algorithms, which is more efficient. TKO is based on HUI-Miner to adopt the utility-list structure. THUI is also a list-based algorithm. However, the differences between THUI and TKO are from the underlying structure and threshold raising strategies.

Recently, HUPM algorithms have taken into consideration the hierarchical datasets [5-9]. The output of these HUPM algorithms can contain patterns at higher abstraction level instead of the patterns that appear the lowest taxonomy level. Some notable algorithms are MLHUI-Miner [5], MLHMiner [7], MCML-Miner [9] for the multi-level HUPM task; CLH-Miner [17], TKC [6], FEACP [8] for the cross-level HUPM task. Among these algorithms, there is only one top-*k* algorithm for the cross-level HUPM task, TKC [6]. There is no top-*k* approach designed to perform the multi-level HUPM task. Thus, the users of this mining direction still have to manually pick their own μ thresholds instead of simply pointing out their number of patterns of interest. The goal of this work is to propose a framework to address this issue.

3 Preliminaries

This section provides the foundation information for the task of mining top-*k* multi-level HUPs. To keep the paper compact, only core definitions are presented.

3.1 Definitions

Given a set of m distinct items, $\mathcal{I} = \{i_1, i_2, \ldots, i_m\}$, a transaction dataset is a set of transaction, $\mathcal{D} = \{T_1, T_2, \ldots, T_n\}$. A transaction $T_q (1 \leq q \leq n)$ has a unique identifier q. Each item $i \in \mathcal{I}$ is linked with a positive integer, called external utility: $eu(i)$, or the unit profit of i in $\mathcal{D}$. Each item i in the transaction T_q has a linked positive integer, called its internal utility in T_q: $iu(i, T_q)$. The utility of item i in T_q is determined as

$u(i, T_q) = eu(i) \times iu(i, T_q)$. The utility of a pattern X in T_q is defined as $u(X, T_q) = \sum_{i \in X, X \subseteq T_q} u(i, T_q)$. Utility of transaction T_q in $\mathcal{D}$ is given as $TU(T_q) = \sum_{i \in T_q} u(i, T_q)$. Transaction Weighted Utilization of a pattern is $TWU(X) = \sum_{X \subseteq T_i \wedge T_i \in D} TU(T_i)$ [18]. In this work, the TWU is used in the EUCP strategy to safely prune unpromising patterns before constructing their utility-list [12].

Table 1. An example transaction dataset

TID	Transaction	Internal utility
T_1	a, b, c, e	2, 2, 6, 2
T_2	a, b, c, d, e, f	1, 2, 1, 6, 1, 5
T_3	a, b, e	1, 4, 1
T_4	a, c, d, e, f	1, 2, 2, 1, 1

Table 2. List of external utilities

i	a	b	c	d	e	f
$eu(i)$	5	2	1	2	3	1

Considering the dataset $\mathcal{D}$ given in Table 1 and Table 2, utility of item a in transaction T_1 is $u(a, T_1) = 5 \times 2 = 10$. Utility of item d in $\mathcal{D}$ is $u(d) = u(d, T_2) + u(d, T_4) = 16$. Utility of the pattern $X = \{a, d\}$ in T_2 is $u(\{a, d\}, T_2) = u(a, T_2) + u(d, T_2) = 5 \times 1 + 6 \times 2 = 17$. Similarly, the utility of $\{a, d\}$ in $\mathcal{D}$ is $u(\{a, d\}) = u(\{a, d\}, T_2) + u(\{a, d\}, T_4) = 17 + 9 = 26$. The TWU of item b in $\mathcal{D}$ is $TWU(b) = TU(T_1) + TU(T_2) + TU(T_3) = 26 + 30 + 16 = 72$.

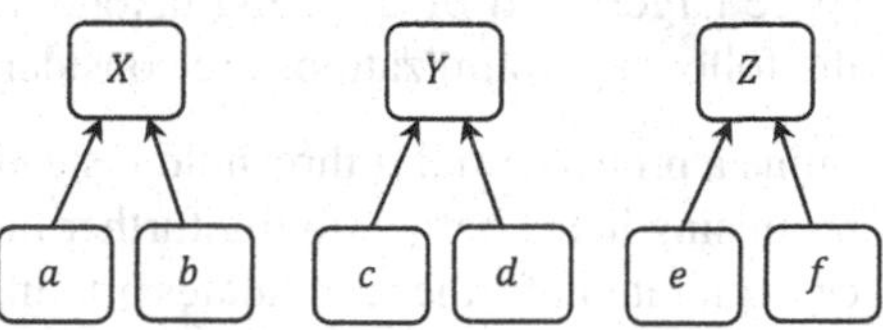

Fig. 1. An example taxonomy of the given transaction dataset

Taxonomy τ of items in $\mathcal{D}$ is a tree. Leaf nodes of τ represent all items $i \in \mathcal{I}$. Internal nodes represent the generalized items (categories) that group items or subcategories into a higher abstraction level in τ. The set of all generalized items is denoted as $\mathcal{G}$. The set $C(g, \tau)$ containing all child nodes (descendants) of g in τ. The level of a generalized item $g \in \mathcal{G}$ is the length of the shortest path between g to any of its leaf nodes. Utility of $g \in \mathcal{G}$ in T_q is $u(g, T_q) = \sum_{i \in C(g,\tau)} u(i, T_q)$. A generalized multi-level pattern is a set of items in $\mathcal{G}$ at the same taxonomy level. Utility of generalized pattern

X in T_q is $u(X, T_q) = \sum_{g \in X} u(g, T_q)$. The utility of generalized pattern X in $\mathcal{D}$ is $u(X) = \sum_{X \subseteq T_q, T_q \in \mathcal{D}} u(X, T_q)$. The TWU is also leveraged to work with generalized patterns [5].

With the dataset $\mathcal{D}$ in Table 1, Table 2 and the taxonomy in Fig. 1, it is observed that a, b are grouped into a generalized item X, the same for c, d to Y and e, f to Z. In turn, X, Y and Z could be grouped into a higher abstraction level if needed. A pattern $\{Y, Z\}$ is considered a multi-level pattern as Y, Z belong to the same level. However, the pattern $\{Y, f\}$ is not. The utility of the generalized item X in $\mathcal{D}$ and τ is $u(X) = u(a) + u(b) = 25 + 16 = 41$. Similarly, $u(Y) = 34$ and $u(Z) = 22$. The utility of generalized pattern $\{Y, Z\}$ is $u(\{Y, Z\}) = u(\{Y, Z\}, T_1) + u(\{Y, Z\}, T_2) + u(\{Y, Z\}, T_4) = 54$.

3.2 Problem Statement

Given a transaction database $\mathcal{D}$ combined with taxonomy τ and a user specified k parameter, the task of mining top-k multi-level high utility patterns is to discover exact k multi-level high utility patterns that have the highest utilities from $\mathcal{D}$.

Considering the dataset $\mathcal{D}$ in Table 1, Table 2, the taxonomy τ in Fig. 1 and $k = 5$, the following five patterns are MLHUPs: $\{a, c, d, e, f\}$,$\{Y, Z\}$,$\{X, Y\}$,$\{X, Z\}$ and $\{X, Y, Z\}$ with the respective utilities are 48, 54, 62, 63 and 82.

4 Top-*k* MLHUP Mining

In this section, we aim to extend the TKO algorithm [14] by adopting the taxonomy-based framework [5] on top of it. TKO is a single-phase top-k HUP mining algorithm [11]. By extending TKO, our proposed algorithm mlTKO would inherit all the efficient strategies used in TKO. Besides, by adopting the multi-level HUPM framework from [5] also make it capable of extracting top-k MLHUPs from hierarchical datasets. Furthermore, mining top-k MLHUPs is a more challenging task than the regular MLHUPs mining, which is based on μ. Since there is no μ provided, the task must initially assume $\mu = 0$. This would make the search space larger. And by adopting taxonomy into the mining task extends it further. Thus, the following optimizations are considered.

- It is important to determine a proper initial μ threshold instead of setting it to 0. This would prune early unpromising items before diving further into the search space.
- During the search space exploration, further candidates pruning is needed to compact the search space more. This could be done by employing the EUCP strategy [12].

4.1 Determining the Initial μ Threshold

Originally, TKO initializes μ to 0, which is inefficient as it would consider the whole set $\mathcal{I}$ to be promising in the DFS phase, especially when $k \ll |\mathcal{I}|$. This would significantly increase the number of candidates to be checked and lengthen the mining time and memory consumption. To reduce the number of unpromising items participating in the pattern construction phase, a proper μ must be determined instead of setting it to 0.

To address this issue, when the *TWUs* and utility of all $i \in \mathcal{I}$ are obtained in the first dataset scans, we sort the list of the obtained pairs $\{TWU(i), u(i)\}$ based on the descending order of *TWUs*. Assuming after sorted, we have the list of TWUs as $pairs = \{twu_1 : u_1, twu_2 : u_2, \ldots, twu_k : u_k, \ldots, twu_m : u_m\}$. Then μ is determined as follows.

- If $k \geq |\mathcal{I}|$, we select the utility of the last element from the list, $\mu = pair[m].u$.
- Otherwise, $\mu = pair[k].u$.

If $k \ll |\mathcal{I}|$, then a large number of unpromising items will be discarded, thus significantly dropping the number of candidates checked.

4.2 Pruning Candidates via EUCP

During the mining phase, TKO does not have any other pruning strategy other than utility-list pruning. If $|\mathcal{I}|$ and $|\mathcal{G}|$ large, the number of combinations checked per taxonomy level is also increased. To reduce the number of unpromising candidates' pattern prior utility-list construction, EUCP strategy from the FHM algorithm [12] is adopted. The structure is extended to work with both set $\mathcal{I}$ and $\mathcal{G}$. The EUCS construction is done in the second dataset scan, which is shown in Algorithm 1. During the DFS phase in Algorithm 2, the EUCS is used to prune candidates. With the help of EUCP, mlTKO could significantly lower utility-list constructing and joining cost.

4.3 The mlTKO Algorithm

The proposed algorithm to handle the top-*k* multi-level HUP mining is named mlTKO (multi-level TKO). The algorithm has the following characteristics.

- mlTKO is a single-phase algorithm.
- mlTKO accepts the *k* parameter as the number of patterns to extract.
- mlTKO is taxonomy driven. It capable of working on both hierarchical datasets and traditional transaction datasets.
- The outputs' of mlTKO are *k* MLHUPs patterns with the highest utility values.
- mlTKO includes both optimizations presented.

Algorithm 1: mlTKO algorithm
Input: transaction dataset $\mathcal{D}$, taxonomy τ, number of patterns k
Output: complete set of k highest utility patterns.

1: $kList \leftarrow \emptyset$.
2: Scan $\mathcal{D}$ to compute $TWU(i), u(i)\ \forall i \in \mathcal{I}$,
3: Store $\{TWU(i), u(i)\}, \forall i \in \mathcal{I}$ in *pairs*.
4: Sort *pairs* in the descending order of TWUs.
5: **IF** $k \geq |\mathcal{I}|$ **THEN** $\mu = pairs[|\mathcal{I}|].u$ **ELSE** $\mu = pairs[k].u$.
6: Computing TWUs for $\forall g \in \mathcal{G}$.
7: $\mathcal{G}' \leftarrow \{g \in \mathcal{G} \mid TWU(g) \geq \mu\}$
8: $\mathcal{I}' \leftarrow \{i \in \mathcal{I} \mid TWU(i) \geq \mu \wedge \forall g \in \mathcal{G}' : i \in \mathcal{C}(g, \tau) \wedge TWU(g) \geq \mu\}$
9: Scan $\mathcal{D}$, τ. Build *EUCS* and utility list for $\forall i \in \mathcal{I}' \cup \mathcal{G}'$.
10: **FOREACH** level $l = |\tau|..0$ **AND** $l \in \tau$ **DO**
11: | $\mathcal{IG}' \leftarrow$ items in $\mathcal{I}' \cup \mathcal{G}'$ at level l.
12: | $kList \leftarrow$ **SEARCH**($\emptyset$, $\mathcal{IG}'$, μ, *EUCS*, *kList*)
13: **RETURN** *kList*

The pseudo-code of the mlTKO algorithm is presented in Algorithm 1. mlTKO uses a min-heap structure to store the currently found top-k MLHUPs, *kList*. This list would be filled by the DFS phase in Algorithm 2. Lines #2 to #5 carry out the first optimization, determining a proper initial μ. Line #9 performs the second optimization, constructing EUCS for use in the *SEARCH* function call at line #12.

Algorithm 2 performs the search space exploration using the DFS mechanism. μ is raised at line #2, using the RUC strategy proposed in [14]. Lines #6 to #9 perform the EUCP strategy [12]. At line #8, the *CONSTRUCT* function that performs the utility-list construction/joining is exactly the one used in work [11].

Algorithm 2: SEARCH - the top-k multi-level HUPs miner
Input: pattern to be extended P, set of extensions of P $E(P)$, current minimum utility threshold μ, EUCS structure, a snapshot of the previously found top-k HUPs *kList*
Output: currently snapshot of the highest k HUPs.

1: **FOREACH** $P_x \in E(P)$ **DO**
2: | **IF** $SUM(P_x.UL.iutils) \geq \mu$ **THEN** $\mu \leftarrow RUC(P, kList)$
3: | **IF** $SUM(P_x.UL.iutils) + SUM(P_x.UL.rutils) \geq \mu$ **THEN**
4: | | $E(P_x) \leftarrow \emptyset$
5: | | **FOREACH** $P_y \in E(P)$ **AND** y succeed x **DO**
6: | | | **IF** $EUCS[x, y] \geq \mu$ **THEN**
7: | | | | $P_{xy} \leftarrow P_x \cup P_y$.
8: | | | | $P_{xy}.UL \leftarrow$ **CONSTRUCT**(P, P_x, P_y)
9: | | | | $E(P_x) \leftarrow E(P_x) \cup P_{xy}$
10: | | | $kList \leftarrow$ **SEARCH**$(P_x, E(P_x), \mu, EUCS, kList)$

5 Evaluation Studies

To evaluate the performance of the mlTKO algorithm, a series of experiments were conducted on four hierarchical datasets. The computer used in the experiments is equipped with an Intel Core i5-5257U @ 2.7 Ghz, 8 GB RAM using macOS Monterey. The algorithm mlTKO is implemented in Java, using JDK 8.0.

Table 3. Characteristics of experiment datasets

Dataset	$\|\mathcal{D}\|$	$\|\mathcal{I}\|$	$\|\mathcal{G}\|$	$\|\tau\|$	T_{max}	T_{avg}	τ_{type}	Size
Fruithut	181,970	1,265	43	4	36	3.59	Real	6.9M
Liquor	90,826	2,626	78	7	15	10.28	Real	11.5M
Foodmart	53,537	1,560	102	5	28	4.60	Real	2.2M
Chess	3,196	75	30	3	37	37.00	Synthetic	0.7M

Table 3 provides the characteristics of the datasets. Whereas $|\mathcal{D}|$ is the dataset size in term of transaction count; $|\mathcal{I}|$ and $|\mathcal{G}|$ are the number of specialized items and generalized item, respectively; $|\tau|$ is the taxonomy's depth; T_{max}, T_{avg} respectively are the maximum and average transaction length; the taxonomy type is given in τ_{type}; finally, *Size* is the actual size on disk of the datasets (in Megabytes). Fruithut, Liquor and their taxonomies are obtained from the SPMF library [19]. Foodmart is obtained from GitHub[1], transformed into SPMF format. Chess is also from the SPMF library, its taxonomy is synthesized to contain 30 categories with a depth of 3.

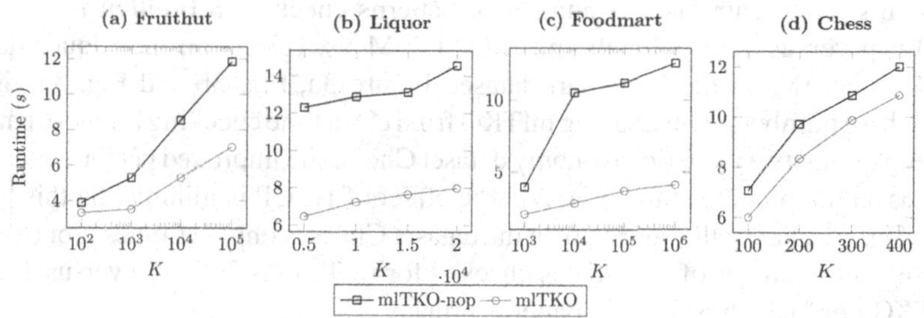

Fig. 2. Execution time comparisons across test datasets

As stated, mlTKO is the first algorithm to mine top-*k* multi-level HUPs. Thus, we compare mlTKO against its own non-EUCP version, referred as mlTKO-nop. The compared factors are runtime, number of candidates checked and peak memory usage. During the test, *k* parameter is varied from within a specific range per dataset to evaluate its influence over the comparison factors.

[1] Source: https://github.com/arunkjn/foodmart-mysql.

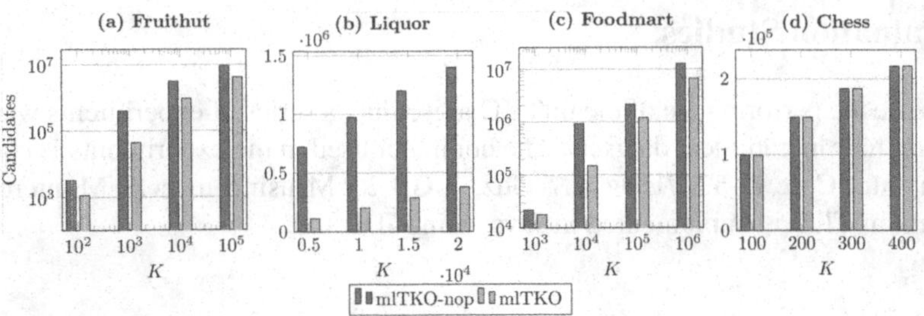

Fig. 3. Number of candidates checked across datasets

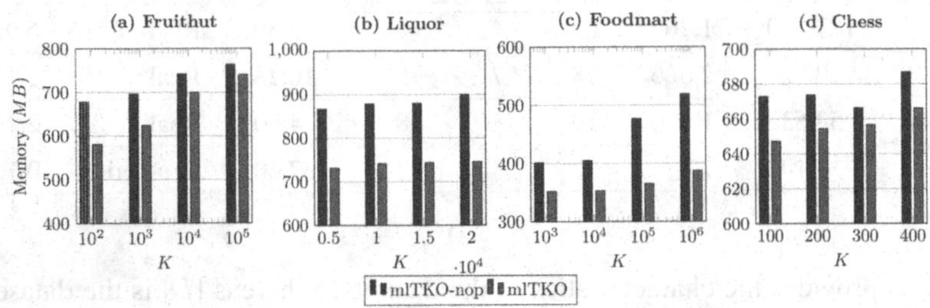

Fig. 4. Peak memory usage across test datasets

In the runtime evaluations on all four datasets, as shown in Fig. 2, the performance of mlTKO is superior to its non-EUCP version, the mlTKO-nop, by up to 3 times in the Foodmart datasets Fig. 2c. Larger k would increase this margin further.

The number of unpromising candidates pruned by EUCP on the first three datasets (Fruithut, Liquor and Foodmart) is huge as we increase k. For example, take the highest k on the first three datasets, the number of patterns checked in Fruithut is 9.27M for mlTKO-nop versus 4.13M for the mlTKO; 1.39M vs. 0.37M on the Liquor dataset; 12.61M vs. 6.63M on the Foodmart dataset. Figure 3a, Fig. 3b and Fig. 3c visually presents these numbers. Thus, saving mlTKO from extra time checking for these patterns. On the high-density, synthetic taxonomy dataset Chess, the improved performance is not as high as on the previous three Fig. 2d. The effect of EUCP is minimal on this type of dataset (Fig. 3d), hence the speed up on the dataset Chess is only 1.1 times. For example, at $k = 400$, the number of candidates checked for mlTKO is 216, 590 versus 216, 607 for mlTKO-nop, which is only 17 patterns pruned.

The peak memory usage evaluations in all four test datasets (Fig. 4) also showed that mlTKO has better memory usage than mlTKO-nop, thanks to the EUCP strategy. EUCP cuts off a large amount of memory needed to store the required information for the unpromising candidates. For the Chess dataset, the amount of memory reduced is marginal. Although EUCP cannot prune many unpromising candidates, the memory usage of mlTKO is still lower than that of mlTKO-nop.

As observed on all the test datasets, when we increase the k parameter, the memory usage of both algorithms is also increased. However, the amount required by mlTKO is always lower than that of mlTKO-nop. Thus, mlTKO is shown to be more memory efficient than mlTKO-nop.

6 Conclusions and Future Works

In this work, we suggest the task of mining top-*k* multi-level high utility patterns. This task allows decision makers to specify their desired number of patterns to be extracted, instead of guessing a proper minimum utility threshold to use. By adopting taxonomy into transaction datasets, mining algorithms can reveal patterns that cannot be found in traditional transaction datasets. An algorithm named mlTKO is also proposed, which is based on the TKO algorithm to solve the task. Evaluations of three real-life datasets and one synthetic dataset, all combined taxonomy data, have shown that mlTKO is efficient. The pruning strategy used in mlTKO also improves its performance further in terms of execution time, memory footprint and number of candidates checked.

In the future, we are planning to implement more efficient pruning strategies to boost performance. Besides, extending this mining task to discover compact pattern representations such as closed or maximal from taxonomy-based transaction datasets is also considered. Parallel frameworks are also our focus to optimize further the mining performance as well as to utilize the modern processors' computing power.

Acknowledgment. This research is funded by Vietnam National University HoChiMinh City (VNU-HCM) under grant number B2023-28-02.

References

1. Fournier-Viger, P., Lin, J.C.W., Vo, B., Chi, T.T., Zhang, J., Le, H.B.: A survey of itemset mining. Wiley Interdiscip. Rev. Data Min. Knowl. Discov. **7**(4) (2017)
2. Agrawal, R., Srikant, R.: Fast algorithms for mining association rules in large databases. In: 20th International Conference on Very Large Data Bases (VLDB'94), Morgan Kaufmann Publishers Inc., pp. 487–499 (1994)
3. Yao, H., Hamilton, H.J., Butz, G.J.: A foundational approach to mining itemset utilities from databases. In: SIAM International Conference on Data Mining, pp. 482–486 (2004)
4. Fournier-Viger, P., Lin, J.C.-W., Truong-Chi, T., Nkambou, R.: A survey of high utility itemset mining. In: High-Utility Pattern Mining: Theory, Algorithms and Applications, Fournier-Viger, P., Lin, J.C.-W., Nkambou, R., Vo, B., Tseng, V.S. (eds.), Springer International Publishing, Cham, pp. 1–45 (2019)
5. Cagliero, L., Chiusano, S., Garza, P., Ricupero, G.: Discovering high-utility itemsets at multiple abstraction levels. In: Kirikova, M., et al. (eds.) European Conference on Advances in Databases and Information Systems, pp. 224–234. Springer International Publishing, Cham (2017)
6. Nouioua, M., Wang, Y., Fournier-Viger, P., Lin, J.C.-W., Wu, J.M.-T.: TKC: Mining top-k cross-level high utility itemsets. In: 2020 International Conference on Data Mining Workshops (ICDMW), pp. 673–682 (2020)
7. Tung, N.T., Nguyen, L.T.T., Nguyen, T.D.D., Vo, B.: An efficient method for mining multi-level high utility Itemsets. Appl. Intell. **52**(5), 5475–5496 (2022)
8. Tung, N.T., Nguyen, L.T.T., Nguyen, T.D.D., Fourier-Viger, P., Nguyen, N.T., Vo, B.: Efficient mining of cross-level high-utility itemsets in taxonomy quantitative databases. Inf. Sci. (Ny) **587**, 41–62 (2022)

9. Nguyen, T. D.D., Nguyen, L.T.T., Kozierkiewicz, A., Pham, T., Vo, B.: An efficient approach for mining high-utility itemsets from multiple abstraction levels. In: Intelligent Information and Database Systems., Springer International Publishing, pp. 92–103 (2021). https://doi.org/10.1007/978-3-030-73280-6_8
10. Baralis, E., Cagliero, L., Cerquitelli, T., D'Elia, V., Garza, P.: Expressive generalized itemsets. Inf. Sci. (Ny) **278**, 327–343 (2014)
11. Liu, M., Qu, J.: Mining high utility itemsets without candidate generation. In: ACM International Conference Proceeding Series, pp. 55–64 (2012)
12. Fournier-Viger, P., Wu, C.W., Zida, S., Tseng, V.S.: FHM: Faster high-utility itemset mining using estimated utility co-occurrence pruning. In: International Symposium on Methodologies for Intelligent Systems, pp. 83–92 (2014)
13. Nguyen, L.T.T., Nguyen, P., Nguyen, T.D.D., Vo, B., Fournier-Viger, P., Tseng, V.S.: Mining high-utility itemsets in dynamic profit databases. Knowledge-Based Syst. **175**, 130–144 (2019)
14. Tseng, V.S., Wu, C.W., Fournier-Viger, P., Yu, P.S.: Efficient algorithms for mining top-K high utility itemsets. IEEE Trans. Knowl. Data Eng. **28**(1), 54–67 (2016)
15. Ryang, H., Yun, U.: Top-k high utility pattern mining with effective threshold raising strategies. Knowledge-Based Syst. **76**, 109–126 (2015)
16. Krishnamoorthy, S.: Mining top-k high utility itemsets with effective threshold raising strategies. Expert Syst. Appl. **117**, 148–165 (2019)
17. Fournier-Viger, P., Yang, Y., Lin, J.C.-W., Luna, J.M., Ventura, S.: Mining cross-level high utility itemsets. In: 33rd International Conference on Industrial, p. 12. Springer, Engineering and Other Applications of Applied Intelligent Systems (2020)
18. Liu, Y., Liao, W.K., Choudhary, A.: A two-phase algorithm for fast discovery of high utility itemsets. In: 9th Pacific-Asia Conference on Advances in Knowledge Discovery and Data Mining, in PAKDD'05, vol. 3518. Springer-Verlag, pp. 689–695 (2005)
19. Fournier-Viger, P., et al.: The SPMF open-source data mining library version 2. In: Joint European Conference on Machine Learning and Knowledge Discovery in Databases, pp. 36–40 (2016)

Lightweight and Efficient Privacy-Preserving Multimodal Representation Inference via Fully Homomorphic Encryption

Zhaojue Li[1], Yingpeng Sang[1](✉), Xinru Deng[1], and Hui Tian[2]

[1] School of Computer Science and Engineering, Sun Yat-sen University, Guangzhou, China
{lizhj33,dengxr3}@mail2.sysu.edu.cn, sangyp@mail.sysu.edu.cn

[2] School of Information and Communication Technology, Griffith University, Brisbane, Australia
hui.tian@griffith.edu.au

Abstract. Machine learning models are now being widely deployed in clouds, but serious data leakage problems are also exposed when dealing with sensitive data. Homomorphic encryption (HE) has been used in the secure inference on unimodal private data because of its ability to calculate encrypted data. Although the privacy protection of multimodal data is of great significance, there is still no privacy-preserving inference scheme for multimodal data. In this work, we propose a lightweight and efficient homomorphic-encryption based framework that enables privacy-preserving multimodal representation inference. Firstly, we propose an HE scheme based on Tensor Fusion Network, which can perform encrypted multimodal feature fusion. Then we propose a pre-expansion method and a packaging method for multimodal data, which can effectively reduce the time delay and data traffic of homomorphic computing. The experimental results show that our encryption inference method has almost no loss of accuracy and obtains an F1 score of 0.7697, while using less than 220KB of data throughput and about 0.91 s of evaluation time.

Keywords: Fully Homomorphic Encryption · Privacy-Preserving Machine Learning · Multimodal Privacy

1 Introduction

Our perception of the environment is inherently multimodal, involving multiple sensory channels such as vision, hearing and taste. A modality refers to the way in which something happens or is experienced, and research is considered multimodal when it involves more than one of these channels. Machine learning approaches based on multimodal data have emerged as a rapidly growing field of research [12,26,27,29]. Substantial research [4,21] has shown that classifiers

N. T. Nguyen et al. (Eds.): ACIIDS 2023, LNAI 13995, pp. 307–321, 2023.
https://doi.org/10.1007/978-981-99-5834-4_25

based on multimodal data outperform those based on unimodal data, which is consistent with the way in which humans perceive and comprehend the world.

Transmitting multimodal sensitive data from client to server typically entails privacy risks, as the server may be malevolent or susceptible to attacks by third parties. Moreover, because multimodal data contains rich cross-information, the privacy damage caused by leakage is more severe than that caused by unimodal data. Thus, there is a significant motivation to design a privacy-preserving strategy for multimodal machine learning that can ensure the privacy of clients who use cloud computing services. A highly relevant technology is Fully Homomorphic Encryption (FHE) [15]. FHE is an encryption system that enables data owners to encrypt their data and authorize third parties to perform calculations on it. Third parties can perform several calculations but are not authorized to access the original data, thus preserving the privacy of consumers. However, practical applications require specialized solutions to reduce the high computing and transmission costs associated with FHE. Previous works have provided specialized homomorphic schemes for unimodal machine learning [2,8,11].

In this paper, we propose, for the first time, a privacy-preserving prediction approach for multimodal representation based on the FHE scheme CKKS [10] and Tensor Fusion Network (TFN) [32]. In this method, the server accepts the encrypted multimodal representation sent by the client, performs fusion and model evaluation, and feeds back the evaluation results to the client. Only the client can decrypt the results and obtain the inference. The main contributions of our study are as follows:

- This research is the first to propose a multimodal representation inference approach based on FHE. Specifically, we implement the multimodal feature fusion network in the ciphertext state.
- We provide a pre-expansion method to reduce the computational complexity of the homomorphic tensor feature fusion layer.
- We utilize the rotation operation of CKKS and the unoccupied slot in ciphertext, to pack data of multiple modes in the same ciphertext. This significantly decreases the amount of data transmission required. To optimize the ciphertext matrix multiplication with the highest latency, we also leverage multithreading to achieve a 3.5-times acceleration.

In the rest of the paper, we summarize the related work in Sect. 2, and present the background knowledge in Sect. 3. In Sect. 4 we present our proposed approach in details, which is followed by performance evaluation and experimental results in Sect. 5. Finally, we conclude the paper in Sect. 6.

2 Related Work

2.1 Multimodal Machine Learning

The employment of multimodal data provides human beings with a comprehensive and multifaceted understanding, thereby facilitating them to make more informed decisions [4,12]. Despite the convenience brought by cloud services, the privacy concern regarding multimodal data is increasingly urgent. To address this issue, Cai et al. [9] proposed implementing differential privacy in the representation after multimodal fusion to protect the privacy of the original training data. However, the introduction of noise by differential privacy inevitably compromises the utility of the data and the accuracy of the model inference. Moreover, it is noteworthy that this method cannot protect user privacy data during the inference process.

2.2 Homomorphic Encrypted Neural Network

With the increasing application of machine learning in education, finance, and other fields that deal with sensitive customer data, there is a growing need for privacy protection in machine learning algorithms that make accurate predictions. To address this issue, several cryptographic techniques have been proposed, such as Trusted Execution Environments [20], Secure Multi-Party Computation (SMPC) [28] and Homomorphic Encryption (HE) [15]. Each method has its own tradeoffs in terms of calculation, accuracy, and security. Among them, schemes based on Fully Homomorphic Encryption (FHE) can generally achieve quantum security, which is the most rigid security model [1].

FHE was first proposed by Rivest et al. [24], and Gentry [15] proposed the first generation of FHE systems. Despite allowing more homomorphic multiplication and addition operations, practical applications still require too much computation. To address this issue, several practical leveled homomorphic encryption schemes have been proposed, such as the integer-based BGV algorithm [7], BFV [14], and the complex scheme CKKS [10]. These schemes can perform homomorphic computation within the pre-set maximum multiplication depth. CKKS permits approximate HE operations for real numbers, making it a suitable choice for the inference task of machine learning with a fixed number of layers.

Dowlin et al. [16] proposed CryptoNets, which proved the feasibility of using HE for private neural network inference. However, CryptoNets have two limitations. The first is on the time cost, although it supports high-throughput prediction, the prediction of a single sample still takes 205 s. The second is on the width of the network. CryptoNet encodes each node of the network into a separate ciphertext, so it needs a lot of memory to support it. In order to solve these problems, LoLa [8] encrypted the entire network layer, significantly reducing memory requirements and achieving single sample prediction in 2.2 s. Furthermore, homomorphic schemes have been proposed for text classification [2] and audio similarity calculation [22] in addition to image classification. However, these schemes only consider the homomorphic implementation of unimodal machine learning. To the best of our knowledge, there is currently no homomorphic encryption scheme available for multimodal machine learning and multimodal feature fusion.

2.3 Homomorphic Representation Inference

The aforementioned work can be used to encrypt shallow networks through homomorphic inference. However, there are two primary limitations when Fully Homomorphic Encryption (FHE) is applied to deep network models: the growth of noise and the growth of ciphertext size. Each ciphertext contains noise that increases with each homomorphic operation. Therefore, too many operations increase the noise to the point where the decryption may not be correct. Secondly, the operation of the HE scheme can double the size of the encryption parameter without bootstrapping, resulting in a large ciphertext that increases memory requirements and causes greater latency. To address these issues, LoLa [8] proposed using deep representations for encrypted inference. Customers convert the original data into deep representations locally through the feature extraction network. Their prediction only requires a shallow network model, which is more suitable for low-latency homomorphic implementation. Chou et al. [11] proposed extracting deep representations from the screenshot of the original phishing website locally, encrypting them, and sending them to the cloud for homomorphic logistic regression calculation, achieving a low-delay and secure homomorphic inference scheme.

3 Preliminaries

3.1 Fully Homomorphic Encryption (FHE)

Fully Homomorphic Encryption (FHE) is a powerful encryption method that allows ciphertext computation with minimal loss of accuracy upon decryption. This capability makes it useful for secure computing outsourcing: the client encrypts the data and sends it to a third party for computation, where the third party cannot access the plaintext data. After receiving the encrypted output, the client decrypts it to obtain the computation result. Specifically, the encryption function is denoted by Enc, and plaintext data by x and y. Then, $Enc(x + y) = Enc(x) \oplus Enc(y)$ and $Enc(x * y) = Enc(x) \odot Enc(y)$, where $\oplus$ and $\odot$ represent homomorphic addition and multiplication, respectively.

While HE allows ciphertext computation to obtain the encrypted output with practically little loss of accuracy, it adds noise to plaintext data, and homomorphic operations increase noise continuously. Once the noise reaches a certain threshold, the correct plaintext result cannot be decrypted. Although a primitive bootstrapping method [15] can refresh noise, it is limited by the massive amount of computation required. A more practical technique is to use leveled homomorphic encryption [5], which allows multiple addition and multiplication operations at a predetermined maximum multiplication depth. Since the calculation times of neural networks are also determined, leveled homomorphic encryption has been widely used in encrypted machine learning inference tasks. In this paper, we choose the CKKS leveled homomorphic encryption method, which is specifically designed to handle real numbers and approximate calculations.

3.2 The Levelled FHE Scheme - CKKS

In the following paragraphs, we will briefly introduce the CKKS scheme. Let N be a power of two, and $R = \mathbb{Z}[X]/(X^N + 1)$ be the ring integer of the $2N$-th cyclotomic polynomial. For some small prime integers p_i, let $q_L = \prod_{i=1}^{L} p_i$ and $R_{q_L} = \mathbb{Z}_{q_L}[X]/(X^N + 1)$ consists of the polynomials whose coefficients are moduloq_L. Here, L represents the preset maximum multiplicative depth, and a message $m \in \mathbb{C}^{N/2}$ is encoded and encrypted into R_{q_L}, followed by homomorphic addition and multiplication. Each multiplication consumes a layer of depth and rescales the ciphertext of R_{q_l} into $R_{q_{L-1}}$. Another essential operation enabled by ciphertext is rotation, which allows encrypted elements to rotate in $N/2$ slots.

One advantage of CKKS is on its SIMD feature, that is, one single homomorphic addition (or multiplication) among ciphertexts can attain a corresponding addition (or element-wise product) of two vectors in plaintexts. Suppose that m is a k-dimensional vector. When $k < N/2$, m will be padded with zeros to size $N/2$. As the homomorphic operation operates bit by bit on all slots, the operation is highly inefficient when $k << N/2$. The optimization methods introduced later in this paper will fully utilize the free spaces in the ciphertext to improve computational efficiency without incurring additional space consumption.

3.3 Homomorphic Linear Layer

Here we introduce the implementation of the homomorphic linear layer [5]. The linear layer is one of the most important network layers in the machine learning model. A linear layer consists of a vector-matrix multiplication and an addition of a bias. Traditionally, each column of the matrix can be packaged and multiplied by the ciphertext vector, and the result can be summed bit by bit. This practice causes the output to be spread across multiple ciphertexts rather than stored under a single ciphertext. Halevi et al. [17] proposed a matrix multiplication that is realized by matrix diagonalization. Let $n = N/2$ denotes the number of slots in ciphertext c, matrix $M \in R^{n \times m}$. Firstly M is decomposed into n vectors in diagonal order, in which the j'th element in the i'th diagonal $diag[i][j] = M[(i+j) mod\, n][j]$. Then we have $M.c = \sum_{i=0}^{n-1} diag(i) \odot Rotate(c, i)$, where $\odot$ denotes the coefficient wise vector multiplication and $Rotate(c, i) = (c_i, c_{i+1}, \ldots, c_0, c_1, \ldots, c_{i-1})$ is the rotation of c by shifting i slots to the left. In practice, in order to get the correct rotation result, it is necessary to copy the encryption vector [19], then $c = (c_0, c_1, \ldots, c_{n-1}, c_0, c_1, \ldots, c_{n-1}, 0, \ldots, 0)$. The complexity of vector-matrix multiplication is $O(n)$, and the computational cost is the largest in the homomorphic linear layer. Since the nonlinear activation function (such as ReLu) cannot be calculated in the ciphertext state, the standard practice [3] is to substitute it with the square activation function. The last layer of the network model is a homomorphic dot product layer, which is first multiplied by a k-dimensional vector. All elements are then added to the first element of the output ciphertext by $log(k)$ rotations. It should be noted that the final activation function is computed locally after decryption by the client [2].

3.4 Multimodal Fusion Representation Learning

The original multimodal data contains a large amount of redundant information, and the feature vectors of each modality are initially located in different subspaces. This can impede the learning of data in subsequent models. To address this issue, representation learning has been proposed as a solution [6]. This technique maps input data to a low-dimensional representation, enabling efficient learning. Currently, unimodal representation learning is widely used in Natural language processing (NLP)[13] and Computer Vision (CV) [18]. In our work, we employ an appropriate encoder network for each modality of the data, and fuse the resulting low-dimensional feature vectors. The validity of the representation is ensured by the convergence of network parameters.

Multimodality Fusion Technology (MFT) [23] involves fusing the feature vectors of each modality to create a more effective representation for subsequent networks. One simple method [33] for feature fusion is direct concatenation, but it can be challenging to capture the interaction information and nonlinear relationships between different modalities. To address this issue, Zadeh et al. proposed Tensor Fusion Network [32]. This approach models each sub-modality feature as different dimensions of a Cartesian space. Therefore, the fusion process between different modes can be achieved through the tensor cross-product, as shown in Eq. (1):

$$z = \begin{bmatrix} v_1 \\ 1 \end{bmatrix} \otimes \begin{bmatrix} v_2 \\ 1 \end{bmatrix} \otimes \cdots \otimes \begin{bmatrix} v_n \\ 1 \end{bmatrix} \tag{1}$$

where z represents the output after TFN, v_i denotes different modalities and $\otimes$ denotes the outer product operator. To ensure that the fusion representation contains both the cross-information from dual mode to n mode and the independent information of each unimodal feature, an element of 1 is spliced at the end of each modality representation.

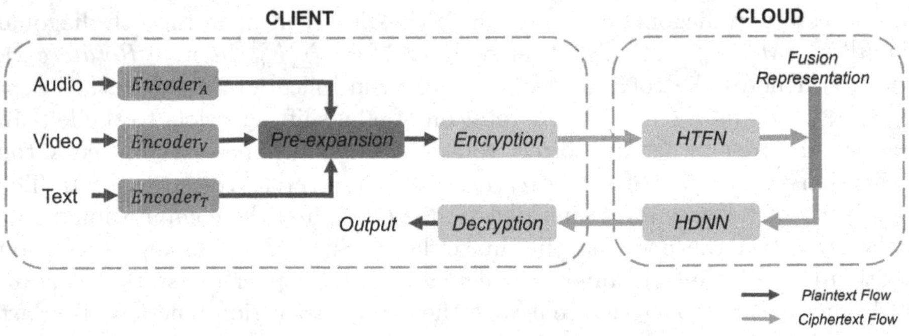

Fig. 1. HE-based Multimodal Representation Inference System

4 The Proposed Approach

4.1 System Model

This section introduces our system model and task settings. A client possesses multimodal raw data and uses the corresponding encoder for feature extraction. Subsequently, the client performs pre-expansion and transmits the ciphertext to the cloud. In the cloud, the encrypted fusion representation is obtained through the homomorphic tensor fusion network (HTFN). The homomorphic dense neural network (HDNN) performs the final evaluation, and the encryption result is delivered back to the client. The system model is illustrated in Fig. 1. Throughout the computing process, the server is limited to operating solely on the ciphertext and cannot access the sensitive data.

Consistent with current private reasoning tasks, such as those described in [8, 11, 22], the complete model network utilizes plaintext data during the common training phase and encrypted data during the inference phase. In this task, we implicitly assume that the client has the computational capacity to preprocess the raw data and execute the encryption/decryption tasks.

4.2 Homomorphic TFN

In TFN, the fusion features of a high-dimensional Cartesian product are obtained by multiplying the eigenvectors of each modality twice. However, in the ciphertext state, only the operation between vectors is supported, and the result of the Cartesian product cannot be retrieved directly. One possible solution, related to CryptoNets, is to encrypt each element of each feature independently and send it to the server for multiplication one by one. However, this approach is computationally expensive since the fusion requires computing hundreds of ciphertexts. An optimization for this method is *packing one modality* (POM) into a single ciphertext and multiply it with the elements of other features one by one. Even with this optimization, the computational complexity and the amount of output ciphertext are still dependent on the length of the feature.

To solve this issue, we propose a *pre-expansion* processing method. As the fusion features of the higher-dimensional Cartesian product will be flattened as input to the linear layer, and the SIMD feature of CKKS makes the bit-by-bit multiplication between ciphertexts efficient, we expand each representation to the fused length before encryption. In the simplest case of two modalities, pre-expansion will repeat each bit of one modality for L_1 times, given the other modality has a length of L_1. The other modality will also be expanded so as to be aligned with the previous modality, and finally both modality will be with a length of $L_1 \cdot L_2$, given L_2 is the length of the first modality. This will ensure that only one ciphertext multiplication is needed in the HTFN.

Let's consider the most widely used fusion of video, audio, and text, as an example. As shown in Fig. 2, the lengths of text, video and audio modality are 2, 2, 3, respectively. Firstly each bit of the video modality is repeated by 2 times, so that each bit can be aligned with the whole length of the text modality. Then correspondingly the lengths of video and text modalities will both be expanded to 4. In the same way, each bit of the audio modality is repeated by 4 times, and finally, all the three modalities will be expanded to be with a length of 12.

Let m denote the number of modalities, and L_i denote the length of the i'th modality feature. After pre-expansion, all modalities will be with a length of $\prod_{i=1}^{m} L_i$, and $m-1$ ciphertext multiplications are required to compute the HTFN. As shown in Table 1, the number of required ciphertext multiplications in HTFN can be reduced using pre-expansion, and the resulting fused representation can be stored in a single ciphertext. In Sect. 3.2, we mentioned that the CKKS ciphertext has $N/2$ slots, which is significantly larger than the length of each feature. Therefore, pre-expansion can take advantage of the vacant slots in the ciphertext without additional space requirements.

Table 1. Performance Comparison of HTFN Under Different Pretreatment Methods.

Pretreatment Method	Multiplications	Output Size
CryptoNets	$\prod_{i=1}^{m} L_i - 1$	$\prod_{i=1}^{m} L_i$
POM	$\prod_{i=1}^{m-1} L_i$	$\prod_{i=1}^{m-1} L_i$
Pre-expansion	$m-1$	1

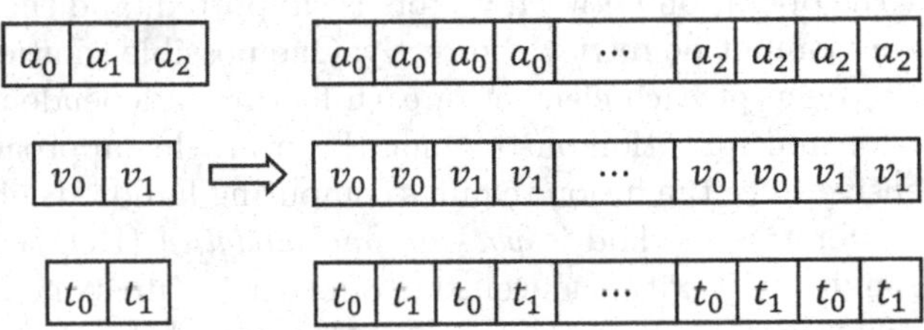

Fig. 2. An Illustration of Pre-expansion on 3 Modalities

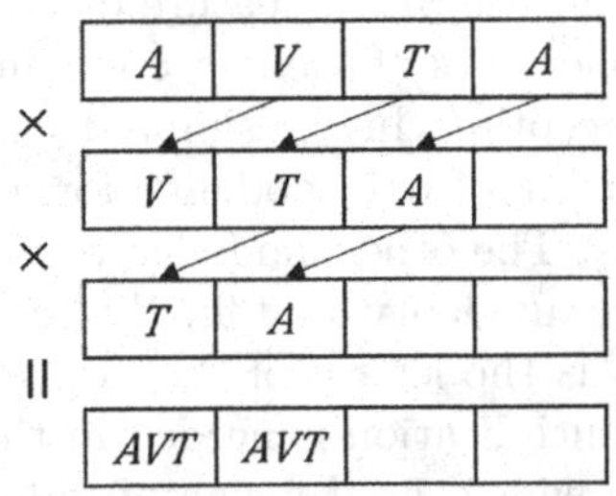

Fig. 3. An Illustration of HTFN with our Packing Method based on Ciphertext Rotations

4.3 Other Optimizations

The client sends a ciphertext for each modality to the cloud. As the number of modalities increases, the size of the ciphertexts also increases, resulting in a linear increase in encryption time and data transmission. To solve this problem, we concatenate all the feature vectors after the pre-expansion and finally encrypt them into a single ciphertext. In the ciphertext state, the ciphertext of each feature can be decoupled by two CKKS rotation operations. As mentioned in Sect. 3.3, the subsequent homomorphic linear layer computation requires a copy of the original vector. This can be achieved by concatenating the features of the first modality.

As shown in Fig. 3, assuming that the features fused by the three modalities have n dimensions, we perform two consecutive rotation operations to obtain two ciphertexts, where the first n bits of each ciphertext store the corresponding modality representation. Through this packaging method, the transmission of the original n ciphertexts can be optimized to a single ciphertext, which significantly decreases the client's encryption computation cost and transmission delay. Finally, to optimize the homomorphic linear layer with the longest computation time, we employ a multi-threaded approach for the n-times rotation and homomorphic multiplication. Algorithm 1 outlines the complete process of HTFN, where $\oplus$ and $\odot$ represent homomorphic addition and multiplication, respectively. For the sake of simplicity, we omit the relinearization and rescale operations that are mandatory after homomorphic multiplication and evaluation.

5 Experiments

5.1 Experimental Setup

We use two commonly used multimodal datasets for our experiments. Multimodal Corpus of Sentiment Intensity (CMU-MOSI) [33] contains 2199 movie reviews on YouTube video blog, which came from 89 narrators of ages among 20 to 30 years. Each has audio, text, and video information available. The videos were labeled $[-3, 3]$ with seven categories ranging from negative to positive affective tendencies. Different from the original CMU-MOSI dataset, we use BERT preprocessing to obtain more accurate text information [31]. CH-SIMS [30] is a Chinese multimodal emotion analysis data set. It cuts out 2281 video clips from different movies and television works and collates the data in audio, text and video modalities. It contains 474 different speakers with a wide range of characters and ages. Each video clip has five categories of emotional intensity values between $[-1, 1]$.

Algorithm 1: Homomorphic Multimodal Representation Inference

```
Input: ciphertext c encrypting vector v ∈ R^{m.k}, where m is the number of
       modalities, linear layer weight W ∈ R^{k×z}, y ∈ R^z, bias b1 ∈ R^z, b2 ∈ R
Output: ciphertext s encrypting the evaluation result
Step 1: Homomorphic Tensor Fusion Layer
r ← c ∈ R^N;
for i = 1 to m − 1 do
    r ← r ⊙ Rotate(c, k ∗ i);
end
Step 2: Homomorphic Linear Layer
s ← 0 ∈ R^N
for i = 0 to k − 1 do
    for j = 0 to z − 1 do
        D[i][j] ← W[(i + j)mod k][j];
    end
    s ← s ⊕ D[i] ⊙ Rotate(r, i);
end
s ← s ⊕ b1;
s ← s ⊙ s;
Step 3: Homomorphic Dot Product Layer
s ← s ⊙ y;
for i = ⌈log(z)⌉ to 1 do
    t ← Rotate(s, 2^i);
    s ← s ⊕ t;
end
s ← s ⊕ b2;
return s
```

For network settings, as in the previous work of [32], we choose LSTM to extract text features. DNN with three hidden layers is used for audio and video features. Following this, we use two hidden layers to make the final prediction of the fusion representation. It is worth noting that the maximum fusion representation length should not exceed 2048. Considering the actual usability, our experiment uses a smaller length, 512. The setting of other hyperparameters and the partitioning of data sets are consistent with the TFN experimental settings in MMSA [31]. After the model training, we use the SEAL library [25] to carry out the inference task of homomorphic encryption.

5.2 Experimental Results

In Table 2, we compare the inference results of the model before and after encryption on the test set using the F1 score and MAE as evaluation metrics, which are commonly used in multimodal machine learning. Results demonstrate that our encryption inference method is highly effective, as our model exhibits no performance loss compared to plaintext inference up to four decimal places of precision.

Table 3 illustrates the enhancements made by the proposed optimization method concerning data throughput and inference time in the cloud. We began with the idea of CryptoNets, which encrypts each element individually. Although it supports batch prediction of multiple samples, the time for a single calculation is approximately 150 s. Lola [8] is a low-delay computing model proposed by CryptoNets for single-sample calculations, but it does not support multimodal feature fusion. In POM, each element of the two features (lengths of 4 and 16) had to be packed into a separate ciphertext since TFN necessitates element-wise multiplication. The computation time has been reduced to 4.43 s, but for scenarios with high feature lengths, this time will increase rapidly and require larger data transfers. With pre-expansion, feature fusion can be completed with only two ciphertext vector multiplications, enabling the use of smaller encryption parameters like $N = 8192$ for further optimization while ensuring security. Furthermore, packing the features of the three modalities into a single ciphertext reduces the amount of data transmission to 1/3, which must be decoupled by two rotation operations on the server. Multithreading can effectively reduce the delay of ciphertext vector-matrix multiplication for the first homomorphic linear layer after feature fusion. Finally, the proposed optimization method yields an inference time of approximately 0.91 s for a single sample and the data transfer volume is 211.88 KB.

We present a detailed comparison of the performance of HTFN implemented under three preprocessing methods in Table 4. The number of multiplications required by our approach is linearly dependent on the number of modes, whereas the other two methods show a rapid increase in the number of multiplications as the feature length increases. Our approach, combined with two rotations and packaging, compactly stores both input and output in a single ciphertext, thus reducing the encryption burden on the client and the computational cost on the server. Experimental results demonstrate that our method can maintain stable and efficient computing performance in practical multimodal scenarios.

Table 2. Encrypted Test Set Inference Results for CMU-MOSI and SIMS.

Dataset	F1 score	MAE	Accuracy Loss
CMU-MOSI	0.7467	1.0400	No
SIMS	0.7697	0.4356	No

Table 3. Model Performance - Data Throughput and Timing.

Method	Data Throughput	Computation Time
CKKS	6.03 MB	149.21 s
CKKS + POM	2.70 MB	4.43 s
CKKS + pre-expansion	637.14 KB	3.12 s
CKKS + pre-expansion + packed	211.88 KB	3.14 s
CKKS + pre-expansion + packed + multi-thread	211.88 KB	0.91 s

Table 4. Performance Comparisons of HTFN.

	Ours	POM	CryptoNets
Multiplication	2	68	544
Rotation	2	-	-
Input Size	427.08 KB	8.77 MB	12.09 MB
Output Size	282.95 KB	17.92 MB	141.56 MB
Computation Time	0.036 s	1.051 s	8.407 s
Encryption Time	0.006 s	0.212 s	0.350 s

6 Conclusion

In this paper, we propose the first multimodal representation inference method based on Fully Homomorphic Encryption (FHE). Our method provides complete protection of multimodal data privacy and enables private prediction tasks on the cloud server. We propose a pre-expansion method and a homomorphic TFN scheme using only two ciphertext multiplications. We also propose a variety of optimization schemes that not only improve the computational efficiency of ciphertext but also reduce the amount of data transmission. The experimental results show that our method has the advantages of low computation and communication costs. In the future, we will continue to improve the performance of the model and extend it to multi-user application scenarios using multi-key FHE.

Acknowledgement. This work was supported by the Key-Area Research and Development Program of Guangdong Province (No. 2020B010164003), China.

References

1. Albrecht, M.R., et al.: Homomorphic encryption standard. IACR Cryptol. ePrint Arch., 939 (2019). https://eprint.iacr.org/2019/939
2. Badawi, A.A., Hoang, L., Mun, C.F., Laine, K., Aung, K.M.M.: PrivFT: private and fast text classification with homomorphic encryption. IEEE Access **8**, 226544–226556 (2020). https://doi.org/10.1109/ACCESS.2020.3045465
3. Badawi, A.A., et al.: Towards the AlexNet moment for homomorphic encryption: HCNN, the first homomorphic CNN on encrypted data with GPUs. IEEE Trans. Emerg. Top. Comput. **9**(3), 1330–1343 (2021). https://doi.org/10.1109/TETC.2020.3014636
4. Baltrusaitis, T., Ahuja, C., Morency, L.: Multimodal machine learning: a survey and taxonomy. IEEE Trans. Pattern Anal. Mach. Intell. **41**(2), 423–443 (2019). https://doi.org/10.1109/TPAMI.2018.2798607
5. Benaissa, A., Retiat, B., Cebere, B., Belfedhal, A.E.: TenSEAL: a library for encrypted tensor operations using homomorphic encryption. CoRR abs/2104.03152 (2021). https://arxiv.org/abs/2104.03152

6. Bengio, Y., Courville, A.C., Vincent, P.: Representation learning: a review and new perspectives. IEEE Trans. Pattern Anal. Mach. Intell. **35**(8), 1798–1828 (2013). https://doi.org/10.1109/TPAMI.2013.50
7. Brakerski, Z., Gentry, C., Vaikuntanathan, V.: (Leveled) fully homomorphic encryption without bootstrapping. ACM Trans. Comput. Theor. **6**(3), 13:1-13:36 (2014). https://doi.org/10.1145/2633600
8. Brutzkus, A., Gilad-Bachrach, R., Elisha, O.: Low latency privacy preserving inference. In: Chaudhuri, K., Salakhutdinov, R. (eds.) Proceedings of the 36th International Conference on Machine Learning, ICML 2019, 9–15 June 2019, Long Beach, California, USA. Proceedings of Machine Learning Research, vol. 97, pp. 812–821. PMLR (2019). http://proceedings.mlr.press/v97/brutzkus19a.html
9. Cai, C., Sang, Y., Tian, H.: A multimodal differential privacy framework based on fusion representation learning. Connect. Sci. **34**(1), 2219–2239 (2022). https://doi.org/10.1080/09540091.2022.2111406
10. Cheon, J.H., Kim, A., Kim, M., Song, Y.: Homomorphic encryption for arithmetic of approximate numbers. In: Takagi, T., Peyrin, T. (eds.) ASIACRYPT 2017. LNCS, vol. 10624, pp. 409–437. Springer, Cham (2017). https://doi.org/10.1007/978-3-319-70694-8_15
11. Chou, E.J., Gururajan, A., Laine, K., Goel, N.K., Bertiger, A., Stokes, J.W.: Privacy-preserving phishing web page classification via fully homomorphic encryption. In: 2020 IEEE International Conference on Acoustics, Speech and Signal Processing, ICASSP 2020, Barcelona, Spain, 4–8 May 2020, pp. 2792–2796. IEEE (2020). https://doi.org/10.1109/ICASSP40776.2020.9053729
12. Deldjoo, Y., Schedl, M., Hidasi, B., Wei, Y., He, X.: Multimedia recommender systems: algorithms and challenges, 3rd edn. In: Recommender Systems Handbook (2020)
13. Devlin, J., Chang, M., Lee, K., Toutanova, K.: BERT: pre-training of deep bidirectional transformers for language understanding. In: Burstein, J., Doran, C., Solorio, T. (eds.) Proceedings of the 2019 Conference of the North American Chapter of the Association for Computational Linguistics: Human Language Technologies, NAACL-HLT 2019, Volume 1 (Long and Short Papers), Minneapolis, MN, USA, 2–7 June 2019, pp. 4171–4186. Association for Computational Linguistics (2019). https://doi.org/10.18653/v1/n19-1423
14. Fan, J., Vercauteren, F.: Somewhat practical fully homomorphic encryption. IACR Cryptol. ePrint Arch., 144 (2012). http://eprint.iacr.org/2012/144
15. Gentry, C.: Fully homomorphic encryption using ideal lattices. In: Mitzenmacher, M. (ed.) Proceedings of the 41st Annual ACM Symposium on Theory of Computing, STOC 2009, Bethesda, MD, USA, 31 May–2 June 2009, pp. 169–178. ACM (2009). https://doi.org/10.1145/1536414.1536440
16. Gilad-Bachrach, R., Dowlin, N., Laine, K., Lauter, K.E., Naehrig, M., Wernsing, J.: CryptoNets: applying neural networks to encrypted data with high throughput and accuracy. In: Balcan, M., Weinberger, K.Q. (eds.) Proceedings of the 33nd International Conference on Machine Learning, JMLR Workshop and Conference Proceedings, ICML 2016, New York City, NY, USA, 19–24 June 2016, vol. 48, pp. 201–210. JMLR.org (2016). http://proceedings.mlr.press/v48/gilad-bachrach16.html
17. Halevi, S., Shoup, V.: Algorithms in HElib. IACR Cryptol. ePrint Arch. **2014**, 106 (2014)

18. He, K., Zhang, X., Ren, S., Sun, J.: Deep residual learning for image recognition. In: 2016 IEEE Conference on Computer Vision and Pattern Recognition, CVPR 2016, Las Vegas, NV, USA, 27–30 June 2016, pp. 770–778. IEEE Computer Society (2016). https://doi.org/10.1109/CVPR.2016.90
19. Huynh, D.: Cryptotree: fast and accurate predictions on encrypted structured data. CoRR abs/2006.08299 (2020). https://arxiv.org/abs/2006.08299
20. McKeen, F., et al.: Innovative instructions and software model for isolated execution. In: Lee, R.B., Shi, W. (eds.) The Second Workshop on Hardware and Architectural Support for Security and Privacy, HASP 2013, Tel-Aviv, Israel, 23–24 June 2013, p. 10. ACM (2013). https://doi.org/10.1145/2487726.2488368
21. Ngiam, J., Khosla, A., Kim, M., Nam, J., Lee, H., Ng, A.Y.: Multimodal deep learning. In: Getoor, L., Scheffer, T. (eds.) Proceedings of the 28th International Conference on Machine Learning, ICML 2011, Bellevue, Washington, USA, 28 June–2 July 2011, pp. 689–696. Omnipress (2011). https://icml.cc/2011/papers/399_icmlpaper.pdf
22. Rahulamathavan, Y.: Privacy-preserving similarity calculation of speaker features using fully homomorphic encryption. CoRR abs/2202.07994 (2022). https://arxiv.org/abs/2202.07994
23. Ramachandram, D., Taylor, G.W.: Deep multimodal learning: a survey on recent advances and trends. IEEE Sig. Process. Mag. **34**(6), 96–108 (2017). https://doi.org/10.1109/MSP.2017.2738401
24. Rivest, R.L., Dertouzos, M.L.: On Data Banks and Privacy Homomorphisms (1978)
25. Microsoft SEAL (release 4.0). Microsoft Research, Redmond, WA, March 2022. https://github.com/Microsoft/SEAL
26. Sun, L., Wang, J., Zhang, K., Su, Y., Weng, F.: RpBERT: a text-image relation propagation-based BERT model for multimodal NER. In: Thirty-Fifth AAAI Conference on Artificial Intelligence, AAAI 2021, Thirty-Third Conference on Innovative Applications of Artificial Intelligence, IAAI 2021, The Eleventh Symposium on Educational Advances in Artificial Intelligence, EAAI 2021, Virtual Event, 2–9 February 2021, pp. 13860–13868. AAAI Press (2021). https://ojs.aaai.org/index.php/AAAI/article/view/17633
27. Wang, D., Xiong, D.: Efficient object-level visual context modeling for multimodal machine translation: masking irrelevant objects helps grounding. In: Thirty-Fifth AAAI Conference on Artificial Intelligence, AAAI 2021, Thirty-Third Conference on Innovative Applications of Artificial Intelligence, IAAI 2021, The Eleventh Symposium on Educational Advances in Artificial Intelligence, EAAI 2021, Virtual Event, 2–9 February 2021, pp. 2720–2728. AAAI Press (2021). https://ojs.aaai.org/index.php/AAAI/article/view/16376
28. Yao, A.C.: Protocols for secure computations (extended abstract). In: 23rd Annual Symposium on Foundations of Computer Science, Chicago, Illinois, USA, 3–5 November 1982, pp. 160–164. IEEE Computer Society (1982). https://doi.org/10.1109/SFCS.1982.38
29. Yu, F., et al.: ERNIE-ViL: knowledge enhanced vision-language representations through scene graphs. In: Thirty-Fifth AAAI Conference on Artificial Intelligence, AAAI 2021, Thirty-Third Conference on Innovative Applications of Artificial Intelligence, IAAI 2021, The Eleventh Symposium on Educational Advances in Artificial Intelligence, EAAI 2021, Virtual Event, 2–9 February 2021, pp. 3208–3216. AAAI Press (2021). https://ojs.aaai.org/index.php/AAAI/article/view/16431

30. Yu, W., et al.: CH-SIMS: a Chinese multimodal sentiment analysis dataset with fine-grained annotation of modality. In: Jurafsky, D., Chai, J., Schluter, N., Tetreault, J.R. (eds.) Proceedings of the 58th Annual Meeting of the Association for Computational Linguistics, ACL 2020, Online, 5–10 July 2020, pp. 3718–3727. Association for Computational Linguistics (2020). https://doi.org/10.18653/v1/2020.acl-main.343
31. Yu, W., Xu, H., Yuan, Z., Wu, J.: Learning modality-specific representations with self-supervised multi-task learning for multimodal sentiment analysis. In: Thirty-Fifth AAAI Conference on Artificial Intelligence, AAAI 2021, Thirty-Third Conference on Innovative Applications of Artificial Intelligence, IAAI 2021, The Eleventh Symposium on Educational Advances in Artificial Intelligence, EAAI 2021, Virtual Event, 2–9 February 2021, pp. 10790–10797. AAAI Press (2021). https://ojs.aaai.org/index.php/AAAI/article/view/17289
32. Zadeh, A., Chen, M., Poria, S., Cambria, E., Morency, L.: Tensor fusion network for multimodal sentiment analysis. In: Palmer, M., Hwa, R., Riedel, S. (eds.) Proceedings of the 2017 Conference on Empirical Methods in Natural Language Processing, EMNLP 2017, Copenhagen, Denmark, 9–11 September 2017, pp. 1103–1114. Association for Computational Linguistics (2017). https://doi.org/10.18653/v1/d17-1115
33. Zadeh, A., Zellers, R., Pincus, E., Morency, L.: Multimodal sentiment intensity analysis in videos: facial gestures and verbal messages. IEEE Intell. Syst. **31**(6), 82–88 (2016). https://doi.org/10.1109/MIS.2016.94

Neural Machine Translation with Diversity-Enabled Translation Memory

Quang Chieu Nguyen[1,2], Xuan Dung Doan[1], Van-Vinh Nguyen[2], and Khac-Hoai Nam Bui[1(✉)]

[1] Viettel Cyberspace Center, Viettel Group, Hanoi, Vietnam
{chieunq,dungdx4,nambkh}@viettel.com.vn
[2] Vietnam National University of Hanoi, Hanoi, Vietnam
vinhvn@vnu.edu.vn

Abstract. Neural machine translation (NMT) using translation memory (TM) has been introduced as an emergent technique for improving machine translation systems (MTS). In this study, we propose an end-to-end NMT model with TM by exploiting the diversity of the retrieval-augmented phase using maximal marginal relevance (MMR). In particular, the proposed model is designed with monolingual TM, which is able to support low-resource scenarios. Furthermore, the memory retriever and translation models are jointly trained to improve translation performance. For the experiment, we use IWSLT15 (En $\longleftrightarrow$ Vi) as a benchmark dataset to evaluate the performance of the proposed method. Accordingly, the experiential results show the effectiveness of the proposed method compared with strong baselines in this research field.

Keywords: Neural Machine Translation · Translation Memory · Maximal Marginal Relevance · Low Resource Language

1 Introduction

Translation Memory (TM) is conceptually regarded as a database of sentence pairs (source and target texts), which is utilized to reuse previously translated content when working on new texts. Recent works have focused on memory augmentation to improve the performance of neural machine translation (TM-augmented NMT) [12].

Technically, a typical TM-augmented NMT model performs the translation process in two phases, as shown in Fig. 1: i) *Retrieval Stage* extracts the candidate sentence memories from training corpus based on calculating similarity; and ii) *Generation Stage* integrates the candidate sentences into translation model for the translation. Subsequently, the trend research focuses on jointly learning models of two phases (retriever and translation models) with remarkable results [4].

N. T. Nguyen et al. (Eds.): ACIIDS 2023, LNAI 13995, pp. 322–333, 2023.
https://doi.org/10.1007/978-981-99-5834-4_26

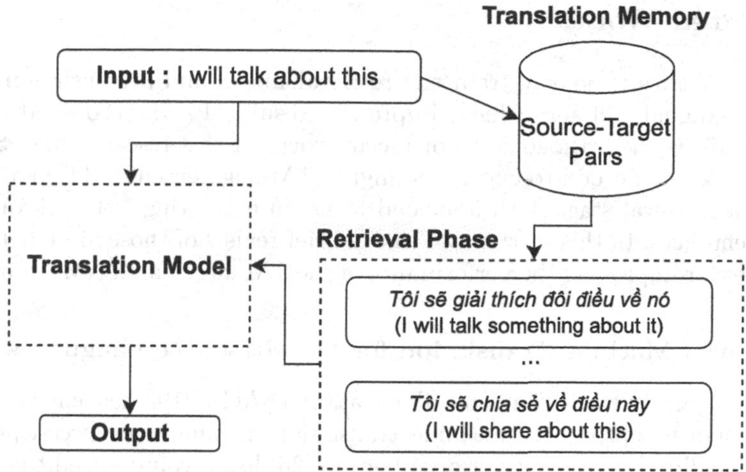

Fig. 1. An example of the neural machine translation using translation memory.

Despite the success of the recent TM-augmented NMT models, there are still two remaining research issues that need take into account: i) The retrieval stage mainly uses a greedy method to extract the top-r nearest sentence pairs, which results in redundant information because the top-r sentence pairs are highly similar to each other [4]. ii) Most previous works use TM with sentence pairs (source-target pairs), which is not able to take advantage of abundant monolingual data [9]. In this regard, this study proposes a new method for TM-based NMT to deal with the aforementioned issues. Specifically, for the retrieval phase, we adopt Maximal Marginal Relevance (MMR) [7] to enable the diversity, guaranteeing the two most challenging properties of candidates: *informativeness* obtained by the distance between query and candidates; *diversity* expressed by the distances among candidates themselves. For the monolingual TM, following the work in [4], a simple dual-encoder framework is adopted for selecting the most relevant sentences. Generally, the main contributions of this study are two folds as follows:

- We present a novel end-to-end model for TM-augmented NMT, which aims to leverage two emergent issues of TM-augmented NMT such as balancing the relevance and diversity of the retrieval phase and using monolingual data. To the best of our knowledge, the proposed method is the first study that integrates two aforementioned issues in a unified framework.
- We execute our approach on IWSLT15, a benchmark English-Vietnamese dataset [8] to demonstrate the effectiveness of the model in low-resource language pair scenarios. Specifically, the reported results show that our model outperforms strong baseline models in this research field.

The rest of the paper is organized as: Sect. 2 presents a brief review of previous works regarding this study. The proposed method is presented in Sect. 3. We report and analyze the evaluated results in Sect. 4. Section 5 is the conclusion and discussion of this study.

2 Related Work

Recent work tries to jointly train the retrieval model and a translation model with monolingual TM and achieve impressive results [4]. The proposed method in this study is the orthogonality of recent works of this research line. Specifically, we present an end-to-end monolingual TM-augmented NMT model that includes a retrieval stage with a special focus on extracting both relevant and diverse sentences. In this section, we take a brief review of those aforementioned techniques for improving the performance of the TM-augmented NMT approach.

2.1 Neural Machine Translation for Low-Resource Languages

In recent years, Neural Machine Translation (NMT) [19] has emerged as a state-of-the-art approach to machine translation, gaining widespread popularity. Specifically, the Transformer architecture [25] has revolutionized the field of NMT by achieving success in multiple language pairs. However, supervised NMT requires large datasets, which are often limited in low-resource languages. To address this issue, several data augmentation techniques have emerged, including back-translation [23], and self-training [15]. Additionally, transfer learning techniques [20,29] show promise in leveraging pre-trained models for improved performance. In cases where parallel data is not available, unsupervised technique NMT [2], pivot-based [10] or multi-NMT-based solution [11] can be employed. Recent studies focus on using TM with monolingual data, as an emergent technique, to improve the translation quality of NMT.

2.2 Translation Memory-Augmented Neural Machine Translation

Augmenting TM has become an emerging research topic for improving NMT. There are primarily two approaches for incorporating translation memory (TM) into neural machine translation (NMT): constraining the decoding process with TM and using TM to train a more powerful NMT model.

The main idea of the first research line is to increase the generation probability of some target words based on TM. Zhang et al. [28] increased the generation probability of target words aligned with the TM. In [13] a bilingual dictionary is used as auxiliary information to tackle infrequent word translation. Khandelwal et al. [17] used kNN-MT to retrieve TMs from dense vectors by creating a key-value datastore and interpolating the generation probability of the NMT model with similar target distributions from the datastore at each time step.

The second research line aims to train the translation model to learn how to deal with the retrieved TMs. A data augmentation way was used by Pham et al. [22] to concatenate the retrieved TMs with input sentences during training. Several studies have explored modifying the architecture of the NMT model to improve integration with TM. Cao et al. [6] introduced a gating mechanism to control the signal from the retrieved TM, and following this, in [5] an additional transformer encoder is designed to incorporate the target sentence of the TM through attention. In Xia et al. [26] work, multiple retrieved TMs are compressed into a graph structure to enhance efficiency and space usage and then integrated into the model through attention.

2.3 Retrieval for Translation Memory-Augmented Neural Machine Translation

Previous works focus on the TM with bilingual sentence pairs [26,27], which used fuzzy matching to retrieve the most similar sentences from the corpus with a query. In TM with monolingual, retrieval task is more challenging due to the cross-lingual setting. To address this challenge, Cai et al. [4] proposed a simple dual-encoder framework pre-trained on two tasks: sentence-level cross-alignment and token-level cross-alignment.

Regarding diversity for the retrieval results, authors in [9] have proven that diverse translation memories are able to improve the performance of the NMT, making it important to ensure diversity in the retrieval stage. There are several methods to enable diversity, including MMR [7], IA-Select [1] or MaxSum Diversification [3]. We employed MMR in this study due to its straightforward implementation and, more importantly, its ease of interpretation.

3 Methodology

3.1 Overview System

Figure 2 depicts the overview structure of the proposed method. In particular, the main contribution of this study focuses on the retrieval stage, which selects the most relevant and diverse sentences from a large monolingual TM in the target language. Specifically, given an input sentence x in the source language and a large monolingual TM $M = \{m_1, m_2, .., m_n\}$, the output of the retrieval stage is a subset (top k) TM and relevance scores $\{f(x, m_i)\}_{i=1}^{k}$ Then, the translation model conditions on both the input x, the retrieved set, and their scores x to generate the output y.

3.2 Retrieval Model

3.2.1 Relevant Monolingual TM: The input sentence x (source sentence) and monolingual TM M of target language are encoded by using two independent Transformers [25], which are sequentially formulated as follows:

$$\begin{aligned} z_x &= W_1 Trans_x(< bos >, x^1, x^2, ..x^{|x|}) \\ z_{m_i} &= W_2 Trans_m(< bos >, m_i^1, m_i^2, ..m_i^{|m_i|}) \end{aligned} \tag{1}$$

where $m_i \in M$ denotes the memory target sentence. W_1 and W_2 are learning parameters. In this regard, the relevance score $f(x, m_i)$ between the source sentence and the candidate sentence can be calculated using the dot product:

$$f(x, m_i) = z_x^T z_{m_i} \tag{2}$$

Subsequently, the top r relevant sentences are extracted using Maximum Inner Product Search (MIPS).

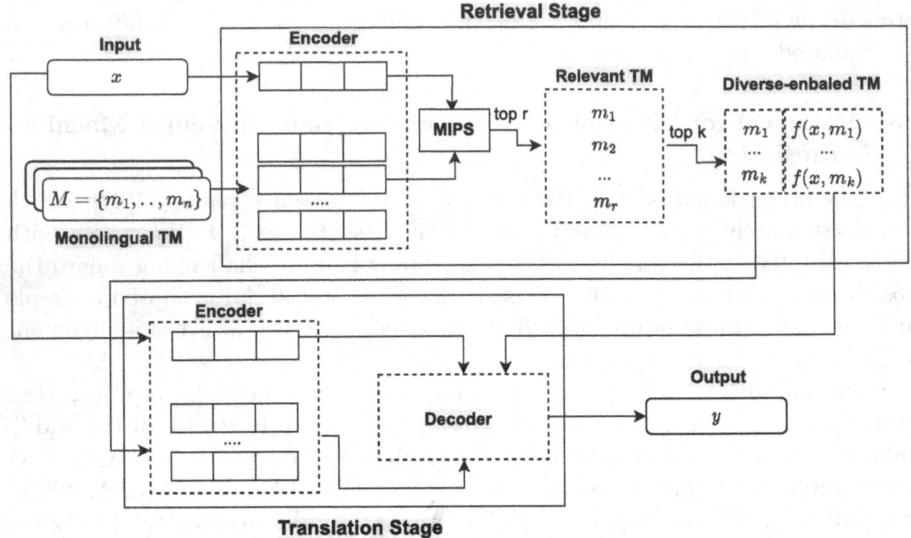

Fig. 2. Overall structure.

3.2.2 Diversity-Enabled TM: After obtaining $R = \{m_1, .., m_r\}$ as the relevant sentences, a subset of size k is selected from R is selected to increase the diversity by using MMR [7], the MMR function can be formulated as follows:

$$MMR(x, R, S) = \underset{m_i \in R \setminus S}{argmax}[\lambda.cosine(x, m_i) - (1 - \lambda).\underset{m_j \in S}{max}(cosine(m_i, m_j))] \quad (3)$$

where S is the current set of chosen candidates. $R \setminus S$ is a set of unselected sentences. The hyperparameter λ, which takes values in the range $[0, 1]$, is used to trade off accuracy and diversity. A high value of lambda corresponds to high accuracy, whereas a low value corresponds to high diversity.

Algorithm 1: Maximal Marginal Relevance (MMR)

---

input : top-r most relevant sentences $R = \{m_i\}_{i=1}^{r}$ and result size k
output: result set $S \subseteq R, |S| = k$

```
S ← ∅
while |S| < k do
    s_s = MMR(x, R, S)
    S ← S ∪ s_s
    R = R \ s_s
end
```

The diverse-enabled TM processed can be described in the Algorithm 1. Specifically, the output of this process is a set of translation memories $S = \{m_1, .., m_k\}$ and its retrieval score $f(x, S) = \{f(x, m_1), .., f(x, m_k)\}$.

3.3 Translation Model

For the translation stage, we follow the work in [4] for the end-to-end model, which is built based on the standard encoder-decoder NMT model [25]. Specifically, given source sentence x, a set of retrieval TM $S = \{m_i\}_{i=1}^{k}$ and its scores $\{f(x, m_i)\}_{i=1}^{k}$ in the previous step, the objective of the translation model is to define the conditional probability as follows:

$$p(y|x, m_1, f(x, m_1), ..., m_k, f(x, m_k)) \tag{4}$$

To incorporate the information of TM contextualized token embeddings $\{z_{m_i,j}\}_{j=1}^{|m_i|}$ $(1 \le i \le k)$, the cross attention is calculated as follows:

$$\alpha_{ij} = \frac{exp(h_t^T W_3 z_{m_i,j} + \beta f(x, m_i))}{\sum_{i=1}^{i=k} \sum_{l=1}^{L_i} exp(h_t^T W_3 z_{m_i,l} + \beta f(x, m_i))} \tag{5}$$

Here, α_{ij} and L_i denote the attention score of the j-th token in z_{m_i} and the length of the sentence z_{m_i}, respectively. W_3 represents the learning parameter. h_t is the decoder's hidden state at time step t. The weighted sum of memory information can be updated as follows:

$$c_t = W_4 \sum_{i=1}^{k} \sum_{j=1}^{L_i} \alpha_{i,j} z_{i,j} \tag{6}$$

where W_4 denotes the learning parameter. Following this, h_t is updated with c_t, i.e., $h_t = h_t + c_t$. In this regard, the next token probabilities can be computed as follows:

$$p(y_t|x, m_1, f(x, m_1), ..., m_k, f(x, m_k)) = (1 - \lambda_t) P_v(y_t) + \lambda_t \sum_{i=1}^{k} \sum_{j=1}^{L_i} \alpha_{i,j} \mathbb{1}_{z_{m_i,j}=y_t} \tag{7}$$

where $\lambda_t = g(h_t, c_t)$ denotes the feed-forward network, $\mathbb{1}$ is the indicator function, the next-token probabilities P_v are obtained by converting the hidden state h_t using a linear projection and then applying the softmax function, which can be formulated as follows:

$$P_v = softmax(W_v h_t + b_v) \tag{8}$$

4 Experiment

4.1 Experiment Setup

4.1.1 Dataset and Evaluation: We use the English-Vietnamese as the evaluated dataset of this study (publicly available in the MT track of the IWSLT 2015 corpus [8]). Specifically, this dataset comprises a collection of parallel sentences in spoken language domains. The detailed data statistic is illustrated in Table 1.

Table 1. Statistics of the evaluated dataset.

# Train Pairs	# Dev Pairs	# Test Pairs
133317	1553	1268

In all experiments, the target language in the training set is utilized as monolingual translation memory data M. Subsequently, different bilingual datasets are generated for later experiments by randomly selecting 60%, 80%, 100% of the training dataset, referred to as D60, D80, and D100 datasets, respectively. For evaluation, we use the BLEU score [21].

4.1.2 Baseline Models: We compare the proposed model with the following baselines:

- **NMT wo TM**: the original NMT model without TM [25].
- **NMT + TM-BM25**: source similarity search method based on BM25, which is used in many recent TM-augmented NMT models [14,26].
- **NMT + Monolingual TM**: The joint training retrieval and translation models by adopting a dual encoder architecture [4].

4.1.3 Implementation Details: Our model utilizes Transformer blocks with the same setup as Transformer Base [25], which includes 8 attention heads, a hidden state with 512 dimensions, and a feed-forward state with 2048 dimensions. We employ 3 Transformer blocks for the retrieval model, 4 blocks for the memory encoder in the translation model, and 6 blocks for the encoder-decoder architecture in the translation model. We set trade-off hyperparameter in MMR $\lambda = 0.5$. The FAISS [16] has been used for indexing the dense representations. The learning rate schedule, dropout, and label smoothing are set following the default settings in [25]. We use Adam optimizer [18] and train models with up to $30K$ steps throughout all experiments. The number of tokens in every batch is 4096. BPE [24] tokenizer is employed with a vocabulary size of 4000. In order to execute the BM25-based method, we used a BM25 search engine[1] to obtain a preliminary set of TM sentences.

4.2 Main Results

Table 2 shows the results of the evaluation on different sizes of the training dataset. Particularly, the reported results are conducted with 3 and 5 retrieval sentences for the TM, respectively. As reported results, we make the following observations: i) NMT by using greedy retrieval (e.g., BM25) does not outperform the original NMT model, which indicates that joint training is an important method for the MT-augmented NMT approach; ii) Our model, which focuses

[1] https://github.com/dorianbrown/rank_bm25.

Table 2. Report results (BLEU scores) with two different values of top k retrieval TM for English ⟶ Vietnamese. Bold texts are the best results of each column.

Dataset	Model	k = 3		k = 5	
		Dev	Test	Dev	Test
D60	NMT wo TM	24.31	26.14	24.31	26.14
	NMT+TM-BM25	23.8	25.1	24.03	25.39
	NMT+Monolingual TM	24.22	26.51	24.38	26.72
	Our Model	**24.45**	**27.3**	**24.59**	**26.87**
D80	NMT wo TM	25.61	28.14	25.61	28.14
	NMT+TM-BM25	25.65	27.5	25.38	28.21
	NMT+Monolingual TM	25.61	28.32	25.7	**28.69**
	Our Model	**25.8**	**28.71**	**25.84**	28.67
D100	NMT wo MT	26.53	29.56	26.53	29.56
	NMT+TM-BM25	26.31	28.99	26.56	29.37
	NMT+Monolingual TM	**26.7**	**30.03**	26.74	29.28
	Our Model	**26.7**	29.6	**26.88**	**29.84**

on improving the retrieval sentence in terms of enabling the diversity for TM, achieves the best results compared with strong baseline models. The reported results indicate that diverse TM is able to improve the performance of NMT, especially with the low-resource scenarios; iii) The results between the number k sentences (k = 3 and k = 5) of the retrieval models are not too different. However, in our opinion, selecting the number of k sentences should be regarded as a hyperparameter and tuned during the training process. We leave this issue as future work regarding this study.

Table 3. Report the BLEU scores obtained when comparing monolingual translation memory (TM) and bilingual TM for English ⟶ Vietnamese.

Dataset	Model	Dev	Test
D60	Cheng et al., 2022 [9]	24.22	26.48
	Our Model+Bilingual TM	24.4	27.17
	Our Model+Monolingual TM	**24.45**	**27.3**
D80	Cheng et al., 2022 [9]	25.53	28.39
	Our Model+Bilingual TM	25.78	28.54
	Our Model+Monolingual TM	**25.8**	**28.71**
D100	Cheng et al., 2022 [9]	26.02	29.34
	Our Model+Bilingual TM	26.62	29.1
	Our Model+Monolingual TM	**26.7**	**29.6**

Furthermore, we also try to evaluate the performance between monolingual TM and bilingual TM. Accordingly, we re-implement the most recent work [9] for bilingual TM and comparing with our method in terms of both monolingual and bilingual, respectively. Table 3 shows the results of the variant of our model and Cheng et al., [9] with $k = 3$. An interesting observation is that the performance of monolingual TM is slightly better than bilingual TM. The evaluated results indicate that taking advantage of abundant monolingual data is able to improve the performance of NMT tasks, especially for low-resource scenarios.

5 Conclusion

In this paper, we propose a new framework for TM-augmented NMT by enabling the diversity of monolingual TM. To the best of our knowledge, the proposed method is the first study of end-to-end TM-augmented NMT that takes both monolingual and diversity-enabled TM into account. Specifically, by adding a non-heuristic module using the MMR algorithm, our proposed framework is able to enable diversity for the retrieval stage. Furthermore, instead of utilizing bilingual sentence pairs for the retrieval stage, we adopt two transformer encoders to exploit the capability of abundant information by monolingual data. Experiments show the effectiveness of the proposed method. Specifically, with varying the number of training datasets, our method is able to increase the performance from 0.5 to 1 Bleu score compared with strong baseline models in this research field. Regarding the future work of this study, we plan to exploit the size of translation memory by integrating this hyperparameter into the learning process in order to improve the performance of TM-augmented NMT tasks.

References

1. Agrawal, R., Gollapudi, S., Halverson, A., Ieong, S.: Diversifying search results. In: Proceedings of the Second ACM International Conference on Web Search and Data Mining, WSDM 2009, pp. 5–14. Association for Computing Machinery, New York (2009). https://doi.org/10.1145/1498759.1498766
2. Artetxe, M., Labaka, G., Agirre, E., Cho, K.: Unsupervised neural machine translation (2018)
3. Borodin, A., Lee, H.C., Ye, Y.: Max-sum diversification, monotone submodular functions and dynamic updates. In: Proceedings of the 31st ACM SIGMOD-SIGACT-SIGAI Symposium on Principles of Database Systems. PODS 2012, pp. 155–166. Association for Computing Machinery, New York (2012). https://doi.org/10.1145/2213556.2213580
4. Cai, D., Wang, Y., Li, H., Lam, W., Liu, L.: Neural machine translation with monolingual translation memory. In: Zong, C., Xia, F., Li, W., Navigli, R. (eds.) Proceedings of the 59th Annual Meeting of the Association for Computational Linguistics and the 11th International Joint Conference on Natural Language Processing, ACL/IJCNLP 2021, (Volume 1: Long Papers), Virtual Event, 1–6 August 2021, pp. 7307–7318. Association for Computational Linguistics (2021). https://doi.org/10.18653/v1/2021.acl-long.567

5. Cao, Q., Kuang, S., Xiong, D.: Learning to reuse translations: guiding neural machine translation with examples. In: Giacomo, G.D., et al. (eds.) 24th European Conference on Artificial Intelligence, ECAI 2020, Santiago de Compostela, Spain, 29 August–8 September 2020, Including 10th Conference on Prestigious Applications of Artificial Intelligence, PAIS 2020. Frontiers in Artificial Intelligence and Applications, vol. 325, pp. 1982–1989. IOS Press (2020). https://doi.org/10.3233/FAIA200318
6. Cao, Q., Xiong, D.: Encoding gated translation memory into neural machine translation. In: Riloff, E., Chiang, D., Hockenmaier, J., Tsujii, J. (eds.) Proceedings of the 2018 Conference on Empirical Methods in Natural Language Processing, Brussels, Belgium, 31 October–4 November 2018, pp. 3042–3047. Association for Computational Linguistics (2018). https://doi.org/10.18653/v1/d18-1340
7. Carbonell, J., Goldstein, J.: The use of MMR, diversity-based reranking for reordering documents and producing summaries. In: Proceedings of the 21st Annual International ACM SIGIR Conference on Research and Development in Information Retrieval, SIGIR 1998, pp. 335–336. Association for Computing Machinery, New York (1998). https://doi.org/10.1145/290941.291025
8. Cettolo, M., Niehues, J., Stüker, S., Bentivogli, L., Cattoni, R., Federico, M.: The IWSLT 2015 evaluation campaign. In: Proceedings of the 12th International Workshop on Spoken Language Translation: Evaluation Campaign, Da Nang, Vietnam, 3–4 December 2015, pp. 2–14 (2015). https://aclanthology.org/2015.iwslt-evaluation.1
9. Cheng, X., Gao, S., Liu, L., Zhao, D., Yan, R.: Neural machine translation with contrastive translation memories. In: Goldberg, Y., Kozareva, Z., Zhang, Y. (eds.) Proceedings of the 2022 Conference on Empirical Methods in Natural Language Processing, EMNLP 2022, Abu Dhabi, United Arab Emirates, 7–11 December 2022, pp. 3591–3601. Association for Computational Linguistics (2022). https://aclanthology.org/2022.emnlp-main.235
10. Cheng, Y., Yang, Q., Liu, Y., Sun, M., Xu, W.: Joint training for pivot-based neural machine translation. In: Proceedings of the Twenty-Sixth International Joint Conference on Artificial Intelligence, IJCAI-17, pp. 3974–3980 (2017). https://doi.org/10.24963/ijcai.2017/555
11. Fan, A., et al.: Beyond English-centric multilingual machine translation. J. Mach. Learn. Res. **22**, 1–48 (2020)
12. Feng, Y., Zhang, S., Zhang, A., Wang, D., Abel, A.: Memory-augmented neural machine translation. In: Palmer, M., Hwa, R., Riedel, S. (eds.) Proceedings of the 2017 Conference on Empirical Methods in Natural Language Processing, EMNLP 2017, Copenhagen, Denmark, 9–11 September 2017, pp. 1390–1399. Association for Computational Linguistics (2017). https://doi.org/10.18653/v1/d17-1146
13. Feng, Y., Zhang, S., Zhang, A., Wang, D., Abel, A.: Memory-augmented neural machine translation. In: Proceedings of the 2017 Conference on Empirical Methods in Natural Language Processing, Copenhagen, Denmark, September 2017, pp. 1390–1399. Association for Computational Linguistics (2017). https://doi.org/10.18653/v1/D17-1146

14. Gu, J., Wang, Y., Cho, K., Li, V.: Search engine guided neural machine translation. In: McIlraith, S.A., Weinberger, K.Q. (eds.) Proceedings of the Thirty-Second AAAI Conference on Artificial Intelligence, AAAI 2018, The 30th Innovative Applications of Artificial Intelligence, IAAI 2018, and the 8th AAAI Symposium on Educational Advances in Artificial Intelligence, EAAI 2018, New Orleans, Louisiana, USA, 2–7 February 2018, pp. 5133–5140. AAAI Press (2018). https://www.aaai.org/ocs/index.php/AAAI/AAAI18/paper/view/17282
15. He, J., Gu, J., Shen, J., Ranzato, M.: Revisiting self-training for neural sequence generation (2020)
16. Johnson, J., Douze, M., Jégou, H.: Billion-scale similarity search with GPUs. IEEE Trans. Big Data **7**(3), 535–547 (2021). https://doi.org/10.1109/TBDATA.2019.2921572
17. Khandelwal, U., Fan, A., Jurafsky, D., Zettlemoyer, L., Lewis, M.: Nearest neighbor machine translation. CoRR abs/2010.00710 (2020). https://arxiv.org/abs/2010.00710
18. Kingma, D.P., Ba, J.: Adam: a method for stochastic optimization. In: Bengio, Y., LeCun, Y. (eds.) 3rd International Conference on Learning Representations, ICLR 2015, San Diego, CA, USA, 7–9 May 2015, Conference Track Proceedings (2015). http://arxiv.org/abs/1412.6980
19. Luong, M.T., Pham, H., Manning, C.D.: Effective approaches to attention-based neural machine translation (2015)
20. Neubig, G., Hu, J.: Rapid adaptation of neural machine translation to new languages. In: Proceedings of the 2018 Conference on Empirical Methods in Natural Language Processing, Brussels, Belgium October–November 2018, pp. 875–880. Association for Computational Linguistics (2018). https://doi.org/10.18653/v1/D18-1103. https://aclanthology.org/D18-1103
21. Papineni, K., Roukos, S., Ward, T., Zhu, W.J.: BLEU: a method for automatic evaluation of machine translation. In: Proceedings of the 40th Annual Meeting on Association for Computational Linguistics, pp. 311–318. Association for Computational Linguistics (2002)
22. Pham, M.Q., Xu, J., Crego, J., Yvon, F., Senellart, J.: Priming neural machine translation. In: Proceedings of the Fifth Conference on Machine Translation, November 2020, Online, pp. 516–527. Association for Computational Linguistics (2020). https://aclanthology.org/2020.wmt-1.63
23. Sennrich, R., Haddow, B., Birch, A.: Improving neural machine translation models with monolingual data (2016)
24. Sennrich, R., Haddow, B., Birch, A.: Neural machine translation of rare words with subword units. In: Proceedings of the 54th Annual Meeting of the Association for Computational Linguistics (Volume 1: Long Papers), Berlin, Germany, August 2016, pp. 1715–1725. Association for Computational Linguistics (2016). https://doi.org/10.18653/v1/P16-1162. https://aclanthology.org/P16-1162
25. Vaswani, A., et al.: Attention is all you need. CoRR abs/1706.03762 (2017). http://arxiv.org/abs/1706.03762
26. Xia, M., Huang, G., Liu, L., Shi, S.: Graph based translation memory for neural machine translation. In: The Thirty-Third AAAI Conference on Artificial Intelligence, AAAI 2019, The Thirty-First Innovative Applications of Artificial Intelligence Conference, IAAI 2019, The Ninth AAAI Symposium on Educational Advances in Artificial Intelligence, EAAI 2019, Honolulu, Hawaii, USA, 27 January–1 February 2019, pp. 7297–7304. AAAI Press (2019). https://doi.org/10.1609/aaai.v33i01.33017297

27. Xu, J., Crego, J.M., Senellart, J.: Boosting neural machine translation with similar translations. In: Jurafsky, D., Chai, J., Schluter, N., Tetreault, J.R. (eds.) Proceedings of the 58th Annual Meeting of the Association for Computational Linguistics, ACL 2020, Online, 5–10 July 2020, pp. 1580–1590. Association for Computational Linguistics (2020). https://doi.org/10.18653/v1/2020.acl-main.144
28. Zhang, J., Utiyama, M., Sumita, E., Neubig, G., Nakamura, S.: Guiding neural machine translation with retrieved translation pieces (2018). https://doi.org/10.48550/ARXIV.1804.02559. https://arxiv.org/abs/1804.02559
29. Zoph, B., Yuret, D., May, J., Knight, K.: Transfer learning for low-resource neural machine translation (2016)

GIFT4Rec: An Effective Side Information Fusion Technique Apply to Graph Neural Network for Cold-Start Recommendation

Tran-Ngoc-Linh Nguyen[1,2], Chi-Dung Vu[1,2(✉)], Hoang-Ngan Le[1], Anh-Dung Hoang[1], Xuan Hieu Phan[2], Quang Thuy Ha[2], Hoang Quynh Le[2], and Mai-Vu Tran[2]

[1] Data Analytics Center - Viettel Telecom, Viettel Group, Hanoi, Vietnam
{linhntn3,dungvc2,nganlh,dungha7}@viettel.com.vn
[2] University of Engineering and Technology - Vietnam National University, Hanoi, Vietnam
{hieupx,thuyhq,lhquynh,vutm}@vnu.edu.vn

Abstract. Recommendation systems are highly interested in technology companies nowadays. The businesses are constantly growing users and products, causing the number of users and items to continuously increase over time, to very large numbers, this leads to the cold-start problem. Recommending purely cold-start users is a long-standing and fundamental challenge in the recommendation systems where systems are unable to recommend relevant items to the users due to unavailability of adequate information about them. To solve this problem, extensive studies have been carried out using the side information techniques (user information, item information, ...). However, we argue that this work will affect the user/product group that had a lot of interaction, using this side information can reduce the performance of the model when just focusing on learning based on the side information. In this paper, we propose a combination of global and local side **I**nformation **F**usion **T**echniques based on attention algorithm applied to **G**raph neural network-based models for cold-start users recommendation, and we call this architecture **GIFT4Rec**.

Keywords: recommendation system · cold-start problem · meta learning · side information fusion · graph neural network

1 Introduction

Personalization is the topic of investment with high returns in recent years. Two typical collaborative filtering (CF) algorithms for the recommendation problem are matrix factorization [1] and two-head DNNs [4,5]. While recent studies focus on the accuracy of the recommended system in the lab and achieve positive results such as BiVAE [2], VASP [3], ... We find these methods facing difficulties in deploying on the product environment because as the number of users increases, the model cannot recommend for new users and new items.

N. T. Nguyen et al. (Eds.): ACIIDS 2023, LNAI 13995, pp. 334–345, 2023.
https://doi.org/10.1007/978-981-99-5834-4_27

Cold-start recommendation is a challenge in recommendation systems where the system needs to make recommendations for new users or new items that have little to no historical interaction data. In other words, it refers to the situation where a recommendation system is presented with a user or item that it has never encountered before. The cold-start problem can occur in two scenarios: new user cold-start, for example, when a new user signs up for a service, there is no historical data available for the system to use to make personalized recommendations for them and new item cold-start, when a new item is added to the system, there is little or no data available about the item's characteristics and how users might interact with it. To address the cold-start problem, recommendation systems can use various techniques such as transfer learning, cross-domain, information fusion. Transfer learning approach involves leveraging knowledge learned from other domains or tasks to improve recommendation accuracy in cold-start scenarios. Transfer learning can be effective in situations where there is limited data available for the target domain or task.

Side information fusion is a technique used in cold-start recommendation systems to address the problem of limited data by incorporating additional information about users and items. The technique involves using side information or auxiliary data, such as demographic information, social network information, or item attributes, to improve the accuracy of recommendations for new users or items. The side information fusion technique works by combining the user-item interaction data with the side information to build a more comprehensive user-item model. This combined model can then be used to generate recommendations for new users or items by leveraging the information in the side information. For example, in a movie recommendation system, side information such as demographics, movie genres, directors, and actors can be used to improve recommendations for new movies or users. By incorporating this information into the recommendation model, the system can make more accurate predictions about which movies a new user might like based on their preferences for specific genres or actors. The side information fusion technique can be applied using various machine learning methods such as matrix factorization, or graph-based approaches. It is particularly useful for cold-start scenarios, where the system lacks sufficient data to make accurate recommendations, and can improve the overall performance of the recommendation system.

Some recent studies [14,15] have shown that the side information fusion is a useful technique for cold-start recommendation systems, there are some disadvantages and limitations to consider:

First, side information can introduce bias into the recommendation model if the side information is biased or incomplete. For example, if the side information is based on user demographics, it may lead to recommendations that are biased towards a particular group or stereotype.

Second, incorporating side information can increase the risk of overfitting, where the model becomes too closely tailored to the training data and performs poorly on new, unseen data. This can happen if the side information is too closely aligned with the training data or if the model is too complex.

In this paper, we try to solve these two non-optimal points, and propose a new architecture that can effectively be implemented in the production environment. Our main contributions include:

- We propose a new technique using side information to learn cold-start users interest and recommend more suitable items for them.
- We propose a new attention-based technique that can control and estimate the priority of each user's information used from different sources to capture their interest for an unbiased and fairness recommendation system.
- We propose a meta learning techniques that can decrease the risk of overfitting, model can learn generally on the unseen data.

2 Related Works

2.1 Cold-Start User Problems

The main issue of the cold start problem is non-availability of information required for making recommendations. In such cases, the only There are two popular methods to address this problem:

First is Cross-Domain Recommendation technique, which uses users behavior of source domain to predict their interests at target domain. Ye Bi et al. [9] and Cheng Zhao et al. [8] both map users behavior embedding of source domain to target domain via MLP layers. However, there aren't always more than one domain sharing the same users in reality.

Second is the side information fusion method. This method is more stable than the first one due to the fact that the side information always exists. DropoutNet [12] aims to maintain recommendation accuracy on non-cold start users while improving model performance on cold-start users, by combining all side information with users interactions to learn to reconstruct output from model just using users interaction. Beside, this model also randomly choosing some data just using side information of users or items to learn to reconstruct, that increases the affection of side information to model output which is very suitable for cold-start recommendation. However, this technique is not designed to control and estimate the affection of side information to each user, which may harm to model performance. To address this limitation of DropoutNet, we propose a new technique called AttentionDropoutNet improving model performance in all users type as active users, warm-start users and cold-start users simultaneously.

2.2 Meta-learning

Meta-learning, also called learning-to-learn, aims to train a model that can rapidly adapt to a new task which is not used during the training with a few examples Meta-learning can be classified into three types: metric-based, memory-based, and optimization-based meta-learning. Previous research as Manqing

Dong et al. [10] or Ye Bi et al. [11] are about applying optimization-based meta-learning to recommendation system that provide a more quickly and efficiently new data learning method for better cold-start recommendation. Inspired by that, we create a new metric-based meta-learning method for an unbiased and fairness recommendation system [13] and the rapidly changing of users preference better learning.

2.3 Graph Neural Network

Graph Neural Network [6] also known as GNN is a deep learning model applied to graph structure for many different problems. GNN learns the higher representation by aggregating their neighbor nodes information and learn them jointly with downstream task as node classification, link prediction or graph classification ...

In recent years, there are many methods applying GNN to recommendation systems by treating interactions between users and items like graph structure. In this graph structure, users and items are defined as nodes, each interaction between them is defined as an edge. GNN aims to learn the relation between each node via available links displaying in graph structure to predict possible relation of each two nodes not displaying of graph for different recommendation tasks. Rex Ying et al. [18] proposed a combination of GCN and hard-negative sampling method for similar items recommendation. Besides, Xiang Wang et al. [7] are about each node neighbors weight learning via their relation that is suitable for recommendation system.

3 Proposal Model

In this section, we will first give an overview about the proposed model, then detail each model component

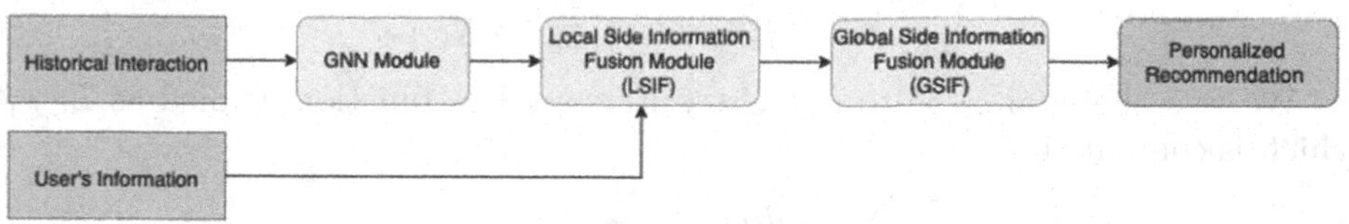

Fig. 1. Overall GIFT4Rec architecture

The architecture of the proposal model is shown in Fig. 1. The model consists of two components: Graph neural network (GNN) module, our global and local side information fusion module. The GNN module learns and extracts the characteristics of the user's behavior and the item's representation. The global and local side information fusion module builds a way to integrate side information into the user's embedding vector, which is the output of GNN module.

Given the items catalog $V = \{v_1, v_2, ..., v_p\}$ with p items. For a sample user u_i, $i \in \{1, 2, \ldots, N\}$ with side information vector X_{info_i}, we have a set of interacted items $S_i = \{s_{i1}, s_{i2}, s_{i3}, ..., s_{iq}; s_{ij} \in T, q \leqslant p\}$.

The GNN module is shown in Fig. 2. A graph is represented as $G = (U, V)$, which is defined as $\{(u_i, s_{i_j}, v_j)|u_i \in U, v_j \in V\}$, where U and V separately denote the user and item sets, and a link $s_{i_j} = 1$ indicates that there is an observed interaction between user u_i and item v_j, otherwise $v_{i_j} = 0$. The neighborhood of a node is denoted as $\mathtt{N}(.)$. Given the graph data, the main idea of GNN is to iteratively aggregate feature information from neighbors and integrate the aggregated information with the current central node representation during the propagation process [19,20]. From the perspective of network architecture, GNN stacks multiple propagation layers, which consist of the aggregation and update operations. The formulation of propagation is

$$\text{Aggregation: } n_{.}^{(\ell)} = \text{Aggregator}_{\ell}(\{h_u^{\ell}, \forall u \in \mathtt{N(.)}\})$$
$$\text{Update: } h_{.}^{(\ell+1)} = \text{Update}_{\ell}(h_{.}^{(\ell)}, n_{.}^{(\ell)})$$

Where $h_{u_i}^{(\ell)}$ denotes the representation of user u_i and $h_{v_j}^{(\ell)}$ denotes the representation of item v_j at ℓ^{th} layer, and Aggregator$_\ell$ and Update$_\ell$ represent the function of aggregation operation and update operation at ℓ^{th} layer, respectively. In the aggregation step, existing works either treat each neighbor equally with the mean-pooling operation [21,22], or differentiate the importance of neighbors with the attention mechanism [23]. In the update step, the representation of the central node and the aggregated neighborhood will be integrated into the updated representation of the central node. After training, the GNN model G will perform interaction embedding to build a vector $X_{u_i} \in R^{1 \times D}$ - the behaviors embedding of user i and a vector X_{i_j} - the representation of item v_{i_j}:

$$X_{u_i}, X_{i_j} \leftarrow G(s_i)$$

The combination of X_{u_i} and $X_{info_{u_i}}$ via our Weight Generated module in Fig. 3 for the last representation of user i are defined as $X_{final_{u_i}}$. Then the final score between u_i and i_j is computed as:

$$y_{u_i, i_j} = sofmtax(X_{final_{u_i}} \cdot X_{i_j})$$

We feed the final score to our Cross Entropy loss function defined as L_{CF}, which is computed as:

$$L_{CF} = \sum_{u_i, i_j, i_{jneg}} [log y_{u_i, i_j} + log(1 - y_{u_i, i_{jneg}})]$$

After learning the relation of each user and item having interactions used via L_{CF}, we use a new technique that make our model learning user representations more efficiently called Global Side Information Fusion also know as GSIF. In GSIF module, X_{final_u} and X_i are generated from the GNN module. Then all parameters are all frozen except the Weight Generated module in local side information fusion module and global side information module. Finally, a_u are generated from local side information fusion module then feed to global side information fusion module along with X_{final_u} and X_i

3.1 General Side Information Module

We propose two side information techniques that support each other by observing each user with from different angles. Those methods aims to control and estimate the impact of each information to each user to combine them efficiently for fairness and unbiased recommendation that focus not only in any source of information, which can't always contain information related to user interest. The first one forces Weight Generated module to learn via optimize L_{CF}, the remaining technique provide this module a general knowledge via indirectly observing unseen interactions. These two modules shared parameters that generate weights for each user side information and behaviors called Weight Generated.

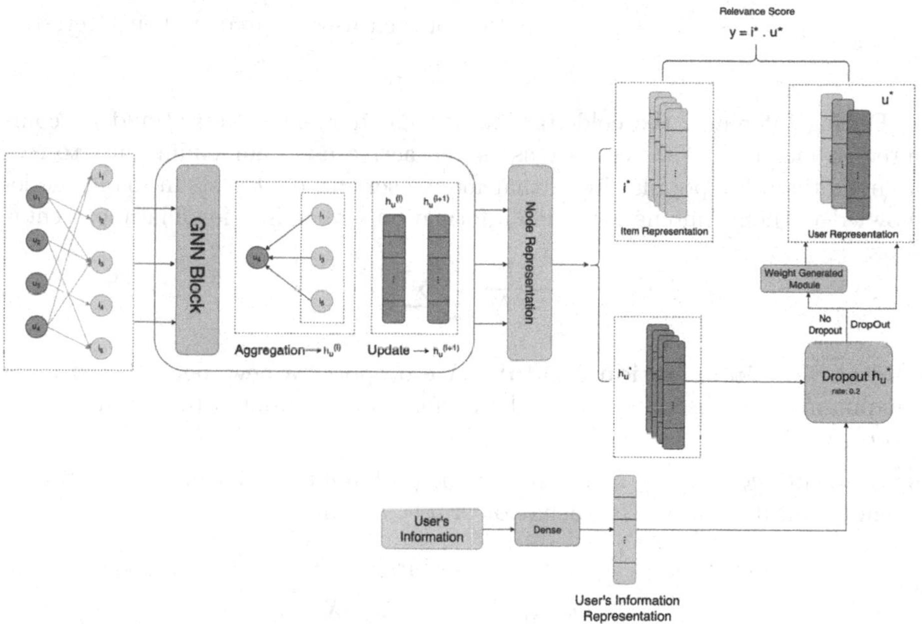

Fig. 2. Local Side Information Fusion module architecture

Local Side Information Fusion Module. We proposed a new technique called Attention DropoutNet also known as ADN that combining the technique used in [12] with out Weigh Generated module controlling side information and behavior of each users to better learning. Our module concatenate X_{u_i} and $X_{info_{u_i}}$ via the last dim as the input of module called $X_{concat_{u_i}}$

$$X_{concat_{u_i}} = concat([X_{u_i}, X_{info_{u_i}}])$$

We apply the MLP model to our Weight Generated module. We feed $X_{concat_{u_i}}$ to Weight Generated module using a Sigmoid activation function in the last layer to get a_{u_i}

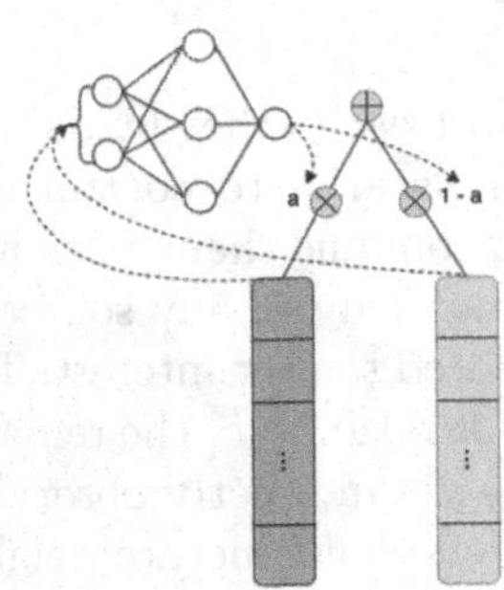

Fig. 3. Weight Generated Module

The last representation of user i:

$$X_{final_{u_i}} = a_{u_i} \cdot X_{u_i} + (1 - a_{u_i}) \cdot X_{info_{u_i}}$$

That is how we estimate the impact of each information to user i and combine them to control the representation. Beside that, we use a technique that sample a random value from a uniform distribution over $[0, 1)$ for each data when training. If that value less than the limit we set, the last representation would just be computed as side information embedding to force our model learning to use more side information of each user to predict their interest:

$$X_{final_{u_i}} = X_{info_{u_i}}.$$

During inference, the cold-start users behavior embedding would be computed as mean of all warm-start users and active users embedding just to recommend them the popular items that many users have interests in to our model knowledge, then combine with side information embedding for final representations:

$$X_{u_i} = \frac{1}{N_u - N_{U_{cold}}} \dot{\sum}_{u_j \notin U_{cold}} X_{u_i}$$

Global Side Information Module. We proposed a new metric-based meta learning method observing the model performance computed by our metrics at two case:

- We define $y_{behavior_{u_i,i}}$ as the list of the probability user i having interest of each item if we just using behavior of user i to model:

$$y_{behavior_{u_i,i}} = \left[y_{behavior_{u_i,i_1}}, y_{behavior_{u_i,i_2}}, \ldots, y_{behavior_{u_i,i_{n_I}}}\right]$$

$$y_{behavior_{u_i,i_j}} = X_{u_i} \cdot X_{i_j}$$

- We use $y_{behavior_{u_i,i}}$ to calculate model performance each metrics and then average them.
- Similar to the case if we just using side information of user i to model

We choose to use the validation set to test our model performance at two case above that help model Weight Generated module indirectly learning more objective knowledge from each user unseen interaction

We define $label_{u_i} = 0$ if the model performance at case one is better. If not, then $label_{u_i} = 1$

We encourage our Weight Generated module to learn more objectively and globally by optimizing a loss function called L_{global} defined as:

$$L_{global} = -\sum_i \left[(1 - label_{u_i}) * log(1 - a_{u_i}) + label_{u_i} * log(a_{u_i})\right]$$

L_{global} would be training separately with L_{CF} in each epoch.

4 Experiments

4.1 Experiment Setting

Dataset. We use Movielen 1M (ML1M) [16], a relatively large and popular data set with the demographic of each user, item ratings and user's interaction in the research field to test our proposed architecture performance. In addition, we use the Douban Dataset [17] to examine the effectiveness of side information fusion techniques (Table 1).

Table 1. Dataset Information

Dataset	Users	Items	Ratings
ML1M [16]	6040	3955	715477
Douban [17]	2442	8349	1053069

We split users into three set:

- Top 80% users having highest number of interactions will be choose as active users set
- Top 10% users having lowest number of interactions will be choose as cold-start users set that their interactions we don't use for training and the first item each user interacts are used during testing.
- The remain users will be choose as warm-start users set

To evaluate model performance efficiently, for each user in the active users set, we hold out the last item for the testing set for active users, treat one random item before the last item as the validation set, and use the remaining items for the training set. For each user in the warm-start users set, we hold out the last item as a testing set for warm-start users, we combine the remaining interactions with the training set to create our graph structure. For each user in the cold-start users set, we just hold out the first item as a testing set for cold-start users.
Baseline Methods. To verify the effectiveness of our method, we compare it with the following representative baselines:

- GAT: a model only using graph neural network module from KGAT [7] paper along with using mean of all non cold-start users embedding for each cold-start users during testing
- GAT + DropoutNet: a model only using combination of graph neural network module from KGAT paper and DropoutNet technique
- GIFT4Rec (w/o Local): our proposed model without updating Weight Generated Module parameters via optimizing L_{CF}
- GIFT4Rec (w/o Global): our proposed model without updating Weight Generated Module parameters via optimizing L_{Global}
- GIFT4Rec: our proposed model

Metrics. We define A_i as top k highest ranking items generated by model for user i, B_i as the real items set that user i interacted, N as the number of users.

Recall@k:

$$\frac{\sum \frac{|A_i \cap B_i|}{|B_i|}}{N}$$

We define overall score as mean of three sets scores. This metrics could evaluate model performance more fairly than just calculating score of combination of three sets which each of them always has different number of users that the more numbers of users one set has, the more its affection to overall score. In our experiment, we choose k as 50.

4.2 Experiment Result

Table 2. Benchmark

Dataset name	Method	Active users	Warm-start user	Cold-start user	Overall
ML1M	GAT [7]	19.64	0.066	24.52	14.94
	GAT [7] + DropoutNet	**20.24**	**0.5**	24.78	15.17
	GIFT4Rec (w/o Global)	**14.76**	**17.22**	**24.83**	**18.94**
	GIFT4Rec (w/o Local)	14.22	17.05	24.81	18.7
	GIFT4Rec	14.66	**18.71**	24.55	**19.3**
Douban	GAT	2.3	**2.95**	5.33	3.53
	GAT + DropoutNet	5.44	0.07	4.07	6.17
	GIFT4Rec (w/o Global)	**10.5**	1.48	7.52	**6.5**
	GIFT4Rec (w/o Local)	8.94	**2.58**	**7.94**	6.49
	GIFT4Rec	**10.18**	2.21	**8.36**	**6.92**

Our experiments results on ML1M dataset show GAT + DropoutNet model having the good performance at cold-start users set and active users simultaneously, which proves the efficient of DropoutNet. But this model has a very bad scores at warm-start users, which is nearly zero. Besides, this model also perform worse at all sets of Douban dataset than ours. That proves the accurate of our insight into the bad affection to model performance caused by uncontrollable side information learning of DropoutNet (Table 2).

GAT performs very good at active users set but worst at almost all of the remains methods on ML1M dataset, that is considered biased. In additions, this model performance at almost all sets of Douban dataset are the worst, compared to other models in experiment that prove the existence of large information about each user interest hidden inside their side information.

Our model without Global module gets a very good result at active users set of each dataset. The lower results of our model proposed and itself without Local module can be explained with the difference of distribution between two tasks we learning which one task is directly observing via each user that is also in active users set. That 's also an open challenge for meta-learning method.

Our model achieves the best result at cold-start users set of Douban dataset as soon as warm-start users set of ML1M dataset. Moreover, it also gets the second best result at active-users set of Douban dataset. But most of all, based on the most important metrics, our model outperforms the remains method on both datasets that is clearly the most fairness and unbiased recommendation system.

5 Conclusion

In this paper, we applied the attention-based side information fusion technique to cold-start users problem resolving and an unbiased and fairness recommendation system. Experiment results on two popular datasets show that our model outperforms the remains method which are variants of our model ore based on many popular algorithms for recommendation systems in recent years.

In future, we will upgrade our model to apply to cold-start items problem resolving. Another directions for future work would be research about how to combine L_{CF} and L_{Global} for less time consuming and efficient knowledge transfer between local and global modules to resolve open challenge that we describe in experiment result section.

Appendix B

We defines Θ and Θ_{WG} as our model parameters and the parameters of Weight Generated module. In addition, $scores_{u_B}$ and $scores_{u_{B_{info}}}$ denote users u_B (batch of U) model performance in validation set if just using their behavior embedding or their side information. Here is a pseudo code for our proposed model learning algorithm.

Algorithm 1. GIFT4Rec

Random initialize Θ and Θ_{WG}
while not converged:
  for u_B in U:
    update Θ via optimizing L_{CF} with each user in u_B
    $X_{final_{u_B}} \leftarrow X_{u_B}$
    calculate $scores_{u_B}$
    $X_{final_{u_B}} \leftarrow X_{info_{u_B}}$
    calculate $scores_{u_{B_{info}}}$
    compare $scores_{u_B}$ and $scores_{u_{B_{info}}}$ to get output labels for optimizing L_{Global}
    update Θ_{WG} via optimizing L_{global} with each user both in u_B and validation set users
return Θ, Θ_{WG}

References

1. Koren, Y., Bell, R., Volinsky, C.: Matrix factorization techniques for recommender systems. Computer **42**(8), 30–37 (2009). https://doi.org/10.1109/MC.2009.263
2. Truong, Q.-T., Salah, A., Lauw, H.: Bilateral variational autoencoder for collaborative filtering, pp. 292–300 (2021). https://doi.org/10.1145/3437963.3441759
3. Vančura, V., Kordík, P.: Deep variational autoencoder with shallow parallel path for top-N recommendation (VASP). In: Farkaš, I., Masulli, P., Otte, S., Wermter, S. (eds.) ICANN 2021. LNCS, vol. 12895, pp. 138–149. Springer, Cham (2021). https://doi.org/10.1007/978-3-030-86383-8_11
4. Krichene, W., et al.: Efficient training on very large corpora via Gramian estimation. In: ICLR 2019 (2019)
5. Mehrotra, R., Lalmas, M., Kenney, D., Lim-Meng, T., Hashemian, G.: Jointly leveraging intent and interaction signals to predict user satisfaction with slate recommendations. In: WWW 2019 (2019)
6. Wu, S., Sun, F., Zhang, W., Xie, X., Cui, B.: Graph neural networks in recommender systems: a survey, 2 April 2022
7. Wang, X., He, X., Cao, Y., Liu, M., Chua, T.-S.: KGAT: knowledge graph attention network for recommendation, 8 June 2019
8. Zhao, C., Li, C., Xiao, R., Deng, H., Sun, A.: CATN: cross-domain recommendation for cold-start users via aspect transfer network, 23 May 2020
9. Bi, Y., Song, L., Yao, M., Wu, Z., Wang, J., Xiao, J.: A heterogeneous information network based cross domain insurance recommendation system for cold start users, 30 July 2020
10. Bi, Y., Song, L., Yao, M., Wu, Z., Wang, J., Xiao, J.: MeLU: meta-learned user preference estimator for cold-start recommendation, 31 July 2019
11. Dong, M., Yuan, F., Yao, L., Xu, X., Zhu, L.: MAMO: memory-augmented meta-optimization for cold-start recommendation, 7 July 2020
12. Volkovs, M., Yu, G., Poutanen, T.: DropoutNet: addressing cold start in recommender systems. In: NIPS (2017)
13. Li, Y., et al.: Fairness in recommendation: a survey, 1 June 2022
14. Xie, Y., et al.: KoMen: domain knowledge guided interaction recommendation for emerging scenarios. In: ACM SIGWEB (2022)
15. Zhu, Z., Sefati, S., Saadatpanah, P., Caverlee, J.: Recommendation for new users and new items via randomized training and mixture-of-experts transformation. In: SIGIR (2020)
16. Harper, F.M., Konstan, J.A.: The MovieLens datasets: history and context. ACM Trans. Interact. Intell. Syst. (TiiS) **5** (2015). Article 19, 19 pages. https://doi.org/10.1145/2827872
17. Zhu, F., Wang, Y., Chen, C., Liu, G., Zheng, X.: A graphical and attentional framework for dual-target cross-domain recommendation. In: Proceedings of the Twenty-Ninth International Joint Conference on Artificial Intelligence, IJCAI 2020, pp. 3001–3008 (2020)
18. Ying, R., He, R., Chen, K., Eksombatchai, P.: Graph convolutional neural networks for web-scale recommender systems (2018)
19. Wu, Z., Pan, S., Chen, F., Long, G., Zhang, C., Yu, P.S.: A comprehensive survey on graph neural networks. TNNLS **32**(1), 4–24 (2020)
20. Zhou, J., et al.: Graph neural networks: a review of methods and applications. AI Open **1**(2020), 57–81 (2020)

21. Hamilton, W.L., Ying, R., Leskovec, J.: Inductive representation learning on large graphs. In: NeurIPS, pp. 1025–1035 (2017)
22. Li, Y., Tarlow, D., Brockschmidt, M., Zemel, R.: Gated graph sequence neural networks. In: ICLR (2016)
23. Veličković, P., Cucurull, G., Casanova, A., Romero, A., Lio, P., Bengio, Y.: Graph attention networks. In: ICLR (2017)

A Decision Support System for Improving Lung Cancer Prediction Based on ANN

Yen Nhu Thi Phan[1], Lam Son Quoc Pham[1], Sinh Van Nguyen[1](✉), and Marcin Maleszka[2]

[1] School of Computer Science and Engineering, International University, Vietnam National University HCM City, Ho Chi Minh City, Vietnam
nvsinh@hcmiu.edu.vn

[2] Department of Applied Informatics, Wroclaw University of Science and Technology, Wrocław, Poland

https://it.hcmiu.edu.vn, https://kis.pwr.edu.pl/en/

Abstract. Recent advancements in artificial intelligence (AI) and big data analysis have shown great potential for improving the diagnosis of lung cancer. Early detection of lung cancer is crucial for increasing patient survival rates. This paper analyze the data BRFSS (Behavioral Risk Factor Surveillance System), conducted from 2017 to 2020 to identify risk factors and symptoms of lung cancer. We develop a decision support system (DSS) based on data mining technique to assist healthcare practitioners and users in early diagnosis of lung cancer. Thirteen risk factors and demographic data are selected as predictors. The ANN and a logistic regression (LR) model are performed to predict the probability of lung cancer and to serve as a prognostic index respectively. The ANN model shown an accuracy of 84.79%, a sensitivity of 79.8%, and a specificity of 89.76%, a 93% of the ROC (AUROC) curve. While the LR model obtained an accuracy of 80.2%, a sensitivity of 80%, and a specificity of 72.2%, with a 76.1% AUROC. The models are trained with a batch size of 100, using stochastic gradient descent (SGD) optimizer. By using data analysis and mining techniques, we discovered new patterns in the health behavioral risk data that are previously unknown. Overall, our proposed method has a potential to significantly improve the early detection and treatment of lung cancer.

Keywords: Lung cancer · ANN Model · Data analysis · Decision support system

1 Introduction

Lung cancer is one of the most deadly diseases in the world, affecting people of all ages and becoming more prevalent in every year. The three stages of analyzing the BRFSS data [1] includes training an LR and ANN models, and creating a web application for the final diagnosis of patients. The input data of the system requires 13 risk factors and demographic attributes from the user and produces

N. T. Nguyen et al. (Eds.): ACIIDS 2023, LNAI 13995, pp. 346–357, 2023.
https://doi.org/10.1007/978-981-99-5834-4_28

a diagnosis within a second. According to cancer statistics 2018 in [2], there is an estimation of 234,030 new cases of lung cancer in the United States and 154,050 cancer-related deaths. In India, there are approximately 70,000 cases per year, with a number is go on to rise. Early diagnosis of lung cancer is critical in improving a patient's chances of survival and recovery, highlight the need for a Machine Learning (ML) based support system.

Computer aided diagnosis (CAD) systems have proven to be useful in supporting medical professionals in the detection and diagnosis of diseases [3]. These systems serve as a second opinion to the physician, offering additional support in the diagnostic process [4]. The AI-based expert systems convert knowledge of an expert in a specific field into software that have been used for many years in the medical field. These systems can answer questions and have ability to accommodate new knowledge, either through traditional rule-based techniques or through more advanced approaches like ANN and neuro-fuzzy systems. Despite of the numerous studies that have been conducted on the treatment and diagnosis of lung cancer, our research focus on an early prediction of the disease through an analysis of risk factors. A decision aid system will be built using a multi-parameter neural network that is fed with the most important risk factors and trained to classify lung cancer. The development of AI-based applications in healthcare has become necessary due to the limited availability of medical expertise globally. In many countries, the problem in recent years is lack of medical professionals, who can serve a large number of citizens. For this reason, a computer-based intelligent system is very important and necessary to support. The expert systems can help address human inconsistencies, lack of qualified experts, and a large amount of easily accessible data to provide better decision-making support for patients [5]. One of the biggest challenges in the diagnosis of lung cancer is that the symptoms in the early stage is often asymptomatic, make it difficult to detect [6]. This is why early detection is critical, as the survival rate of lung cancer is the second lowest of all cancers, with only 18%. The goal of this study is to improve the early diagnosis of lung cancer and provide a more efficient and accessible tool for patients and medical professionals alike.

The remaining of the paper is structured as follows: Sect. 2 presents the related works. Our proposed method is presented in Sect. 3. Section 4 shows the implementation and evaluation of the proposed method. The last section (Sect. 5) presents conclusion and future work.

2 Related Works

With the development of modern technologies, the field of medical diagnosis has been significantly impacted by AI and ML. While the image processing techniques combined with the ML methods are widely studied and applied to recognize objects (e.g. face recongnition [7]) and medical diagnosis on medical image datasets [8,9], this section discuss several related works which focus on lung cancer diagnosis and DSS. The most noteworthy work in the area of lung cancer diagnosis was conducted by Ardila et al. (2019) which demonstrated the potential to improve the consistency of lung cancer diagnoses using deep-learning models [10]. The method is developed based on deep-learning model for lung cancer screening on the CT scanner image. The method achieved higher accuracy to the state-of-the-art methods when compared to radiologists. It showed an absolute reduction of 11% in false positives and 5% in false negatives. The method proposed by Linh et al. [8] to determine the tumor region on the brain MRI images based on 3D generative adversarial network (3D-GAN). The method proved the performance and accuracy are better than the existing methods. Sharmila et al. [9] presented an effective approach for predicting and classifying lung cancer using machine learning and image processing techniques. The method proved superior accuracy for lung cancer prediction. However, further details on classifier selection and future research directions would be improved. Sinh et al. [11,12] suggested a method for building 3D models of MRI image objects based on geometric modeling. This methods not only visualize the 3D medical image objects but also support for doctors and medical staffs in diagnostic and treatment. Singla et al. (2013) [13] built a DSS for lung disease diagnosis using a rule-based inference engine to diagnose various lung diseases. The system components includes a knowledge base, a fact base, an inference engine, an explanation module, and an interface for both users and developers. Although the system obtained promising results in diagnosing certain lung diseases, it also highlighted some limitations of using rule-based expert systems, such as the lack of human inconsistencies and the limited knowledge base. Even so, the study set the stage for future research in this field and offered a fundamental perspective on how expert systems can support decision-making. Because of the capacity to handle uncertainty and imprecise information of the existing methods, fuzzy logic has been applied in medical diagnosis recently. Rodiah et al. (2020) presented a web-based fuzzy logic inference engine that is implemented to support lung cancer diagnosis [14]. The system used a knowledge base contained rules for assisting decision-making and the output was determined by a defuzzification step. The method has addressed to overcome the drawbacks of previous works. The study showed promising results in classifying patients based on their age, anamnesis, and heavy smoking. A research for lung cancer prediction using data mining techniques suggested by E. Yatish et al. (2019) [15]. The various data mining techniques like decision trees, k-nearest neighbor, logistic regression, random forest, and SVM are used to predict lung cancer tumor. The study found that KNN and logistic regression had a higher rate of prediction accuracy compared to other techniques. In recent researches, ANN is widely used to analyze patient data obtained from a web survey related to lung cancer symptoms such as age, gender, and six risk factors. The "Age" attribute is found as the greatest impact on the results. However, the imitation of a small dataset and incompleted consideration of all the risk factors and symptoms could impact the generalizability of the findings (2019) [22]. Adrian Cassidy et al. (2007) [16] built a tool for early detection and treatment of lung cancer. The valuable review is to compare the proposed risk models and investigate the works of others. The paper shown the importance of system validation for accurate risk prediction models and identify risk factors beyond age and smoking. G. Chada [17] proposed a method for lung cancer risk prediction based on a multi-parameterized ANN. The study used the datasets from 1997–2015 of National Health Interview Survey to create an ANN based on personal health clinical and demographic information. Despite the limited data is used, the model performed very well in predicting patients at risk of lung cancer, with a sensitivity of 75.3% and specificity of 80.6%. The model identified 649 cancer and 488,418 non-cancer cases using attributes such as age, gender, BMI, diabetes, smoking status, heart disease, and history of stroke. The AUC of 0.86 (0.85–0.88) for the training and validation sets indicates a promising tool for lung cancer risk prediction. Another method for identifying lung cancer risk factors based on the deep neural networks that is presented in [18]. The method is implemented on the web-based using Behavioral Risk Factor Surveillance System data source from 1997 to 2017 to realize the lung cancer risk factors. They analyzed more than 7 million records and excluded missing factors to obtain over 230,000 records for further selection. By leveraging the weights of DNN models, they identified notable lung cancer risk factors and quantitatively analyzed their degree of influence on lung cancer incidence in the elderly. For men at age of 65 years or older, the risk factor was smoking frequency, followed by time since quitting, use of e-cigarettes, and having smoked at least 100 cigarettes in their lifetime.

In general, the several studies focused on the development of expert systems and rule-based inference engines; the other methods used ML techniques such as deep learning, fuzzy logic, and data mining to improve diagnosis accuracy. Additionally, the models varied in their inputs, with some studies used patient data that obtained from the surveys, while others used imaging data.

3 Proposed Method

3.1 Data Selection and Analysis

Our method aim to build a model that could serve as a DSS for individuals at high risk of lung cancer. The BRFSS data is a join project of all 50 states participating in the US territories, as well as the CDC and Prevention. It aimed at collecting uniform state-specific data on health risk behaviors, chronic diseases and conditions, access to health care, and utilization of preventive medical services related to the prominent causes of illness and death in the US. The survey is conducted in both landline and mobile phone-based surveys with individuals over the age of 18, and in 2020 (see Fig. 1). It assessed health status and healthy days, exercise, sleep problems, chronic diseases, dental health, tobacco use, cancer screenings, and healthcare accessibility as general factors. An important part of building a model with predictive accuracy is the used data for training, testing and evaluating. Therefore, data selection is a critical step in our research. Our data selection process is started with the exclusion of individuals who did not satisfy the age requirement of being 18 or older. We then excluded missing demography data and behavioral factors that might harm the model performance. To address the issue of imbalanced data, we excluded non-lung cancer

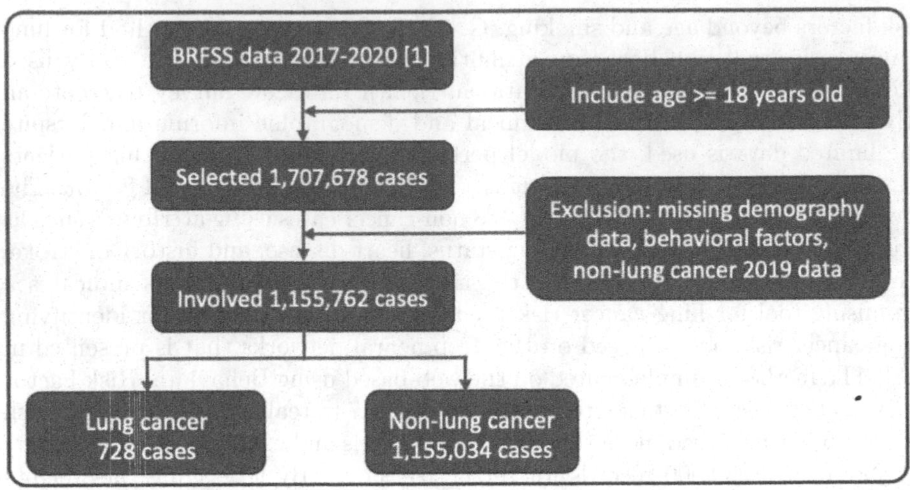

Fig. 1. Data selection plan of BRFSS from 2017–2020 [1]

data in year 2019, since the ratio of respondents with and without lung cancer is 1:30000, potentially causing significant imbalance issues in our final data. We use the Synthetic Minority Oversampling Technique (SMOTE) [19], as a powerful pre-processing technique to address the issue of imbalanced data and enhance the performance of our predictive models. The final dataset contains information unique to survey participants. We selected the 13 most referenced lung cancer risk factors out of more than 279 different features. These features include demographic data such as age, sex, and body mass index (BMI), as well as smoking history, general health status, and lung cancer screening history. The age feature is categorized into two levels, $18 \leq AGE \leq 64 : 0$ and $65 \leq AGE \leq 99 : 1$. The sex feature is represented by two levels, male: 1 and female: 0. The BMI feature is computed based on four categories: underweight, healthy weight, overweight, and obese. Notably, the most frequent category in this column is overweight, which is consistent with the trend observed in the US population from 2017 to 2020. To provide an additional feature for the intelligent support system, we presented a BMI calculator. Patients who do not know their exact body mass index can input their respective weight and height. The system will automatically calculate and categorize them into one of the four ranges. The BMI encoding will then entered the model as another feature, encoded as 1 to 4, respectively, at the backend.

The input values for underlined variables have been modified from using 1 and 2 to binary encoding with 0 and 1, as it offers certain advantages in terms of interpretation and model fitting. While changing a small dataset from 0 to 2 and keeping 1 as it is may not result in significant changes in model performance, consistent encoding becomes crucial for larger datasets. This approach facilitates better recall and benefits loss functions like binary cross-entropy, which is used to evaluate the neural network model in this study. For instance, to consider the height data of women and men with a regression equation of $height = a + b * gender + residual$. With the dummy variable of 0 and 1, the average height of women can be estimated to be 170 and the difference between the average height of men and women is 10. However, with the dummy variable of 1 and 2, estimating the average height of women would be harder, resulting in a less interpretable model. The general health feature consists of five categories: excellent, very good, good, fair, and poor. The smoking history features include two fields: "smoked at least 100 cigarettes" and "four-level smoker status." Finally, we selected all the questions in the lung cancer screening section added after the year 2016, as they are the only lung-specific questions available in the BRFSS data.

3.2 Learning Methodology

BMI Calculator and Classification: The first function of the system is simple to automatically calculate the BMI of patients through their weight and height inputs. They are classified into four ranges for encoding feature feeding model training and serving purpose. The formula to compute BMI and the logic to classify them are presented as in [21]: $BMI = weight/((height * 0.01) * (height * 0.01))$. If (BMI $<$ 18.5) then BMI $=$ 1 (you are underweight); If (BMI $\geq$ 18.5 and BMI $<$ 24.9) then BMI $=$ 2 (you are healthy weight); If (BMI $\geq$ 25 and BMI $<$ 29.9) then BMI $=$ 3 (you are overweight); If (BMI $>$ 30) then BMI $=$ 4 (you are obese); else BMI $=$ 0.

Logistic Regression: LR is a statistical method used to analyze and model the relationship between a categorical dependent variable and one or more independent variables. In medical applications, it can be used as a prognostic or diagnostic index to classify patients based on their risk of developing a certain disease. The LR model is based on the *logit* transformation, which computes the probability of output with the explanatory variables taken into account. The transformation is written as $logit(p)$, where p is the proportion of objects that have a certain characteristic. In medical applications, p can be seen as the probability of a patient having a specific disease. The following formula [23] shall be used in order to transform the value of the dependent variable into a binary one: $logit(p) = ln\frac{p}{1-p} = b_0 + b_1 * x_1 + b_2 * x_2 + ... + b_m * x_m$. The above formula obtains the value α so as to calculate for p basic algebra is used to resolve the equation: $p = \frac{1}{1+e^{-\alpha}}$. In order to compute the intercept and the regression coefficients, the matrix form of the model is used: $logit(p) = X * b + \epsilon$, where:

$$logit(p) = \begin{pmatrix} ln(\frac{p_1}{1-p_1}) \\ \vdots \\ ln(\frac{p_n}{1-p_n}) \end{pmatrix} \quad and \quad X = \begin{pmatrix} 1 \; x_1^1 \dots x_m^1 \\ \vdots \; \vdots \; \ddots \; \vdots \\ 1 \; x_1^n \dots x_m^n \end{pmatrix} \tag{1}$$

thus: $b = (X' * X)^{-1} * X' * ln(\frac{p}{1-p})$. Logistic regression is often used as a binary classification algorithm, meaning it classifies a patient as either having or not having a certain disease. In the case of lung cancer, for example, the logistic regression model can be used to classify patients who are at higher risk of developing lung cancer compared to others within the group.

To compute the risk of lung cancer in a patient, the logistic regression model takes into account various risk factors such as age, gender, smoking habits, body mass index, and medical history. These risk factors are selected based on their association with lung cancer. The first step in developing the model is to construct the model and define its intercept and coefficients after training the model to get the complete equation.

Interpreting the results of the logistic regression model involves understanding the equation and the values of the coefficients. For example, a patient who is 55 years old, male, a daily smoker, has a healthy weight range, started smoking at age 18, stopped smoking at age 0, smokes an average of 20 cigarettes per day, has fair general health status, has smoked more than 5 packs of cigarettes in their lifetime, attempted to quit smoking, and does not have any respiratory diseases or asthma would have a $logit(p)$ of −2.437. This means that the probability of this patient having lung cancer is about 8.32%.

3.3 Artificial Neural Network

The ANN model in [24] is a powerful machine learning algorithm used for the early detection of lung cancer. Our proposed process of design and implementation is presented as in Fig. 2. We uses the feed-forward backpropagation algorithm, which adjusts the weights in the network through each iteration to reduce errors. The architecture of the feed-forward backpropagation consists of 13 input neurons in the input layer representing significant symptoms of lung diseases, 100 hidden neurons in the first and second hidden layers, and one output neuron in the output layer representing the presence or absence of lung disease (Table 1).

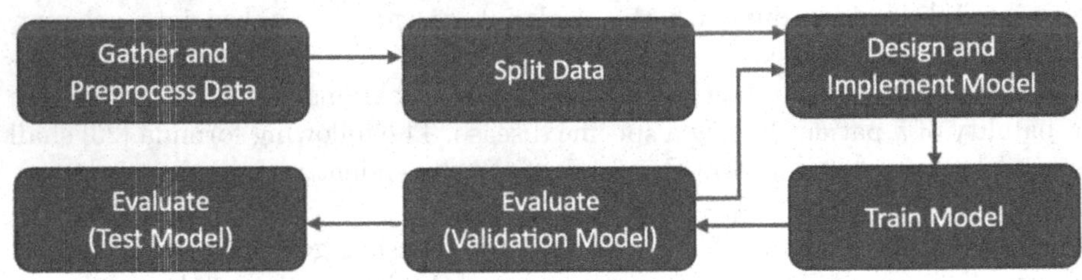

Fig. 2. Our process of design and implementation

After obtaining and pre-processing data, we design and setup information for the ANN architecture as follow:

Table 1. Configuration of ANN architecture.

Layers	Parameter	Value
1 (Input)	Neuron	13
2–5 (Hidden)	Neuron	24
	Activation	Relu
	Dropout	50%
6 (Output)	Neuron	1
	Activation	Sigmoid
Training times = 50 Batch size = 100 Optimizer = SGD Learning rate = 0.05 Train type = Feed forward Loss = Binary crossentropy		

As we known, the deep neural network is a modern variant of the ANN, which replaces the step activation function with something better, such as ReLU, and applies Sigmoid to the output. The neural net consists of three layers, the input layer, the hidden layer, and the output layer. Each neuron receives a signal from the neurons in the previous layer, and each of those signals is multiplied by a separate weight value that is improved through each epoch. The decision is made as to whether the user suffered from lung cancer or not based on the output. The activation function of each layer is computed by applying a ReLU and a sigmoid function. When the sigmoid activation function is applied to the output, the final output of the neural network is converted into probabilities, from which one can choose the classification with the highest probability. The neural network's training method is a variant of gradient descent called backpropagation, which enables the adjustment of all weights at the same time. Binary cross-entropy is used as the loss function, as it is appropriate for binary classification. The cost function can be calculated for each epoch as the whole dataset is fed to the neural network, and the weights are updated until the goal is reached, which is to minimize the cost function. Stochastic Gradient Descent, on the other hand, takes the row one by one and then runs the neural network, and then adjusts the weights right after. This technique helps in finding the global minimum rather than getting stuck in the local one and is faster than batch gradient descent since there is no need to load all data into memory. The binary cross-entropy loss function will be used for system evaluation, where unseen data will be fed into the models ten times to see how the system will perform on real-world data. TensorFlow 2 will be used for model training, and visualization will be made using popular libraries such as seaborn and matplotlib. Backpropagation is driven by very interesting and sophisticated mathematics that enable the adjustment of all the weights at the same time, making it an efficient method for training neural networks.

4 Implementation and Evaluation

Data pre-processing and analysis is a critical phase in any data-driven research, and Spark has proven to be a valuable tool in this regard. Its application in this work allowed for the implementation of big data processing techniques, which led to a modern, interactive, and user-friendly approach to data pre-processing. The chosen database covers various behavioral risk factors for general health, including lung cancer. However, when pre-processing for logistic regression and neural network models for early prediction purposes, an issue of imbalance between respondents with and without lung cancer emerged. To address this, various techniques were used to balance the data and improve model performance. The data in the database comprises multiple explanatory variables of different levels of data, each encoded according to the status of each respondent. Therefore, mode imputation was employed during the pre-processing stage to fill in missing data and improve model performance. The process of data integration for each dataset was unique, with the inferSchema tool automatically defining the datatype for each field, leading to inconsistencies between datasets. For instance, the 2017 and 2018 datasets required the conversion of null values to numbers, which helped standardize the data types across all datasets. Compared to the original dataset, the processed data was cleaned, balanced, and ready for training and fitting machine learning models. This dataset served as the foundation for the models that produced the predictions in this study. Hyperparameter tuning is a critical step in optimizing ANN models, as the performance of the model significantly depends on the values of hyperparameters. However, determining the optimal values for hyperparameters is a challenging task, and trying all possible combinations manually can be time-consuming and resource-intensive. To automate this process, we used GridSearchCV, which is an iterative tuning process that helps to determine the optimal hyperparameters for ANN models. In our study, it took more than three hours to find the optimal hyperparameters for the ANN model using GridSearchCV. We also learned from our previous attempts to optimize the model manually. The first step in our predictive model is to input the significant symptoms from the subject's test data into the neural network and compute the net output value from the output layer. If the output value is less than the predefined threshold, the output neuron inhibits and displays "Low risk of lung cancer" along with the probability of lung cancer presence. If the output value is greater than the threshold, the output neuron fires and displays "High risk of lung cancer". As a result, ROC curves were developed and used in various applications and research. The ROC curve is useful because one point in space is considered better than another if it is positioned more to the northwest of the square. Predictions are better if the false positives rate is low and true positives rate is high, and the area under the curve (AUC) was introduced as a measure of performance. The AUC value falls within a specific range and shows the quality of classification. For example, a range of 0.9 to 1.00 is considered excellent classification, 0.8 to 0.9 is considered good, and 0.7 to 0.8 is considered fair. In contrast, the rate is considered a failure if it is below 0.5 [24]. To visualize the models, we used the ROC curve and the accuracy of the training and validation data. First, we split the data into training and testing sets and then performed imputation and oversampling on the training data (see Fig. 3) (Table 2).

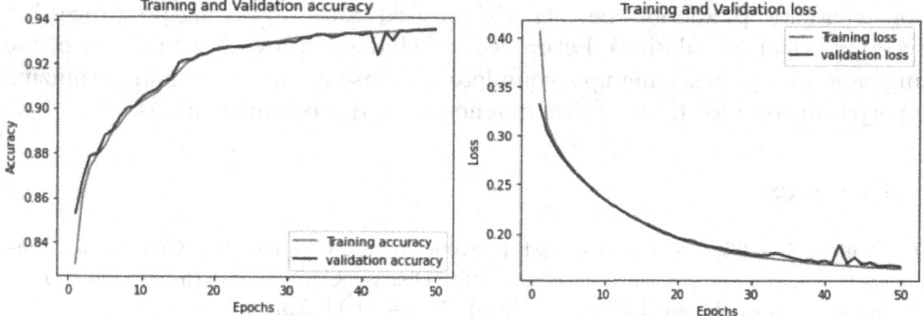

Fig. 3. The obtained results of our training and validation ANN model

Table 2. Comparison of the metric index between the models.

Model	Accuracy	Sensitivity	Specificity	AUROC
Our ANN Model	84.79%	79.80%	89.76%	93.00%
Our logistic regression Model	80.20%	80.00%	72.20%	76.10%
Multi-parameterrized ANN [20]	No	75.30%	80.60%	86.00%

The models are trained 50 times with a batch size of 100 and optimized using stochastic gradient descent. Our ANN model showed (accuracy: 84.79%, sensitivity: 79.80%, specificity: 89.76% and AUROC: 93.00%). While the LR model obtained (accuracy: 80.20%, sensitivity: 80.00%, specificity: 72.20% and AUROC: 76.10%). Comparing to the Multi-parameterrized ANN (the accuracy is not shown, sensitivity: 75.30%, specificity: 80.60% and AUROC: 86.00%).

5 Conclusion and Future Work

This study developed the ANN and LR models to aid an early detection of lung cancer. The proposed models utilized data mining techniques to discover new patterns in health behavioral risk data that had not been previously discovered. The contribution in this study is answered the questions: how the data mining techniques can be used to support detecting lung cancer; and how to increase the interpretability and trustworthiness of the developed models. A decision support system is developed to aid user in decision-making and recommended next steps in possible treatment plans. To address the training problem, the lightweight model with high accuracy is recommended and training on the computer with limited resources. The visual insights are utilized to address model failure and improve the architecture. A web application is developed for lung cancer detection through risk factors, utilizing deep algorithms and state-of-the-art evaluation metrics. The obtained results have been shown better data for more synthetic prediction, dealing with skewed data, improving pre-processing steps for better calculations. Future work includes improving the dataset of the lung cancer centric; setting up a complete database architecture, and optimizing hyperparameters for more precise diagnostics and recommendations.

References

1. Centers for Disease Control and Prevention. CDC - BRFSS, Centers for Disease Control and Prevention. Centers for Disease Control and Prevention (2023). https://www.cdc.gov/brfss/index.html. Accessed 11 Apr 2023
2. Siegel, R.L., Miller, K.D., Jemal, A.: Cancer statistics (2018). CA Cancer J. Clin. **68**, 7–30 (2018). https://doi.org/10.3322/caac.21442
3. Tiwari, A.: Prediction of lung cancer using image processing techniques: a review. Adv. Comput. Intell. Int. J. (ACII) **3**, 1–9 (2016). https://doi.org/10.5121/acii.2016.3101
4. Donoso, L.: Europe's looming radiology capacity challenge: a comparative study. For me the main threat is the shortage of radiologists, ESR President 2015/2016 Healthmanagement.org, vol. 16, no. 1 (2016)
5. Turban, E., Sharda, R., Delen, D.: Decision Support and Business Intelligence Systems, 9th edn. Pearson Education, New Jersey (2011). ISBN: 978-0136107293
6. Hamilton, W., Peters, T.J., Round, A., Sharp, D.: What are the clinical features of lung cancer before the diagnosis is made? A population based case-control study. Thorax **60**(12), 1059–65 (2005). PMID: 16227326; PMCID: PMC1747254. https://doi.org/10.1136/thx.2005.045880
7. Nguyen, L.D.V., Chau, V.V., Nguyen, S.V.: Face recognition based on deep learning and data augmentation. In: Dang, T.K., Küng, J., Chung, T.M. (eds.) FDSE 2022. CCIS, vol. 1688, pp. 560–573. Springer, Singapore (2022)
8. Phung, L.K., Nguyen, S.V., Le, T.D., Maleszka, M.: A research for segmentation of brain tumors based on GAN model. In: Nguyen, N.T., et al. (eds.) ACIIDS 2022. LNAI, vol. 13758, pp. 369–381. Springer, Cham (2022)
9. S Nageswaran, et al.: Lung cancer classification and prediction using machine learning and image processing. BioMed Res. Int. **2022**, Article ID 1755460, 8 pages (2022). https://doi.org/10.1155/2022/1755460
10. Diego, A., et al.: End-to-end lung cancer screening with three-dimensional deep learning on low-dose chest computed tomography. J. Nat. Med. **25**(6), 954–961 (2019)
11. Nguyen, V.S., Tran, M.H., Le, S.T.: Visualization of medical images data based on geometric modeling. In: Dang, T.K., Küng, J., Takizawa, M., Bui, S.H. (eds.) FDSE 2019. LNCS, vol. 11814, pp. 560–576. Springer, Cham (2019). https://doi.org/10.1007/978-3-030-35653-8_36
12. Nguyen, V.S., Tran, M.H., Vu, H.M.Q.: An improved method for building a 3D model from 2D DICOM. In: Proceedings of International Conference on Advanced Computing and Applications (ACOMP), pp. 125–131, IEEE (2018). ISBN: 978-1-5386-9186-1

13. Singla, J.: The diagnosis of some lung diseases in a prolog expert system. Int. J. Comput. Appl. **78**, 37–40 (2013)
14. Rodiah, E.H., Fitrianingsih, Susanto, H.: Web based fuzzy expert system for lung cancer diagnosis. In: International Conference on Science in Information Technology (ICSITech), p. 142 (2016)
15. Yatish Venkata Chandra, E., Ravi Teja, K., Hari Chandra Siva Prasad, M., Mohammed Ismail, B.: Lung cancer prediction using data mining techniques. Int. J. Recent Technol. Eng. (IJRTE) **8**(4), 12301–12305 (2019). ISSN: 2277–3878
16. Cassidy, A., Duffy, S.W., Myles, J.P., Liloglou, T., Field, J.K.: Lung cancer risk prediction: a tool for early detection. Int. J. Cancer **120**(6), 1–6 (2007)
17. Chada, G.: Using 3D convolutional neural networks with visual insights for classification of lung nodules and early detection of lung cancer (2019)
18. Songjing-Chen, S.W.: Identifying lung cancer risk factors in the elderly using deep neural networks: quantitative analysis of web-based survey data. J. Med. Internet Res. **22**(3), e17695 (2020)
19. Maldonado, S., López, J., Vairetti, C.: An alternative SMOTE oversampling strategy for high-dimensional datasets. Appl. Soft Comput. **76**, 380–389 (2019)
20. Hart, G., Roffman, D., Decker, R., Deng, J.: A multi-parameterized artificial neural network for lung cancer risk prediction. PLoS ONE **13**, e0205264 (2018). https://doi.org/10.1371/journal.pone.0205264
21. What Is Lung Cancer? The American Cancer Society. https://www.cancer.org/cancer/lung-cancer/about/what-is.html. Accessed April 2023
22. Nasser, I.: Lung cancer detection using artificial neural network. Int. J. Eng. Inf. Syst. (IJEAIS) **3**(3), 17–23 (2019). https://ssrn.com/abstract=3700556
23. S. Belciug and F. Gorunescu. Intelligent Decision Support Systems - Journal Smarter Healthcare, 1st edn. Springer, Cham (2020). https://doi.org/10.1007/978-3-030-14354-1
24. Peto, R., Darby, S., Deo, H., Silcocks, P., Whitley, E., Doll, R.: Smoking, smoking cessation, and lung cancer in the UK since 1950: combination of national statistics with two case-control studies. BMJ **321**(7257), 323–329 (2000)

Emotion Detection from Text in Social Networks

Barbara Probierz[1,2(✉)], Jan Kozak[1,2], and Przemysław Juszczuk[1]

[1] Department of Machine Learning, University of Economics in Katowice, 1 Maja 50, 40-287 Katowice, Poland
{barbara.probierz,jan.kozak,przemyslaw.juszczuk}@ue.katowice.pl

[2] Łukasiewicz Research Network – Institute of Innovative Technologies EMAG, Leopolda 31, 40-189 Katowice, Poland

Abstract. Emotion detection from text in social networks is an important tool for monitoring and analyzing discussions on social networks. However, the difficulty of the problem related to emotional analysis from texts is caused by the lack of additional features such as facial expressions or tone of voice, and the use of colloquial language and various signs and symbols in the texts contributes to misinterpretation of the content of the statement. For this reason, this article proposes an approach to automatically detect emotions from real text data from social networking sites and classify them into appropriate emotion categories. The aim of the article is to develop a model for detecting and classifying emotions through the use of natural language processing techniques and machine learning algorithms. In addition, it was decided to analyze the impact of the number of types of emotions on the classification results. Our research was tested for six classifiers (CART, SVM, AdaBoost, Bagging, Random Forest and K-NN) and three word weighting measures (TF, TF-IDF and Binary). Four measures of classification quality were used to evaluate the classifiers, i.e. accuracy, precision, recall and F1-score. Two datasets from different social networking sites were used for the research - one is a collection of comments from a social networking site and the other is tweets from Twitter. The analysis of the results confirmed that the proposed solution allows detecting emotions from the actual content of tweets, and the results obtained are better for a smaller number of emotion categories.

Keywords: emotion detection · natural language processing · classification · machine learning · social networks

1 Introduction

Emotion detection (ED) is an issue derived from the field of sentiment analysis (SA), which is based on solutions developed in the field of natural language processing (NLP) and artificial intelligence (AI). The main purpose of sentiment analysis is to determine the degree of positivity or negativity of a given statement, where the evaluation of the content is most often presented as a positive,

N. T. Nguyen et al. (Eds.): ACIIDS 2023, LNAI 13995, pp. 358–370, 2023.
https://doi.org/10.1007/978-981-99-5834-4_29

negative or neutral evaluation. Emotion detection, on the other hand, is a branch of sentiment analysis that deals with the extraction and analysis of emotions [5].

Emotions are strong and subjective feelings that influence our decisions, attitudes and reactions. Emotion detection is about identifying those feelings and then analyzing their emotional meaning. The purpose of this analysis is to understand what is the emotional intention of the author of the text and what emotions are evoked in his audience. However, the range of emotions extends far beyond the three-level scale of SA, so additional techniques are needed to detect emotions and classify them into more categories [7].

Emotion recognition and sentiment analysis are particularly important not only in industries such as marketing and advertising, but are also used in customer service processes, content supervision or in medicine. In marketing and advertising, sentiment analysis allows companies to understand how people perceive their products and services, so that they know how to align their marketing strategies with user sentiment [6]. In customer service, sentiment analysis allows you to quickly identify potential problems and better meet customer needs [16]. In content supervision, sentiment analysis allows for early detection and response to image crises [17,18]. Sentiment analysis and emotion detection is also important in the medical sector, where it can be used to detect patients' emotional needs and understand their attitudes towards treatment [14,19]. It can also be used to monitor and understand the emotional state of patients with mental illnesses such as depression and anxiety [13].

The aim of our research is to develop an approach to automatically detect emotions from texts by applying natural language processing techniques and machine learning algorithms. Our approach consists in analyzing the content posted on social networking sites, identifying words that describe emotional overtones, training learning algorithms, and finally properly classifying the entire statement into one of several predefined categories of emotions. The main problem for which we are looking for a solution is detecting not only two types of emotions - positive or negative, as two decision classes, but also detecting and classifying content into many categories, i.e. anger, fear, sadness, joy, shame, disgust, neutral and surprise.

This paper is organized as follows. Section 1 comprises an introduction to the subject of this article. Section 2 provides an overview of related works on the analysis sentiments and emotion detection and we present natural language processing methods (NLP) and machine learning (ML). Section 3 we explain our methodology for emotion detection. In Sect. 4, we present our experiments and discuss results. Finally, in Sect. 5 we conclude with general remarks on this work, and a few directions for future research are pointed out.

2 Emotion Detection

Sentiment analysis is especially important for content posted on social networking sites, as it allows you to quickly and effectively understand the emotions and opinions of users. On social networks such as Twitter, Facebook and others, sentiment analysis and emotion detection from text can be used to monitor and analyze discussions about specific products, services and events [11,12]. This allows you to get information about what people think about these issues, which is especially important for companies that want to understand how their brand is perceived and what are the potential disadvantages of their products or services [9]. Sentiment analysis and emotion detection from text in social networks is also important for governments and organizations that want to monitor and understand public sentiment regarding specific events and political issues. This allows to obtain feedback from the society, thanks to which it is possible to make the right political and social decisions [3].

Emotion detection from text involves identifying words or phrases in natural language and then analyzing their emotional overtones [12]. Emotion detection from texts is a difficult problem because texts may not provide specific emotional cues, such as facial expressions and postures, and additional sounds such as crying, laughing, and tone of voice [2]. Difficulties are also associated with extracting emotions from grammatically incorrect texts, short messages, sarcasm in written documents, contextual information, etc. Research [1] shows that over the last decade (2010–2020), only a small percentage of articles dealt with detecting emotions from text. In addition, of all research papers on emotion detection, only about 10% deal with emotion detection from texts. Therefore, insufficient knowledge of effective text extraction methods in this field due to insufficient research is a major obstacle to the successful detection of emotions from written texts [4,8,15].

Emotion detection from the text is most often divided into two approaches: the rule-based approach and the machine learning approach. The rule-based approach consists in extracting and applying predetermined rules and criteria for text analysis, which are based on manually, automatically or semi-automatically created dictionaries or language corpora. The machine learning approach, on the other hand, is based on machine learning algorithms, the application of which consists in learning classifiers on a small set of training data, and then performing predictions on the remaining data.

2.1 Rule-Based Approach

Dictionary analysis uses emotion and sentiment dictionaries that contain phrases and words related to specific emotions and sentiments. Dictionary analysis algorithms search the text and evaluate what words and phrases from the text are included in the dictionary. These dictionaries allow you to identify words expressing emotions by checking the occurrence of these words in the appropriate sets of words with positive or negative connotations. In order to determine the emotional overtones of an utterance, the frequency of using positive and negative words is counted and the overtone of the entire utterance is determined by applying statistical methods. Among the Natural Processing Language (NLP) techniques are mechanisms that can be used to count the occurrence of words in tweets, and these are three measures of word weighting:

- Binary measure checks whether the word is in the tweet or not and each value is 0 or 1;
- Term Frequency measure (TF) checks tat the number of occurrences of a word in a tweet against the number of occurrences of all words in that tweet. TF measure is calculated using the formula:

$$TF_{i,j} = \frac{n_{i,j}}{\sum_k n_{k,j}}, \tag{1}$$

  where $n_{i,j}$ is the raw count of a word t_i in the tweet d_j and the denominator is the sum of the raw count of all $n_{k,j}$ words in the tweet d_j.
- Term Frequency-Inverse Document Frequency (TF-IDF) measure, is used to reduce the weights of less significant words in a set of words. In addition to the term weight in relation to the complete set of tweets, the TF-IDF measure also determines the corresponding word weight locally. In this way, the weights of words appearing several times in one document are lower than the weights of words appearing in multiple documents. TF-IDF measure is calculated using the formula:

$$(TF - IDF)_{i,j} = TF_{i,j} \times IDF_i \tag{2}$$

  where Inverse Document Frequency (IDF) is the ratio of the number of processed n_d tweets to the number of tweets containing at least one occurrence of the given word $\{d : t_i \in d\}$. IDF is expressed by the formula:

$$IDF_i = \log \frac{n_d}{\{d : t_i \in d\}} \tag{3}$$

2.2 Machine Learning Approach

The use of statistical methods to analyze content from social networking sites has become insufficient due to the very rapid increase in information and the frequent use of colloquial language, for which there are no appropriate dictionaries. Therefore, to perform sentiment analysis in social networking sites, various approaches based on NLP and ML are being developed. To detect emotions in a text, techniques based on many factors such as keywords, phrases and context are used to determine the degree of positivity or negativity of a given text. In addition, NLP uses techniques such as text classification and feature extraction. Text classification consists in determining whether a given text is positive, negative or neutral, while feature extraction consists in searching for keywords and phrases that are associated with specific emotions. For this reason, the use of the machine learning approach enables quick and effective analysis of large data sets, which is crucial for the effectiveness and efficiency of activities on social networking sites.

In sentiment analysis and detecting emotions from text, it is also crucial to choose the right database of words and phrases that are used as a reference point. For example, a database of positive and negative words may include words such as *happiness* and *sadness*, and a database of emotions may include emotions such as *joy* and *fear*.

In machine learning, the method used is text classification [10]. Machine learning algorithms are used to learn and identify distinctive features of text that are associated with specific emotions and sentiments. Machine learning models rely on big data and learn from examples to determine what text features are best associated with specific emotions and sentiments.

3 Proposed Approach

Emotion detection from text is one of the challenging problems in natural language processing. The reason is the unavailability of tagged datasets and the multi-class nature of the problem. People have different emotions, and it is difficult to collect enough records for each emotion, hence the problem of class imbalance. Therefore, our goal is to build a model for detecting emotions by applying NLP techniques and known machine learning algorithms for classification, and to analyze the impact of the number of types of emotions on the classification results.

The proposed approach is to automatically detect emotions from texts shared by users on the Twitter social network and classify them into appropriate categories. Emotion detection is based on the analysis of texts from the content of tweets through the use of NLP techniques and machine learning algorithms. The scheme of operation of the proposed approach is presented in the Fig. 1.

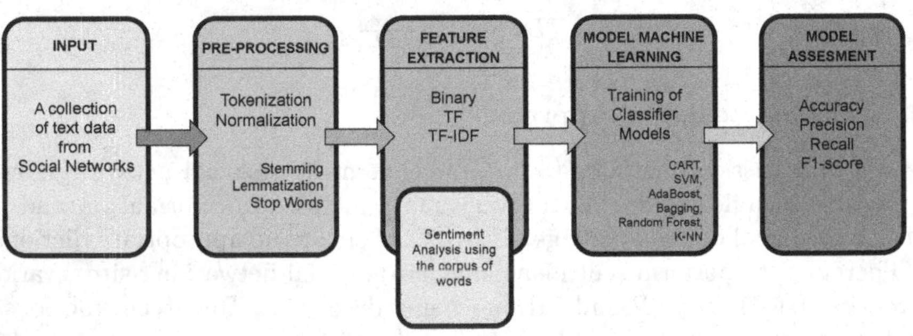

Fig. 1. The scheme of operation of the proposed approach

The first step of our approach is to read the content of all tweets along with the correct categories, and then pre-process the data. For this purpose, a cleaning process is carried out to remove unnecessary punctuation marks as well as common words that do not give insight into the specific subject of the tweet. A set of such words creates a Stopwords list, and words such as "the", "and", "me", etc. are also an example. Additionally, the tweet content is tokenized, i.e. the content of tweets is divided into individual words. Finally, a normalization process is carried out, which consists in converting the words from tweets to the basic form of the word, consistent with the dictionary definition, by means of the stemming process.

The next step in our approach is to use NLP techniques to perform content analysis and build a vector representation of the tweets under study. For this purpose, a sentimental analysis of the content of tweets is carried out, which consists in marking words with emotional overtones. The occurrence of each word in the tweet is checked against the corpus of words. Statistical measures were used to count the occurrences of words in tweets: TF-IDF, TF, Binary. We transform the resulting word vectors into a decision table that we use in the classification process. The columns in the decision table are the words that appear in the content of the tweets, and the rows are the subsequent tweets. At the intersection of the row and the column, there is a weight value of the occurrence of individual words in tweets, which depends on the selected measures. The last column is a decision in which the appropriate category is entered as a type of emotion.

In order to carry out the classification, we selected several classic classifiers such as CART, SVM, AdaBoost, Bagging, Random Forest and K-NN described in the Sect. 2.2. In the case of machine learning algorithms used for classification, it is necessary to use a training set and a test set. The training set includes tweets (all their content) and categories (serving as decision classes), while the test set contains only the content of new tweets, for which the algorithm assigns the emotion category as a decision class.

The last step in our method is to evaluate the classification quality measures. For this purpose, we use measures such as accuracy, precision, recall and F-1 score. All these measures are based on the confusion matrix defined for the multi-criteria problem.

4 Experiments and Results

The main goal of our experiments was to build a model for detecting emotions from real text data from social networking sites and to analyze the impact of the number of types of emotions on the classification results. Therefore, in order to test the proposed approach, we conducted two types of experimental studies.

In the first experiments, we wanted to test the effectiveness of the proposed solution to the problem of detecting and classifying emotions from texts, as well as identify words that describe emotional overtones. In the second experiment, we classified the tweets into a different number of emotion categories and analyzed the impact of this number on the classification results.

Two datasets were prepared for the experiments and evaluated against three text-based word weighting measures (TF-IDF, TF and Binary) and six machine learning algorithms (CART, SVM, AdaBoost, Bagging, Random Forest and K-NN). The obtained results are described in the Sect. 4.3.

4.1 Identifying Words as Emotional Overtones

The first research attempt related to achieving the assumed goal of detecting emotions from selected texts from social networking sites was testing the proposed solution on a small set of comments from a social networking site. The

Table 1. The results of the a sentiment analysis of the content of comments from social network.

Number of words	Decision classes			
	joy	fear	anger	sadness
joy	2180	925	816	823
confidence	570	546	567	623
happiness	2121	805	713	671
beauty	315	151	120	117
fear	271	2243	494	358
anger	921	2571	2400	1147
sadness	880	2528	1350	2021
repulsion	625	2443	926	750
surprise	122	485	194	181
ugliness	756	666	364	328
unhappiness	777	2926	2151	1959
total number of words	12887	18795	13534	13092
total number of comments	823	1147	857	786

dataset consisted of 3,613 records that were assigned to one of four emotion categories, i.e. *joy, fear, anger* and *sadness.* The exact division of records into emotion categories is presented in the Table 1 in the last row as *number from documents.*

In addition, a sentimental analysis of the content of comments was carried out on the same dataset, which consisted in marking words with emotional overtones. In order to correctly identify such words, the sentimental analysis corpus for the English language was used, which consisted of a list of words describing emotions and categories of emotions. For each of the four categories of emotions, the occurrence of words describing different emotions was calculated, and the results are presented in the Table 1.

It should be noted that many words in the comment set do not define any emotions, and other words are assigned several emotions, which, depending on the presence in the content, may indicate positive or negative emotions, e.g. the word *love* has the following emotions: *anger, joy, sadness, disgust, happiness* and the word *nice* has associated emotions: *joy, sadness, disgust, anger, trust, happiness.*

4.2 Number of Emotion Categories

The second research attempt was related to the application of the proposed solution and testing it on real text data from Tweeter. For this purpose, experiments were carried out on two different data sets. The dataset consisted of 33,795 tweets that were assigned to one of eight emotion categories: *anger,fear, joy, sadness,*

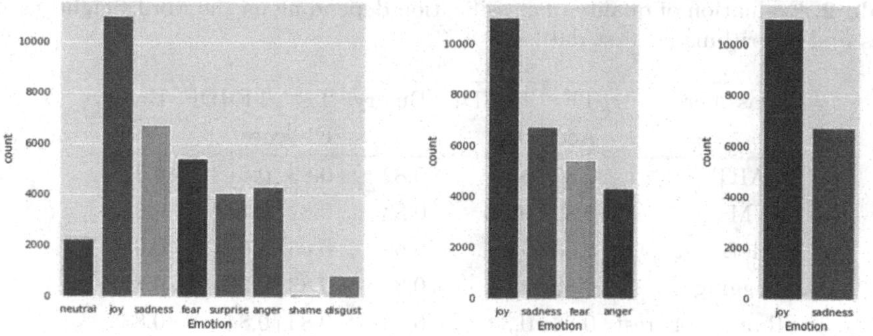

Fig. 2. Division of records into emotion categories for the first set into three stages

surprise, neutral, disgust and *shame.* During the experiments, it was decided to test the impact of the number of emotion categories on the results of tweet content classification. For this purpose, tests were carried out for the 8th, 4th and 2nd categories of emotions for the first set. In the first stage, classification was carried out for the entire data set (a total of 33,795 tweets), in the second, for records assigned to the four most numerous categories (a total of 27,474 tweets), while in the last stage, a classification was carried out only for the two largest classes (a total of 17,767 tweets). The exact division of records into emotion categories is presented in the Fig. 2.

4.3 Results of Experiments

This section presents the classification results for six classifiers (CART, SVM, AdaBoost, Bagging, Random Forest and K-NN) and three word weighting measures (TF-IDF, TF, and Binary). Four measures of classification quality were used to evaluate the classifiers, i.e. accuracy, precision, recall and F1-score. Two sets of data were prepared for the experiments (described in Sects. 4.1 and 4.2), for which the classification process was carried out. Due to the non-deterministic operation of machine learning classifiers, all test trials were carried out in 30 iterations, and the averages were calculated from the obtained results. All results presented in this chapter were obtained with a 70–30 split, where 70% is the training set and 30% is the test set. Due to the large number of combinations of test conditions, the results were aggregated. In fact, more experiments were conducted with different settings of the proposed approach.

The obtained classification results for the first set of data are presented in the Table 2. The best results are marked in bold. The obtained results indicate that detecting emotions from texts is a difficult task, and the classification accuracy reached a level above 80% for all tested classifiers. The best results were obtained using the SVM classifier and the TF-IDF word importance measure, for which the classification accuracy is 85%. In the case of the second data set, the research was divided into three stages due to the number of emotion categories – 8 categories, 4 categories and 2 categories, respectively, which are described in

Table 2. Evaluation of quality of classification depending on the word weighting measures and algorithms for first data set.

Classifier	TF	TF-IDF	Binary	TF	TF-IDF	Binary
	Accuracy			F1-score		
CART	0.82	0.81	0.82	0.83	0.81	0.82
SVM	0.82	**0.85**	0.83	0.82	**0.85**	0.83
AdaBoost	0.68	0.68	0.68	0.68	0.67	0.68
Bagging	0.83	0.82	0.81	0.83	0.82	0.81
Random Forest	0.84	0.84	0.83	0.84	0.84	0.83
K-NN	0.56	0.58	0.55	0.55	0.56	0.55
	Precision			Recall		
CART	0.83	0.81	0.83	0.83	0.80	0.81
SVM	0.84	**0.87**	**0.87**	0.80	**0.84**	0.83
AdaBoost	0.80	0.81	0.80	0.66	0.65	0.66
Random Forest	0.86	0.86	0.86	0.83	0.83	0.83
Bagging	0.85	0.84	0.84	0.83	0.81	0.80
K-NN	0.58	0.64	0.60	0.55	0.56	0.55

the Sect. 4.2. The purpose of such a division was to analyze the impact of the number of emotion categories on the classification results.

The obtained results are presented in the Table 3, 4, 5 and Table 6 for various measures of classification quality, respectively. This way of presenting the results obtained from the experiments allows us to notice the influence of the number of emotion categories on the classification results. The best results for the second set of data are marked in bold, separately for each of the three stages of the study. The obtained results indicate that detecting emotions from texts is a difficult task, and the classification accuracy decreases with the increase in the number of emotion categories. In all three stages, the best results were obtained using the SVM classifier and the TF-IDF word importance measure. The classification accuracy for two emotion categories is 82%, for four emotion categories it is 72%, and for eight emotion categories it is 64%. The worst results were obtained using the K-NN classifier and all three measures of word importance. Similar dependencies can be seen when analyzing the results for precision (Table 4), recall (Table 5) and F1-score (Table 6).

From the analysis of the results obtained, it can be concluded that the greater the number of emotion categories, the more difficult it is to properly classify the content of tweets. This is undoubtedly a problem of multiclass, in which classes are different categories of emotions, often similar to each other, e.g. *joy* and *happines* or *fear* and *worry*. An additional difficulty is the feature selection problem, in which the features are single words appearing in the content of tweets. Often, the use of the same words in combination with other words can express different conflicting emotions. For these reasons, detecting emotions from

Table 3. Accuracy of classification depending on the word weighting measures and algorithms for second data set.

Classifier	for 8 classes			for 4 classes			for 2 classes		
	TF	TF-IDF	Binary	TF	TF-IDF	Binary	TF	TF-IDF	Binary
CART	0.54	0.55	0.54	0.64	0.64	0.65	0.76	0.76	0.75
SVM	0.63	**0.64**	0.63	0.70	**0.72**	0.71	0.80	**0.82**	0.81
AdaBoost	0.33	0.34	0.33	0.59	0.58	0.59	0.75	0.76	0.74
Bagging	0.56	0.55	0.56	0.67	0.67	0.67	0.77	0.79	0.78
Random Forest	0.59	0.62	0.59	0.70	0.70	0.70	0.80	0.81	0.80
K-NN	0.28	0.17	0.26	0.47	0.44	0.48	0.68	0.65	0.70

Table 4. Precision of quality of classification depending on the word weighting measures and algorithms for second data set.

Classifier	for 8 classes			for 4 classes			for 2 classes		
	TF	TF-IDF	Binary	TF	TF-IDF	Binary	TF	TF-IDF	Binary
CART	0.54	0.53	0.54	0.63	0.63	0.63	0.74	0.75	0.73
SVM	0.73	0.73	**0.75**	0.73	0.74	0.74	0.81	**0.82**	**0.82**
AdaBoost	0.24	0.26	0.22	**0.78**	0.77	0.76	0.78	0.78	0.75
Bagging	0.59	0.60	0.61	0.66	0.67	0.65	0.76	0.78	0.76
Random Forest	0.70	0.67	0.68	0.71	0.73	0.71	0.80	0.81	0.79
K-NN	0.55	0.52	0.48	0.51	0.50	0.53	0.70	0.68	0.72

Table 5. Recall of quality of classification depending on the word weighting measures and algorithms for second data set.

Classifier	for 8 classes			for 8 classes			for 2 classes		
	TF	TF-IDF	Binary	TF	TF-IDF	Binary	TF	TF-IDF	Binary
CART	0.52	0.53	0.54	0.62	0.61	0.62	0.74	0.75	0.73
SVM	0.54	0.55	0.55	0.64	**0.67**	0.65	0.76	**0.78**	0.77
AdaBoost	0.17	0.23	0.14	0.48	0.48	0.49	0.69	0.70	0.68
Bagging	0.53	0.55	0.55	0.64	0.64	0.63	0.74	0.77	0.75
Random Forest	0.54	**0.56**	**0.56**	0.65	0.65	0.65	0.77	0.78	0.77
K-NN	0.25	0.19	0.23	0.43	0.37	0.43	0.60	0.54	0.61

Table 6. F1-score measure depending on the word weighting measures and number of classes for second data set.

Classifier	for 8 classes			for 8 classes			for 2 classes		
	TF	TF-IDF	Binary	TF	TF-IDF	Binary	TF	TF-IDF	Binary
CART	0.52	0.53	0.52	0.62	0.62	0.62	0.74	0.75	0.75
SVM	0.57	**0.58**	**0.58**	0.67	**0.69**	0.68	0.77	**0.79**	0.78
AdaBoost	0.13	0.17	0.09	0.52	0.59	0.52	0.70	0.71	0.69
Bagging	0.53	0.55	0.54	0.64	0.65	0.64	0.75	0.77	0.75
Random Forest	0.55	0.57	0.56	0.67	0.68	0.67	0.78	**0.79**	0.77
K-NN	0.21	0.15	0.19	0.43	0.37	0.42	0.58	0.49	0.60

real text data, such as the content of tweets, is difficult, but as shown in the research, it is possible to perform.

5 Conclusions

Emotion detection from text in social networks is an important tool for monitoring and analyzing discussions on social networks. It allows you to get quick and up-to-date information on what people think and feel about specific products, services, events and more. However, there are some limitations and challenges associated with this process. First of all, the text in social networks is often short and incomplete, which can make it difficult to analyze. Also, people often use informal and humorous language in their posts, which can cause erroneous sentiment analysis results.

The main purpose of the article was to develop a model for automatic emotion detection based on texts appearing on social networking sites. An additional objective was to analyze the impact of the number of emotion categories on the classification results. The sentiment corpus and three weight measures based on natural language processing were selected for content analysis. Machine learning algorithms based on supervised learning were used for classification, while accuracy, precision, recall and F1-score were selected to assess the quality of classification. By combining methods from various fields and processing complex data, we wanted to show a scheme by which emotions can be detected from real texts and classified into appropriate categories.

Our experimental research was carried out on two different datasets to show how important and difficult it is to properly define the types of emotions we want to detect. The analysis of the obtained results showed that the proposed solution allows detecting emotions from the actual content of tweets, and the obtained results are better for a smaller number of emotion categories. Unfortunately, too many categories of emotions (as a multi-class problem) that are often referred to by similar words, as well as class imbalance caused by not enough records for each emotion, make detecting emotions from the texts themselves very difficult.

In the future, it is worth exploring different corpora for sentiment analysis, as well as the use of different measures depending on the size of the set and the number of emotion categories. The range of words sufficient to detect emotion should also be explored. In addition, the preprocessing of the data set should be carried out well, in particular the analysis of emoticons, which often reflect various emotions, such as laughter, anger or crying, which can be useful in detecting emotions from the text. On the other hand, for classification algorithms, it is worth additionally checking the impact of the division of training and test data on the classification results.

References

1. Acheampong, F.A., Wenyu, C., Nunoo-Mensah, H.: Text-based emotion detection: advances, challenges, and opportunities. Eng. Rep. **2**(7), e12189 (2020)
2. Agbehadji, I.E., Ijabadeniyi, A.: Approach to sentiment analysis and business communication on social media. In: Bio-inspired Algorithms for Data Streaming and Visualization, Big Data Management, and Fog Computing, pp. 169–193 (2021)
3. Arbieu, U., Helsper, K., Dadvar, M., Mueller, T., Niamir, A.: Natural language processing as a tool to evaluate emotions in conservation conflicts. Biol. Cons. **256**, 109030 (2021)
4. Calvo, R.A., D'Mello, S.: Affect detection: an interdisciplinary review of models, methods, and their applications. IEEE Trans. Affect. Comput. **1**(1), 18–37 (2010)
5. Cambria, E., White, B.: Jumping NLP curves: a review of natural language processing research [review article]. IEEE Comput. Intell. Mag. **9**(2), 48–57 (2014). https://doi.org/10.1109/MCI.2014.2307227
6. Dash, P., Mishra, J., Dara, S.: Sentiment analysis on social network data and its marketing strategies: a review. ECS Trans. **107**(1), 7417 (2022)
7. Garcia-Garcia, J.M., Penichet, V.M., Lozano, M.D.: Emotion detection: a technology review. In: Proceedings of the XVIII International Conference on Human Computer Interaction, pp. 1–8 (2017)
8. Gosai, D.D., Gohil, H.J., Jayswal, H.S.: A review on a emotion detection and recognization from text using natural language processing. Int. J. Appl. Eng. Res. **13**(9), 6745–6750 (2018)
9. Kaur, J., Saini, J.R.: Emotion detection and sentiment analysis in text corpus: a differential study with informal and formal writing styles. Int. J. Comput. Appl. **101**, 1–9 (2014). ISSN 0975–8887
10. Li, Q., et al.: A survey on text classification: from traditional to deep learning. ACM Trans. Intell. Syst. Technol. (TIST) **13**(2), 1–41 (2022)
11. Madhuri, S., et al.: Detecting emotion from natural language text using hybrid and NLP pre-trained models. Turkish J. Comput. Math. Educ. (TURCOMAT) **12**(10), 4095–4103 (2021)
12. Nandwani, P., Verma, R.: A review on sentiment analysis and emotion detection from text. Soc. Netw. Anal. Min. **11**(1), 1–19 (2021). https://doi.org/10.1007/s13278-021-00776-6
13. Panicker, S.S., Gayathri, P.: A survey of machine learning techniques in physiology based mental stress detection systems. Biocybernet. Biomed. Eng. **39**(2), 444–469 (2019)
14. Saffar, A.H., Mann, T.K., Ofoghi, B.: Textual emotion detection in health: advances and applications. J. Biomed. Inf. **137**, 104258 (2022)

15. Santini, S., Schettini, R.: Internet imaging iv. Internet Imaging IV 5018 (2003)
16. Saxena, A., Khanna, A., Gupta, D.: Emotion recognition and detection methods: a comprehensive survey. J. Artif. Intell. Syst. **2**(1), 53–79 (2020)
17. Zad, S., Heidari, M., James Jr., H., Uzuner, O.: Emotion detection of textual data: an interdisciplinary survey. In: 2021 IEEE World AI IoT Congress (AIIoT), pp. 0255–0261. IEEE (2021)
18. Zad, S., Heidari, M., Jones, J.H., Uzuner, O.: A survey on concept-level sentiment analysis techniques of textual data. In: 2021 IEEE World AI IoT Congress (AIIoT), pp. 0285–0291. IEEE (2021)
19. Zhang, T., Schoene, A.M., Ji, S., Ananiadou, S.: Natural language processing applied to mental illness detection: a narrative review. NPJ Dig. Med. **5**(1), 46 (2022)

Finite Libby-Novick Beta Mixture Model: An MML-Based Approach

Niloufar Samiee[1(✉)], Narges Manouchehri[1,2], and Nizar Bouguila[1]

[1] Concordia Institute for Information Systems Engineering (CIISE), Concordia University, Montreal, QC, Canada
niloufar.samiee@mail.concordia.ca, narges.manouchehri@ieee.org, nizar.bouguila@concordia.ca

[2] Algorithmic Dynamics Lab, Unit of Computational Medicine, Karolinska Institute, Stockholm, Sweden

Abstract. We propose an unsupervised algorithm for learning the optimal number of clusters in a finite Libby-Novick Beta mixture model. In unsupervised learning, it is crucial to determine the number of clusters that best describes the data. By extending the minimum message length (MML) principle, we are able to determine the number of clusters in Libby-Novick Beta mixtures. Our model has been evaluated on three publicly available and real-world medical datasets.

Keywords: Finite mixture models · MML · Libby-Novick Beta · Medical image analysis · Healthcare

1 Introduction

The development of artificial intelligence and machine learning has accelerated tremendously in recent years, and we now use machine learning techniques in almost all fields, including healthcare, to perform tasks more efficiently and effectively [1]. Biomedical records are increasingly being analyzed with these approaches to find latent patterns and extract valuable information. When it comes to healthcare, the first and most important issues are the diagnosis and prevention of diseases. Due to the fact that medical science is full of unknowns like all other sciences, it is important to determine the number of clusters identified in a medical record. This will enable experts to diagnose different types of diseases, including diseases that have yet to be discovered. For example, machine learning algorithms can be used to identify patterns of symptoms in a medical record that have not been previously associated with any disease, allowing for the detection of new disease states. Data annotation in healthcare is carried out only by medical experts, resulting in a lengthy and expensive process. Therefore, supervised learning algorithms such as classification are not suitable for use in medical applications. Unsupervised machine learning algorithms based on statistical models are among the most widely used and important algorithms.

N. T. Nguyen et al. (Eds.): ACIIDS 2023, LNAI 13995, pp. 371–383, 2023.
https://doi.org/10.1007/978-981-99-5834-4_30

There are many fields in which statistical models are used, such as image processing and pattern recognition. These models describe the possible behavior of data using probabilities. In analyzing data using statistical models, the first challenge is determining a distribution that can adequately explain the data. In terms of probability models, finite mixture models are one of the most widely used models [2–8]. As a result of their wide application in a wide range of fields, these models have received increasing interest in recent years. Clustering based on these models has many advantages, including the ability to select the number of clusters or to assess the validity of a given model in a formal manner. In fact, determining how many consistent components are necessary to describe the data is a critical part of the modeling problem. There have been numerous approaches proposed for this purpose, including the minimum message length (MML) [9], the partition coefficient (PC) [10], fuzzy hypervolume (FHV) [11], Akaike's information criterion (AIC) [12], the minimum description length (MDL) [13], and the mixture MDL (MMDL) [14]. Based on various studies conducted on this topic, the minimum message length (MML) appears to be one of the most effective approaches for this purpose [15–20].

In this paper, we employ the finite Libby-Novick Beta mixture model and MML to determine the optimal number of components in our data. It is noteworthy that we have demonstrated in a previous work that the Libby-Novick Beta mixture model can provide a good fit to different datasets [21], yet we supposed that the number of components is known in advance. In this paper, we go a step further by developing a method, based on MML, for automatically estimating the optimal number of components. The MML method evaluates statistical models according to their capacity to compress a message containing data. A mixture's optimal cluster number is the number that minimizes the amount of information required from a sender to a receiver to efficiently transmit data X. For high compression, it is necessary to develop appropriate models of the data to be encoded. The message contains two parts for each model in the model space. In the first part of the encoder, only prior knowledge about the model is used, and no information about the data is provided. As the second part encodes only data, it utilizes the model encoded in the first part.

Subsequently, we describe the specification of our model in Sect. 2 of our paper. A brief description of the Libby-Novick Beta mixture model (LNBMM) is provided, followed by the formulation of MML and a complete algorithm of estimation and selection. The results of the experiments are discussed in Sect. 3. A conclusion and future research directions are given in Sect. 4.

2 Model Specification

2.1 Finite Libby-Novick Beta Mixture Model

We assume that $\vec{X}_i = (x_{i1}, \ldots, x_{iD})$, a D-dimensional vector such that $0 \leq x_{id} \leq 1$ and $d = 1, \ldots, D$, follows a Libby-Novick Beta mixture model [22,23]. We consider $\mathcal{X} = \{\vec{X}_1, \ldots, \vec{X}_N\}$ as a dataset with N independent and identically distributed D-dimensional observations. In the following equation, we have the Libby-Novick Beta mixture model:

$$p(\mathcal{X} \mid \vec{\pi}, \vec{\theta}) = \prod_{i=1}^{N}\left[\sum_{j=1}^{M}\pi_j p(\vec{X}_i \mid \vec{\theta}_j)\right]$$
$$= \prod_{i=1}^{N}\left[\sum_{j=1}^{M}\pi_j \prod_{d=1}^{D}\frac{\lambda_{jd}^{a_{jd}} {x_{id}}^{a_{jd}-1}(1-x_{id})^{b_{jd}-1}}{B(a_{jd}, b_{jd})\{1-(1-\lambda_{jd})x_{id}\}^{a_{jd}+b_{jd}}}\right] \quad (1)$$

The parameters of component j^{th} and its weight are given by $\vec{\theta}_j = (\vec{a}_j, \vec{b}_j, \vec{\lambda}_j)$ and π_j, where $j = 1, \ldots, M$. $\vec{\pi} = (\pi_1, \ldots, \pi_M)$, $\vec{\theta} = (\vec{\theta}_1, \ldots, \vec{\theta}_M)$, and $\Theta = \{\vec{\pi}, \vec{\theta}\}$, the complete set of mixture parameters that $\sum_{j=1}^{M}\pi_j = 1$ and $\pi_j >= 0$ for $j = 1, \ldots, M$. $\vec{a} = (\vec{a}_1, \ldots, \vec{a}_M)$, $\vec{b} = (\vec{b}_1, \ldots, \vec{b}_M)$, $\vec{\lambda} = (\vec{\lambda}_1, \ldots, \vec{\lambda}_M)$ are the parameters of mixture model, such that $a_{jd} > 0$, $b_{jd} > 0$, $\lambda_{jd} > 0$ for $d = 1, \ldots, D$. In the following subsection, we give a brief summary of the maximum likelihood approach that we developed in [21] to estimate the parameters of the proposed mixture model. The estimation of these parameters is an important step towards developing the MML criterion.

2.2 Maximum Likelihood and EM Algorithm

As part of the estimation process, our model parameters are calculated using ML estimation within an EM framework, which determine the parameters that maximize the model's likelihood function. In fact, ML estimates mixture model parameters in order to maximize the log-likelihood [21]. We define a vector $\vec{Z}_i = (Z_{i1}, ..., Z_{ij})$ such that $Z_{ij} = 1$ if $\vec{X}_i$ belongs to component j and 0 otherwise and $\sum_{j=1}^{M} Z_{ij} = 1$. As a result, we define a set of membership vectors $\mathcal{Z} = \{Z_1, \ldots, Z_N\}$ for the set $\mathcal{X} = \{\vec{X}_1, \ldots, \vec{X}_N\}$. The posterior probability of each vector $\vec{X}_i$ determines its assignment to one of the M clusters, as follows:

$$\hat{Z}_{ij} = p(j \mid \vec{X}_i, \vec{\theta}_j) = \frac{\pi_j p(\vec{X}_i, \vec{\theta}_j)}{\sum_{j=1}^{M}\pi_j p(\vec{X}_i, \vec{\theta}_j)} \quad (2)$$

This leads to the following formulation for log-likelihood:

$$L(\Theta, \mathcal{Z}, \mathcal{X}) = \sum_{j=1}^{M}\sum_{i=1}^{N}\hat{Z}_{ij}\Bigg[\log \pi_j \quad (3)$$
$$+ \log \prod_{d=1}^{D}\frac{\lambda_{jd}^{a_{jd}} {x_{id}}^{a_{jd}-1}(1-x_{id})^{b_{jd}-1}}{B(a_{jd}, b_{jd})\{1-(1-\lambda_{jd})x_{id}\}^{a_{jd}+b_{jd}}}\Bigg]$$
$$= \sum_{j=1}^{M}\sum_{i=1}^{N}\hat{Z}_{ij}(\log \pi_j + \sum_{d=1}^{D}\Bigg[a_{jd}\log \lambda_{jd} + a_{jd}\log x_{id}$$

$$- \log x_{id} + b_{jd} \log(1 - x_{id}) - \log(1 - x_{id})$$
$$+ \log \Gamma(a_{jd} + b_{jd}) - \log \Gamma(a_{jd}) - \log \Gamma(b_{jd})$$
$$- a_{jd} \log(1 - (1 - \lambda_{jd}) x_{id}) - b_{jd} \log(1 - (1 - \lambda_{jd}) x_{id}) \Big]$$

In order to maximize the complete log-likelihood, we compute the gradient of log-likelihood with respect to the parameters. In this case, Newton-Raphson is used as an iterative method to update the parameters [21], as there is no closed-form solution to following equation:

$$\frac{\partial L(\Theta, \mathcal{Z}, \mathcal{X})}{\partial \Theta} = 0 \tag{4}$$

For more details about the parameters estimation, the reader is referred to [21].

2.3 The MML Criterion for a Finite Libby-Novick Beta Mixture

As a model selection technique, MML (minimum message length) is used in this section. According to information theory, the optimal number of clusters transmits data from sender to receiver efficiently with the least amount of information. As a result, MML can be defined as follows for a mixture of distributions:

$$\text{MML} = -\log\left(\frac{h(\Theta) p(\mathcal{X} \mid \Theta)}{\sqrt{\mid F(\Theta) \mid}}\right) + N_p\left(-\frac{1}{2}\log(12) + \frac{1}{2}\right) \tag{5}$$

where $h(\Theta)$ is prior probability distribution, $p(\mathcal{X} \mid \Theta)$ is the complete data log-likelihood, $F(\Theta)$ is the expected Fisher information matrix computed by taking the second derivative of the negative log-likelihood, and $|F(\Theta)|$ is its determinant. N_p represents the number of free parameters and equals $(M(2D + 1)) - 1$.

2.4 Fisher Information for Libby-Novick Beta Mixture Model

As the expected value of the negative of the Hessian matrix, the Fisher matrix, also known as the curvature matrix, describes the curve of the likelihood function around its maximum. Since MML is based on a Hessian matrices, it takes on a sophisticated analytical form that is difficult to reproduce. As a result, we will approximate this matrix using these two following assumptions: First, it is important to keep in mind that $\vec{\theta}$ and the vector $\vec{\pi}$ are independent, as one's preconceived notions about the value of the mixing parameter vector $\vec{\pi}$ do not typically influence one's notions about $\vec{\theta}$. Moreover, we presume that the $\vec{\theta}$ components are also independent. Fisher information matrix can be calculated after clustering data vectors according to a mixture model. A Fisher information matrix has the following determinant:

$$|F(\Theta)| = |F(\vec{\pi})| \prod_{j=1}^{M} |F(\vec{\theta_j})| \tag{6}$$

$|F(\vec{\pi})|$ is the determinant of Fisher information of mixing parameters π_j and $|F(\vec{\theta_j})|$ is the determinant of Fisher information with regard to the vector $\vec{\theta_j} = (\vec{a_j}, \vec{b_j}, \vec{p_j})$ of a single Libby-Novick Beta distribution. As a result, we can compute the Fisher information matrix determinant by assuming a generalized Bernoulli process where there are M possible results for M clusters for each trial as follows:

$$|F(\vec{\pi})| = \frac{N^{M-1}}{\prod_{j=1}^{M} \pi_j} \tag{7}$$

The Fisher information for our mixture is as follows:

$$\log(|F(\Theta)|) = (M-1)\log(N) - \sum_{j=1}^{M} \log(\pi_j) + \sum_{j=1}^{M} \log(|F(\vec{\theta_j})|) \tag{8}$$

2.5 Determinant of the Fisher Information

Assuming the mixture model procedure is followed, $X_j = (\vec{X_t}, \ldots, \vec{X}_{t+n_j-1})$ samples of data are allocated to cluster jth, such that $t \leq N$ and n_j is the number of samples assigned to cluster j.

$$\begin{aligned} -\log p(\mathcal{X} \mid \Theta) &= -\log(\prod_{n=t}^{t+n_j-1} p(\vec{X} \mid \vec{\theta}_M)) \\ &= -(\sum_{n=t}^{t+n_j-1} \log p(\vec{X} \mid \vec{\theta}_M)) \end{aligned} \tag{9}$$

$F(\vec{\theta_j})$ is defined as the negative of the second derivative of complete log-likelihood. In accordance with the parameters $a_{jd}, b_{jd}, \lambda_{jd}$, we calculate the second and mixed derivatives:

- Derivatives with respect to a_{jd}, a_{jd}:

$$F_{a_{jd},a_{jd}} = -\frac{\partial^2 \log p(\mathcal{X} \mid \Theta)}{\partial a_{jd}^2} = -n_j(\psi'(a_{jd}+b_{jd}) - \psi'(a_{jd})) \tag{10}$$

$$-\frac{\partial^2 \log p(\mathcal{X} \mid \Theta)}{\partial a_{jd_s} \partial a_{jd_t}} = 0, d_s \neq d_t \tag{11}$$

- Derivatives with respect to a_{jd}, b_{jd}:

$$F_{a_{jd},b_{jd}} = -\frac{\partial^2 \log p(\mathcal{X} \mid \Theta)}{\partial a_{jd} \partial b_{jd}} = -n_j(\psi'(a_{jd}+b_{jd})) \tag{12}$$

$$-\frac{\partial^2 \log p(\mathcal{X} \mid \Theta)}{\partial a_{jd_s} \partial b_{jd_t}} = 0, d_s \neq d_t \tag{13}$$

– Derivatives with respect to a_{jd}, λ_{jd}:

$$F_{a_{jd},\lambda_{jd}} = -\frac{\partial^2 \log p(\mathcal{X} \mid \Theta)}{\partial a_{jd} \partial \lambda_{jd}} = -\sum_{i=1}^{N} \hat{Z}_{ij}[\frac{1}{\lambda jd} - \frac{x_{id}}{(1-(1-\lambda_{jd})x_{id})})] \quad (14)$$

$$-\frac{\partial^2 \log p(\mathcal{X} \mid \Theta)}{\partial a_{jd_s} \partial \lambda_{jd_t}} = 0, d_s \neq d_t \quad (15)$$

– Derivatives with respect to b_{jd}, a_{jd}:

$$F_{b_{jd},a_{jd}} = -\frac{\partial^2 \log p(\mathcal{X} \mid \Theta)}{\partial b_{jd} \partial a_{jd}} = -n_j(\psi'(a_{jd} + b_{jd})) \quad (16)$$

$$-\frac{\partial^2 \log p(\mathcal{X} \mid \Theta)}{\partial b_{jd_s} \partial a_{jd_t}} = 0, d_s \neq d_t \quad (17)$$

– Derivatives with respect to b_{jd}, b_{jd}:

$$F_{b_{jd},b_{jd}} = -\frac{\partial^2 \log p(\mathcal{X} \mid \Theta)}{\partial b_{jd}^2} = -n_j(\psi'(a_{jd} + b_{jd}) - \psi'(b_{jd})) \quad (18)$$

$$-\frac{\partial^2 \log p(\mathcal{X} \mid \Theta)}{\partial b_{jd_s} \partial b_{jd_t}} = 0, d_s \neq d_t \quad (19)$$

– Derivatives with respect to b_{jd}, λ_{jd}:

$$F_{b_{jd},\lambda_{jd}} = -\frac{\partial^2 \log p(\mathcal{X} \mid \Theta)}{\partial b_{jd} \partial \lambda_{jd}} = -\sum_{i=1}^{N} \hat{Z}_{ij}[-\frac{x_{id}}{(1-(1-\lambda_{jd})x_{id})}] \quad (20)$$

$$-\frac{\partial^2 \log p(\mathcal{X} \mid \Theta)}{\partial b_{jd_s} \partial \lambda_{jd_t}} = 0, d_s \neq d_t \quad (21)$$

– Derivatives with respect to λ_{jd}, a_{jd}:

$$F_{\lambda_{jd},a_{jd}} = -\frac{\partial^2 \log p(\mathcal{X} \mid \Theta)}{\partial \lambda_{jd} \partial a_{jd}} = -\sum_{i=1}^{N} \hat{Z}_{ij}[\frac{1}{\lambda_{jd}} - \frac{x_{id}}{(1-(1-\lambda_{jd})x_{id})}] \quad (22)$$

$$-\frac{\partial^2 \log p(\mathcal{X} \mid \Theta)}{\partial \lambda_{jd_s} \partial a_{jd_t}} = 0, d_s \neq d_t \quad (23)$$

– Derivatives with respect to λ_{jd}, b_{jd}:

$$F_{\lambda_{jd},b_{jd}} = -\frac{\partial^2 \log p(\mathcal{X} \mid \Theta)}{\partial \lambda_{jd} \partial b_{jd}} = -\sum_{i=1}^{N} \hat{Z}_{ij}[-\frac{x_{id}}{(1-(1-\lambda_{jd})x_{id})}] \quad (24)$$

$$-\frac{\partial^2 \log p(\mathcal{X} \mid \Theta)}{\partial \lambda_{jd_s} \partial b_{jd_t}} = 0, d_s \neq d_t \quad (25)$$

– Derivatives with respect to $\lambda_{jd}, \lambda_{jd}$:

$$F_{\lambda_{jd},\lambda_{jd}} = -\frac{\partial^2 \log p(\mathcal{X} \mid \Theta)}{\partial \lambda_{jd}^2} \tag{26}$$

$$= -\sum_{i=1}^{N} \hat{Z}_{ij}[\frac{a_{jd}x_{id}^2}{(1-(1-\lambda_{jd})x_{id})^2} - \frac{a_{jd}}{\lambda_{jd}^2} + \frac{b_{jd}x_{id}^2}{(1-(1-\lambda_{jd})x_{id})^2}]$$

$$-\frac{\partial^2 \log p(\mathcal{X} \mid \Theta)}{\partial \lambda_{jd_s} \partial \lambda_{jd_t}} = 0, d_s \neq d_t \tag{27}$$

The $F(\vec{\theta_j})$ is a $3D$ by $3D$ matrix as follows:

$$F_j = \begin{bmatrix} F_{(a_{jd},a_{jd})} & F_{(a_{jd},b_{jd})} & F_{(a_{jd},\lambda_{jd})} \\ F_{(b_{jd},a_{jd})} & F_{(b_{jd},b_{jd})} & F_{(b_{jd},\lambda_{jd})} \\ F_{(\lambda_{jd},a_{jd})} & F_{(\lambda_{jd},b_{jd})} & F_{(\lambda_{jd},\lambda_{jd})} \end{bmatrix} \tag{28}$$

2.6 Prior Distribution

We must choose the model's parameters' prior distribution $h(\Theta)$ in order to calculate the MML criterion. We define $h(\Theta)$ as follows:

$$h(\Theta) = h(\vec{\pi})h(\vec{a})h(\vec{b})h(\vec{\lambda}) \tag{29}$$

For modelling proportional vectors, we assume a Dirichlet distribution for $h(\vec{\pi})$ where $\vec{\eta} = (\eta_1, \eta_2, \ldots, \eta_M)$:

$$h(\pi_1, \pi_2, \ldots, \pi_M) = \frac{\Gamma(\sum_{j=1}^{M} \eta_j)}{\prod_{j=1}^{M} \Gamma(\eta_j)} \prod_{j=1}^{M} \pi_j^{\eta_j - 1} \tag{30}$$

By calculating a uniform prior for the parameter η, $(\eta_1 = 1, ..., \eta_M = 1)$, we can simplify (30) as follows:

$$h(\vec{\pi}) = (M-1)! \tag{31}$$

We assume that dimensions are independent, so we have:

$$h(\vec{a}) = \prod_{j=1}^{M} \prod_{d=1}^{D} h(a_{jd}) \tag{32}$$

The assumption is that we do not have any prior information regarding parameter a_{jd} in this case. Consequently, we use a simple uniform prior in accordance with Ockham's razor, which has shown to be effective in producing effective results, to ensure that its effect on the posterior is minimal [24]. $h(b_{jd})$ and $h(\lambda_{jd})$ will be chosen in the same way:

$$h(a_{jd}) = e^{-6} \frac{a_{jd}}{||a_j||}, h(b_{jd}) = e^{-6} \frac{b_{jd}}{||b_j||}, h(\lambda_{jd}) = e^{-6} \frac{\lambda_{jd}}{||\lambda_j||} \tag{33}$$

Log of prior is provided by:

$$\log(h(\Theta)) = -D\sum_{j=1}^{M}\log(||a_j||) + \sum_{j=1}^{M}\sum_{d=1}^{D}\log(a_{jd}) \tag{34}$$
$$- D\sum_{j=1}^{M}\log(||b_j||) + \sum_{j=1}^{M}\sum_{d=1}^{D}\log(b_{jd})$$
$$- D\sum_{j=1}^{M}\log(||\lambda_j||) + \sum_{j=1}^{M}\sum_{d=1}^{D}\log(\lambda_{jd}) + \sum_{j=1}^{M-1}\log(j) - 18MD$$

2.7 Full Learning Algorithm

Here is a summary of all the steps in our method:

Algorithm 1. Full Learning Algorithm

1. Input $\mathcal{X}$ and the number of clusters M.
2. Use K-Means algorithm to initialize the M clusters.
3. Initialize the parameters.

   Repeat
4. EM algorithm [21]
5. MML
   (a) Calculate the criterion of MML(M).
   (b) Find the optimal M^* i.e. $M^* = argmin_M$ MML(M).

---

3 Experimental Results

In this part of our research, we begin by extracting features from images and preprocessing our data. We employed Scale-invariant Feature Transform (SIFT) and Bag of Visual Words (BoVW) as feature extraction methods [25,26]. Using a bag of visual words, images are presented as patches, with distinctive patterns derived from each one. SIFT is used as a feature detector for the extraction of these visual characteristics. Based on the assumption that all input values fall between 0 and 1, we normalized our datasets using the min-max method:

$$X = \frac{X - X_{min}}{X_{max} - X_{min}} \tag{35}$$

In the following, three real-world medical imaging applications were used to evaluate our algorithm, including images of lung tissue samples, histological malaria images, and histological images of breast tissue.

3.1 Malaria Detection

Malaria is a parasitic disease. Individuals can become infected with a parasite when bitten by a mosquito. Using a microscope is the most common method of diagnosing malaria. Microscopists diagnose malaria by identifying malaria-infected blood cells in blood smears. Unlike uninfected cells, infected cells have a small clot. Humans find it difficult to detect positive results among a large number of smears and to verify each sample precisely. For this reason, using a dependable machine learning technique to diagnose this disease can be extremely helpful. As part of this study, we evaluated our model by analyzing images of 2,000 cells from the National Institutes of Health and determined the optimal number of clusters for modeling this dataset [27,28]. Our algorithm was able to determine the optimal number of clusters as shown in Fig. 1.

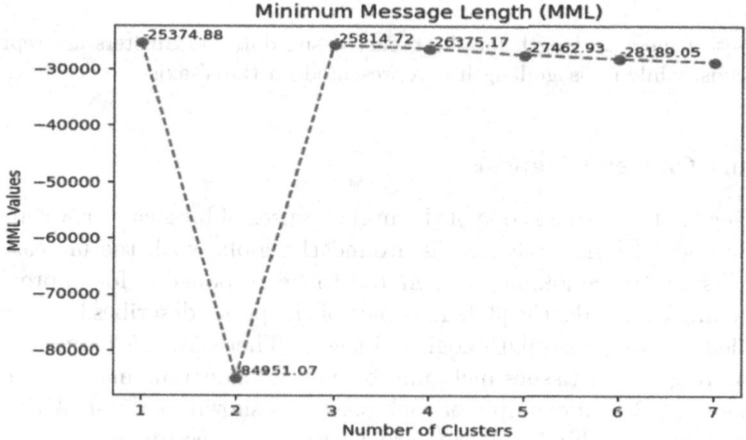

Fig. 1. Plot of message length for the Malaria detection dataset. Clusters are represented on the X-axis, while message length is represented on the Y-axis.

3.2 Breast Tissue Analysis

In the world, breast cancer is the leading cause of cancer-related death. According to current estimates, 12.9% of American women will suffer from breast cancer during their lifetimes [29]. In order to increase the chance of survival, it is essential to identify breast cancer cells at an early stage and start treatment. During the diagnostic process, an expert pathologist evaluates tissues and identifies whether they are benign or malignant (cancerous) based on their histological characteristics. It is important to note that cancer diagnosis by specialists is not error-free. Therefore, machine learning techniques could be used to reduce the number of erroneous diagnoses as well as improve accuracy in breast cancer diagnosis. A publicly available dataset containing 500 samples, each with malignant and benign labels was used to evaluate our algorithm [30]. Figure 2 illustrates how our algorithm was able to determine the optimal number of clusters.

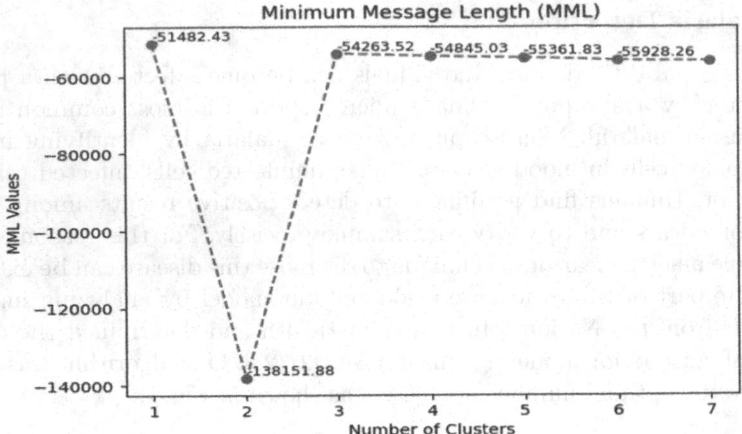

Fig. 2. Plot of message length for the breast tissue dataset. Clusters are represented on the X-axis, while message length is represented on the Y-axis.

3.3 Lung Cancer Diagnosis

Considering that smoking is one of the major causes of lung cancer, we are seeing a large number of lung cancer cases around the globe with the increase in the number of smokers. Smoking is estimated to be responsible for approximately 80% of all lung cancer deaths [31]. This part of the paper describes how our model was applied to lung histopathological images. There are 2500 images in this dataset of lung cancer tissues including benign, adenocarcinoma, and squamous cell carcinoma [32]. An example of each cluster is shown in Fig. 4. A timely and accurate diagnosis of lung cancer is crucial to its successful treatment, as with any other type of cancer. We were able to determine the optimal number of clusters using our algorithm, which can be seen in Fig. 4 (Fig. 3).

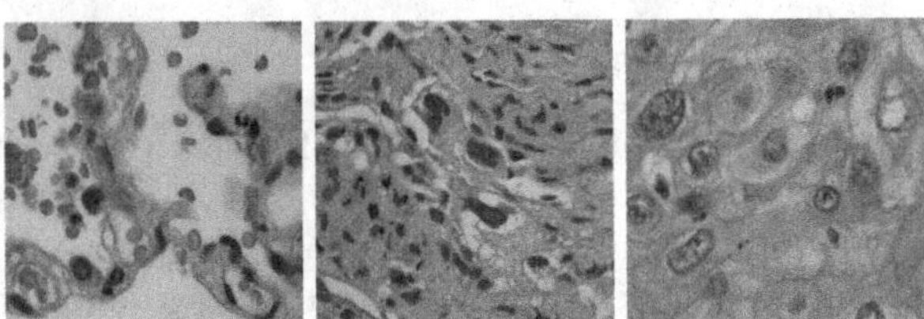

Fig. 3. An illustration of benign lung tissue, adenocarcinoma, and squamous cell carcinoma.

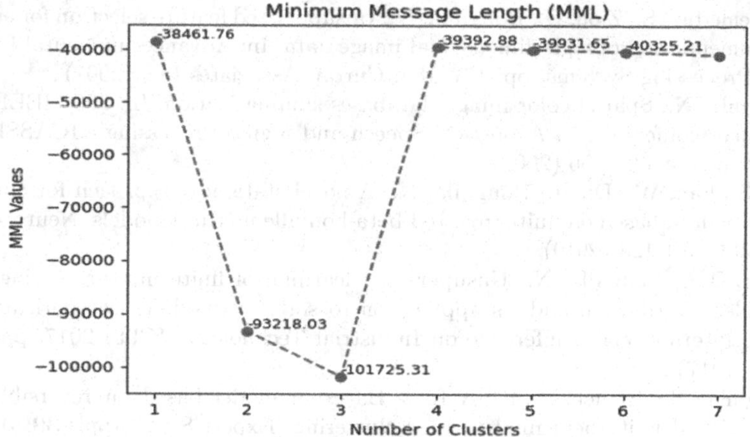

Fig. 4. Plot of message length for lung cancer dataset. Clusters are represented on the X-axis, while message length is represented on the Y-axis.

4 Discussion and Conclusion

In the context of clustering as an unsupervised method, one of the most challenging aspects of the process is determining the optimal number of components. In this paper, we have discussed a method for selecting the number of components in Libby-Novick Beta mixtures based on MML. According to the results, the MML model selection method performs well on real-world data. The reason for this can be attributed to the fact that the prior term in this criteria is present in a manner that is not present in the other criteria. Finally, we conducted an evaluation of our algorithm for three medical applications, including detection of malaria, diagnosis of breast cancer, and lung tissue analysis. As part of our subsequent work, we plan to extend the finite Libby-Novick Beta mixture model to the infinite scenario, which will enable the model to have even greater flexibility.

Acknowledgment. The completion of this research was made possible thanks to the Natural Sciences and Engineering Research Council of Canada (NSERC).

References

1. Bouguila, N.: Hybrid generative/discriminative approaches for proportional data modeling and classification. IEEE Trans. Knowl. Data Eng. **24**(12), 2184–2202 (2012)
2. Bouguila, N., Fan, W.: Mixture Models and Applications. Springer, Cham (2020). https://doi.org/10.1007/978-3-030-23876-6
3. Bouguila, N.: A model-based approach for discrete data clustering and feature weighting using map and stochastic complexity. IEEE Trans. Knowl. Data Eng. **21**(12), 1649–1664 (2009)

4. Boutemedjet, S., Ziou, D., Bouguila, N.: Unsupervised feature selection for accurate recommendation of high-dimensional image data. In: Advances in Neural Information Processing Systems, pp. 177–184. Curran Associates Inc. (2007)
5. Bouguila, N.: Spatial color image databases summarization. In: 2007 IEEE International Conference on Acoustics, Speech and Signal Processing - ICASSP 2007, vol. 1, pp. I-953–I-956 (2007)
6. Hu, C., Fan, W., Du, J., Bouguila, N.: A novel statistical approach for clustering positive data based on finite inverted beta-liouville mixture models. Neurocomputing **333**, 110–123 (2019)
7. Oboh, B.S., Bouguila, N.: Unsupervised learning of finite mixtures using scaled dirichlet distribution and its application to software modules categorization. In: IEEE International Conference on Industrial Technology (ICIT) 2017, pp. 1085–1090 (2017)
8. Bouguila, N., Elguebaly, T.: A fully Bayesian model based on reversible jump MCMC and finite beta mixtures for clustering. Expert Syst. Appl. **39**(5), 5946–5959 (2012)
9. Wallace, C.S., Boulton, D.: An Information Measure for Classification. Comput. J. **11**(2), 185–194 (1968)
10. Bezdek, James C..: Selected applications in classifier design. In: Pattern Recognition with Fuzzy Objective Function Algorithms. AAPR, pp. 203–239. Springer, Boston, MA (1981). https://doi.org/10.1007/978-1-4757-0450-1_6
11. Gath, I., Geva, A.B.: Unsupervised optimal fuzzy clustering. IEEE Trans. Pattern Anal. Mach. Intell. **11**(7), 773–780 (1989)
12. Akaike, H.: A new look at the statistical model identification. IEEE Trans. Autom. Control **19**(6), 716–723 (1974)
13. Rissanen, J.: Modeling by shortest data description. Automatica **14**, 465–471 (1978)
14. Figueiredo, M., Jain, A.: Unsupervised learning of finite mixture models. IEEE Trans. Pattern Anal. Mach. Intell. **24**(3), 381–396 (2002)
15. Bouguila, N., Ziou, D.: Unsupervised selection of a finite dirichlet mixture model: an mml-based approach. IEEE Trans. Knowl. Data Eng. **18**(8), 993–1009 (2006)
16. Roberts, S.J., Husmeier, D., Rezek, I., Penny, W.D.: Bayesian approaches to gaussian mixture modeling. IEEE Trans. Pattern Anal. Mach. Intell. **20**(11), 1133–1142 (1998)
17. Baxter, R.A., Oliver, J.J.: Finding overlapping components with mml. Stat. Comput. **10**(1), 5–16 (2000)
18. Bouguila, N., Ziou, D.: MML-based approach for finite dirichlet mixture estimation and selection. In: Perner, P., Imiya, A. (eds.) MLDM 2005. LNCS (LNAI), vol. 3587, pp. 42–51. Springer, Heidelberg (2005). https://doi.org/10.1007/11510888_5
19. Bouguila, N., Ziou, D.: On fitting finite dirichlet mixture using ECM and MML. In: Singh, S., Singh, M., Apte, C., Perner, P. (eds.) ICAPR 2005. LNCS, vol. 3686, pp. 172–182. Springer, Heidelberg (2005). https://doi.org/10.1007/11551188_19
20. Bouguila, N., Ziou, D.: Mml-based approach for high-dimensional unsupervised learning using the generalized dirichlet mixture. In: IEEE Conference on Computer Vision and Pattern Recognition, CVPR Workshops 2005, San Diego, CA, USA, 21–23 September, 2005, p. 53. IEEE Computer Society (2005)
21. Samiee, N., Manouchehri, N., Bouguila, N.: Maximum likelihood-based estimation of finite multivariate Libby-Novick beta mixture models in medical applications. In: IEEE International Conference on Industrial Technology (ICIT) **2023**, 1–6 (2023)

22. Ketabchi, K., Manouchehri, N., Bouguila, N.: Fully Bayesian Libby-Novick beta mixture model with feature selection. In: IEEE International Conference on Industrial Technology, ICIT 2022, Shanghai, China, 22–25 August 2022, pp. 1–6. IEEE (2022)
23. Cordeiro, G., Santana, L., Ortega, E., Pescim, R.: A new family of distributions: Libby-Novick beta. Int. J. Stat. Probability **3**, 63–80 (2014)
24. Autzen, B.: Bayesian Ockham's razor and nested models. Econ. Philos. **35**(2), 321–338 (2019)
25. Lowe, D.G.: Distinctive image features from scale-invariant keypoints. Int. J. Comput. Vision **60**(2), 91–110 (2004)
26. Rao, L.J., Neelakanteswar, P., Ramkumar, M., Krishna, A., Basha, C.Z.: An effective bone fracture detection using bag-of-visual-words with the features extracted from sift. In: International Conference on Electronics and Sustainable Communication Systems (ICESC) 2020, pp. 6–10 (2020)
27. https://ceb.nlm.nih.gov/repositories/malaria-datasets
28. Tangpukdee, N., Duangdee, C., Wilairatana, P., Krudsood, S.: Malaria diagnosis: a brief review. Korean J. Parasitol. **47**, 93–102 (2009)
29. https://www.cancer.gov/types/breast
30. https://www.kaggle.com/paultimothymooney/breast-histopathology-images
31. https://www.cancer.org/cancer/lung-cancer/causes-risks-prevention/risk-factors.html
32. Lung dataset (2018). https://www.kaggle.com/andrewmvd/lung-and-colon-cancer-histopathological-images

Artificial Intelligences on Automated Context-Brain Recognition with Mobile Detection Devices

Ja-Hwung Su[1], Wei-Jiang Chen[1], Ming-Cheng Zhang[1], and Yi-Wen Liao[2(✉)]

[1] Department of Computer Science and Information Engineering, National University of Kaohsiung, Kaohsiung, Taiwan

[2] Department of Intelligent Commerce, National Kaohsiung University of Science and Technology, Kaohsiung, Taiwan

pinkwen923@gmail.com

Abstract. In the past few decades, lots of studies were proposed on investigations of brain circuits for physical health, mental health, educational learning, controlling system and so on. However, very few studies concentrate their attention on contextual brain recognition. Actually, the brain neural system is the main bridge between mental senses and physical behaviors, and the mental senses such as emotions and preferences are heavily impacted by the context, which can be represented by the brain signals. Hence, how to recognize the contextual brain signal for mental senses is the aimed issue of this paper. To deal with this issue, in this paper, we examine a number of machine learning methods on a real dataset, including traditional classifiers and LSTM (Long Short-Term Memory Neural Network). Also, a free Convolutional Neural Network called BrainCNN is proposed to recognize the contextual brain signal using a mobile brain-signal detection device. Through the brain-signal recognition, the context can be identified and thereby facilitates the prediction of mental senses in the future. The experimental results show the proposed BrainCNN is more promising than the other methods in recognizing the context brain signal. Further, this work provides a basic idea for future interests in investigating the contextual brain.

Keywords: BrainCNN · context-brain recognition · Convolutional Neural Network · Artificial Intelligence · machine learning

1 Introduction

Recent Artificial Intelligence techniques has been successfully used in the field of biomedical sciences, such as disease recognition and risk assessment. In this field, the brain neural system attracts much attention because it is highly related to physical health and mental health. Most related researches are proposed to investigate the associations between the brain neural system and the health with recognitions of CT images and brain signals. However, in addition to the health investigation, there exists another important issue for how to imply the mental senses by the brain neural system. Actually, a number

N. T. Nguyen et al. (Eds.): ACIIDS 2023, LNAI 13995, pp. 384–394, 2023.
https://doi.org/10.1007/978-981-99-5834-4_31

of user preferences are hidden in the mental senses and the mental senses are impacted by the environment (context). Imagine that you are taking an exercise, the sense could trigger you prefer the light music. Figure 1 shows the motivated idea in this paper. The preference changes as the mental sense changes, and the mental sense changes as the context changes. If the context can be predicted successfully, the user preference can also be implied for many applications such as recommendation to reading, listening and watching. Therefore, the main challenge of this paper is how to effectively predict the context by the contextual brain signal. To aim at this issue, in this paper, a number of classifiers including Support Vector Machine (SVM), Linear Discriminant Analysis (LDA), Random Forest (RF) and Long Short-Term Memory Neural Network (LSTM) are employed as the baselines. Accordingly, a free CNN architecture named BrainCNN is proposed to improve the accuracy. The evaluation results on a real dataset reveals that, the proposed method performs better than the baselines. In overall, the primary contributions can be listed as follows.

I. From novelty point of view, although many ubiquitous multimedia applications are lifted, very few studies are made on associating the brain neural system with context recognition. This paper provides a considerable idea for future researches interested in this topic or even extensions.
II. From technical point of view, the contributions are: 1) because no related data is publicly released, we propose a robust brain-collection scheme to touch the real circumstances; 2) because traditional classifiers are not sensitive to the contextual brain, we construct a manual CNN named BrainCNN to enhance the context recognition.
III. From practical point of view, it is applicable. The experimented equipment is a headband-like mobile detection device which is convenient and easy to wear. In the future, the proposed BrainCNN can be embedded into a wearable device such as over-ear headphones with WiFi to link the online web system for recommendations of listening, reading or watching.

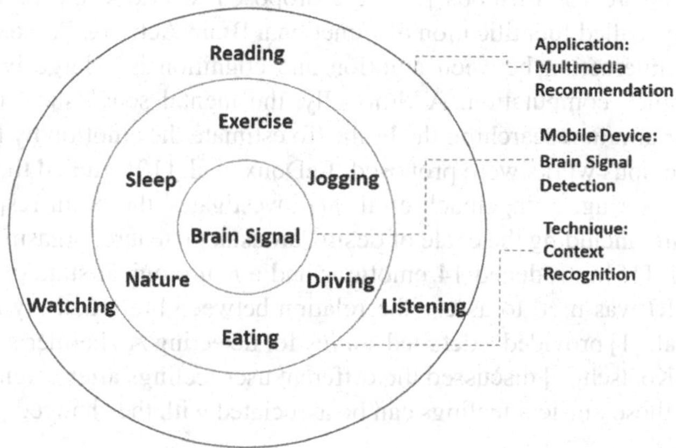

Fig. 1. Motivated idea for contextual brain recognition.

The rest of this paper is organized as follows. A review of previous works is made in Sect. 2. Next, the proposed approach including data collection, data preprocessing, model training and brain-signal recognition is interpreted in Sect. 3. Then, the empirical analysis is shown in Sect. 4. Finally, the summary and future works are conducted in Sect. 5.

2 Related Work

In the field of Bioinformatics, the brain neural system has been studied [12] for many decades. In general, the related works can be briefly categorized into three domains, namely brain detection, brain awareness and applications. These domains are actually three levels in exploring the brain neural system, where the detection provides the awareness with sources and thereby the awareness supports the applications. In this section, the literatures are reviewed category by category.

2.1 Brain Detection

Brain detection indicates detections of the brain activity by invasive or non-invasive methods [3, 17, 18]. For invasive methods, Electrocorticography (ECoG) and Intracortical inhibition are two main techniques, while Electroencephalogram (EEG), Electrooculography (EOG), Magnetoencephalography (MEG), Functional near-infrared spectroscopy (FNIRS) and Functional magnetic resonance imaging (FMRI) are five main techniques for non-invasive methods. In overall, the extracted signals are represented by two paradigms, namely time series and visual images. Whatever the paradigm is, they are used as the features for brain awareness.

2.2 Brain Awareness

Most of brain awareness methods [2] were proposed to bridge the brain activity to physical activity, called Identification of Functional Brain Activity. Pessoa scholar [14] stated that the interaction between emotion and cognition is a large brain network, needing a complex computation. Additionally, the mental sense such as emotion is another popular aim in researching the brain. To estimate the emotion by brain signals, a number of previous works were proposed. LeDoux et al. [10] studied the brain-signal response while fearing. Kringelbach et al. [7] investigated the brain responses to the emotion pleasure, including the cycle of desire, arousal, plateaure, orgasm to refraction. Saarimäki et al. [15] introduced 14 emotions and a neurological state of the brain. In this study, FMRI was used to assess the relation between brain activity and emotion. Al-Shoukry et al. [1] provided a detailed survey for detecting Alzheimer's diseases with deep learning. Koelsch [6] discussed the different user feelings after listening to music. Through MRI, these musical feelings can be associated with the changed parts of brain.

2.3 Application

Besides above, the other popular topic in studying the brain neural system is the application. Navarro et al. [13] extended the traditional ambient intelligence for context awareness by enhancing the brain computer interface. Koban et al. [8] proposed a self-in-context model for connecting the mental and physical health. Maren et al. [11] clearly defined context and provided details of associations between context and brain. Jamil et al. [4] reviewed the brain computer interfaces and eye-tracking techniques to evaluate the effectiveness of online learning. Su et al. [16] proposed a content-based music retrieval by calculating the context-brain similarity. Kim [5] used brain signals to switch music while listening. In this work, the α signal on the EEG is adopted as the threshold for music switching. Kohli et al. [9] provided a survey of combining Brain Computer Interface and Extended Reality for smart cities.

3 Proposed Method

3.1 Primary Concept

Up to the present, more and more evidences approve the relation between context and brain [11]. This is the main motivation to propose a method for regularizing the relation in this paper. However, materializing this motivation will encounter three challenges: 1) how to collect the nearly real data by a mobile detection device, 2) how to model the regulation between the context and brain, and 3) how to apply the regulation to the reality. Actually, the solutions can be echoed with the contributions mentioned in Sect. 1. Here, a more detailed concept is presented. For the first challenge, it is really not easy to collect the real context brain signals. Hence, a simulation embedded in a virtual reality box (called VR Box) is conducted for this challenge. With the VR box, the user will obtain the immersive experience. Thereupon, the collected brain signals will be close to the truth. For the second challenge, because no baseline can be referred to, a number of well-known classifiers are assessed in this paper. Yet, it is because the brain signals from the experimented device are represented in a one-dimension sequence that an adaptive data preprocess for different types of methods is necessary. For CNN-based classifiers, a two-dimension matrix is used, while a one-dimension vector is used for the other ones. The main intent is to approximate a best classifier for recognizing the contextual brain signal. Note that, because the original brain signal is uninformative (only 10 attributes), data enhancement and data augmentation are performed ahead of training. For the third challenge, the proposed method can be materialized in a fashion wearable device, like current Wearable Health Monitoring Systems. Based on the context, the user preference could be implied further for listening, reading, watching and so on. Moreover, the health could be linked to the context brain for effective treatments.

3.2 Framework

As mentioned above, the goal of this paper is to propose a context awareness method by recognizing contextual brain signals. For this purpose, the proposed method is conducted by offline training and online prediction stages, referred to Fig. 2.

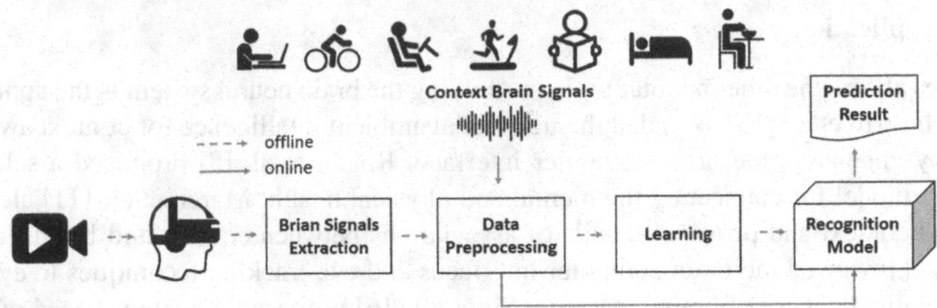

Fig. 2. Framework of proposed method.

I. Offline training stage: In this stage, the main intent is to train a recognition model for contexts. To this end, techniques can be classified into three components, namely data collection, data preprocessing and model training. For data collection, a nearly real simulation is designed. For data preprocessing, an adaptive process for 1-Dimension and 2-Dimsion learning is performed. For model training, the classifiers based on 1D and 2D data are trained, respectively.
II. Online prediction stage: For the contextual brain signals, they will be processed and recognized by the recognition model. In this paper, a number of classifiers are evaluated and the best performance is approximated by the proposed BrainCNN.

3.3 Offline Training

Even though the training is claimed to contain data collection in above, it could be better to show the data collection in the experimental setting section, not here. However, no existing public data is released and the real data is not easy to collect. Thus, the proposed data collection is regarded as one of the contributions claimed in Sect. 1. In the following, three training components, including data collection, data preprocessing and model training will be presented in detail.

A *Data collection*

To make the data closer to the truth, a creative simulation is proposed in this paper. In this simulation, first, we collected four types of 3D context videos, including reading, biking, diving and sleeping. Second, a VR box and a mobile detection device were ready for the collection. Third, a number of volunteers were invited to watch the videos by the VR box and the mobile detection device. Finally, the brain signals for each context are gathered as a vector sequence with 10 attributes. Figure 3 shows the scenarios of proposed simulation, VR Box and mobile detection device. The mobile detection device is a wearable equipment and the signal structure is a vector $S = \{\{\text{Attention, Meditation}\}, \{\text{Delta, Theta, Low Alpha, High Alpha, Low Beta, High Beta, Low Gamma, High Gamma}\}\}$. Therefore, a transaction is a matrix which can be defined as:

$$F_T \rightarrow S[f_t, s],$$

where T stands for the time sequence set, S stands for the signal set and $f_{t,\,s}$ stands for the sth signal value in the tth second [16]. Therefore, for each user, a related transaction matrix is derived.

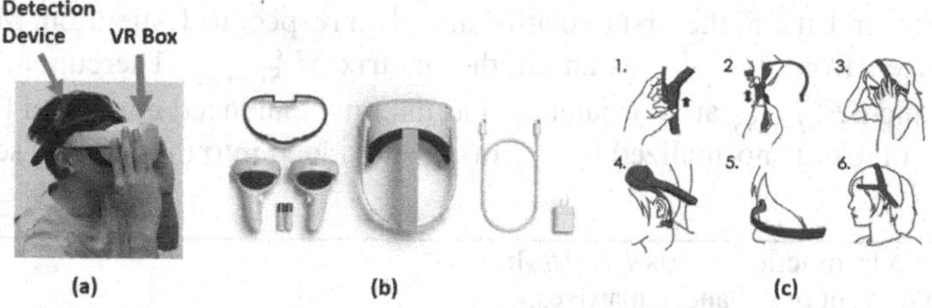

Fig. 3. (a): Example of proposed simulation; (b): VR Box; (c): Mobile detection device.

B *Data preprocessing*

It is because no former research is proposed for contextual brain recognition that we test a number of well-known classifiers, including SVM, LDA, RF, LSTM and our proposed BrainCNN. In overall, SVM, LDA, RF and LSTM taking the 1D vector as the input can be viewed as 1D-based classifiers, while the proposed BrainCNN adopting the 2D matrix as the input can be viewed as 2D-based classifiers. Therefore, the data preprocessing generates two types of data, namely 1D vector and 2D matrix. For a 1D vector, it is calculated from the transaction matrix $F_{T\to S}[f_{t,s}]$, which further contains three paradigms: average (*avg*), difference to average (*davg*) and standard deviation (*std*). The related definitions of three 1D vector paradigms for the user u can be defined as:

$$avg_u = \{a_{u,1}, a_{u,2}, \ldots, a_{u,s}, \ldots, a_{u,10}\}, \text{where} a_{u,s} = \frac{\sum_{1\leq t\leq |T|} f_{t,s}}{|T|}, \quad (1)$$

$$davg_u = \{v_{u,1}, v_{u,2}, \ldots, v_{u,s}, \ldots, v_{u,10}\}, \text{where} v_{u,s} = \frac{\sum_{1\leq t\leq |T|} (f_{t,s} - a_s)}{|T|}, \quad (2)$$

$$\text{and} std_u = \{d_{u,1}, d_{u,2}, \ldots, d_{u,s}, \ldots, d_{u,10}\}, \text{where} d_{u,s} = \sqrt{\frac{\sum_{1\leq t\leq |T|} (f_{t,s} - a_s)^2}{|T|}}. \quad (3)$$

Obviously, it is a level-based representation with respect to avg., davg., and std., respectively. Afterwards, the vector will be normalized by the Min-Max normalization, which can be defined as: $Navg_u$, $Ndavg_u$ and $Nstd_u$, respectively. In addition to 1D vector, 2D matrix is another structure to be processed for the proposed BrainCNN. In fact, BrainCNN is a deep learning method needing more data for high accuracy. Moreover, an original 2D matrix composed of only 10 attributes is too poor to encode a rich feature map. For these two concerns, data augmentation and data enhancement are necessary. In this paper, data augmentation indicates the operation increasing the data amount, while data enhancement indicates the one enlarging the matrix size. Figure 4 shows the procedure of preprocessing for the proposed BrainCNN. In this process, Lines 2–9 show the data augmentation with a sliding window where the size is 32. For each window, in Line 3, the sub-matrix $SF^i_{ST\to S}$ is fetched first where the matrix size is 32 × 10. Next, in Line 4, the $SF^i_{ST\to S}$ is copied into two same matrixes and then concatenate three as a new matrix $N^i_{ST\to 2S}$. That is, the size of $N^i_{ST\to 2S}$ is 32

× 32. Then, in Line 5, the first two attributes with respect to {Attention, Meditation} are extracted from $SF^{i}_{ST \to S}$ as an another matrix $SF^{i}_{ST \to SS}$. Thereupon, in Line 6, $N^{i}_{ST \to 2S}$ and $SF^{i}_{ST \to SS}$ are concatenated as the final enhanced matrix. In line 7, each enhanced matrix is normalized by z-scoring and added into the training set B at last.

Input: a transaction matrix $F_{T \to S}[f_t, s]$;
Output: a set of enhanced matrixes;
Process: Augmentation&Enhancement
1. set window size win=32;
2. **for** i=1 to (|T|-win) **do**
3. fetch the sub-matrix $SF^{i}_{ST \to S}$ where $i \le |ST| \le (i+win-1)$;
4. concatenate $SF^{i}_{ST \to S}$ with two $SF^{i}_{ST \to S}$ as a new matrix $N^{i}_{ST \to 2S}$;
5. fetch a sub-matrix as $SF^{i}_{ST \to SS}$ from $N^{i}_{ST \to 2S}$, where $1 \le |SS| \le 2$;
6. generate a new matrix $NN^{i}_{ST \to 2SS}$ by concatenating $SF^{i}_{ST \to SS}$ with $N^{i}_{ST \to 2S}$;
7. normalize the $NN^{i}_{ST \to 2SS}$ as $Z^{i}_{ST \to 2SS}$ by z-scoring.
8. let $B = \cup Z^{i}_{ST \to 2SS}$;
9. **end for**
10. **return** B;

Fig. 4. Preprocessing for the proposed BrainCNN.

C *Model training*

In this paper, the main contributed technique is a free CNN. As shown in Fig. 5, it consists of 3 blocks for feature filtering where each block contains 2 convolutions (called Conv2D) and 1 pooling. The parameters for the numbers of filtering are called F1 and F2, respectively. After the feature filtering, the features are flattened and the result is derived by 3 Dense networks with the activation functions of Relu and Softmax. In overall, the depth of the proposed BrainCNN is 9, which is light for a input small matrix (size = 32 × 32).

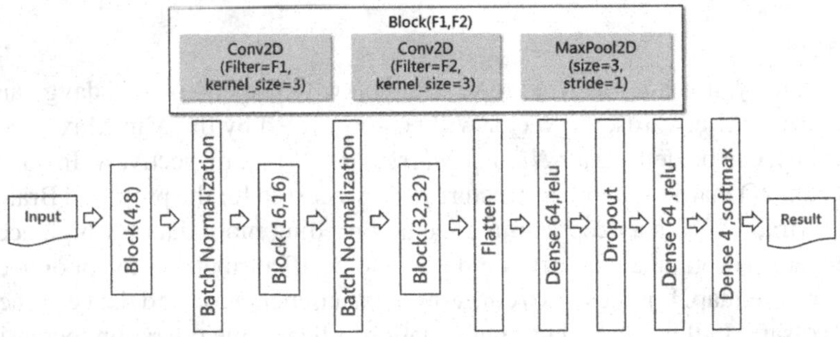

Fig. 5. Architecture of BrainCNN.

In this paper, for the robustness concern, the other 1D classifiers such as SVM, RF and LDA are examined. Also, another important deep learning method LSTM is examined. For LSTM, the input is a 1D vector sequence extracted from the transaction matrix $F_{T \to S}[f_t, s]$.

3.4 Online Prediction

Based on the training model, the potential context for a contextual brain signal can be predicted. Note that, the input for testing is just a sub-matrix fetched from the transaction matrix $F_{T \to S}[f_{t,\,s}]$, not all sequential sub-matrixes. By referring to definitions in Fig. 4, the testing sub-matrix is further defined as: $TF_{TT \to 2SS}$ where $|TT|$ is 32. That is, the time window size is 32, ranged from $(|T|/2–16)$ to $(|T|/2 + 16)$ in $F_{T \to S}[f_{t,\,s}]$. Here, to clarify the practical concern, an application example is lifted in Fig. 6. In this example, the brain signal detection device could be embedded into the over-ear headphone as a new design of online listening devices. Therefore, the proposed prediction method can be combined with an online music recommender system in this new device.

Fig. 6. Applications for the proposed method.

4 Experiment

4.1 Experimental Setting

To evaluate the proposed method, a set of experiments were made by two settings: experimental data and parameter settings. For the experimental data, 30 volunteers were invited for this experiment. Each volunteer worn both VR box and mobile detection device first. Then, the related videos were played in the VR box and the brain signals were detected by the device. In the video, to make the tester calm before testing, black scenes are presented in the first 10 s. Afterwards, the brain signals were caught in the succeeding 60 s. For the robustness concern, the experiments were conducted by a 5-fold cross validation. One served for test, and the others were for training. For BrainCNN settings, the learning rate is 0.001, batch size is 32, dropout rate is 0.2 and epoch is 25. There were 4 methods compared with the proposed method, and all methods were measured by *Accuracy*. For LSTM, the related architecture is shown in Fig. 7, which contains 2 LSTMs with activations of Relu and Sigmoid. Moreover, the related learning rate is 0.001, batch size is 32 and epoch is 25.

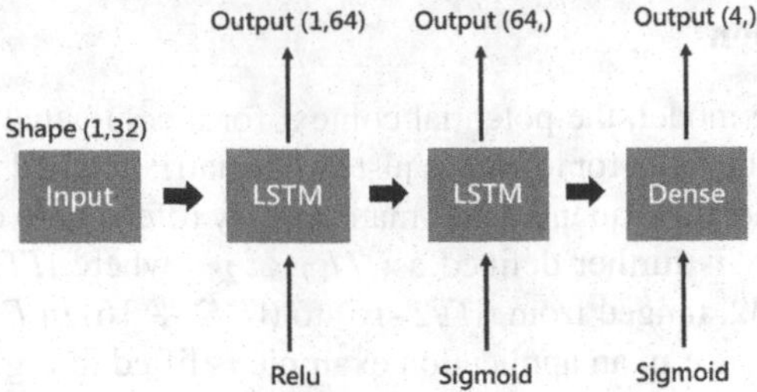

Fig. 7. Architecture of experimented LSTM.

4.2 Experimental Result

For effectiveness, the evaluations were conducted by two aspects: effectiveness for data preprocessing and comparisons for all evaluated methods. In terms of effectiveness for data preprocessing, Fig. 8 depicts the results for different processing features, referring to Subsect. 3.3.B. It delivers some points. First, the best performance is generated by RF with feature *std*. However, *std* is not sensitive for the other two methods. Second, on average, RF performs the best. Third, in contrast, the worst method is SVM. Fourth, from the feature viewpoint, the most sensitive feature is *avg* and the insensitive one is *std*. An insightful analysis is that, the original feature is better than the difference feature, and the difference feature is better than the variance feature. The performance is corresponding to the gradient level of processing.

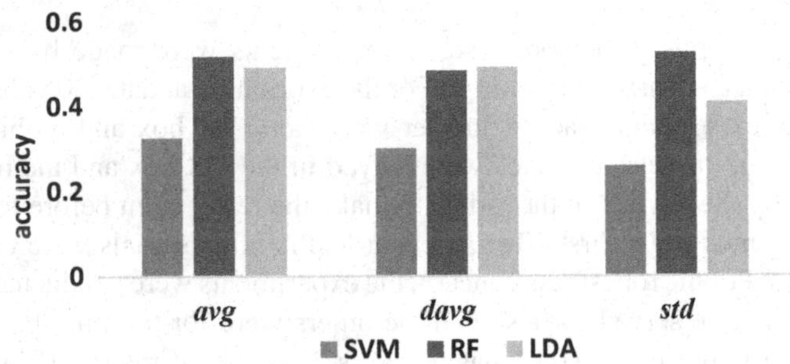

Fig. 8. Accuracies of compared methods for different processing features.

In addition to the feature analysis, the other issue for experiments is the comparison for all evaluated methods. In this comparison, the best accuracies for SVM, RF and LDA are selected from Fig. 8. Figure 9 shows the overall comparison for all evaluated methods. In Fig. 9, a set of observations are listed here. First, for 1D-based methods, the frame-level method LSTM is better than brain-level ones of SVM and LDA. The potential reason is that LSTM considers the memory/sequence characteristics, but the others do not. However, without memory, RF performs equally as well as LSTM. This is because the processing feature *std* of RF presents the same characteristics as the original feature of LSTM. Second, although LSTM considers the memory/sequence characters, it cannot be better than BrainCNN. This could be explained by that, BrainCNN is richer than LSTM due to the data enhancement. Third, in summary, the best method is the proposed method BrainCNN, which delivers an aspect that big data and rich information bring out the better encoding features for the better recognition.

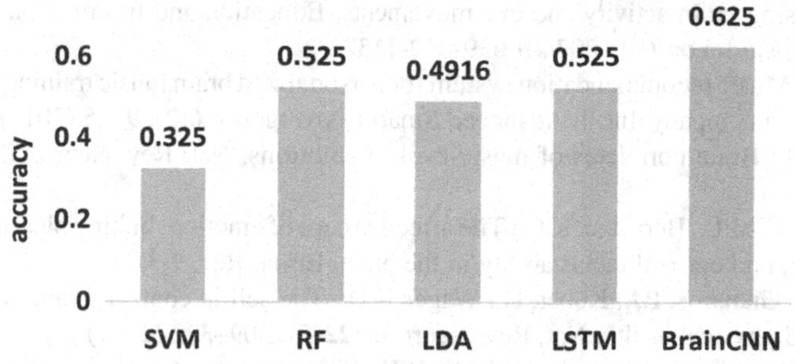

Fig. 9. Accuracies of all evaluated methods.

5 Conclusion

Although there are a lot of researches proposed for exploring the brain neural system, very little attention is focused on the recognition of contextual brain signals. Indeed, the context recognition by brain signals is helpful to many applications such as recommendations and health care. For these needs, in this paper, a creative recognition model called BrainCNN is proposed. Because no public data was posted and no baseline method was proposed ever, we further provide a robust data collection paradigm and a robust examination for a number of well-known classifiers. Finally, a detailed empirical study is lifted and the evaluation result reveals the proposed method performs more promising than the compared fashion ones. Actually, this paper is just a start for this topic. In the future, there remain several investigations to do. First, the tester amount and context category size will be enlarged. Second, the evaluation results provide much valuable information for conducting a more effective method from viewpoints of feature processing, data enhancement and data augmentation. Finally, as shown in Fig. 6, further collaborations with commercial companies will be pursued for materializing the proposed method.

Acknowledgement. This research was supported by Ministry of Science and Technology, Taiwan, R.O.C. under grant no. MOST 111–2221-E-390–014 and NSTC 111–2410-H-230 -003 -MY2.

References

1. Al-Shoukry, S., Rassem, T.H., Makbol, N.M.: Alzheimer's diseases detection by using deep learning algorithms: a mini-review. IEEE Access **8**, 77131–77141 (2020)

2. Finnegan, S.L., Browning, M., Duff, E., et al.: Brain activity measured by functional brain imaging predicts breathlessness improvement during pulmonary rehabilitation. Thorax (2022). https://doi.org/10.1136/thorax-2022-218754
3. Gallego, J.A., Makin, T.R., McDougle, S.D.: Going beyond primary motor cortex to improve brain–computer interfaces. Trends Neurosci. **45**(3), 176–183 (2022)
4. Jamil, N., Belkacem, A.N., Lakas, A.: On enhancing students' cognitive abilities in online learning using brain activity and eye movements. Education and Information Technologies (2022). https://doi.org/10.1007/s10639-022-11372-2
5. Kim, B.J.: Music recommendation system for personalized brain music training research with jade solution company. Int. J. Advanced Smart Convergence **6**(2), 9–15 (2017)
6. Koelsch, S.: Brain correlates of music-evoked emotions. Nat. Rev. Neurosci. **15**, 170–180 (2014)
7. Kringelbach, M.L., Berridge, K.C.: The affective core of emotion: linking pleasure, subjective well-being, and optimal metastability in the brain. Emot. Rev. **9**(3), 1–9 (2017)
8. Koban, L., Gianaros, P.J., Kober, H., Wager, T.D.: The self in context: brain systems linking mental and physical health. Nat. Rev. Neurosci. **22**(5), 309–322 (2021)
9. Kohli, V., Tripathi, U., Chamola, V., Rout, B.K., Kanhere, S.S.: A review on virtual reality and augmented reality use-cases of brain computer interface based applications for smart cities. Microprocess. Microsyst. **88**, 104392 (2022)
10. LeDoux, J.E.: Emotion circuits in the brain. Annu. Rev. Neurosci. **23**, 155–184 (2000)
11. Maren, S., Phan, K.L., Liberzon, I.: The contextual brain: implications for fear conditioning, extinction and psychopathology. Nat. Rev. Neurosci. **14**(6), 417–428 (2013)
12. Moreno, J.D., Schulkin, J.: The Brain in Context: A Pragmatic Guide to Neuroscience. *Columbia University Press* (2019)
13. Navarro, A.A., et al.: Context-awareness as an enhancement of brain-computer interfaces. In: Proceedings of International Workshop on Ambient Assisted Living, pp. 216–223 (2011)
14. Pessoa, L.: Understanding emotion with brain networks. Curr. Opin. Behav. Sci. **19**, 19–25 (2018)
15. Saarimäki, H., et al.: Distributed affective space represents multiple emotion categories across the human brain. Social Cognitive and Affective Neuroscience **13**(5), 471–482 (2018)
16. Su, J.-H., Liao, Y.-W., Wu, H.-Y., Zhao, Y.-W.: Ubiquitous music retrieval by context-brain awareness techniques. In: Proceedings of IEEE International Conference on Systems, Man, and Cybernetics (2020)
17. Zabcikova, M., Koudelkova, Z., Jasek, R., Lorenzo Navarro, J.J.: Recent advances and current trends in brain-computer interface research and their applications. International Journal of Developmental Neuroscience **82**(2), 107–123 (2021)
18. Zhang, X., Yao, L., Wang, X., Monaghan, J., McAlpine, D., Zhang, Y.: A survey on deep learning-based non-invasive brain signals: recent advances and new frontiers. Journal of Neural Engineering **18**(3) (2021)

A Novel Meta-heuristic Search Based on Mutual Information for Filter-Based Feature Selection

Bui Quoc Trung, Duong Viet Anh, and Bui Thi Mai Anh(✉)

School of Information and Communication Technology,
Laboratory of Data Analysis and Operational Research,
Hanoi University of Science and Technology, Hanoi, Vietnam
anhbtm@soict.hust.edc.in

Abstract. Recent research has shifted its focus towards feature selection to improve model performance and reduce training costs, as high dimensional and noisy datasets have become ubiquitous. Filter-based feature selection, which uses external criteria to filter features based on relevance scores, has become increasingly popular in this research area. However, existing studies that use a greedy search strategy to find optimal feature subsets suffer from suboptimal solutions, as the selected features cannot be changed or removed later. In this paper, we propose a novel method for filter feature selection that combines a local search strategy with meta-heuristic techniques to improve the exploitation of the most promising candidates for the global search. We aim to use evolutionary computing with mutual information-based criteria instead of relying solely on classification performance like existing state-of-the-art methods. We conducted empirical experiments on various benchmark datasets that differ in volume and dimensionality to evaluate the performance of our proposed approach. Our findings demonstrate that our method outperforms other state-of-the-art work by requiring less time to reach the optimal solution.

Keywords: feature selection · local search · meta-heuristic search

1 Introduction

The use of massive volume and high dimensional datasets is becoming increasingly prevalent in various domains, such as bioinformatics [6], medicine [1], and others. However, existing machine learning algorithms that were designed for low-dimensional space may face significant challenges when dealing with high dimensional data, including high computational complexity, overfitting, and degradation of performance. Research has demonstrated that the presence of irrelevant and redundant information in high-dimensional data is one of the main reasons for low prediction on unseen data [14]. To address this issue, a common solution is to apply a feature selection process to select the most important

N. T. Nguyen et al. (Eds.): ACIIDS 2023, LNAI 13995, pp. 395–407, 2023.
https://doi.org/10.1007/978-981-99-5834-4_32

information from the original dataset. The primary goal of feature selection is to reduce the number of features by eliminating those that are irrelevant or redundant. To accomplish this, a feature selection approach typically involves two essential elements: a searching mechanism and an assessment method. The searching mechanism refers to the process of exploring the feature space to identify the most relevant feature subsets, while the assessment method is used to evaluate the quality of the identified feature subsets based on specific evaluation criteria.

Feature selection algorithms are categorized into two directions, depending on how the relevance of a feature subset candidate is assessed: (i) filter-based approaches and (ii) wrapper-based approaches [14]. The former approach employ external criteria to evaluate the relevance of features. Features with low relevance score are considered as irrelevant and will be filtered out. Filter-based feature selection approaches can works as a pre-processing step prior to the learning model. Wrapper-based techniques explore the search space over several iterations until some stopping criteria are satisfied. A heuristic function built based on the precision of the targeted learning algorithm is utilized to assess the quality of candidate solutions at each iteration. Wrapper-based methods, however, require excessive computations to train the model with numerous feature subsets and thus, are impractical in reality. Inversely, filter-based feature selection is more practical since it is straightforward, computationally fast and independent of any learning approach. The main disadvantage of filtering techniques is that the generated feature subset might not be the best for the intended learning model. A trade-off between these two techniques is embedded methods, which are less time-consuming than wrapper techniques but are more tailored to the learning algorithm than filter ones [17]. Nevertheless, the design of an embedded feature selection is intricate and necessitates altering the target model's training process [14]. Due to the computational efficiency, filter-based feature selection gains more research focus and is also the subject of this study.

Filter-based feature selection approaches rely on a wide range of criteria to evaluate the importance of features. Given an evaluation criterion f, a filter feature selection method can be formulated as an optimization problem that aims to find an optimal subset $\mathcal{S}^*$ of k features from the original set $\mathcal{S}$ ($\mathcal{S}^* \subset \mathcal{S}$) such that $f(\mathcal{S}^*)$ is as maximal as possible. Since it is impractical to conduct an exhaustive search to consider all possible combinations of features, numerous searching strategies are therefore employed and may affect significantly the effectiveness of a filter feature selection. Forward and backward sequential searching strategies are widely used due to their simplicity and low computational cost [9]. However, such approaches suffer from the problem of becoming stuck in a local optimum because selected features cannot be later changed or removed. Random searching strategies have been considered to overcome this problem however they may come with a small probability of failure [3]. In this research, we combine a local search approach with meta-heuristic methods to enhance the efficiency of filter-based feature selection. In order to assess the importance of feature subsets more effectively, our proposed method combines evolutionary computing with theoretical information-based criteria. The limitations of current systems that exclusively rely on classification performance can be addressed by including mutual information-based criteria into the process of solution assessment. Otherwise, the introduction of a local search strategy assists to increase the exploration of the searching space, which can help overcome the local optimum issue typically seen in heuristic searching methods.

The remaining of this paper is structured as follows. The state-of-the-art works along this research line are depicted in Sect. 2. Section 3 introduces briefly the fundamental concepts of information theory and some representative criteria. We present our proposed approach in Sect. 4 where two main algorithms for filtered based feature selection are detailed. Section 5 describes the benchmark datasets and our evaluation plan. The results and discussion are analyzed in Sect. 6. Finally, Sect. 7 concludes with some future directions.

2 Related Work

This section gives a brief overview of state-of-the-art filter-based feature selection approaches with the focus on those that rely on theoretical information-based criteria.

The literature on filter-based feature selection has identified two main categories of ranking criteria: similarity-based and theoretical information-based [14]. Similarity-based criteria rank features based on their ability to discriminate data samples with respect to the target variable in the dataset. Examples of similarity-based criteria include Laplacian Score [12], Fisher Score [11], and others. The similarity-based criteria take into consideration only the information content of a single feature to a given class label and ignore the relationship among features. Theoretical information-based criteria focus not only on the relevance of features but also the redundancy among them, thus, are more preferred for filter feature selection [2,10,16]. The mutual information is the fundamental concept of this research direction, based on which numerous variants are built for feature ranking including CIFE [15], MIFS [4] etc. Due to more than two relations between features, theoretical information-based criteria may lead to an extremely increased computational cost, proportionate to the number of features in the dataset.

Because feature selection based on ranking scores is a greedy heuristic, it is impossible to adequately explore the feature space. In order to completely explore the feature space and find an exact solution, Nie et al. [16] have converted the feature selection into a binary integer programming problem in which a mutual-information based criterion is reformulated as a sequence of linear constraints. Bui et al. [2] have generalized the set of linear constraints for different theoretical information based criteria and enhanced the binary integer programming model to improve searching performance.

Transforming feature selection into an integer programming model can improve the efficiency of exploring promising candidate solutions and prevent sub-optimal solutions. However, this approach can also become computationally expensive and even result in a "no-solution found" problem. To address the limitations of greedy search techniques and integer programming models, evolutionary heuristic search has been investigated as an alternative method for feature selection. Cervante et al. [5] combined Partial Swarm Optimization (PSO) with mRmR criterion to select features. Candidate feature subsets are evaluated based on maximizing their relevance to the class label and minimizing the redundancy between selected features. In the recent research of Ge and Hu [7], mutual information is also examined as an objective function of the filter selection method based on an genetic algorithm. Hancer et al. [10] have proposed a differential evolutionary method combined with three criteria including mutual information, ReliefF and Fisher Score to take into consideration the highly ranked features (given by ReliefF and Fisher Score) while maximizing the relevance of selected features. We have argued that the majority o research have focused on using a single evolutionary algorithm to explore the feature space. However, these traditional algorithms alone may suffer from a weak local search capability and a slow convergence speed. In this paper, we also investigate a hybrid approach for filter based feature selection, which combines a meta-heuristic search and local search strategy to more efficiently reach optimal solutions.

3 Background

3.1 Theoretical Information Based Criteria

Mutual information, which measures the amount of information exchanged by two random discrete variables, is the key concept of theoretical information. The mutual information represents the relevance of a feature X to the class variable Y and is computed as in Eq. 1.

$$I(X;Y) = \sum_{x_i \in X} \sum_{y_i \in Y} P(x_i, y_i) log \frac{P(x_i, y_i)}{P(x_i)P(y_i)} \quad (1)$$

where $X = \{x_i\}$ denotes the feature X with all the sample values from the dataset, $P(x_i)$ denotes the probability of x_i over all possible values of X, $P(x_i, y_i)$ is the joint probability of x_i and y_i. Based on the mutual information, a wide range of criteria have been studied for feature selection such as CIFE [15], MIFS [4] and others. In this paper, we study CIFE [15] as the theoretical information based criterion for feature selection. Given a feature, the CIFE criterion aims to not only minimize the redundancy among features but also maximize the conditional redundancy between this feature and all the others with respect to the class label [15].

$$CIFE(X_i) = I(X_i;Y) - \sum_{X_j \in \mathcal{S}} I(X_i; X_j) + \sum_{X_j \in \mathcal{S}} I(X_j; X_i|Y) \quad (2)$$

where the third term introduces the conditional redundancy among X_i and the other features in $\mathcal{S}$ given the class label Y.

3.2 Filter-Based Feature Selection Formulation

The filter-based feature selection can be considered as an optimization problem in which we have to find a subset of features to maximize a specific criterion. Given a dataset $\mathcal{D}$ with N samples and characterized by n attributes. $\mathcal{S} = \{X_1, X_2, ..., X_n\}$ denotes the set of n features (i.e., attributes or variables). We represent a candidate solution for feature selection as a binary vector $d = \{d_1, d_2, ..., d_n\}$ in which $d_i \in \{0, 1\}$. The value 1 denotes that the feature X_i is selected and vice versa. Given a criterion function $\mathcal{F}$ (e.g., CIFE), the objective of filter feature selection is to find a subset $\mathcal{S}^*$ of k optimal features from the original set such that $\mathcal{F}(\mathcal{S}^*)$ is as maximal as possible.

$$\mathcal{S}^* = arg\max_{d} \mathcal{F}(\mathcal{S}|d), s.t. |\mathcal{S}^*| = \sum_{i=1}^{n} d_i = k \tag{3}$$

where $|\mathcal{S}^*|$ indicates the size of the selected feature subset. Otherwise, maximizing the score of selected features is equivalent to minimizing the impact of unselected features, it is then possible to transform the optimization problem in Eq. 7 to the following objective function:

$$\min \sum_{X_i \in \mathcal{S} \setminus \mathcal{S}^*} \mathcal{F}(X_i) + \lambda \sum_{i=1}^{n} d_i \tag{4}$$

where $\mathcal{F}(X_i)$ is the individual score of the feature X_i with respect to all the unselected features in $\mathcal{U} = \mathcal{S} \setminus \mathcal{S}^*$. $\mathcal{F}(X_i)$ is computed as in Eq. 2. λ is a predefined parameter to emphasize on the number of the selected features.

4 The Proposed Solution

In this section, we introduce a novel approach for feature selection that combines a local search technique and a meta-heuristic search algorithm in order to enhance the exploration and exploitation of the search space. The meta-heuristic approach plays the role of a global search strategy which aims to locate a global optimum while preventing premature convergence to a local solution. We have chosen the Tabu search algorithm [8] as the meta-heuristic mechanism for feature selection. This method uses a short-term memory called a Tabu-list, along with a set of systematic rules, to guide and maintain the search process. Although Tabu Search is effective at preventing the search from getting stuck in local optima or revisiting already-explored solutions, it may converge slowly because it could fail to detect promising search directions in the neighborhood of a global minimum. To address this issue, we propose incorporating a local search strategy into the main algorithm in this paper. The objective of the local search strategy is to intensify the search in promising areas of the search space.

4.1 The Local Search Strategy

A local search strategy aims to improve the quality of solutions by exploring the search space in the vicinity of the current solution. We first define the set of neighbor candidates of an under-consideration solution in the context of feature selection, then introduce our proposed neighborhood search strategy.

Algorithm 1: $NeighborhoodSearch(S)$

Input : S as a candidate solution; $[k_1, k_2]$ as the range of desirable numbers of selected features
Output: An optimal subset of features $\hat{\mathcal{S}}$ in the vicinity of S $(\hat{\mathcal{S}} \subset \mathcal{S}), k_1 \leq |\hat{\mathcal{S}}| \leq k_2$

$\hat{\mathcal{S}} \leftarrow S$
$nbIterations \leftarrow 0$
while $nbIterations \leq maxIterations$ **do**
  Generate three neighborhoods S_1, S_2, S_3 of S
  $S_{best} \leftarrow getBestCandidate(S_1, S_2, S_3)$
  if $\mathcal{F}(\mathcal{S}_{best}) < \mathcal{F}(\hat{\mathcal{S}})$ **then**
    $\hat{\mathcal{S}} \leftarrow \mathcal{S}_{best}$
  end
  $S \leftarrow S_{best}$
  $nbIterations \leftarrow nbIterations + 1$
end
return $\hat{\mathcal{S}}$

Neighbor-Candidate Formulation. Given a candidate solution $\mathcal{S}^* \subset \mathcal{S}$ is the set of selected features X_i, we define three neighbor candidates of $\mathcal{S}^*$ as follows.

1. A *removal-based neighborhood* of $\mathcal{S}^*$ is created by removing a feature from $\mathcal{S}^*$ such that
$$S_1 = \mathcal{S}^* \setminus \{f^*\}, f^* = \arg \min_{f_i \in \mathcal{S}^*} \mathcal{F}(f_i) \tag{5}$$
2. An *insertion-based neighborhood* of $\mathcal{S}^*$ is created by inserting an un-selected feature which has not been involved in the current solution $\mathcal{S}^*$ such that
$$S_2 = \mathcal{S}^* \cup \{f^*\}, f^* = \arg \max_{f_i \in \mathcal{S} \setminus \mathcal{S}^*} \mathcal{F}(f_i) \tag{6}$$
3. A *swap-based neighborhood* of $\mathcal{S}^*$ is created by swapping a feature in $\mathcal{S}^*$ with another one from $\mathcal{S} \setminus \mathcal{S}^*$ such that
$$S_3 = \mathcal{S}^* \cup \{f_1\} \setminus \{f_2\}, f_1 = \arg \max_{f_i \in \mathcal{S} \setminus \mathcal{S}^*} \mathcal{F}(f_i), f_2 = \arg \min_{f_i \in \mathcal{S}^*} \mathcal{F}(f_i) \tag{7}$$

Based on these neighbor candidates, a neighborhood search strategy which is introduced in the next section will perform to explore promising areas or the feature search space.

Algorithm 2: *NBHTabuSearch*($\mathcal{S}$)

Input : $\mathcal{S}$ as the set of features; $[k_1, k_2]$ as the range of desirable number of selected features

Output: An optimal subset of features $\mathcal{S}^* (\mathcal{S}^* \subset \mathcal{S})$

```
1  S* ← ∅
2  tabuIterations ← 0
3  while tabuIterations ≤ maxTabuIterations do
4      lsIterations ← 0
5      S ← initializeSolution(𝒮)
6      Ŝ ← S
7      tabuList ← ∅
8      while lsIterations ≤ maxLSIterations do
9          Generate three neighborhoods S1, S2, S3 of S that are not in tabuList
10         S_best ← getBestCandidate(S1, S2, S3)
11         S ← S_best
12         if ℱ(S_best) < ℱ(Ŝ) then
13             Ŝ ← S_best
14         end
15         updateTabuList(tabuList)
16         lsIterations ← lsIterations + 1
17     end
18     if ℱ(S*) < ℱ(Ŝ) then
19         S* ← Ŝ
20     end
21     tabuIterations ← tabuIterations + 1
22 end
23 return S*
```

Neighborhood Search Strategy. The objective of our proposed neighborhood search strategy is to explore the vicinity of an under-consideration solution to find a more promising candidate. As being depicted in Algorithm 1, the neighborhood search process takes as input a given solution $\hat{\mathcal{S}}$ ($|\hat{\mathcal{S}}| \in [k_1, k_2]$) and outputs a more optimal candidate solution $\mathcal{S}^*$. This is performed by first building three neighbor candidates of the initial solution (lines 4–6). All the neighbor candidates are then evaluated using the objective function $\mathcal{F}$ (see Sect. 3.2). Based on their objective value, the best candidate is identified using the function *getBestCandidate* (line 57). Given two candidates S_i, S_j, S_i is better than S_j if $\mathcal{F}(S_i) < \mathcal{F}(S_j)$. In the case that $\mathcal{S}_{best}$ is better than $\mathcal{S}^*$, the optimal candidate will be updated (lines 8–10). This procedure is repeated until the maximum number of iterations is reached (specified by *maxIterations* (line 3)).

4.2 The Meta-heuristic Search for Feature Selection

Tabu-Search. Given the local search strategy which has been introduced in the previous section, we propose a global optimization searching process for fea-

Algorithm 3: *IteratedLocalSearch*($\mathcal{S}$)

Input : $\mathcal{S}$ as the set of n features; $[k_1, k_2]$ as the range of desirable number of selected features

Output: The optimal subset of features $\mathcal{S}^*(\mathcal{S}^* \subset \mathcal{S})$

1 $\mathcal{S}^* \leftarrow \emptyset$
2 $lsIterations \leftarrow 0$
3 **while** $lsIterations \leq maxLSIterations$ **do**
4 $\quad S \leftarrow initializeSolution(\mathcal{S})$
5 $\quad S \leftarrow NeighborhoodSearch(S)$
6 $\quad$ **if** $\mathcal{F}(S) < \mathcal{F}(\mathcal{S}^*)$ *or* $\mathcal{S}^* = \emptyset$ **then**
7 $\quad\quad \mathcal{S}^* \leftarrow S$
8 $\quad$ **end**
9 $\quad lsIterations \leftarrow lsIterations + 1$
10 **end**
11 **return** $\mathcal{S}^*$

ture selection based on the Tabu-Search algorithm [8]. The neighborhood search strategy is incorporated into the main procedure of Tabu-Search to maintain efficiently the tabu list. The iterative process of our proposed Tabu-Search is depicted in Algorithm 2. To begin the process, a solution S is generated randomly with the number of selected features within the range of $[k_1, k_2]$ (line 5). The tabu list is initialized to keep track of explored solutions during the search (line 7). The neighborhood search process executes the exploration phase of Tabu Search (lines 8–17), identifying neighbor candidates for a given candidate solution as long as they are not in the tabu list (lines 9). The best neighbor candidate is identified (line 10) that serves as a new initial trial point for the next iteration (lines 11–14). The tabu list inserts already-explored candidates (S_1, S_2, S_3) and keeps track of them for five iterations (line 15), after which they may be considered as promising trial search regions. At the final stage of the algorithmn, the global optimal solution will be updated (lines 18–20). The Tabu-Search procedure is repeated after a specific number of iterations (i.e., $maxTabuIterations$).

Iterated Local Search. In order to demonstrate the effectiveness of our proposed neighborhood search strategy and tabu-search approach, we also study another metaheuristic search technique, called *iterated local search.* This method performs the neighborhood search from multiple starting points to sample different regions of the search space and to prevent returning a low-quality local minimum. The pseudo-code of our proposed iterated local search is illustrated in Algorithm 3. The search process executes a specific number of neighborhood search runs, denoted by $maxLSIterations$ and returns the best solution found during the search. In each iteration, a local search is conducted (line 5) and a new starting point is generated by either perturbing the local minimum solution or by generating a new initial solution.

5 Experimental Design

5.1 Benchmark Datasets

To evaluate the effectiveness of our proposed approach, we conducted experiments on a diverse set of datasets used in prior research [2,16]. Table 1 presents the key characteristics of the benchmark datasets. Eight of the datasets were sourced from the UCI Machine Learning Repository[1]. The remaining datasets are highly dimensional in terms of both attributes and volume, with a feature count ranging from 13 to 856 and a sample size ranging from 32 to over one million[2] [13].

Table 1. Benchmark Datasets

Dataset	#features	#samples	#classes
Heart	13	1,025	2
Congress	16	435	2
Parkinsons	22	195	2
Breast	30	569	2
CreditCard	30	284,807	2
Ionosphere	34	351	2
Waveform	40	5,000	3
FinancialRisk	42	1,048,575	2
Donor	49	19,372	2
Sensor	55	220,320	3
Lungcancer	56	32	3
Sonar	60	208	2
BIG Malware	68	10,868	9
Company Bankruptcy	95	6,819	2
HomeCredit	120	307,513	2
Musk	168	6,598	2
ECG Heartbeat	187	21,892	5
CreditScore	304	58,194	2
CNAE-9	856	1,080	9

5.2 Evaluation Plan

We compare the performance of the proposed approach which combines Tabu-Search and the neighborhood search (TabuFS) for feature selection to three methods including:

[1] https://archive.ics.uci.edu/ml/datasets.php.

[2] https://github.com/JLZml/Credit-Scoring-Data-Sets.

1. The integer programming model (IPFS) for feature selection proposed by Bui et al. [2]. This approach reformulate the filter feature selection as a binary integer programming problem in which the mutual information based criterion is examined as a sequence of linear constraints. We re-implement the model and compare to our approach on the same dataset configurations and scoring criterion (i.e., CIFE).
2. The neighborhood search method (NBHS) which has been introduced in Sect. 12.
3. The iterated local search approach (ILS) which applies the neighborhood search strategy for a pre-defined number of iterations.

We evaluate the performance of all methods using two metrics: (i) the total computation time and (ii) the objective value of the best solution discovered.

5.3 Algorithm Settings

We have implemented all the algorithms using the programming language `Python`. The solvers for integer programming model is based on `scip`[3] and `OR-Tools`[4]. The weight λ is set to 1 to balance the significance of the number of selected features within the objective function. We determined the percentage of selected features to be between 20 and 30 percent of all features. The experiments have been performed using the criterion CIFE for filter based feature selection. To evaluate the robustness and convergent capability of the neighborhood search and meta-heuristic search, we configured both to run for 50 iterations. Experiments were conducted on a computer with Intel(R) Core(TM) i7-8565U CPU @ 1.80 GHz 1.99 GHz, 8.00 GB RAM, running Windows Operation System version 10. The time limit for each experiment was set to three hours.

6 Results and Discussion

6.1 Algorithm Performance Analysis

Table 2 reports the computational time in *seconds*) and the objective value returned from all the algorithms. The summarized data from this table leads to the following observations. Firstly, comparing to the heuristic and meta-heuristic algorithms (i.e., NBHS, ILS and TabuFS), the TabuFS method has achieved the highest performance in terms of the objective value (i.e., the quality of the resulting subset of selected features). In particular, TabuFS completely outperforms NBHS and ILS with regard to high-dimensional datasets (from Company Bankruptcy to CNAE-9). On the other hand, it is apparent that while switching from NBHS to ILS and from ILS to TabuFS, the quality of the solutions improves consistently. Despite this, the required computational time displays a different pattern. The NBHS approach demonstrates a fast convergence rate, while the

[3] https://www.scipopt.org.

[4] https://developers.google.com/optimization/.

Table 2. Comparing the computational time and objective values of 4 methods

Dataset	Computational Time				Objective Value			
	IPFS	NBHS	ILS	TabuFS	IPFS	NBHS	ILS	TabuFS
Heart	1.72	**0.99**	1.01	1.09	**7.2**	**7.2**	**7.2**	**7.2**
Congress	25.21	**0.09**	0.13	0.12	**356.2**	**356.2**	**356.2**	**356.2**
Park.	662.84	**1.01**	1.19	1.22	**−71.4**	−69.2	**−71.4**	**−71.4**
Breast	TL	**2.81**	4.12	6.35	−184.6	−180.2	−188.9	**−191.3**
Cre.Card	286.35	**2.04**	4.12	4.29	**3.9**	4.8	**3.9**	**3.9**
Ionos.	8.65	**0.88**	1.01	1.13	**−325.4**	−320.9	**−325.4**	**−325.4**
Waveform	6279.47	**5.66**	6.90	8.67	**259.7**	270.3	266.4	**259.7**
Fin. Risk	3605.73	**4.73**	6.03	9.38	**−70.6**	−63.9	−65.5	**−70.6**
Donor	10.14	**0.60**	0.78	1.13	**24.0**	**24.0**	**24.0**	**24.0**
Sensor	3699.49	**2.56**	5.82	6.68	**−99.8**	−83.3	−94.7	**−99.8**
Lung Can.	38.12	**1.00**	1.08	1.17	**21.8**	**21.8**	**21.8**	**21.8**
Sonar	TL	**2.56**	3.73	4.53	−960.9	−954.2	−968.1	**−972.5**
BIG Mal.	4919.40	**1.83**	5.23	5.73	**−258.2**	−249.4	**−258.2**	**−258.2**
Com.Bank.	6003.94	**5.34**	7.90	10.81	**−436.8**	−406.6	−417.5	**−436.8**
HomeCre.	267.11	**3.11**	4.47	6.35	**74.3**	83.2	80.6	**74.3**
Musk	TL	**5.66**	9.10	13.22	118.4	130.2	118.4	**115.9**
ECGHeart.	5009.12	**4.03**	7.64	9.01	**−4532.8**	−4491.4	−4520.2	**−4532.8**
Cre.Score	2208.01	**4.17**	5.22	7.79	**−14.9**	−8.2	−9.6	**−14.9**
CNAE-9	9237.94	**3.91**	8.20	12.67	**25.4**	40.5	36.7	**25.4**

TabuFS algorithm exhibits comparable or slower performance compared to the ILS method. Secondly, in comparison to the exact approach (IPFS), the approximate methods are more efficient as they provide better solutions while minimizing computational costs. Indeed, the TabuFS method achieves as good solutions as IPFS for the majority of datasets (16 over 19 datasets as showed in Table 2). However, for the remaining datasets, IPFS fails to provide optimal solutions within the time limitation (TL) of three hours per run, while TabuFS achieves the optimal solution in significantly less computational time. These observations suggest that the meta-heuristic algorithms proposed in this paper demonstrated excellent performance by generating high-quality solutions in comparison to the state-of-the-art exact method [2].

6.2 Algorithm Robustness Analysis

In order to evaluate the robustness of all the methods, we examine their objective value through 50 runs for each dataset. Table 3 shows the minimum, maximum and standard deviation values of the objective function computed over 50 running observations. It is evident that the difference between the maximum and

Table 3. Summary of objective values obtained by each algorithm after 50 runs

Dataset	Min			Max			Std		
	NBHS	ILS	TabuFS	NBHS	ILS	TabuFS	NBHS	ILS	TabuFS
Heart	**7.2**	**7.2**	**7.2**	**7.2**	**7.2**	**0.0**	**0.0**	**0.0**	**0.0**
Congress	**356.2**	**356.2**	**356.2**	**356.2**	**356.2**	**356.2**	**0.0**	**0.0**	**0.0**
Park.	**−71.4**	**−71.4**	**−71.4**	−68.0	**−71.4**	**−71.4**	0.7	**0.0**	**0.0**
Breast	−183.2	−190.5	**−192.7**	−172.3	−185.4	**−189.6**	3.1	2.3	**1.1**
Cre.Card	4.1	**3.9**	**3.9**	4.6	**3.9**	**3.9**	1.2	**0.0**	**0.0**
Ionos.	−321.6	**−325.4**	**−325.4**	−319.4	**−325.4**	**−325.4**	0.6	**0.0**	**0.0**
Waveform	268.1	263.3	**259.7**	271.6	268.4	**259.7**	0.5	0.9	**0.0**
Fin. Risk	−65.8	−68.3	**−70.6**	−62.2	−64.1	**−70.6**	1.4	1.5	**0.0**
Donor	**24.0**	**24.0**	**24.0**	**24.0**	**24.0**	**24.0**	**0.0**	**0.0**	**0.0**
Sensor	−87.2	**−99.8**	**−99.8**	−80.1	−93.6	**−99.8**	2.5	2.1	**0.0**
Lung Can.	**21.8**	**21.8**	**21.8**	**21.8**	**21.8**	**21.8**	**0.0**	**0.0**	**0.0**
Sonar	−958.4	−970.7	**−973.6**	−951.0	−965.9	**−971.1**	3.4	2.9	**0.8**
BIG Mal.	−253.8	**−258.2**	**−258.2**	−247.0	**−258.2**	**−258.2**	2.9	**0.0**	**0.0**
Com.Bank.	−415.0	−433.2	**−436.8**	−400.4	−415.0	**−436.8**	6.2	5.8	**0.0**
HomeCre.	81.5	76.1	**74.3**	84.8	82.0	**74.3**	3.3	2.5	**0.0**
Musk	126.0	117.3	**112.4**	134.7	121.8	**116.3**	2.5	2.0	**1.4**
ECGHeart.	−4502.0	−4526.4	**−4532.8**	−4486.4	−4518.3	**−4532.8**	3.4	2.8	**0.0**
Cre.Score	−9.6	−12.3	**−14.9**	−6.7	−7.3	**−14.9**	2.2	1.9	**0.0**
CNAE-9	37.4	32.1	**25.4**	44.0	37.4	**25.4**	7.6	3.5	**0.0**

minimum values for each dataset is quite small for all the methods NBHS, ILS and TabuFS. Otherwise, the standard deviation value of 50 runs shows that the TabuFS is able to produce the most consistent results comparing to NBHS and ILS. Based on these observations, it can be concluded that the proposed algorithms demonstrate robustness and the ability to converge quickly.

7 Conclusion

This paper presents a hybrid approach that incorporates a local search strategy and Tabu-search for feature selection using information-theoretic criteria. The performance of the Tabu-search process is enhanced by identifying the promising neighbor candidates of a given solution and using them to maintain the Tabu-list during the search. Empirical results demonstrate that our proposed method surpasses the state-of-the-art approach in producing optimal solutions while significantly reducing computational time.

References

1. Alyass, A., Turcotte, M., Meyre, D.: From big data analysis to personalized medicine for all: challenges and opportunities. BMC Med. Genom. **8**(1), 1–12 (2015)

2. Anh, B.T.M., Anh, D.V., Trung, B.Q.: A filter approach based on binary integer programming for feature selection. In: 2022 RIVF International Conference on Computing and Communication Technologies (RIVF), pp. 677–682. IEEE (2022)
3. Arai, H., Maung, C., Xu, K., Schweitzer, H.: Unsupervised feature selection by heuristic search with provable bounds on suboptimality. In: Proceedings of the AAAI Conference on Artificial Intelligence, vol. 30 (2016)
4. Battiti, R.: Using mutual information for selecting features in supervised neural net learning. IEEE Trans. Neural Networks **5**(4), 537–550 (1994)
5. Cervante, L., Xue, B., Zhang, M., Shang, L.: Binary particle swarm optimisation for feature selection: a filter based approach. In: 2012 IEEE Congress on Evolutionary Computation, pp. 1–8. IEEE (2012)
6. Chen, Z., et al.: Feature selection may improve deep neural networks for the bioinformatics problems. Bioinformatics **36**(5), 1542–1552 (2020)
7. Ge, H., Hu, T.: Genetic algorithm for feature selection with mutual information. In: 2014 Seventh International Symposium on Computational Intelligence and Design, vol. 1, pp. 116–119. IEEE (2014)
8. Glover, F., Laguna, M.: Tabu search. In: Du, D.Z., Pardalos, P.M. (eds.) Handbook of Combinatorial Optimization. Springer, Boston (1998). https://doi.org/10.1007/978-1-4613-0303-9_33
9. Guyon, I., Elisseeff, A.: An introduction to variable and feature selection. J. Mach. Learn. Res. **3**, 1157–1182 (2003)
10. Hancer, E., Xue, B., Zhang, M.: Differential evolution for filter feature selection based on information theory and feature ranking. Knowl.-Based Syst. **140**, 103–119 (2018)
11. Hart, P.E., Stork, D.G., Duda, R.O.: Pattern Classification. Wiley, Hoboken (2000)
12. He, X., Cai, D., Niyogi, P.: Laplacian score for feature selection. In: Advances in Neural Information Processing Systems, vol. 18 (2005)
13. Lessmann, S., Baesens, B., Seow, H.V., Thomas, L.C.: Benchmarking state-of-the-art classification algorithms for credit scoring: an update of research. Eur. J. Oper. Res. **247**(1), 124–136 (2015)
14. Li, J., et al.: Feature selection: a data perspective. ACM Comput. Surv. (CSUR) **50**(6), 1–45 (2017)
15. Lin, D., Tang, X.: Conditional infomax learning: an integrated framework for feature extraction and fusion. In: Leonardis, A., Bischof, H., Pinz, A. (eds.) ECCV 2006. LNCS, vol. 3951, pp. 68–82. Springer, Heidelberg (2006). https://doi.org/10.1007/11744023_6
16. Nie, S., Gao, T., Ji, Q.: An information theoretic feature selection framework based on integer programming. In: 2016 23rd International Conference on Pattern Recognition (ICPR), pp. 3584–3589. IEEE (2016)
17. Trung, B.Q., Duc, L.M., Anh, B.T.M.: A hybrid approach based on genetic algorithm with ranking aggregation for feature selection. In: Fujita, H., Fournier-Viger, P., Ali, M., Wang, Y. (eds.) IEA/AIE 2022, vol. 13343z, pp. 226–239. Springer, Cham (2022). https://doi.org/10.1007/978-3-031-08530-7_19

Discovering Prevalent Co-location Patterns Without Collecting Co-location Instances

Vanha Tran[1(✉)], Caodai Pham[2], Thanhcong Do[1], and Hoangnam Pham[1]

[1] FPT University, Hanoi 155514, Vietnam
hatv14@fe.edu.vn, {congdthe150385,namphhe160714}@fpt.edu.vn
[2] Le Quy Don Technical University, Hanoi 11355, Vietnam
daipc.isi@lqdtu.edu.vn

Abstract. Discovering prevalent co-location patterns (PCPs) is a process of finding a set of spatial features in which their instances frequently occur in close geographic proximity to each other. Most of the existing algorithms collect co-location instances to evaluate the prevalence of spatial co-location patterns, that is if the participation index (a prevalence measure) of a pattern is not smaller than a minimum prevalence threshold, the pattern is a PCP. However, collecting co-location instances is the most expensive step in these algorithms. In addition, if users change the minimum prevalence threshold, they have to re-collect all co-location instances for obtaining new results. In this paper, we propose a new prevalent co-location pattern mining framework that does not need to collect co-location instances of patterns. First, under a distance threshold, all cliques of an input dataset are enumerated. Then, a co-location hashmap structure is designed to compact all these cliques. Finally, participation indexes of patterns are efficiently calculated by the co-location hashmap structure. To demonstrate the performance of the proposed framework, a set of comparisons with the previous algorithm which is based on collecting co-location instances on both synthetic and real datasets is made. The comparison results indicate that the proposed framework shows better performance.

Keywords: Prevalent co-location pattern · Co-location instance · Clique · Hashmap

1 Introduction

The unknown and valuable knowledge mined from spatial data sets can be applied to many domains, thus spatial data mining has received more and more attention recently. Prevalent co-location pattern (PCP) mining, which discovers a set of spatial features whose instances frequently appear together in proximity space, is an important branch of spatial data mining. For example, shopping centers and restaurants are co-located commonly in cities, thus {Shopping center, Restaurant} is called a prevalent co-location pattern. The information about the pattern can be provided to businessmen where they should set up new shops or restaurants to get the best benefit. The PCP mining technology has been applied widely in location-based services [1], environment [2], public safety [3], socio-economics [4], ecology [5], urban transportation [6], and so on.

N. T. Nguyen et al. (Eds.): ACIIDS 2023, LNAI 13995, pp. 408–420, 2023.
https://doi.org/10.1007/978-981-99-5834-4_33

Different from association rule mining [7], objects in spatial data carry spatial locations and are distributed in a continuous space. They have complex neighbor relationships. Hence, discovering PCPs from spatial data sets is nontrivial.

1.1 Related Work

Many mining algorithms have been proposed and they can be roughly divided into three categories. The first type can be called the conventional mining algorithm which effectively mines all correct prevalent patterns. These algorithms belonging to this type use the time-consuming join operator to generate co-location instances [8, 9], check neighborhoods [10], and construct a prefix tree structure to find co-location instances [11, 12]. However, these algorithms are hard to deal with big data sets, thus based-MapReduce [13], Hadoop [14], and GPU [15–17] parallel mining algorithms have been proposed. The second type can be named the special spatial data type mining algorithm which discoveries co-location patterns from special spatial data such as spatiotemporal data [18], rare event data [19], uncertain data [20], fuzzy data [21], data with network constraints [1] or data with considering density-weighted distance [22], and so on. The third type can be labeled as the compression PCP mining algorithm which represents concisely the mining results by top-k closed patterns [23], maximal patterns [24], and redundant patterns from the mining results [25, 26].

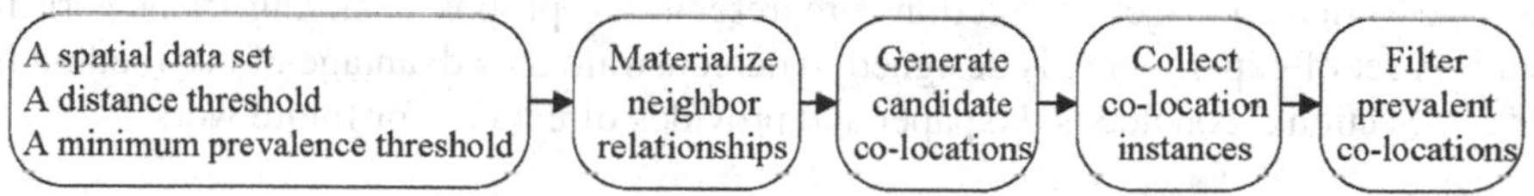

Fig. 1. The common framework of prevalent co-location pattern mining.

All of the above algorithms are developed based on a common framework shown in Fig. 1 [10] with four phases. The first phase materializes neighbor relationships of instances of the spatial data set. A set of candidate co-location patterns is generated in the second phase. The third phase collects all co-location instances of each candidate. After that, the participation index of each candidate pattern is calculated based on the co-location instances and PCPs are filtered in the fourth phase. However, the framework has two drawbacks. First, it employs a time-consuming generate-test candidate model. If the number of features is large or the data set is big/dense, the number of candidates becomes very huge and the mining process will take a lot of execution time [14].

For example, Fig. 2 shows the execution time of each phase [27] in joinless [10] and iCPI-tree [11]. It can be seen that most of the execution time is devoted to collecting co-location instances. Second, if users change the minimum prevalence threshold, this framework has to re-collect the co-location instances of all candidates. Thus, this mining framework is poorly flexible.

1.2 Contributions

This paper proposes a new prevalent co-location pattern mining framework that tackles the two drawbacks. The key contributions are as follows.

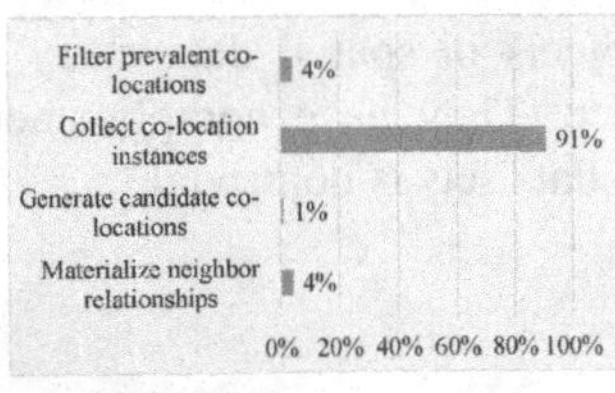

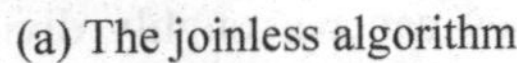

(a) The joinless algorithm

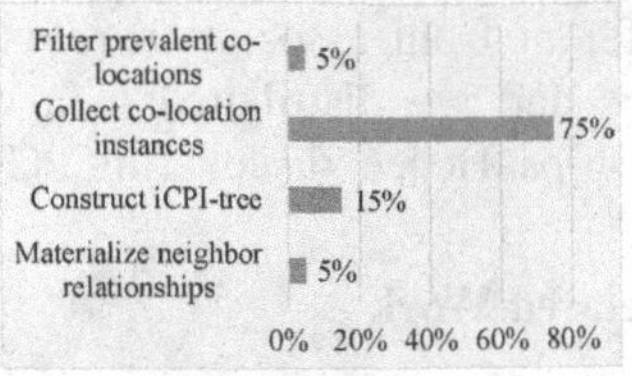

(b) The iCPI-tree algorithm

Fig. 2. The execution time in each phase of the mining framework shown in Fig. 1.

(1) We design a fast enumerating clique approach. The neighbor relationships of instances are represented by using cliques.
(2) A co-location hashmap structure, which is constructed from the cliques, is designed to store compactly neighbor relationships of instances. All information about co-location instances of patterns is located under the hashmap structure.
(3) Our algorithm is no longer using the time-consuming generate-test model. Based on the co-location hashmap structure, the participation indexes of patterns are efficiently and easily calculated. When the minimum prevalence threshold is changed, the proposed method can quickly and adaptively give new results.

The rest of the paper is organized as follows. The basic concept of PCP mining is described briefly in Sect. 2. Section 3 represents the proposed mining framework in detail. A set of experiments is designed to demonstrate the advantage of our method in Sect. 4. Section 5 concludes the paper and provides directions for future work.

2 The Basic Concept

$F = \{f_1,\ldots,f_n\}$ is a set of spatial features and S is a set of their instances. Each instance in S is a vector $<$ feature type, ID, location $>$. A co-location pattern $c = \{f_1,\ldots,f_k\}$ is a subset of F whose instances have neighbor relationships to each other. The number of features in c, k, is called the size of c. If the distance between instances is smaller than a distance threshold d, the two instances have a neighbor relationship, e.g., $\overline{\text{A.1B.3}}$. A co-location instance of c, I, is a set of instances, $I \subseteq S$, which includes the instances of all features in c and forms a clique (all instances have neighbor relationships with each other). A set of all co-location instances is called the table instance of c, denoted $T(c)$.

The participation ratio (PR) of feature f_i in a pattern c is denoted as $PR(f_i, c) = \frac{\text{Number of instances of} f_i \text{in} T(c)}{\text{Total number of instances of} f_i \text{in} S}$. The participation index (PI) of c is defined as $PI(c) = \min\{PR(f_i, c)\}, f_i \in c$. Users give a minimum prevalence threshold, *min_prev*, if the participation index of c is larger than *min_prev*, c is called a prevalent co-location pattern.

Lemma 1: The participation ratio and the participation index are monotonically non-increasing with the size of the co-location pattern.

Proof: Please refer to [8] in detail.

Lemma 1 shows that if pattern c' is a super pattern of c, $c \subseteq c'$, we have a relationship $PI(c) \geq PI(c')$. If c is not prevalent, c' is also not a prevalent co-location. Lemma 1 is employed to quickly discover PCPs in our algorithm.

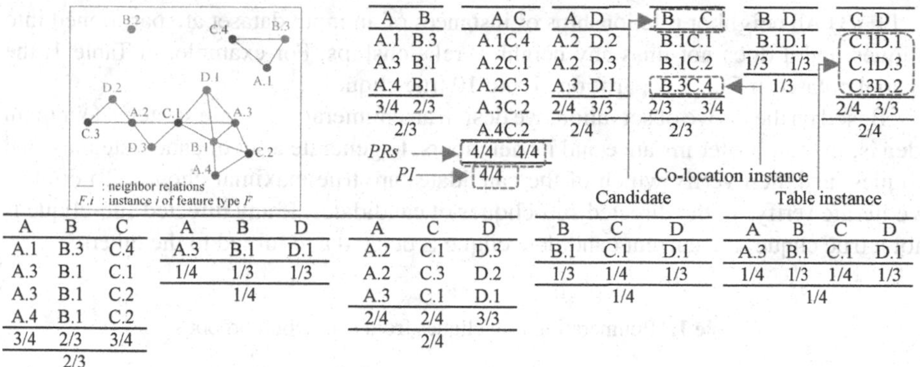

Fig. 3. An example of prevalent co-location pattern mining.

Definition 1 (Participating instance, ParI): The distinguish instance set of each feature in $T(c)$ is denoted as the participating instances of the feature in c.

For example, for candidate $c = \{A, B, C\}$, all co-location instances of it are $T(c) = \{\{A.1, B.3, C.4\}, \{A.3, B.1, C.1\}, \{A.3, B.1, C.2\}, \{A.4, B.1, C.2\}\}$. We obtain $ParI(A, c) = \{A.1, A.3, A.4\}$, $ParI(B, c) = \{B.1, B.3\}$, and $ParI(C, c) = \{C.1, C.2, C.4\}$. Thus, $PR(A,c) = \frac{|\{A.1, A.3, A.4\}|}{|\{A.1, A.2, A.3, A.4\}|} = \frac{|ParI(A,c)|}{4} = \frac{3}{4} = 0.75$, $PR(B, c) = 0.67$, and $PR(C, c) = 0.75$. Finally, $PI(c) = \min\{0.75, 0.67, 0.75\} = 0.67$. Assuming a user sets $min\_prev = 0.5$, since $PI(c) = 0.67 > min\_prev = 0.5$, thus $\{A, B, C\}$ is a PCP.

It can be seen that if the participating instances of each feature in a pattern are obtained first, there is no need to collect all co-location instances of the pattern.

3 The New Prevalent Co-location Pattern Mining Framework

3.1 Enumerating Cliques

After materializing neighbor relationships of instances and converting to a star neighborhood structure which is developed by Yoo et al. [10], we design a finding clique strategy to quickly enumerate all cliques of an input data set. Table 1 shows the star neighborhoods of each center instance constructed from Fig. 3. Note that the neighbors of each center instance in the star neighborhood are sorted in descending order. Based on Fig. 3 and Table 1, we observe three conclusions: (1) The information of neighbor relationships of all sub-cliques is included in the maximal clique. For example, clique $\{A.1, B.3, C.4\}$ includes all neighbor information of sub-cliques $\{A.1, B.3\}$, $\{A.1, C.4\}$, and $\{B.3, C.4\}$; (2) Each center instance combines with its star neighbors to construct several maximal cliques. For example, center instance A.2 and its neighbors $\{C.1, C.3, D.2, D.3\}$ can generate two maximal cliques $\{A.2, C.1, D.3\}$ and $\{A.2, C.3, D.2\}$. Note that the maximal notion here is local, only relative to one center instance. One maximal clique, that is constructed by one center instance, must be a clique (and maybe a maximal clique) in global. For example, $\{B.1, C.1, D.1\}$ is a maximal clique constructed by center instance B.1, however, it is a clique in global since it has a super-clique $\{A.3, B.1, C.1,$

D.1}; (3) All neighbor relationships of instances of an input dataset are partitioned into cliques and it does not miss any neighbor relationships. For example, in Table 1, the spatial dataset in Fig. 3 is partitioned into 10 true cliques.

Based on the above observation, we design an enumerating clique strategy. Our main idea is, in each center instance and its neighbors, to generate a set of candidate maximal cliques, and then verify which of the candidates are true maximal cliques. To do this, we iterate verifying the directed sub-cliques of candidates. If any directed sub-clique is not a true clique, the current candidate clique is deleted and turned to the others.

Table 1. Enumerating true cliques from star neighborhoods

Instance	Star neighbors	Candidate maximum cliques	True cliques
A.1	B.3, C.4	{A.1, B.3, C.4}	{A.1, B.3, C.4}
A.2	C.1, C.3, D.2, D.3	{A.2, C.1, D.2}, {A.2, C.1, D.3} {A.2, C.3, D.2}, {A.2, C.3, D.3}	{A.2, C.1, D.3} {A.2, C.3, D.2}
A.3	B.1, C.1, C.2, D.1	{A.3, B.1, C.1, D.1} {A.3, B.1, C.2, D.1}	{A.3, B.1, C.1, D.1}
A.4	B.1, C.2	{A.4, B.1, C.2}	{A.4, B.1, C.2}
B.1	C.1, C.2, D.1	{B.1, C.1, D.1}, {B.1, C.2, D.1}	{B.1, C.1, D.1}
B.2	-	-	-
B.3	C.4	{B.3, C.4}	{B.3, C.4}
C.1	D.1, D.3	{C.1, D.1}, {C.1, D.3}	{C.1, D.1}, {C.1, D.3}
C.2	-	-	-
C.3	D.2	{C.3, D.2}	{C.3, D.2}
C.4	-	-	-
D.1	-	-	-
D.2	-	-	-
D.3	-	-	-

-: having no neighbors or cliques

For example, Table 2 lists the iterator steps when enumerating maximal cliques of B.1. The star neighbors of B.1 are {C.1, C.2, D.1} and they generate two candidate maximal cliques {B.1, C.1, D.1} and {B.1, C.2, D.1}. First, we obtain the directed sub-clique of the candidate, {C.1, D.1} and {C.2, D.1}. Next, getting the star neighborhood of the first instance in the directed sub-clique, for C.1 be {D.1, D.3} and for C.2 be $\emptyset$. After that, finding the intersection of the directed sub-clique with the star neighbors of C.1, $\{D.1\} \cap \{D.1, D.3\} = \{D.1\}$ and C.2, $\emptyset \cap \{D.1\} = \emptyset$. If the size of the intersecting result (denoted as *Flag*) is equal to the size of the directed sub-clique subtracting 1, it means all instances in the directed sub-clique may have neighbor relationships to each other and the current candidate maximal clique may be a true clique. If not, the current candidate is not a true clique and is deleted immediately.

Table 2. Enumerating maximal cliques of B.1

Candidate maximal cliques	Iterator step				True clique
	Directed sub-clique	Star neighborhoods of the first instance	Intersection	*Flag*	
{B.1, C.1, D.1}	{C.1, D.1}	{D.1, D.3}	{D.1}	1	Yes
{B.1, C.2, D.1}	{C.2, D.1}	∅	∅	0	No

The pseudo-code of enumerating cliques is plotted in Algorithm 1. The first phase generates star neighborhoods of a given dataset under a distance threshold (Step 1). The second phase iterates each item (instance) and generates a set of candidate maximal cliques of the current item (Steps 2–3). The third phase checks which candidate is a true clique. To do this, directed sub-cliques of the candidate are obtained, *subCand* (Step 9). Next, the star neighbors of the first element in directed sub-cliques are also acquired, *starNei* (Step 10). After that, the intersection of *subCand* and *starNei* is found (Step 11). If the size of the intersection is not equal to the size of the directed sub-clique, it means the current candidate maximal clique is not a true clique (Step 12). Then all directed sub-cliques of the current candidate are generated and added to the candidate maximal clique set (Step 14). Else the current candidate may be a truc clique and the process go to the next iteration (Step 18). The final phase returns a set of true cliques (Step 26).

Algorithm 1. Discovering clique algorithm

Inputs: a spatial dataset *S*, and a neighbor distance threshold *d*.
Output: a set of all cliques *SoTC*.
Variables: *SN*: a hashmap structure; *SoCC*: a set of candidate cliques; *s*: the size of *SoCC*; *flag*: marking if a candidate clique is a true/false clique.

```
SN ← generate_star_neighbors(S, d)
for item ∈ SN do
    SoCC ← generate_maximal_candidate_cliques(item)
    while SoCC not empty do
        cand ← SoCC.PopFront(); // get a candidate
        s ← cand.size()
        flag ← True //assuming the current candidate clique is a true clique
        while (s > 2) do
            subCand ← get_directed_sub_clique(cand)
            starNei ← SN.find(subCand.first())
            interSet ← subCand ∩ starNei
            if (interSet.size() != subCand.size()-1) do // not a true clique
                flag ← False
                SoCC.add(generate_direct_sub_candidate_cliques(cand))
                SoCC.remove(cand)
                break
            else // the current sub-candidate clique may be is a true clique, next iteration
                s--
            end if
        end while
        if (flag == True) do
            SoTC.add(cand)
        end if
    end while
end for
return SoTC
```

3.2 Constructing a Co-location Hashmap Structure

As can be seen in Sect. 3.1, all neighbor relationships of instances are partitioned into a set of cliques. Participating instances of all patterns can be discovered from these cliques. We design a co-location hashmap structure to compact these cliques so that information about participating instances of all patterns can be quickly obtained from the structure.

Definition 2 (Co-location hashmap structure): A co-location hashmap structure is a two-level nested hashmap structure whose key and value are denoted as:

(1) The key is a set of feature types of instances in the cliques.
(2) The value is a hashmap structure whose key and value are the feature type and the instance ID of instances in the cliques, respectively.

Figure 4 shows the co-location hashmap structure which is constructed from the cliques enumerated in Table 1. It can be seen that all cliques are compressed compactly in the co-location hashmap structure.

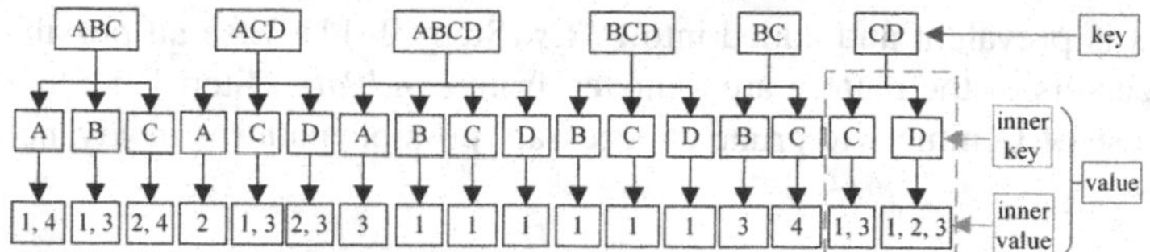

Fig. 4. The co-location hashmap structure of the data set in Fig. 3.

Algorithm 2. Building a co-location hashmap structure

Inputs: a set of cliques *SoTC*.
Output: a co-location hashmap structure *CoLHM*.
Variables: *key*, *value*.
1: **for** *clique* ∈ *SoTC* **do**
2: *key* ← get_all_features(*clique*)
3: *value* ← construct_inner_hashmap(*clique*)
4: *CoLHM* ← update(*key*, *value*, *CoLHM*)
5: **end for**
6: **return** *CoLHM*

Algorithm 2 shows the pseudo-code of building the co-location hashmap structure. The first phase creates the key and the value based on each clique (Steps 2–3). The second phase updates the co-location hashmap structure by the created key and value (Step 4). If the key has already existed in *CoLHM*, update the value; else directly add the key and the value into *CoLHM*. Finally, Algorithm 2 returns a co-location hashmap structure and it will be used to calculate the participation indexes of all patterns in Sect. 3.3.

3.3 Calculating Participation Indexes and Filtering PCPs

Based on the co-location hashmap structure, any patterns can be extracted from the keys and the information about participating instances of a pattern is obtained by the values. The participating instances of a pattern are embedded into two parts: one is the pattern itself and the other is the super-patterns of the pattern.

To describe simply, we use < feature type: {instance ID} > to represent instances of a feature. For example, the participating instances of pattern {B, C} can be acquired form key BC < B: {3}, C: {4} > and its super keys, BCD < B: {1}, C: {1} >, ABCD < B: {1}, C: {1} > and ABC < B: {1, 3}, C: {2, 4} >. Hence, the participating instances of {B, C} are < B: {1, 3}, C: {1, 2, 4} >. This result is exactly the same as Fig. 3.

To make full use of Lemma 1, we first generate all possible patterns from the key set, and then start mining from the size 2 patterns. If a pattern is not prevalent, all its supersets can be deleted directly.

Algorithm 3 is designed to quickly calculate the participation indexes of patterns based on the co-location hashmap structure. The first phase takes all keys in the co-location hashmap structure and generates the power sets of these keys (Steps 1–3). All possible patterns, *candPatts,* are generated and sorted by their size (Step 4). The second phase gets each pattern, *patt* and finds its participating instances, *ParI*, by gathering the values of all keys that are supersets of *patt* (Steps 6–7). The third phase calculates PI of *patt* based on its participating instances (Step 8). If the PI of *patt* is larger than *min_prev*, it is prevalent and added into *PSCs* (Steps 9–11). Else all possible patterns that are supersets of the pattern are removed from *candPatts* (Step 12). Our algorithm makes full use of Lemma 1 to prune unnecessary possible patterns in advance.

Algorithm 3. Calculating PIs and filtering PCPs

Inputs: a co-location hashmap structure *CoLHM* and a prevalence threshold *min_prev*.
Output: all prevalent spatial co-location patterns *PSCs*.
Variables: *key*, *value*.

```
1: for each key in CoLHM.keys() // get all keys in the co-location hashmap structure
2: |   candPatts ← gen_power_sets(key)
3: end for
4: candPatts ← sort_by_size(candPatts)
5: while candPatts is not empty do
6:     patt ← candPatts.PopFront()
7:     ParI ← (patt, CoLHM)
8:     PI ← calculate_PI(ParI)
9:     if (PI ≥ min_prev) do
10:    |   PSCs.add(patt)
11:    else
12:    |   candPatts.delete_all_superset(patt)
13:    end if
14: end while
15: return PSCs
```

Figure 5 shows the proposed mining framework. Different from the framework shown in Fig. 1, our framework has no time-consuming collecting co-location instance phase.

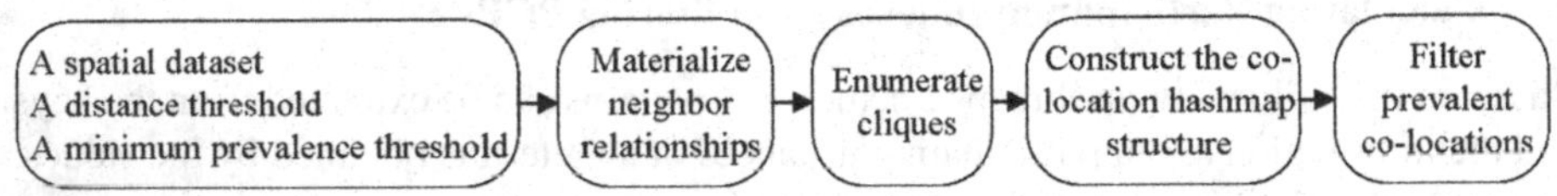

Fig. 5. The proposed prevalent co-location pattern mining framework.

3.4 The Time Complexity Analyses

As shown in Fig. 5, our mining algorithm has four phases. The first phase materializes neighbor relationships and converts to the star neighborhood structure and the computational complexity of this phase is about $O(n^2 \times \frac{d^2}{A})$ [25] where d is the distance threshold, n is the number of instances and A is the area of the input dataset space.

Assuming s_{avg} is the average size of neighbors of each instance. The computational complexity of enumerating cliques is about $O\left(n \times s_{avg}^2\right)$. The third phase constructs the co-location hashmap structure and the computational complexity of this phase is about $O(L)$ where L is the number of cliques. The final phase calculates PIs and filters prevalent co-location patterns and its computational complexity is about $O(l \times m)$ where l and m are the numbers of items in the co-location hashmap structure and numbers of all possible

Table 3. Parameters of the synthetic dataset

Parameters	Values
Number of features	15
Number of instances	20000
Frame size (D × D)	1000, 10000
d	15
min_prev	0.2
clumpy	1
across	0
overlap	0

* *clumpy, across,* and *overlap* refer to [10] in detail

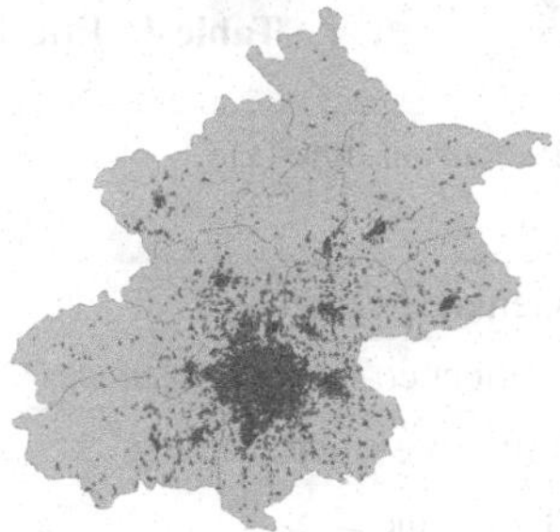

Fig. 6. The distribution of the real dataset.

patterns. Therefore, the total computational complexity of our mining framework is about $O\left(n^2 \times \frac{d^2}{A}\right) + O\left(n \times s_{avg}^2\right) + O(L) + O(l \times m)$.

4 Experiment Evaluations

We design a set of experiments to demonstrate the proposed mining framework is efficient. We chose the joinless algorithm [10] which is known as a correct and efficient algorithm for finding PCPs based on the framework in Fig. 1. Our experiments are coded by C++ and performed on a PC machine with 16 GB main memory.

4.1 Experimental Datasets

Both synthetic and real datasets are used in our experiments. Table 3 shows a summary of the synthetic dataset which is generated by a generator developed by Shekhar et al. [8]. We use the real data set that is a facility point data set from Beijing, China, that contains 25,276 items (instances) belonging to 12 feature types such as residential area, company, and restaurant. The distribution of the real data set is shown in Fig. 6.

4.2 Mining Comparisons

Comparison of the Costs of Computation Complexity Factors: We show the execution time in each phase of our algorithm and the joinless algorithm in Table 4. Since the two algorithms use star neighborhoods, Table 4 only lists the execution time of the last three phases in each framework. In the sparse dataset, since the number of co-location instances is small, PIs of patterns are also small and many candidates are pruned, thus the execution time of joinless is a little faster than our algorithm. However, in the dense dataset, our algorithm is faster than joinless. It can be seen that most of the execution time of joinless is devoted to collecting co-location instances. Our algorithm without collecting co-location instances shows less execution time.

Table 4. Effects of the decision by the two mining frameworks

Execution time (s)	Joinless		Execution time (s)	Our algorithm	
	Sparse	Dense		Sparse	Dense
Generate candidates	0.005	0.051	Enumerate cliques	0.044	22.006
Collect co-location instances	4.017	47.172	Construct hashmap	0.374	0.95
Filter prevalent patterns	0.026	0.294	Filter prevalent patterns	6.332	6.03
Total time	4.048	47.517	Total time	6.75	28.986

Effect of the Minimum Prevalence Threshold: Figure 7 shows the effect of the proposed algorithm when changing minimum prevalence thresholds on the real data set with the distance threshold set to 300m. It is clear to see that the execution time of joinless is very expensive when *min_prev* is small. In this case, many candidates become PCPs and it must be collect co-location instances of all these patterns. With the increase of *min_prev*, the execution time in joinless decreases. While our algorithm is robust because it is designed to avoid the effect of changing *min_prev*. When changing *min_prev*, our algorithm only needs to calculate the PIs of patterns from the co-location hashmap structure without other redundant operations. Thus, it can quickly give new mining results to users.

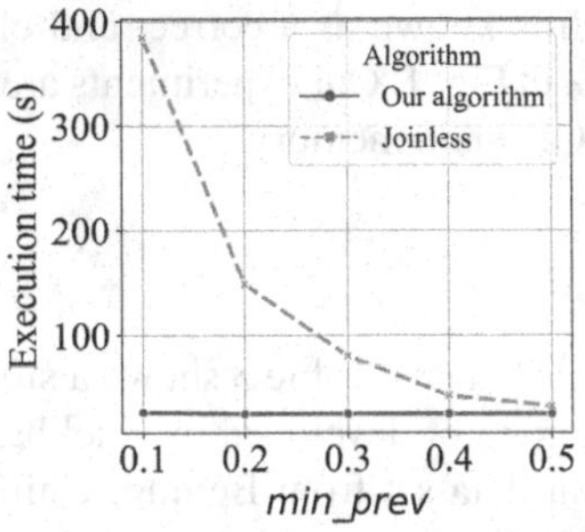

Fig. 7. The effect of the *min_prev* threshold.

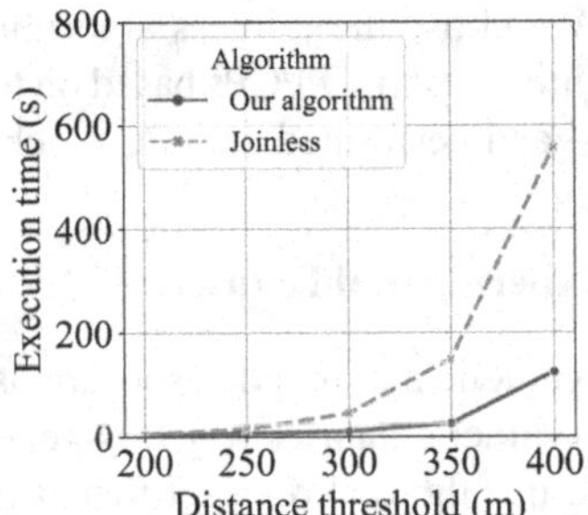

Fig. 8. The effect of the distance threshold.

Effect of the Distance Threshold: The effect of distance thresholds also is evaluated in the real dataset with *min_prev* is set to 0.2. As shown in Fig. 8, with the increase of the distance threshold, the execution time of the two algorithms also increase. However, our algorithm shows better performance.

Effect of Numbers of Instances: The final experiment evaluates the effect of the number of instances of the compared algorithms. We set the synthetic data sets with a frame size is 5000×5000, the number of features is 15, $d = 30$, $min\_prev = 0.2$, and the number of instances is changed. As shown in Fig. 9, with the increase in the number of instances (data sets become larger), the execution time of the two algorithms also increases. However, the performance of Joinless degenerates quickly in the situation of large data sets, while the proposed algorithm shows better scalability.

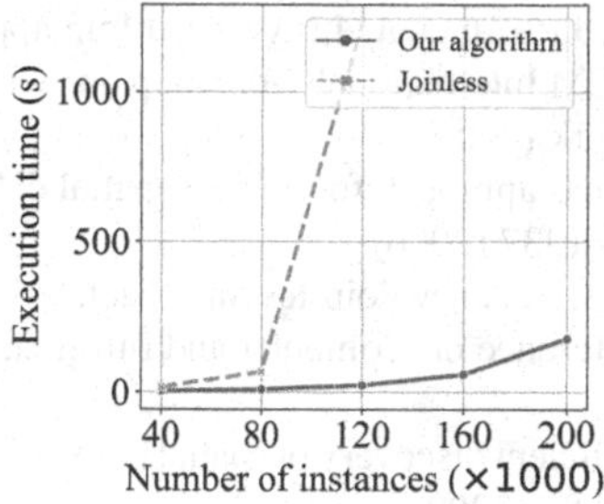

Fig. 9. The effect of numbers of instances.

5 Conclusion

A new PCP mining framework is proposed in this paper. The proposed framework partitions neighbor relationships of instances into cliques and constructs a co-location hashmap structure based on these cliques that compact the neighboring instances. By using the structure, the participation indexes of patterns can be calculated quickly without collecting their co-location instances. If users change the minimum prevalence threshold, our algorithm only needs to re-calculate the participation indexes and directly give new mining results. Thus, the proposed algorithm is robust with the minimum prevalence threshold. The experimental result indicates that the performance of our algorithm is better than collecting co-location instance-based mining algorithms.

It can be seen that a heavy job in our algorithm is enumerating cliques. However, this step is quite easy to implement in parallel computing. Thus, our future work focuses on transforming the proposed algorithm into an efficient parallel mining method.

References

1. Yu, W.: Spatial co-location pattern mining for location-based services in road networks. Expert Syst. Appl. **46**, 324–335 (2016)
2. Akbari, M., Samadzadegan, F., Weibel, R.: A generic regional spatio-temporal co-occurrence pattern mining model: a case study for air pollution. J Geogr Syst. **17**, 249–274 (2015)
3. Mohan, P., Shekhar, S., Shine, J.: A neighborhood graph based approach to regional co-location pattern discovery: a summary of results. In: 19th ACM SIGSPATIAL, pp. 122–132. ACM, NY (2011)
4. Cai, J., Deng, M., Liu, Q.: Nonparametric significance test for discovery of network-constrained spatial co-location patterns. Geogr. Anal. **51**, 3–22 (2019)
5. Deng, M., He, Z., Liu, Q.: Multi-scale approach to mining significant spatial co-location patterns. Trans. GIS **21**, 1023–1039 (2017)
6. Wang, S., Huang, Y., Wang, X.: Regional co-locations of arbitrary shapes. In: Advances in Spatial and Temporal Databases, pp. 19–37. Springer, Berlin (2013). https://doi.org/10.1007/978-3-642-40235-7_2
7. Kishor, P., Porika, S.: An efficient approach for mining positive and negative association rules from large transactional databases. In: ICICT, pp. 1–5. IEEE, India (2016)
8. Shekhar, S., Huang, Y.: Discovering spatial co-location patterns: a summary of results. In: Advances in Spatial and Temporal Databases, pp. 236–256. Springer, Berlin (2001). https://doi.org/10.1007/3-540-47724-1_13

9. Yoo, J., Shekhar, S., Smith, J., Kumquat, J.: A partial join approach for mining co-location patterns. In: 12th Annual ACM International Workshop on Geographic Information Systems, pp. 241–249. ACM, New York (2004)
10. Yoo, J., Shekhar, S.: A joinless approach for mining spatial colocation patterns. IEEE Trans. Knowl. Data Eng. **18**, 1323–1337 (2006)
11. Wang, L., Bao, Y., Lu, J., Yip, J.: A new join-less approach for co-location pattern mining. In: 8th IEEE International Conference on Computer and Information Technology, pp. 197–202. Sydney (2008)
12. Wang, L., Bao, Y., Lu, Z.: Efficient discovery of spatial colocation patterns using the iCPI-tree. The Open Inf. Syst. J. **3**, 69–80 (2009)
13. Yoo, J., Boulware, D., Kimmey, D.: A parallel spatial co-location mining algorithm based on mapreduce. In: International Congress on Big Data, pp. 25–31 (2014)
14. Yoo, J., Boulware, D., Kimmey, D.: Parallel co-location mining with MapReduce and NoSQL systems. Knowl Inf. Syst. (2019)
15. Andrzejewski, W., Boinski, P.: Efficient spatial co-location pattern mining on multiple GPUs. Expert Syst. Appl. **93**, 465–483 (2018)
16. Sainju, A., Aghajarian, D., Jiang, Z., Prasad, S.: Parallel grid-based colocation mining algorithms on GPUs for big spatial event data. IEEE Trans Big Data, pp. 1–1 (2018)
17. Andrzejewski, W., Boinski, P: Parallel approach to incremental co-location pattern mining. Information Sci. **496**, 485–505 (2019)
18. Leibovici, D., Claramunt, C., Guyader, D., Brosset, D.: Local and global spatio-temporal entropy indices based on distance-ratios and co-occurrences distributions. Int. J. Geogr. Inf. Sci. **28**, 1061–1084 (2014)
19. Huang, Y., Pei, J., Xiong, H.: Mining co-location patterns with rare events from spatial data sets. GeoInformatica **10**, 239–260 (2006)
20. Wang, L., Wu, P., Chen, H.: Finding probabilistic prevalent colocations in spatially uncertain data sets. IEEE Trans. Knowl. Data Eng. **25**, 790–804 (2013)
21. Ouyang, Z., Wang, L., Wu, P.: Spatial co-location pattern discovery from fuzzy objects. Int. J. Artif Intell. Tools **26**, 1750003 (2016). https://doi.org/10.1142/S0218213017500038
22. Yao, X., Chen, L., Peng, L., Chi, T.: A co-location pattern-mining algorithm with a density-weighted distance thresholding consideration. Inf. Sci. **396**, 144–161 (2017)
23. Yoo, J., Bow, M.: Mining top-k closed co-location patterns. In: International Conference on Spatial Data Mining and Geographical Knowledge Service, pp. 100–105. IEEE, Fuzhou (2011)
24. Wang, L., Zhou, L., Lu, J., Yip, J.: An order-clique-based approach for mining maximal co-locations. Inf. Sci. **179**, 3370–3382 (2009)
25. Wang, L., Bao, X., Zhou, L.: Redundancy reduction for prevalent co-location patterns. IEEE Trans. Knowl. Data Eng. **30**, 142–155 (2018)
26. Wang, L., Bao, X., Chen, H., Cao, L.: Effective lossless condensed representation and discovery of spatial co-location patterns. Inf. Sci. **436–437**, 197–213 (2018)
27. Boinski, P., Zakrzewicz, M.: Collocation pattern mining in a limited memory environment using materialized iCPI-tree. In: Data Warehousing and Knowledge Discovery, pp. 279–290. Springer, Berlin (2012). https://doi.org/10.1007/978-3-642-32584-7_23

Integrating Geospatial Tools for Air Pollution Prediction: A Synthetic City Generator Framework for Efficient Modeling and Visualization

Krystian Wojtkiewicz[1(✉)], Filip Litwinienko[1], Rafał Palak[1,2], and Marek Krótkiewicz[1]

[1] Faculty of Computer Science and Telecommunication Technologies, Wrocław University of Science and Technology, Wybrzeże Stanisława Wyspiańskiego 27, 50-370 Wrocław, Poland
krystian.wojtkiewicz@pwr.edu.pl

[2] Department of Computer Engineering, Yeungnam University, Daegu, Korea

Abstract. Air pollution is a significant public health and environmental concern that requires accurate prediction and monitoring. This paper introduces a framework that establishes a city-wide abstraction layer for air pollution prediction. The authors present contemporary advancements in air pollution modeling, including research approaches and technologies. The framework promotes a streamlined learning process and improves efficiency by generating a simulated representation of the Earth's surface for air pollution forecasting using the Land-Use Regression (LUR) model and facilitating data visualization. The authors aim to establish a platform for exchanging research experiences and replicating findings to improve air pollution prediction and control. The framework can help policymakers, researchers, and environmentalists monitor air pollution levels and develop effective strategies to mitigate its adverse effects.

Keywords: air pollution prediction · land-use regression · chemical transport model · pollution modeling · LUR · CTM · Synthetic City Generator

1 Introduction

Air pollution is a pervasive problem that significantly affects the environment and people. Extensive academic studies have analyzed the dangers of pollution and its impact on the planet's ecosystem at all levels. The adverse effects of pollution on human health, particularly respiratory and cardiovascular systems, are of great concern [3,16,20]. However, pollution also affects other living beings, including flora. Exposure to pollutants can disrupt vital processes like photosynthesis, respiration, and carbon allocation, leading to stunted growth

N. T. Nguyen et al. (Eds.): ACIIDS 2023, LNAI 13995, pp. 421–435, 2023.
https://doi.org/10.1007/978-981-99-5834-4_34

and decreased survival rates of plants [7]. Furthermore, pollution also harms non-living things, leading to acid rain, soil degradation, water quality deterioration, and damage to rock formations.

To mitigate the harmful consequences of pollution, tracking and predicting its levels accurately and understanding its underlying causes is essential. This can be accomplished by studying pollution behavior in the context of the geospatial environment and identifying critical areas that require concentrated efforts to reduce pollution agents [21,22].

The primary contribution of this research is the proposed framework that focuses on analyzing and predicting air pollution using tile-grid abstraction in cities. This unique approach allows for a more specified and localized understanding of pollution patterns, going beyond the traditional models commonly used in existing research.

The framework is designed to be robust and user-friendly, facilitating the creation of Land Use Regression (LUR) and Chemical Transport Model (CTM) models more reflecting city environments' specific needs and characteristics. These models can potentially offer more precise and actionable insights to researchers and decision-makers, thereby contributing to developing more effective pollution mitigation strategies.

The paper is organized into several sections. The Related Works section provides a comprehensive review of tile generation, a method used in this paper to generate tile grids. The subsequent sections describe the proposed framework and discuss its potential applications and issues. Finally, the paper concludes with a summary of the findings and future research directions.

2 Related Works

2.1 Traditional Computational Models for Air Pollution Prediction

Air pollution prediction is a complex task involving many possible variables to consider. Currently, the research focuses on three main approaches, namely Land Use Regression (LUR) [12], Chemical Transport Model (CTM) [6,14] and Gaussian Plume Model (GPM). Additionally, we can find research focusing on the indeterministic, computational approach involving, among others, Artificial Neural Networks [5,10,19] or genetic algorithms [8]. Below, we will briefly characterize each of them, focusing on the data needed for them to be applied.

LUR – Land Use Regression is a modeling approach derived from statistical linear regression models that aims to estimate air pollution concentrations in urban areas based on land use characteristics and other environmental factors. Anderson was among the first to investigate the relationship between land use and air pollution in 1970 [2]. Since then, LUR has become an increasingly popular method for estimating air pollution concentrations, mainly in urban areas. Advances in geographic information systems (GIS) technology and the availability of high-resolution spatial data have made it possible to create detailed maps of land use and other environmental factors as inputs to LUR models. Based on that, we can identify two elements needed to build a dependable LUR model: environmental properties and data collected from the region of interest.

The environmental properties indicate the possible impact a given area has on the pollution production or distribution, e.g., a heavily industrialized area produces lots of pollution passively. At the same time, a forest might reduce air pollution significantly. The indentation of the terrain is crucial, as it helps to decide which pollution dispersion pattern should be used in prediction.

One of the crucial aspects of the application of LUR models is the spatial approach. We can build a semi-continuous model or divide the map into small regions. The latter method is often referred to as tiling. Both directions allow scaling to create more accurate approximations in desired areas. The quality of predictions the LUR model produces depends on the sensor data provided as an input. In an ideal situation, we would have data available from sensors evenly distributed in the area and a short distance from each other. However, such a situation is highly unbelievable since the maintenance cost of such a sensor grid would be tremendous. Real-life scenarios include several sensors in the area. Each sensor acquires data in a fixed time interval and provides it to the model.

CTM - Chemical Transport Model is a computer model used to simulate the transport and transformation of air pollutants in the atmosphere. CTMs are based on mathematical equations describing the physical and chemical processes involved in the movement of contaminants through the atmosphere. They can estimate the dispersion of secondary pollutants such as ozone or NO_2 and particulate matter secondary components [11,17,18,25].

CTMs provide insights into the sources and causes of regional air pollution episodes to evaluate the effectiveness of air pollution control measures and to provide input to air quality management and policy decisions. The critical components of CTM include:

- atmospheric chemistry while it considers the chemical reactions that occur between pollutants and other atmospheric constituents, such as ozone and nitrogen oxides,
- meteorology that gives means to model atmospheric circulation to simulate the movement of air masses and the transport of pollutants over large distances,
- emissions sources as it relies on accurate inventories of pollutant emissions from, e.g., industry, transportation, and agriculture.

Moreover, we rarely use CTM without boundary conditions set. Thus, we need to specify boundary conditions that define the concentrations of pollutants entering and leaving the simulation domain. The evaluation of the model is based on comparing the predictions with observations from ground-based and satellite-based monitoring systems. The general idea of the CTM approximation is derived from simulating fluid mechanics to track the atmospheric dispersion of pollutants. Therefore, the main issue with using CTM and obtaining high-quality prediction is the computational power needed to run all simulations. The computational complexity of CTM relies on the 3D resolution and simulation period.

GPM - Gaussian Plume Model is a mathematical model used to estimate the dispersion of air pollutants emitted from a source point. According to a mathematical formula based on Gaussian distributions, the pollutant concentration decreases with a growing distance from the source. The computations are true for a plume, e.g., it is not equal in all directions. We estimate primary pollutants' spatial and temporal distribution with meteorological data, emissions data, and dispersion parameters. That model is reliable for modeling traffic-related air pollutants at local (up to 1 km) and urban (up to 10 km) scales [9].

From the meteorological point of view, GPM requires input data on atmospheric stability, wind speed, wind direction, and other meteorological variables that influence the movement and dispersion of pollutants. It also utilizes information on the rate and type of pollutants emitted from the source, as well as the height and location of the emission. The most important, however, is the knowledge of the dispersion parameters, including the plume rise, the horizontal and vertical dispersion coefficients, and the decay rate of the pollutant as it moves away from the source. The quality of dispersion parameters significantly influences the output of this method. GPM can also consider the effects of terrain on the dispersion of pollutants, such as the impact of hills and valleys on wind patterns and atmospheric stability.

$$C(x,y,z) = \frac{Q}{U} \frac{e^{(\frac{-y^2}{2\sigma_y^2})}}{2\pi\sigma_y\sigma_z}(e^{(\frac{-(z-H)^2}{2\sigma_z^2})} + e^{(\frac{-(z+H)^2}{2\sigma_z^2})}), \tag{1}$$

where:

C - concentration of pollutants in coordinates x (meters downwind from source), y (meters crosswind from plume center), z (meters above ground) g/m^3,
Q - rate of emissions in g/s,
U - horizontal wind velocity parallel to plume centerline,
H - the height of the plume centerline,
$e^{(\frac{-y^2}{2\sigma_y^2})}$ - crosswind dispersion,
$e^{(\frac{-(z-H)^2}{2\sigma_z^2})}$ - vertical dispersion not including reflection of emission from the ground,
$e^{(\frac{-(z+H)^2}{2\sigma_z^2})}$ - vertical dispersion of reflection of emission from the ground,
σ_y - standard deviation of horizontal emission distribution,
σ_z - standard deviation of vertical emission distribution.

With exemplary Eq. 1 [1], we can observe that with the number of parameters and their specificity, the computational complexity of the prediction increases. Due to this method's small scale, the primary evaluation is by comparing its predictions with observations from ground-based monitoring systems.

2.2 Technologies Utilized in Computer-Aided Air Pollution Prediction

Geographic Information System (GIS) is a computer-based system designed to capture, store, manipulate, analyze, and present spatial and geographic data [23]. It integrates such data types as maps, satellite imagery, and tabular data to understand spatial relationships and patterns comprehensively. The roots of GIS can be traced back to the 1960s and 1970s when Dr. Roger Tomlinson established the foundation for spatial data handling and analysis [24]. However, Geographic Information Systems (GIS) have evolved over decades, with the development and integration of various technologies, concepts, and standards, e.g., Open Geospatial Consortium (OGC) Standards, ISO 191xx Series, Federal Geographic Data Committee (FGDC) Standards or INSPIRE Directive. These standards help ensure that GIS data and services are compatible and seamlessly integrated across different platforms and systems. They effectively promote data sharing, interoperability, and use of geospatial information in various domains, including government, environmental management, emergency response, and urban planning. There are parallel solutions that either utilize the GIS approach, enhance it, or are data sources for GIS-based systems. They are, among others, remote sensing, GPS, mapping platforms, or CAD software. The scope of application of GIS includes but is not limited to, urban planning and management, environment management, emergency management, public health, transportation, logistics, or natural resource management.

Map Tiling refers to dividing a digital map into smaller, fixed-size sections called tiles. Each tile represents a specific map area at a particular zoom level and shape, e.g., square, rectangular, or hexagonal (see Fig. 1). Tiling is used in web mapping and GIS applications to efficiently display and navigate large-scale maps. The concept is based on the principle of progressive loading and rendering. Instead of loading and displaying the entire map at once, only the tiles necessary for the current view are loaded and shown. Tiling offers advantages in performance, scalability, caching, and flexibility. Tiling allows for the easy integration of map layers, overlays, and different data sources. In particular, it enables the combination of multiple map tiles, such as base maps, thematic layers, and markers, to create rich and informative map visualizations. Tiling schemes often follow established standards such as the Slippy Map Tilenames (used by OpenStreetMap) or the Google Maps XYZ scheme. These standards define the numbering and naming conventions for tiles at different zoom levels and provide a consistent way to reference specific tiles within a tile pyramid. The choice between different shapes of tiles should be made based on the purpose of the application. Our solution will use hexagonal division to allow all adjacent cells to be at the same distance.

Geospatial Indexing is a technique that efficiently organizes and retrieves geospatial data based on their spatial properties. It involves creating a data

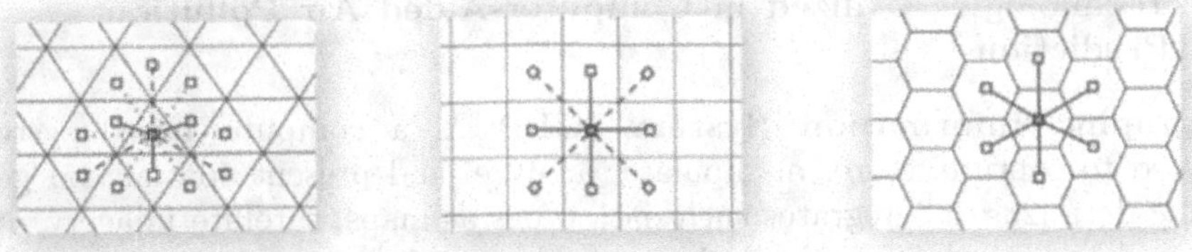

Fig. 1. The comparison between different shapes of tiles and their distance between neighbors (source: https://h3geo.org).

structure or index that allows for quick spatial queries, such as finding objects within a specified area or identifying the nearest neighbors to a given location. Geospatial indexes optimize spatial data retrieval and analysis, reducing the need for exhaustive searches through large datasets. There are diverse types of geospatial indexes, each with advantages and best use cases. Commonly used geospatial indexing methods include, among others, Quadtree, R-tree, grid index, k-d tree, and Geohash. Commercially available solutions include the H3 index, which uses hexagonal tiles with 16 levels of hierarchy and can support analysis on vast and local scales, and Google S2 (Spherical Geometry) system – Fig. 2. The usage of geospatial indexes depends on the specific requirements of the application or system, such as geospatial databases, web mapping, location-based services, or spatial data analysis.

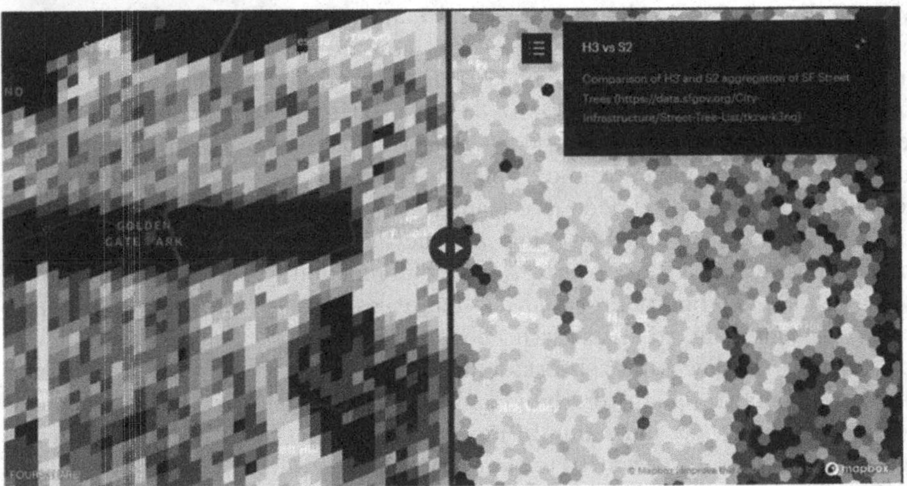

Fig. 2. The comparison between Uber's H3 and Google's S2 tiling systems (source https://h3geo.org) [4].

Consequently, existing models have limitations in terms of scalability and performance due to their high computational needs. It applies mainly to CTM, which requires much computational power for simulations. Conversely, LUR models require detailed environmental properties and data from the region of interest, which may not be readily available or accurate. The GPM's requirement of technical input parameters and relatively small scale makes it less applicable for large-scale predictions.

In this paper, we address these problems by providing a framework that focuses on integrating tiling and geospatial characteristics, thus optimizing the computational complexity and improving the spatial granularity of predictions. This method also offers a more interactive and user-friendly approach, providing real-time feedback and facilitating the reproducibility of the prediction process.

3 The Synthetic City Generator Framework

The methodology employed in this study is called the Synthetic City Generator. Its primary objective is to construct a virtual and simplified representation of the targeted city, which can subsequently be utilized for prediction tasks. The approach integrates the concept of tiling, building upon the geospatial characteristics derived from a geographic information system (GIS) about the specific area of interest. The process of generating the synthetic city encompasses four distinct stages, depicted in Fig. 3, namely:

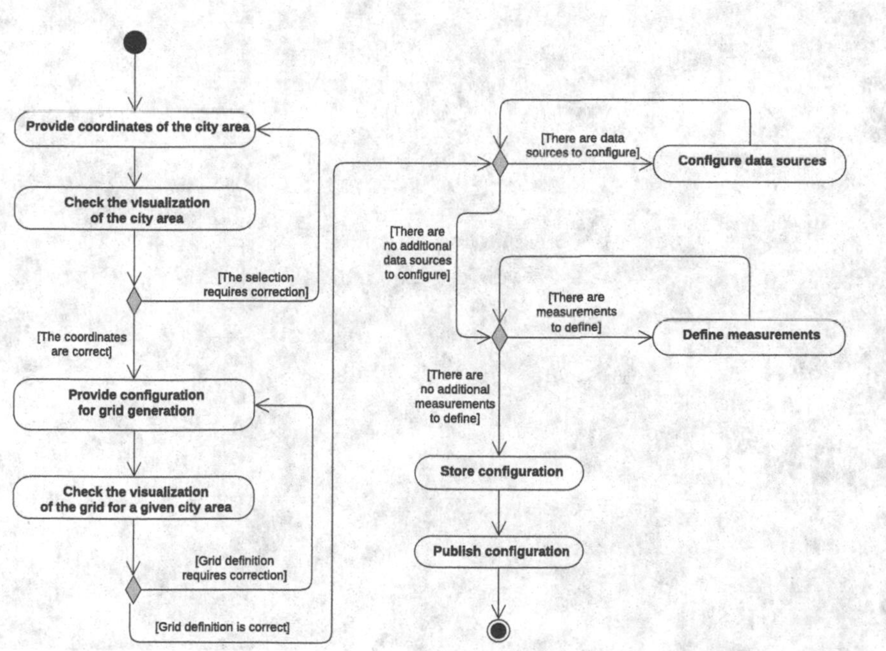

Fig. 3. Diagram of SCG's dataset generation process

step 1 – City selection
step 2 – Grid definition
step 3 – Data sources configuration
step 4 – Measurements definition

Each step involves the specification of user-defined input parameters. Notably, the generator offers real-time feedback to the user throughout the process, ensuring that each configuration is meticulously stored and visually presented, thereby facilitating future reproductions of the entire procedure. We are also providing the ability to publish SCG's configurations for other research teams to reproduce the effort. Below, we will describe each step in more detail.

City Selection. Users can delineate a specific area for subsequent generation within the Synthetic City Generator (SCG) through two approaches. Firstly, they can search for the desired city and employ the selection tool to designate the exact location of interest. Alternatively, users may directly input the geographic coordinates themselves. In the latter scenario, the user specifies two points, denoted as the Starting and Ending points, to define a bounding rectangle for the selection process undertaken by the SCG. An exemplary result of this step is presented in the Fig. 4.

Fig. 4. Area prepared for user input (source: kepler.gl) [13,15].

Grid Definition. Upon acceptance of the chosen area, the user can specify various grid parameters employed in the generation process. These attributes encompass:

- The size of the generated grid.
- The parameters related to land use.

Determining the size of the grid entails defining the radius of the hexagonal units. In theory, the radius can be arbitrarily large; however, it is preferable to maintain a size that does not exceed the dimensions of the selected area. An exemplary result of grid generation is presented in Fig. 5.

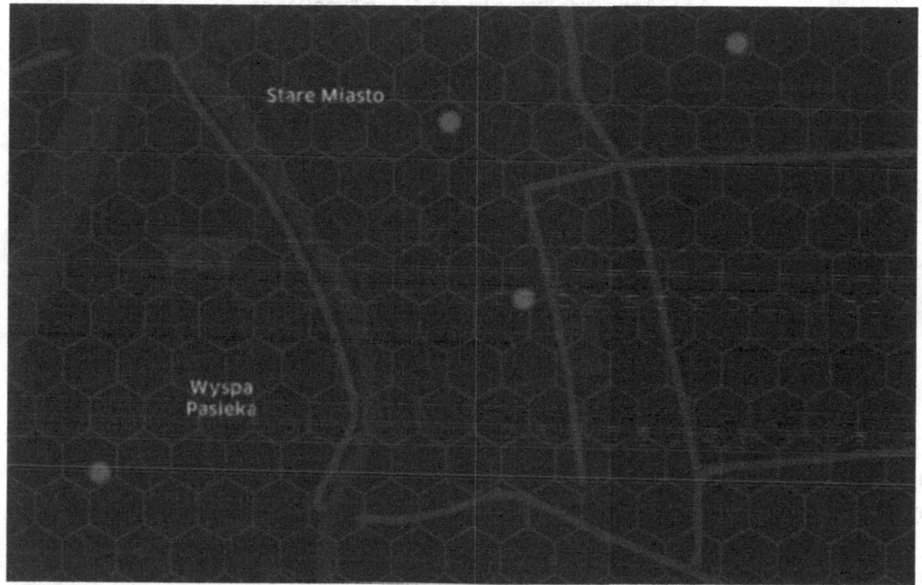

Fig. 5. Area with generated grid and sensors (source: kepler.gl) [13,15].

The tile mesh generation process in the SCG incorporates the utilization of the following libraries:

GeoPy. A Python library specifically developed to streamline the utilization of geocoding web services. Within the system, it plays a crucial role in defining the geographic coordinates of points and calculating distances between them. Its functionality is essential for accurately projecting a flat mesh onto a global coordinate system.

GeoJSON. A Python library that facilitates the encoding and decoding of GeoJSON data. In the Synthetic City Generator context, GeoJSON is employed to generate the necessary files that visually represent the tile mesh. These files enable the effective visualization and manipulation of the mesh within the system.

By incorporating these libraries, the system enhances its capability to handle geospatial data, perform geocoding operations, project meshes onto global coordinates, and generate visually appealing representations of the tile mesh using the GeoJSON format.

Regarding the land use parameters, the user is afforded the flexibility to select the types of land use to be considered during the subsequent calculations. These land use types are enumerated and elaborated upon within the accompanying Table 1. However, these types might be redefined as the list is not arbitrarily closed.

Table 1. List of land use types defined within the SCG.

`fillingTypeSymbole`	`fillingTypeDescr`
0	`residentials`
1	`commercials`
2	`industrial`
3	`forest`
4	`grass`
5	`agriculture`
6	`parking`
7	`shools`
8	`water`

The acquisition of tile-filling data is currently accomplished by utilizing the OpenStreetMap (OSM) application programming interface (API). OSM is a freely accessible and open-source database continuously updated and maintained by a community of volunteers operating under the Open Database License. The OSM database provides a comprehensive and detailed description of geospatial data, encompassing valuable information about land use.

An evaluation is conducted based on the OSM Purpose/Category taxonomy to determine the filling level of the SCG tiles. By categorizing all polygons encompassed by a given map and calculating their respective areas, the filling ratio for each tile can be inferred. In the context of road data, the road filling of a particular tile is determined solely by the length of the roads present within it.

The City Fixed Parameters (CFP) is a structure employed to quantitatively assess the composition of different land use types within a tile (Listing 1.1). It encompasses several key components, including the total length of roads within the tile, the degree of undulation in the terrain, and a timestamp indicating the calculation time.

The $GISLandUse$ structure is utilized further to elucidate the land use characteristics of a specific tile. By specifying the tile index, this structure facilitates the retrieval of filling information about various land use types ($GISFillingType$) from a corresponding Table 1.

By leveraging the *CFP* and *GISLandUse* structures, the system can effectively quantify the representation of different land use categories within a tile, considering factors such as road length, terrain undulation, and the specific calculation time.

Listing 1.1. Definition of CFP parameters used by SCG.

```
struct CFP {
    uint64 index; // H3 index of a tile
    float roads; // Total length of road routes
    float undulationDegree;
    uint64 timestamp; // timestamp
};

struct GISLandUse {
    uint64 tileIndex; // H3 index of a tile
    byte fillingTypeSymbol; // filling type for the tile
    float fillingLevel;
}; // invariant: total sum of fillingLevel equal to 1 for each tileIndex

struct GISFillingType {
    byte fillingTypeSymbol; // filling type for the tile
    char fillingTypeDescr[32];
};
```

The Synthetic City Generator (SCG) incorporates several libraries to classify the tile mesh and process its characteristics. These libraries include:

OSMNX. A Python package for retrieving spatial data from OpenStreetMap. Within the SCG, OSMNX is employed to extract land use and road use information for individual tiles. This information is obtained using the polygon coordinates derived from the generated tiles.

Shapely. A Python package offering functionalities for manipulating and analyzing planar geometric objects. The SCG utilizes Shapely to create polygons based on the coordinates of the tiles. Moreover, this library plays a critical role in the classification of tiles by determining intersections between OSMNX data and the tile polygons.

GeoPandas. An open-source Python project facilitating geospatial data handling. In the SCG, GeoPandas projects the intersected polygon data obtained in the previous step. This projection calculates each polygon's area relative to the corresponding tile.

By leveraging these libraries, the SCG enhances its capabilities in classifying the tile mesh, extracting relevant spatial information from OpenStreetMap, creating and analyzing polygons, and accurately measuring each polygon relative to its associated tile.

Data Sources Configuration. Following selecting the area and generating the grid, the configuration of data sources becomes feasible. A data source comprises various sensors, which can be classified into movable and fixed-position. For movable units, it is imperative that they update their tile index based on the unit's current tile. Optionally, they may also update their precise location to enhance accuracy. These coordinates can be transmitted alongside each measurement performed by any sensor within the unit. On the other hand, fixed-position units do not require positional updates. Once their location is defined, it remains constant. They solely necessitate the knowledge of the tile index they occupy. In terms of sensor configuration within a unit, each sensor can be tailored using the following parameters:

- **sensor index**, representing the identifier of the measured pollutant along with its default units (Table 2),
- **the range** encompassing the measurements,
- **the precision** of the measurement,
- and **the interval** at which readings are taken.

Multiple units can be defined for each map, with their respective definitions visualized on the map. It is crucial to note that every unit must be defined concerning the grid. Consequently, units will not be generated beyond the bounds of the grid.

Table 2. Pollutant types with their symbols and the corresponding units.

fillingTypeSymbole	fillingTypeDescr	baseunit
0	NO_2 concentration	$\frac{\mu g}{m^3}$
1	SO_2 concentration	$\frac{\mu g}{m^3}$
2	CO concentration	$\frac{\mu g}{m^3}$
3	CO_2 concentration	$\frac{\mu g}{m^3}$
4	$PM_2.5$ concentration	$\frac{\mu g}{m^3}$
5	PM_10 concentration	$\frac{\mu g}{m^3}$
6	Wind velocity	$\frac{m}{s}$
7	Wind direction	deg
8	Temperature	$^\circ C$
9	Air pressure	hPa
10	Rainfall	mm
11	Insolation (sunlight)	$\frac{W}{m^2}$
12	Cloudiness	%
13	Relative humidity	%

Measurements Definition. Specifying a measurand within SCG necessitates the inclusion of three essential parameters: symbol, name, and base unit of measurement. The symbol serves as a unique identifier and must be distinct within the context of map generation. In SCG, a measurand denotes a value the system calculates as a spatial or temporal prediction. However, the system does not confine the prediction methodologies. Measurands can encompass extensions to previously defined data sources or values expected to be present within a given tile. The precise definition of measurands enables the execution of prediction tasks using diverse approaches, facilitating the comparison of results derived from different methodologies.

4 Discussion

During the research, several discussions have emerged that we aim to present here:

The utility of SCG. It can be argued that the work presented in this paper may be perceived as an unnecessary exercise or mere intellectual exploration. However, the primary objective of SCG is to integrate tools and solutions that establish a stable and replicable environment for researching air pollution. SCG represents a complex yet approachable solution that individuals with a reasonable understanding of GIS development can reproduce. Nonetheless, following the development of SCG, it becomes unnecessary for researchers to invest valuable time and effort in tool preparation. Moreover, the versatility of SCG allows for seamless integration into various systems, enabling a wide range of applications.

Information updates. To ensure the accuracy of the system's results in light of the ever-changing Earth's surface, regular updates of land use data are conducted. These updates serve to prevent any potential inaccuracies arising from outdated information. The intervals between these updates are typically estimated to be one year. However, if required, the length of these intervals can be adjusted to better accommodate users' specific needs and preferences.

5 Conclusions

In this paper, the authors have presented a framework highlighting the significance of integrating multiple toolsets into a cohesive and intricate solution for air pollution prediction. The approach facilitates a smoother learning curve and enhances overall efficiency by enabling users to specialize in a single tool rather than struggling to coordinate numerous tools simultaneously. The proposed system generates a simulated representation of the Earth's surface and uses the Land-Use Regression (LUR) model for air pollution forecasting while providing data visualization capabilities. While the system has certain limitations, it exhibits exceptional performance in its intended tasks.

Future work includes publishing the solution as an open-source library and developing a service for exchanging Synthetic City Generator (SCG) configurations and research tasks. This will promote collaboration and knowledge sharing among researchers, enabling them to replicate findings and exchange research experiences more efficiently. The authors' framework can help policymakers, researchers, and environmentalists monitor air pollution levels and develop effective strategies to mitigate its adverse effects.

Funding. This work was supported by the National Centre for Research and Development Grant NOR/POLNOR/HAPADS/0049/2019-00.

References

1. Abdel-Rahman, A.A.: On the atmospheric dispersion and Gaussian plume model. In: Proceedings of the 2nd International Conference on Waste Management, Water Pollution, Air Pollution, Indoor Climate, Corfu, Greece, vol. 26 (2008)
2. Anderson, P.M.: The uses and limitations of trend surface analysis in studies of urban air pollution. Atmos. Environ. (1967) **4**, 129–147 (1970). https://doi.org/10.1016/0004-6981(70)90003-X
3. Bernstein, J.A., et al.: Health effects of air pollution. J. Allergy Clin. Immunol. **114**, 1116–1123 (2004). https://doi.org/10.1016/J.JACI.2004.08.030
4. Brodsky, I.: H3: Uber's hexagonal hierarchical spatial index, p. 30 (2018). Available from Uber Engineering website https://enguber.com/h3/. Accessed 22 June 2019
5. Chae, S., Shin, J., Kwon, S., Lee, S., Kang, S., Lee, D.: PM10 and PM2.5 real-time prediction models using an interpolated convolutional neural network. Sci. Rep. **11**, 11952 (2021). https://doi.org/10.1038/s41598-021-91253-9
6. Chipperfield, M.P.: Multiannual simulations with a three-dimensional chemical transport model. J. Geophys. Res. Atmos. **104**, 1781–1805 (1999). https://doi.org/10.1029/98JD02597. https://onlinelibrary.wiley.com/doi/full/10.1029/98JD02597
7. Darrall, N.M.: The effect of air pollutants on physiological processes in plants. Plant Cell Environ. **12**, 1–30 (1989). https://doi.org/10.1111/J.1365-3040.1989.TB01913.X
8. Espinosa, R., Jiménez, F., Palma, J.: Multi-objective evolutionary spatio-temporal forecasting of air pollution. Future Gener. Comput. Syst. **136**, 15–33 (2022). https://doi.org/10.1016/J.FUTURE.2022.05.020
9. Forehead, H., Huynh, N.: Review of modelling air pollution from traffic at street-level - the state of the science. Environ. Pollut. **241**, 775–786 (2018). https://doi.org/10.1016/J.ENVPOL.2018.06.019
10. Gardner, M.W., Dorling, S.R.: Artificial neural networks (the multilayer perceptron) - a review of applications in the atmospheric sciences. Atmos. Environ. **32**, 2627–2636 (1998). https://doi.org/10.1016/S1352-2310(97)00447-0
11. Gariazzo, C., et al.: A gas/aerosol air pollutants study over the urban area of Rome using a comprehensive chemical transport model. Atmos. Environ. **41**, 7286–7303 (2007). https://doi.org/10.1016/J.ATMOSENV.2007.05.018
12. Habermann, M., Billger, M., Haeger-Eugensson, M.: Land use regression as method to model air pollution. Previous results for Gothenburg/Sweden. Procedia Eng. **115**, 21–28 (2015). https://doi.org/10.1016/J.PROENG.2015.07.350
13. He, S.: From beautiful maps to actionable insights: introducing Kepler. gl, uber's open-source geospatial toolbox. Uber Eng. **29** (2018)
14. Kamiński, J.W., McConnell, J.C., Boville, B.A.: A three-dimensional chemical transport model of the stratosphere: midlatitude results. J. Geophys. Res. Atmos. **101**, 28731–28751 (1996). https://doi.org/10.1029/96JD01550

15. Koch, P.R.: Using Kepler. gl to visualize weather data. Ph.D. thesis, Universidade de Passo Fundo (2018)
16. Krutmann, J., et al.: Pollution and skin: from epidemiological and mechanistic studies to clinical implications. J. Dermatol. Sci. **76**, 163–168 (2014). https://doi.org/10.1016/J.JDERMSCI.2014.08.008
17. Kukkonen, J., et al.: Modelling the dispersion of particle numbers in five European cities. Geoscientific Model Dev. **9**, 451–478 (2016). https://doi.org/10.5194/GMD-9-451-2016
18. Kukkonen, J., et al.: A review of operational, regional-scale, chemical weather forecasting models in Europe. Atmos. Chem. Phys. **12**, 1–87 (2012). https://doi.org/10.5194/ACP-12-1-2012
19. Li, Y., Sha, Z., Tang, A., Goulding, K., Liu, X.: The application of machine learning to air pollution research: a bibliometric analysis. Ecotoxicol. Environ. Saf. **257** (2023). https://doi.org/10.1016/j.ecoenv.2023.114911
20. Losacco, C., Perillo, A.: Particulate matter air pollution and respiratory impact on humans and animals. Environ. Sci. Pollut. Res. **25**, 33901–33910 (2018). https://doi.org/10.1007/S11356-018-3344-9/METRICS
21. Palak, R., Wojtkiewicz, K., Merayo, M.G.: An implementation of formal framework for collective systems in air pollution prediction system. In: Nguyen, N.T., Iliadis, L., Maglogiannis, I., Trawiński, B. (eds.) ICCCI 2021. LNCS (LNAI), vol. 12876, pp. 508–520. Springer, Cham (2021). https://doi.org/10.1007/978-3-030-88081-1_38
22. Palak, R., Wojtkiewicz, K.: A centralization measure for social networks assessment. Cybern. Syst. 1–14 (2023). https://doi.org/10.1080/01969722.2022.2162737
23. Raju, P.: Fundamentals of geographical information system. Satell. Remote Sens. GIS Appl. Agric. Meteorol. **103** (2006)
24. Tomlinson, R.: Geographical information systems, spatial data analysis and decision making in government. Doctoral thesis, University of London (1974)
25. Zhang, Y., Bocquet, M., Mallet, V., Seigneur, C., Baklanov, A.: Real-time air quality forecasting, part i: history, techniques, and current status. Atmos. Environ. **60**, 632–655 (2012). https://doi.org/10.1016/J.ATMOSENV.2012.06.031

Design of an Automated CNN Composition Scheme with Lightweight Convolution for Space-Limited Applications

Feng-Hao Yeh[1], Ding-Chau Wang[2], Pi-Wei Chen[1], Pei-Ju Li[1], Wei-Han Chen[3], Pei-Hsuan Yu[4], and Chao-Chun Chen[1(✉)]

[1] IMIS/CSIE, NCKU, Tainan City, Taiwan
{P96114183,NF6111015,P96104154,chaochun,E34081155}@gs.ncku.edu.tw
[2] MIS, Southern Taiwan University of Science and Technology, Tainan City, Taiwan
dcwang@stust.edu.tw
[3] Department of Chemical Engineering, NCKU, Tainan City, Taiwan
[4] Department of Performing Art, LEE-MING Institute of Technology, New Taipei City, Taiwan

Abstract. The emergence of the CNN network has enabled many networks for image object recognition, object segmentation, etc., and has brought amazing results to image processing tasks, including MaskRCNN [4] and YOLO [8]. These networks can achieve comparable performance by stacking Convolutional Layers, as layers go deeper, the performance is improved as well. Although deeper convolution layers make the performance of the entire network better, the huge parameters of the networks makes it difficult to implement the network on embedded systems with constrained hardware resources. Therefore, if these networks are to be run on devices with resource constraint, the structure of the network must be lightweight. Usually the most prevalent way to reduce the weight of the network is to modify the network structure, but the design of the network structure has its own philosophy. Any changes to the structure of the network will compromise the performance of the network. We propose a method to automatically substitute the Convolution architecture in the network without changing the network architecture, thereby reducing the parameter of the network while ensuring the performance of the network.

1 Introduction

As the development of deep learning has seen a drastic improvement, there are more and more industrial applications appearing in the decade. Especially when

This work was supported by National Science and Technology Council (NSTC) of Taiwan under Grants 111-2221-E-006-202. This work was financially supported by the "Intelligent Manufacturing Research Center" (iMRC) in NCKU from The Featured Areas Research Center Program within the framework of the Higher Education Sprout Project by the Ministry of Education in Taiwan.

N. T. Nguyen et al. (Eds.): ACIIDS 2023, LNAI 13995, pp. 436–446, 2023.
https://doi.org/10.1007/978-981-99-5834-4_35

Yann LeCun et al. first bring out the convolutional neural neutral (CNN). The debut of the CNN network has enabled the machine to effectively process the image. This was a huge leap for machines to engage in human society, since it can facilitate humans with any industrial usage. However, the performance of traditional CNN networks heavily rely on stacking the convolutional layer to achieve better performance, which accompanies the constraints of high computational complexity and memory requirements. Such constraints will hinder the implementation of CNN in the frontline industry, since the most of the deployment of the network is usually on resource-constrained devices (e.g, edge computing device), which only has limited computational and memory resources. As to tackle this problematic deployment issue, several researchers have turned to develop light-weighted architecture for CNN. Howard et al. has proposed MobileNet [6] which has reduced the computation and memory cost of the model to one-tenth the original CNN by combining the Depthwise Separable Convolution and Point Wise Separable Convolution, meanwhile remaining comparable accuracy on recognition performance. Several works also made progress in developing light-weighted architecture which reduced the computation and memory demand of the CNN model. Although the aforementioned work has shown extinguishing results in lightening the CNN model, it still faces the challenge of how to maximally optimize the efficiency of model deployment on different resource-constrained devices. manually designing high-performing neural architectures is an intuitive way to solve the problems, but it is undoubtedly burdensome, let alone taking account of the resource-constraint for the specific platform. Fortunately, the success of Neural Architecture Search (NAS) techniques came with hope recently. In this paper we have proposed an automated CNN composition scheme with lightweight depth-correlated convolution. The proposed scheme aims to automatically compose a CNN network by different light weight convolutional architectures which can reach optimal usage of resources while having the beat model performance. The algorithm will select the best composition from a lightweight convolutional architecture pool, in which each architecture shares different properties (The amount of the parameter or capability of feature extraction). In addition, we have proposed a novel lightweight convolution, which enriches the channel information of the feature map that helps provide more robust performance on feature extraction compared with Mobilenet.

In this paper, our main contribution are listed as follow:

- We proposed an automated CNN composition scheme which can automatically compose the CNN model based on the resource of the device, thereby maximizing the computation and memory efficiency of the device.
- We proposed a novel lightweight convolution architecture which is based on Depthwise Separable Convolution and Pointwise Separable Convolution. But we design the architecture in a way that the channel relation can be strengthened during the inference procedure, thereby outperforming the Mobilenet with only a slight increase in parameter.

2 Related Works

2.1 Neural Architecture Search

Neural Architecture Search (NAS) is a technique that automates the process of searching for the best neural network architecture for a given task. In the deep learning domain, instead of manually selecting the number of layers, neurons, activation functions, and other hyperparameters, NAS uses an algorithm to search through the search space of possible architectures to find the one that performs the best. In [10], Zoph et al. first utilized reinforcement learning to guid NAS procedure. The idea is to iteratively choose the model architecture from an infinite search space, and a trainer (environment) will train the selected model with some dataset to return the reward, eventually obtaining the best model architecture. However, it is time consuming for reinforcement-based method to do NAS. in [9] Zoph et al. proposed the progressive NAS, which proposed to compose the model architecture layer by layer. For each layer, it will utilize an additional controller to evaluate which architecture from the finite search space is most suitable for the current layer. Another concept supernet is proposed in [1], we composed all model architecture into one complex and intricate supernet. Once we have trained the supernerd, we can easily estimate the performance of different combinations of the architecture in supernet.

2.2 Model Compression

As the demand of embedding the deep neural network into resource-constrained devices increases in the decade, the research field of model compression has also gained place. Several approach has been proposed to reduce the size and computational complexity meanwhile retain the performance of the model, including network pruning [2], knowledge distillation [3], etc. one of the popular directions is redesign the architecture to lighten the model. [6] proposed MobileNet which utilizes Depth wise separable and point-wise separable convolution to substitute traditional Convolution, which has drastically decreased the parameter to one-tenth while retaining the performance. Although the MobileNet has gained great success in model compression, we observed that there is still a constraint. Since the number of Depth wise convolution is restricted by the channel number of the feature, when we implement it before point-wise separable convolution, the extent of local information extraction will also be limited, which results in the risk of losing important information. In our work, we proposed a new convolution architecture which can amend this problem.

3 Method

3.1 Automatically CNN Composition

Keep CNN Architecture. The current development of neural networks enables the model to become very deep. In the ResNet [5] paper, a neural network with a depth of more than one thousand layers was even proposed. And it is such a deep neural network that results in huge demand for hardware resources, which can't not easily fit to lightweight devices such as embedded devices. Even if the network is applied to devices, It will still fail to operate due to insufficient computational resources. Therefore, the most worthwhile approach to lightweight deep neural networks is to change the network structure, such as reducing the depth or deleting elements. However, in the neural network, there are various design philosophies behind the structure design of the Deep network. If the structure of the neural network is changed arbitrarily, the goal of the network will change, which will undoubtedly be a devastating blow to the network structure.

Modifying Convolutional Layers Struct. Therefore, it is not feasible to simply change the structure to reduce the weight of the network. In order to maintain the integrity of the network structure, we must ensure that "the purpose and behavior of each layer of the network do not change" and "the input and output dimensions of each layer of the network and size does not change", it is undoubtedly a difficult challenge to lightweight CNN under the above conditions. In MobileNet [6], the authors mentioned that the convolution kernel occupies the most parameters and calculations in the CNN network. Therefore, our proposed method is based on the idea to redesign the convolution kernel, thereby achieving the lightweight architecture. Taking Depthwise Separable Convolution (DSC) proposed by MobileNet as an example, this convolution structure fundamentally reduces the amount of parameters and calculations of the convolution kernel by modifying the structure, while maintaining the purpose of the convolution layer and maintaining the input/output size. In order to maintain the original structure of the network, this method is inspired by the way MobileNet changes the convolution structure, and changes the basic requirements of the network for the amount of parameters and computation by replacing the convolution kernel. It fundamentally solves the problem that the network is too large to be implemented by lightweight devices.

Maximize Parameters. In [7], M. Hu et al. mentioned that, the amount of parameter is determinative for the capacity of the model, and the capacity of the network directly affects the fitting of the data set. The larger the number of parameters, the greater the capacity of the network. Therefore, the goal of our method is to reduce the weight of the network while retaining the performance of the inference under limited computing resources.

3.2 Technical Details

Evaluating Parameters of Convolution Layer. In CNN, the amount of parameters occupied by the Convolutional Layer will vary according to the number of input/output channels/convolution kernel size and structure. Therefore, it is necessary to evaluate how many parameters are required by all Convolutional Layers before starting CNN lightweight. Amount, in order to select the Convolutional Layer.

Choosing Architecture of Convolutional Layers. In order to remain the overall architecture unchanged when doing replacement over the convolutional kernel, it is important not to delete a certain convolution layer and keep the input and output size identical. By fulfilling the aforementioned requirement, the overall architecture can remain consistent no matter how we substitute the convolutional architecture (Fig. 1).

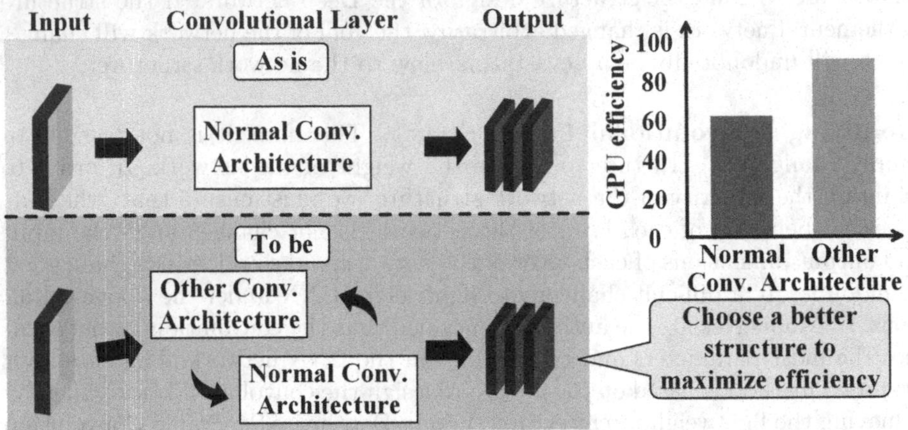

Fig. 1. Replacing convolutional layer to make convolutional neural network lightly.

Finding the Best Network Architecture. As mentioned earlier, in order to maintain the network structure, all convolutional layers must place at least a convolutional structure, and each layer has a different combination of convolutional architecture. In order to find the optimal combination of the network that satisfies the hardware resource while having the optimal network capacity, it is necessary to search all combinations of all available convolutional structures and compose the network with the largest number of parameters among all combinations. If the number of network parameters exceeds the number of available parameters during placement, this combination is skipped.

3.3 Efficient Way to Find Network Architecture - Horizontal Calculation

If there are N layers of Convolutional Layers and M different kernel types to choose from in a CNN, then there are N * M different possibilities in this CNN by changing the convolutional layer. When looking for the best combination, an intuitive way is to lock the convolutional architecture in the current layer, and continue to find the best solution in the next layer. It is necessary to restore the state of the previous layer when changing the convolutional structure of the current layer, so as to make a comprehensive consideration while substituting the convolution. We name this top-down selection method as Vertical Calculation (Fig. 2). In programming terms, this could lead to problems such as Stack Overflow and Out of memory.

This method proposes an efficient way to solve this problem, the core idea of which is "processing one layer at a time". After putting all the possibilities in the first layer, the next layer can be built based on the results of the previous layer. If the amount of available parameters is exceeded during replacement, the selected convolutional structure is discarded. The method of horizontally ordering the convolution kernels of a fixed layer from left to right is called Horizontal Calculation (Fig. 2).

It's the characteristic of replacing the convolutional structure layer by layer, thereby solving the Stack Overflow problem. The proposed Horizontal Calculation has the advantage that the state of the layer will not be changed when the calculation of each layer is completed, so the state of each layer can be stored through the method of Dynamic Programming. When each layer is placed, it will record the "Total Occupied Quantity", "Remaining Parameters" and "Last Kernel Type". By directly recording the results of each layer for different kernel types, it is not necessary to copy the previous results when determining the structure of each layer, so as to save memory and reduce the probability of Out of memory.

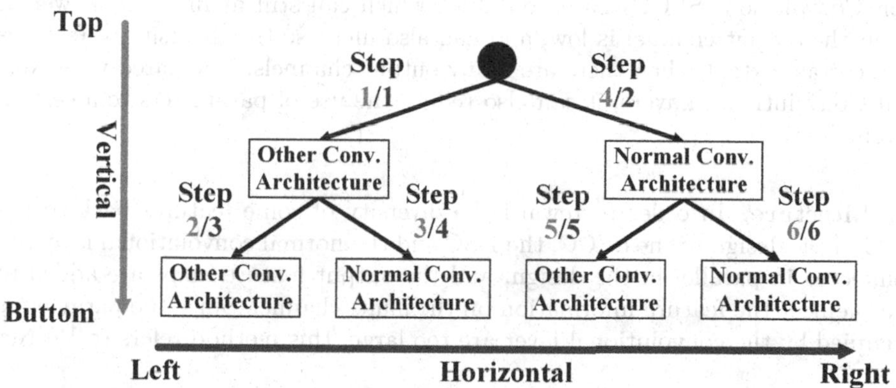

Fig. 2. Schematic diagram of Vertical Calculation and Horizontal Calculation.

3.4 Rollback Architecture

After listing all the combinations of the network, we select the network setting with the largest amount of parameters. It is very simple to push back to the original convolution combination through the parameter quantity record. Each record is composed of the convolution result of the previous layer plus the convolution result of this layer, so the convolution result of this layer is taken out first when pushing back. Then, the result of the previous layer is obtained by subtracting the consumption of this layer of convolution structure, and the original convolution combination can be obtained by repeating. Since the convolution combination is obtained by pushing back and forth, the correct convolution structure combination is to invert the result.

3.5 A New Kernel Convolution Architecture - Sneaking Feature Compensation Convolution (SFCC)

In order to further extend the benefits of this method, we proposed a new convolutional structure "Sneaking Feature Compensation Convolution (SFCC)" (Fig. 3). The structure is based on MobileNet. Different from the ordinary convolutional layer, MobileNet first uses Depthwise Convolution to allow all input channels to have an independent Filter, and allows all kernels to share these Filters to achieve the purpose of reducing the number of parameters, and then changes the output channel through Pointwise Convolution Quantity to complete a convolution process. However, MobileNet has a potential risk that the number of filters will not change regardless of the number of output channels. That is because Depthwise Convolution allows all kernels to share a Filter for individual Input Channels, Which means Even if the number of output channels is increased, the diversity of feature extraction cannot be improved. For this reason, we propose a new convolution structure Sneaking Feature Compensation Convolution (SFCC) based on DSC, which can still maintain light weight when the output channel is low, and can also increase the diversity of features to a certain extent when there are many output channels. Compared with Normal Convolutional Layers, it can also reduce the use of parameters to a certain extent.

Architecture. In order to reward the diversity of some features back to the DSC in the design of the SFCC, the DSC and the normal convolutional layer are connected in parallel in the design, and the output feature maps are added to complement the feature information on the multi-channel. Since the parameters occupied by the convolutional layer are too large, this method refers to ResNet

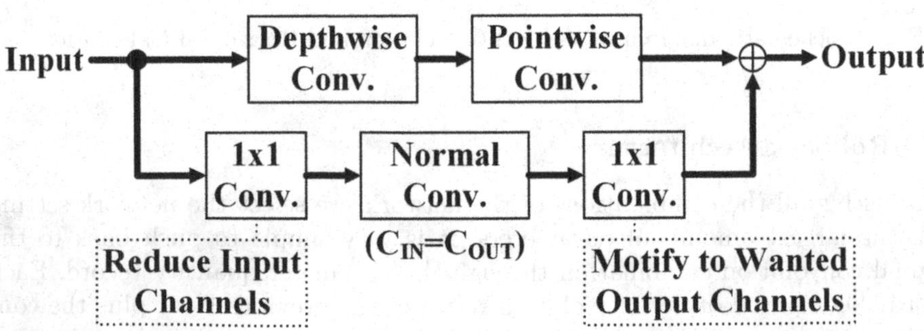

Fig. 3. The architecture of SFCC

[5] to design a bottleneck architecture. First, a 1 * 1 convolution is used to reduce the number of input channels, and then extract features by normal convolution from the image. Finally, The number of feature channels is scaled to the number of output channels by 1 * 1 Convolution which can moderately compensate for feature diversity under the premise of a lightweight convolution structure.

For each convolution layer, the output channel has a convolution kernel, and in the ordinary convolution structure design, each convolution kernel has an independent filter (Fig. 4). On the contrary, due to the design of Depthwise Convolution on the DSC, so that each kernel shares the same filter (Fig. 5). This leads to the fact that the feature information extracted by the DSC structure will remain the same even when the number of output channels increases, thus the diversity of the feature decreases. In order to extract more diverse features as the output channels increase, we must augment the filter. This proposed structure increases the number of Filters by adding Normal Convolution in parallel with DSC. In order not to make the parameter amount of Normal Convolution expand too quickly when the number of output channels increases, this method reduces the number of channels in Normal Convolution by adding 1 * 1 Convolution before and after Normal Convolution. The first 1 * 1 Convolution is used to reduce the number of channels and the second 1 * 1 Convolution is used to restore the number of channels. In this way, the number of Filters will be rewarded.

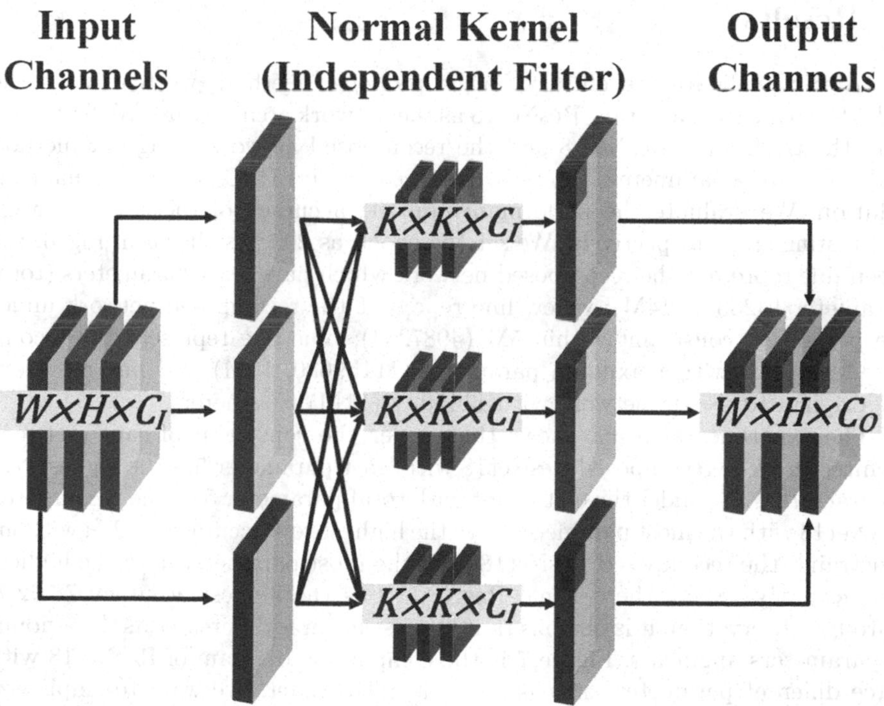

Fig. 4. Schematic diagram of the convolution kernel structure that each convolution kernel has an independent filter.

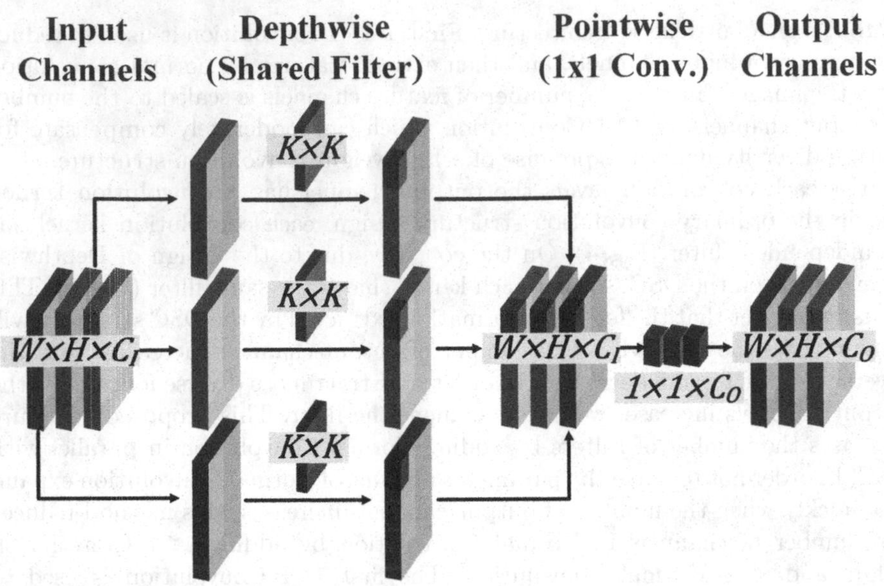

Fig. 5. The Architecture of Depthwise Separable Convolution.

4 Results

In order to validate the efficacy of the proposed method, we test it on the CIFAR100 dataset, and take ResNet18 as the network architecture. We first compare the traditional ResNet18 and the recomposed network using our method under the same parameters. The search place contains DSC and traditional convolution. We evaluate the performance on the accuracy of the training stage and testing stage respectively. We set the epoch as 200. As shown in Fig. 6, the green line represent the recomposed network which have least parameters (total parameter 1423387, 14M); yellow line represent the recomposed network under the parameter constraint within 5M (4987264); blue line represent the recomposed network with maximum parameters(11159680, 11M); red line represent the original ResNet18 network with 11159232(11M) parameter.

The experimental result shows that under the constraint of same network architecture, the recomposed ResNet18 with least parameter has the highest test accuracy 76.21%; under the constraint with total parameter 5M, the recomposed ResNet18 with the most parameter have the highest test accuracy 77.2%; without constraint, the recomposed ResNet18 with the most parameter have the highest test accuracy 78.8%; the original ResNet18 has the highest accuracy 78.62%. Through observation, it is obvious that the test accuracy increases as the amount of parameters augments. Figure 7 is the comparison diagram of ResNet18 with three different parameter settings and original ResMet. We want to emphasize that recomposed ResNet18 with the least parameter has a dramatic disparity

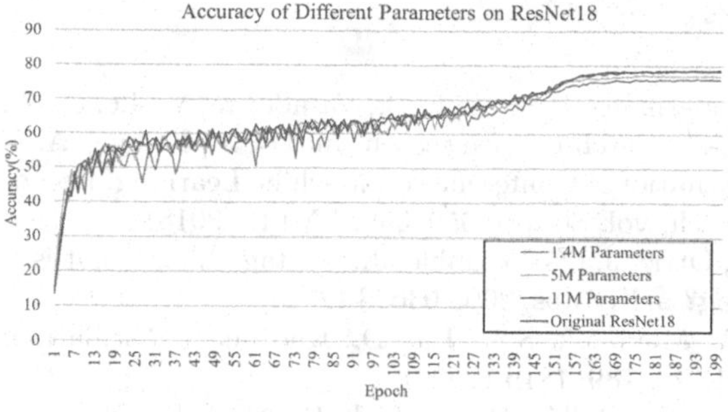

Fig. 6. Accuracy of Different Parameters on ResNet18. (Color figure online)

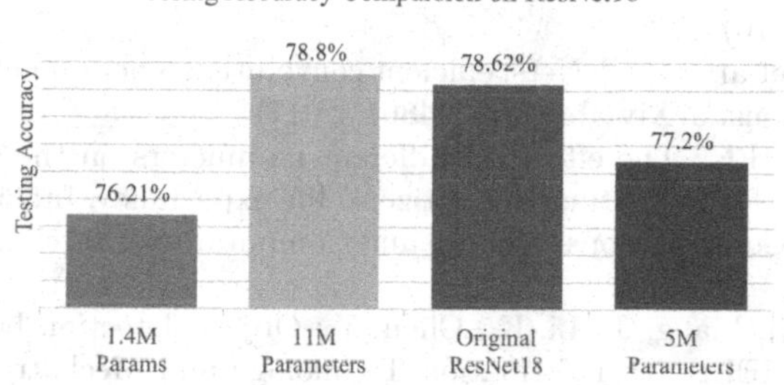

Fig. 7. Testing Accuracy Comparison on ResNet18.

with that of original ResNet18 by 87.25%, yet the test accuracy has only a slight drop with 2.41%. We have validated the method proposed in this paper in lightening the network while retaining the model performance.

5 Conclusions

In this paper we proposed a method which keeps the neural network accuracy when lightweight a convolutional neural network. For keeping the design philosophy of neural networks, our method keeps the neural network architecture and replaces convolutional layers architecture to make the neural network lighter. In experiment, we compared testing accuracy in CIFAR100 dataset between ResNet18 and the lightweight architecture optimize by our method, the result shows the lightweight ResNet18 has test accuracy 76.21%, compared with original ResNet18 architecture only loss 3.29% test accuracy when saves 87.25% parameters. We also proposed a new convolutional architecture to increase more features to Depthwise Separable Convolution when output channels are greater. Our method increases features to DSC by adding normal convolutional in parallel and lightweight the architecture via bottleneck.

References

1. Bender, G., Kindermans, P.J., Zoph, B., Vasudevan, V., Le, Q.: Understanding and simplifying one-shot architecture search. In: Dy, J., Krause, A. (eds.) Proceedings of the 35th International Conference on Machine Learning. Proceedings of Machine Learning Research, vol. 80, pp. 550–559. PMLR (2018)
2. Blalock, D.W., Ortiz, J.J.G., Frankle, J., Guttag, J.V.: What is the state of neural network pruning? arXiv abs/2003.03033 (2020)
3. Gou, J., Yu, B., Maybank, S.J., Tao, D.: Knowledge distillation: a survey. Int. J. Comput. Vis. **129**, 1789–1819 (2020)
4. He, K., Gkioxari, G., Dollár, P., Girshick, R.: Mask R-CNN. In: 2017 IEEE International Conference on Computer Vision (ICCV), pp. 2980–2988 (2017)
5. He, K., Zhang, X., Ren, S., Sun, J.: Deep residual learning for image recognition. In: 2016 IEEE Conference on Computer Vision and Pattern Recognition (CVPR), pp. 770–778 (2016)
6. Howard, A.G., et al.: MobileNets: efficient convolutional neural networks for mobile vision applications. arXiv abs/1704.04861 (2017)
7. Hu, M., Hu, Y.H.F.: The effects of different parameters on the accuracy of deep learning models for predicting U.S. citizen's life expectancy. In: 2021 International Conference on Computational Science and Computational Intelligence (CSCI), pp. 105–109 (2021)
8. Liu, C., Tao, Y., Liang, J., Li, K., Chen, Y.: Object detection based on yolo network. In: 2018 IEEE 4th Information Technology and Mechatronics Engineering Conference (ITOEC), pp. 799–803 (2018)
9. Liu, C., et al.: Progressive neural architecture search. In: Ferrari, V., Hebert, M., Sminchisescu, C., Weiss, Y. (eds.) ECCV 2018. LNCS, vol. 11205, pp. 19–35. Springer, Cham (2018). https://doi.org/10.1007/978-3-030-01246-5_2
10. Zoph, B., Le, Q.: Neural architecture search with reinforcement learning. In: International Conference on Learning Representations (2017)

Author Index

N. T. Nguyen et al. (Eds.): ACIIDS 2023, LNAI 13995, pp. 447–449, 2023.
https://doi.org/10.1007/978-981-99-5834-4

Printed in the United States
by Baker & Taylor Publisher Services

Printed in the United States
by Baker & Taylor Publisher Services